S. Chand's IIT Foundation Series

A Compact and Comprehensive Book of

IIT Foundation Mathematics

CLASS – VII

S.K. GUPTA
ANUBHUTI GANGAL

EURASIA PUBLISHING HOUSE
(An imprint of S. Chand Publishing)
A Division of S. Chand And Company Pvt. Ltd.
(An ISO 9001 : 2008 Company)
7361, Ram Nagar, Qutab Road, New Delhi-110055
Phone: 23672080-81-82, 9899107446, 9911310888; Fax: 91-11-23677446
www.schandpublishing.com; e-mail : helpdesk@schandpublishing.com

Branches :

Ahmedabad	:	Ph: 27541965, 27542369, ahmedabad@schandpublishing.com
Bengaluru	:	Ph: 22268048, 22354008, bangalore@schandpublishing.com
Bhopal	:	Ph: 4274723, 4209587, bhopal@schandpublishing.com
Chandigarh	:	Ph: 2725443, 2725446, chandigarh@schandpublishing.com
Chennai	:	Ph: 28410027, 28410058, chennai@schandpublishing.com
Coimbatore	:	Ph: 2323620, 4217136, coimbatore@schandpublishing.com (Marketing Office)
Cuttack	:	Ph: 2332580; 2332581, cuttack@schandpublishing.com
Dehradun	:	Ph: 2711101, 2710861, dehradun@schandpublishing.com
Guwahati	:	Ph: 2738811, 2735640, guwahati@schandpublishing.com
Hyderabad	:	Ph: 27550194, 27550195, hyderabad@schandpublishing.com
Jaipur	:	Ph: 2219175, 2219176, jaipur@schandpublishing.com
Jalandhar	:	Ph: 2401630, 5000630, jalandhar@schandpublishing.com
Kochi	:	Ph: 2378740, 2378207-08, cochin@schandpublishing.com
Kolkata	:	Ph: 22367459, 22373914, kolkata@schandpublishing.com
Lucknow	:	Ph: 4026791, 4065646, lucknow@schandpublishing.com
Mumbai	:	Ph: 22690881, 22610885, mumbai@schandpublishing.com
Nagpur	:	Ph: 6451311, 2720523, 2777666, nagpur@schandpublishing.com
Patna	:	Ph: 2300489, 2302100, patna@schandpublishing.com
Pune	:	Ph: 64017298, pune@schandpublishing.com
Raipur	:	Ph: 2443142, raipur@schandpublishing.com (Marketing Office)
Ranchi	:	Ph: 2361178, ranchi@schandpublishing.com
Siliguri	:	Ph: 2520750, siliguri@schandpublishing.com (Marketing Office)
Visakhapatnam	:	Ph: 2782609, visakhapatnam@schandpublishing.com (Marketing Office)

First Published in 2012
Reprints 2013, 2014, 2015, 2016 (Thrice)

ISBN : 978-81-219-3898-3 **Code :** 1014 638

PRINTED IN INDIA
By Vikas Publishing House Pvt. Ltd., Plot 20/4, Site-IV, Industrial Area Sahibabad, Ghaziabad-201010
and Published by S. Chand And Company Pvt. Ltd., 7361, Ram Nagar, New Delhi -110 055.

PREFACE AND A NOTE FOR THE STUDENTS

ARE YOU ASPIRING TO BECOME AN ENGINEER AND BECOME AN IIT SCHOLAR ?

Here is the book especially designed to motivate you, to sharpen your intellect and develop the right attitude and aptitude and lay solid foundation for your success in various entrance examinations like **IIT, AIEEE, EAMCET, WBJEE, MPPET, SCRA, Kerala PET, OJEE, Raj PET, AMU**, etc.

SALIENT FEATURES

1. Content based on the curriculum of the classes for ***CBSE, ICSE, Andhra Pradesh*** and ***Boards of School Education of Other States.***
2. Full and comprehensive coverage of all the topics.
3. Detailed synopsis of each chapter at the beginning in the form of ***'Key Concepts'***. This will not only facilitate thorough ***'Revision'*** and ***'Recall'*** of every topic but also greatly help the students in understanding and mastering the concepts besides providing a ***back-up*** to classroom teaching.
4. The books are enriched with an exhaustive range of hundreds of thought provoking objective questions in the form of solved examples and practice questions in Question Banks which not only offer a great variety and reflect the modern trends but also invite, explore, develop and put to test the ***thinking, analysing*** and ***problem solving skills of the students***.
5. **Answers, Hints** and **Solutions** have been provided to boost up the morale and increase the confidence level.
6. **Self Assessment Sheets** have been given at the end of each chapter to help the students to assess and evaluate their understanding of the concepts and learn to attack the problems independently.

   We hope the series will be able to fulfil its aims and objectives and will be found immensely useful by the students aspiring to become topclass engineers.

   Suggestions for improvement and also the feedback from various sources would be most welcome and gratefully acknowledged.

AUTHORS

CONTENTS

UNIT 1 : NUMBER SYSTEM

UNIT 2 : ALGEBRA

UNIT 3 : COMMERCIAL MATHEMATICS

UNIT 4 : GEOMETRY

UNIT 5 : MENSURATION

UNIT 6 : STATISTICS

UNIT 7 : SETS

UNIT–1

NUMBER SYSTEM

- *Numbers*
- *Fractions*
- *Decimals*
- *HCF and LCM*
- *Powers and Roots*

Chapter

NUMBERS

KEY FACTS

1. TYPES OF NUMBERS

(i) **Natural numbers :** Counting numbers, *i.e.,* 1, 2, 3, 4, 5, ... are called natural numbers.

(ii) **Whole numbers :** Counting numbers and 0, *i.e.*, 0, 1, 2, 3, 4, 5, ... are called whole numbers.

(iii) **Integers :** All natural numbers, zero and opposites of natural numbers, *i.e.*, –3, – 2, – 1, 0, 1, 2, 3, ... are called integers.

(iv) **Rational numbers :** All numbers of the form $\frac{p}{q}$, where p and q are integers and $q \neq 0$ are called rational numbers.

A number like $\frac{5}{0}$ or $\frac{-6}{0}$ whose denominator is zero is not defined and hence is not a rational number.

Note :

(a) Every natural number is a rational number but a rational number need not be a natural number.

(b) Zero is a rational number.

(c) Every integer is a rational number, but every rational number need not be an integer.

(d) Every fraction is a rational number but a rational number need not be a fraction.

(e) The operations of addition, subtraction, multiplication are performed on rational numbers as on fractions.

(f) A rational number y is called the **reciprocal** (multiplicative inverse) of a rational number x, if $x \times y = y \times x = 1$. The reciprocal of a **non-zero** rational number $\frac{p}{q}$ is $\frac{q}{p}$ and vice versa.

(g) To divide a rational number x by another (non-zero) rational number y, multiply x by the reciprocal of y.

(h) **Every rational number can be expressed as a decimal**. The decimal representation of a rational number is either ***terminating*** or ***non-terminating*** and ***repeating***. A rational number $\frac{p}{q}$ is a terminating decimal if in its lowest form, the denominator has only 2 and 5 as factors.

(i) A rational number lying between two given rational numbers $\frac{p}{q}$ and $\frac{r}{s}$ is $\frac{1}{2}\left(\frac{p}{q}+\frac{r}{s}\right)$. There exist infinite rational numbers between any two given rational numbers.

(v) **Even and odd numbers :** Natural numbers divisible by 2, *i.e.*, 2, 4, 6, 8, ... etc are called *even numbers*, whereas, those natural numbers which are not divisible by 2 are called *odd numbers*.
For example, 1, 3, 5, 7....... etc., are odd numbers.

(vi) **Prime and composite numbers :** A natural number which has only two factors, *i.e.*, 1 and itself is called a *prime number* while the numbers that are not prime are called *composite numbers*.

For example, 2, 3, 5, 7, 11, ... etc., are called prime numbers, whereas, 4, 6, 9, 10, 12, 14, 15, ... etc., are called composite numbers.

Note that : 1 is neither a prime number nor a composite number. Two natural numbers x and y are said to be **co-primes** if they do not have any common factor except 1, *i.e.*, whose HCF is 1.

For example, (2, 3), (4, 5), (6, 7), (7, 8), (9, 11) etc., are co-primes.

2. DIVISIBILITY RULES

(i) A number is **divisible by 2** if its units digit is any of the numbers 0, 2, 4, 6, 8.

(ii) A number is **divisible by 3** if the sum of its digits is divisible by 3.

(iii) A number is **divisible by 4** if the number formed by its last two digits is divisible by 4.

(iv) A number is **divisible by 5** if its unit's digit is either 5 or 0.

(v) A number is **divisible by 6** if it is divisible by both 2 and 3.

(vi) A number is **divisible by 8** if the number formed by its last three digits, *i.e.*, hundred's, ten's and units' digit is divisible by 8.

(vii) A number is **divisible by 9** if the sum of its digits is divisible by 9.

(viii) A number is **divisible by 10** if its units' digit is 0.

(ix) A number is **divisible by 11**, if the difference between sum of its digits at even places (standing from the rightmost digit) and the sum of its digits at odd places is 0 or a multiple of 11.

(x) A number a is divisible by a number b if it is divisible by the co-prime factors of b.

Solved Examples

Ex. 1. ***What is the product of the greatest prime number that is less than 50 and the smallest prime number that is greater than 50?***

Sol. Required product = 47 × 53 = **2491**.

Ex. 2. ***Find the remainder when $7^{21} + 7^{22} + 7^{23} + 7^{24}$ is divided by 25.***

Sol. $7^{21} + 7^{22} + 7^{23} + 7^{24} = 7^{21}\,(1 + 7 + 7^2 + 7^3)$

$= 7^{21} \times (1 + 7 + 49 + 343) = 7^{21} \times 400$, which is completely divisible by 25. Hence remainder = 0.

Ex. 3. ***If n = 1 + x, where x is the product of four consecutive positive integers, then which of the following is/are true.***

A. n is odd *B. n is prime* *C. n is a perfect square*

(a) A and C only *(b) A and B only* *(c) A only* *(d) None of these*

Sol. Let us take $x = 1 \times 2 \times 3 \times 4 = 24$. Then, $n = 1 + 24 = 25$, *i.e.*, an odd number and a perfect square.

Again let $x = 2 \times 3 \times 4 \times 5 = 120$. Then, $n = 1 + 120 = 121$, *i.e.*, an odd number and a perfect square.

∴ Option (*a*) is correct.

Ex. 4. ***Which one of the following is the rational number lying between $\frac{6}{7}$ and $\frac{7}{8}$?***

(a) $\frac{3}{4}$ *(b)* $\frac{99}{122}$ *(c)* $\frac{95}{112}$ *(d)* $\frac{97}{112}$

Sol. Required rational number $= \frac{1}{2}\left(\frac{6}{7}+\frac{7}{8}\right) = \frac{1}{2}\left(\frac{48+49}{56}\right) = \frac{97}{112}$

Hence option (*d*) is correct.

Ex. 5. ***If 1.525252 .... is converted to a fraction, then what is the sum of its numerator and denominator?***

Sol. 1.525252 .... $= 1.\overline{52} = 1\frac{52}{99} = \frac{151}{99}$

$\therefore$ Required sum = 151 + 99 = **250.**

Ex. 6. ***A number when divided by 136 leaves 36 as remainder. If the same number is divided by 17, what will be the remainder?***

Sol. Given number $= 136K + 36 = 17 \times 8 \times K + (17 \times 2 + 2)$
$= 17(8K + 2) + 2$

$\therefore$ Required remainder = **2.**

Ex. 7. ***Three natural numbers are said to be tri-prime if they are pair-wise co-prime. Then one triplet which is not tri-prime is***

(a) (2, 3, 7) *(b) (2, 9, 11)* *(c) (3, 5, 7)* *(d) (3, 4, 9)*

Sol. In the triplet (3, 4, 9) the numbers 3 and 9 are not pairwise prime so (3, 4, 9) is not tri-prime.

Ex. 8. ***If the sum of the digits of any number between 100 and 1000 be subtracted from the number, by what number is the result divisible?***

Sol. A three digit number is represented by $100x + 10y + z$, where x is the digit at the hundreds' place, y at tens' place and z at ones' place.

Now $100x + 10y + z - (x + y + z) = 99x + 9y = 9(11x + y)$

$\therefore$ The result is always divisible by 9.

Ex. 9. ***Find the least five digit number which is divisible by 666.***

Sol. On dividing 10000 (the least 5-digit number) by 666 we get 10 as the remainder.

$\therefore$ Required number = 10000 + (666 – 10)
= 10656

```
       15
666 | 10000
      666
      3340
      3330
        10
```

Question Bank–1

1. The units' digit of every prime number (other than 2 and 5) must be necessarily
   (a) 1, 3 or 5 (b) 1, 3, 7 or 9
   (c) 7 or 9 (d) 1 or 7

2. Consider the following statements :
   A. The sum of two prime numbers is a prime number.
   B. The product of two prime numbers is a prime number.
   Which of these statements is/are correct ?
   (a) Neither A nor B (b) A alone
   (c) B alone (d) Both A and B

3. If a, $a+2$, $a+4$ are prime numbers, then the number of possible solutions for a is
   (a) one (b) two
   (c) three (d) more than three

4. Let x and y be positive integers such that x is prime and y is composite. Then,
   (a) $y - x$ cannot be an even integer
   (b) xy cannot be an even integer
   (c) $(x + y)/x$ cannot be an even integer
   (d) None of these

5. Consider the following statements :
   The number 23 is
   A. a prime number B. a real number
   C. an irrational number D. a rational number
   Of these statements :
   (a) A, B, D are correct (b) A, B, C are correct
   (c) B, C, D are correct (d) A, C, D are correct

6. What is the units' digit of the product of all prime numbers between 1 and 100 ?
   (a) 0 (b) 1
   (c) 2 (d) 3

7. For what value of n are $2^n - 1$ and $2^n + 1$ prime ?
(a) 7 (b) 5
(c) 2 (d) 1

8. In a six-digit number, the sum of the digits in the even places is 9 and the sum of the digits in the odd places is 20. All such numbers are divisible by
(a) 7 (b) 9
(c) 6 (d) 11

9. Which of the following numbers is divisible by 15 ?
(a) 30560 (b) 29515
(c) 23755 (d) 17325

10. The difference of the number consisting of two digits and the number formed by interchanging the digits is always divisible by
(a) 5 (b) 7
(c) 9 (d) 11

11. The number 89715938* is divisible by 4. The unknown non-zero digit marked as (*) will be
(a) 2 (b) 3
(c) 4 (d) 6

12. An integer is divisible by 16 if and only if its last X digits are divisible by 16. The value of X would be
(a) Three (b) Four
(c) Five (d) Six

13. $(4^{61} + 4^{62} + 4^{63} + 4^{64})$ is divisible by
(a) 3 (b) 11
(c) 13 (d) 17

14. If n is any natural number, then $n^2(n^2 - 1)$ is always divisible by
(a) 12 (b) 24
(c) $12 - n$ (d) 6

15. The sum of all possible two-digit numbers formed from three different one-digit natural numbers when divided by the sum of the original three numbers is equal to
(a) 36 (b) 22
(c) 18 (d) 24

16. A six digit number is formed by repeating a three digit number; for example 256256 or 678678 etc. Any number of this form is always exactly divisible by
(a) 7 only (b) 11 only
(c) 13 only (d) 1001

17. The smallest number to be added to 1000 so that 45 divides the sum exactly is
(a) 35 (b) 80
(c) 20 (d) 10

18. In the product $459 \times 46 \times 28 * \times 484$, the digit in the units' place is 2. The digit to come in place of * is
(a) 3 (b) 5
(c) 7 (d) 4

19. How many of the following numbers are divisible by 3 but not by 9 ?
2133, 2343, 3474, 4131, 5286, 5340, 6336, 7347, 8115, 9276
(a) 5 (b) 6
(c) 7 (d) 4

20. When a number is divided by 893, the remainder is 193. What will be the remainder when it is divided by 47 ?
(a) 3 (b) 5
(c) 25 (d) 33

21. What is the least natural number which leaves no remainder when divided by all the digits from 1 to 9 ?
(a) 1800 (b) 5040
(c) 1920 (d) 2520

22. If we divide a positive integer by another positive integer, what is the resulting number ?
(a) Always a natural number
(b) Always an integer
(c) A rational number
(d) An irrational number

23. Consider the following statements :
A. The product of an integer and a rational number can never be a natural number.
B. The quotient of division of an integer by a rational number can never be an integer.
Which of the statements given above is/are correct ?
(a) A only (b) B only
(c) Both A and B (d) Neither A nor B.

24. The rational number lying between $\frac{5}{6}$ and $\frac{6}{7}$ is
(a) $\frac{1}{2}$ (b) $\frac{15}{21}$
(c) $\frac{35}{42}$ (d) $\frac{71}{84}$

25. The rational number $-\frac{18}{5}$ lies between the consecutive integers
(a) – 2 and – 3 (b) – 3 and – 4
(c) – 4 and 5 (d) – 5 and – 6

26. A rational number equivalent to $-\frac{24}{20}$ with denominator 25 is

(a) $-\frac{30}{25}$ (b) $\frac{28}{25}$

(c) $-\frac{29}{25}$ (d) $-\frac{19}{25}$

27. The pair of rational numbers lying between $\frac{1}{4}$ and $\frac{3}{4}$ is

(a) $\frac{262}{1000}, \frac{752}{1000}$ (b) $\frac{63}{250}, \frac{187}{250}$

(c) $\frac{13}{50}, \frac{264}{350}$ (d) $\frac{9}{40}, \frac{31}{40}$

28. Which of the following is correct ?

3.292929.......... is

(a) an integer (b) a rational number

(c) an irrational number (d) not a real number

29. A rational number whose reciprocal does not exist is

(a) 1 (b) –1

(c) 0 (d) 10

30. A number that can be expressed in the form $\frac{a}{b}$ where a and b are integers and b is not equal to zero is called

(a) a fraction (b) an integer

(c) a rational number (d) a real number.

31. $\frac{3}{4}-\frac{4}{5}$ is not equal to

(a) $-\frac{4}{5}+\frac{3}{4}$ (b) $-\frac{1}{20}$

(c) $\frac{4}{5}-\frac{3}{4}$ (d) $-\frac{4}{5}-\left(-\frac{3}{4}\right)$

32. The rational number which can be expressed as a terminating decimal is

(a) $\frac{1}{6}$ (b) $\frac{1}{12}$

(c) $\frac{1}{15}$ (d) $\frac{1}{20}$

33. Arrange the following rational numbers in descending order.

$$\frac{1}{5}+\frac{1}{-6}, \left|-\frac{7}{6}+1\right|, \left(-\frac{1}{2}\right)^3, \frac{5}{8}\div\frac{-15}{4}$$

(a) $\frac{1}{5}+\frac{1}{-6}, \left(-\frac{1}{2}\right)^3, \frac{5}{8}\div\frac{-15}{4}, \left|-\frac{7}{6}+1\right|$

(b) $\left(-\frac{1}{2}\right)^3, \frac{1}{5}+\frac{1}{-6}, \frac{5}{8}\div\frac{-15}{4}, \left|-\frac{7}{6}+1\right|$

(c) $\left|-\frac{7}{6}+1\right|, \frac{1}{5}+\frac{1}{-6}, \left(-\frac{1}{2}\right)^3, \frac{5}{8}\div\frac{-15}{4}$

(d) $\frac{5}{8}\div\frac{-15}{4}, \left|-\frac{7}{6}+1\right|, \frac{1}{5}+\frac{1}{-6}, \left(-\frac{1}{2}\right)^3$

34. Evaluate $\frac{9|3-5|-5|4|\div 10}{-3(5)-2\times 4\div 2}$

(a) $\frac{9}{10}$ (b) $-\frac{8}{17}$

(c) $-\frac{16}{19}$ (d) $\frac{4}{7}$

35. The value of

$25 - 5\,[2 + 3\,\{2 - 2\,(5 - 3) + 5\} - 10] \div 4$ is

(a) 5 (b) 23.25

(c) 23.75 (d) 25

Answers

1. (b)	**2.** (a)	**3.** (a)	**4.** (d)	**5.** (a)	**6.** (a)	**7.** (c)	**8.** (d)	**9.** (d)	**10.** (c)
11. (c)	**12.** (b)	**13.** (d)	**14.** (a)	**15.** (b)	**16.** (d)	**17.** (a)	**18.** (c)	**19.** (b)	**20.** (b)
21. (d)	**22.** (c)	**23.** (d)	**24.** (d)	**25.** (b)	**26.** (a)	**27.** (b)	**28.** (b)	**29.** (c)	**30.** (c)
31. (c)	**32.** (d)	**33.** (c)	**34.** (c)	**35.** (c)					

Hints and Solutions

1. (b) The prime numbers are 2, 3, 5, 7, 11, 13, 17, 19, 23, ... So the units' digit can be 1, 3, 7 or 9. The units' digit in any prime number cannot be 5 as this number will be divisible by 5.

2. (a) Neither the sum nor the product of two prime numbers is a prime number.

3. (a) If $a = 3$, then the prime numbers are 3, 5 and 7.

4. (d) Let us take $x = 2, y = 6$. Then,

$y - x = 4$, $xy = 12$, $\frac{x+y}{x} = \frac{8}{2} = 4$. So all the three statements (a), (b), and (c) are false.

6. (a) Since $2 \times 5 = 10$.

7. (c) We check by substituting each of the given values.

On putting $n = 2$, we get,

$2^2 - 1 = 3$ and $2^2 + 1 = 5$, which are both prime numbers.

8. (d) Since, sum of digits at odd places – sum of digits at even places = 20 – 9 = 11, the number is divisible by 11.

9. (d)

Number	Sum of digits	Divisibility by 3
30560	$3+5+6=14$	Not divisible
29515	$2+9+5+1+5=22$	Not divisible
23755	$2+3+7+5+5=22$	Not divisible
17325	$1+7+3+2+5=18$	Divisible

All the numbers are divisible by 5 as they end in 0 or 5.

$\therefore$ 17325 is divisible by 3 and 5

$\Rightarrow$ 17325 is divisible by $3 \times 5 = 15$.

10. (c) A two digit number can be written as $10x + y$, where x is the digit at tens' place and y is the digit at ones' place.

On interchanging the digits, the number formed = $10y + x$.

$\therefore$ $10x + y - (10y + x) = 9x - 9y = 9(x - y)$

$\Rightarrow$ Given difference is always divisible by 9.

11. (c) A number is divisible by 4, if the number formed by last two digits are divisible by 4.

13. (d) $4^{61} + 4^{62} + 4^{63} + 4^{64} = 4^{61}(1 + 4 + 4^2 + 4^3)$

$= 4^{61} \times (1 + 4 + 16 + 64) = 4^{61} \times 85$,

which is divisible by 17.

14. (a) For $n = 2$, Given exp. $= 2^2(2^2 - 1) = 4 \times 3 = 12$

$n = 3$, Given exp. $= 3^2(3^2 - 1) = 9 \times 8 = 72$

$n = 4$, Given exp $= 4^2(4^2 - 1) = 16 \times 15 = 240$

etc.

Thus, we can see that $n^2(n^2 - 1)$ is always divisible by 12 for any natural number.

15. (b) Let us take the three different one-digit numbers as 5, 6 and 7.

The two-digit numbers formed from these numbers are 56, 65, 67, 76, 57 and 75.

$\therefore$ Required quotient

$$= \frac{56+65+67+76+57+75}{5+6+7} = \frac{396}{18} = 22$$

You can check with another set of numbers also.

16. (d) Since the number of repeating digits is three, therefore, it will be divisible by the number which has one less, *i.e.*, two zeroes between the same digit, *i.e.*, 1001.

17. (a) On dividing 1000 by 45, the remainder = 10.

$\therefore$ Smallest number to be added = (45 – 10) = 35.

18. (c) Multiplying the units' digit of 459, 46 and 484 we get $9 \times 6 \times 4 = 216$ whose units' digit is 6. So * should be replaced by 7 ($\because 6 \times 7 = 42$) to get 2 as the units' digit in the final product.

19. (b) A number is divisible by 3, if the sum of its digits is divisible by 3.

Such numbers are 2133, 2343, 3474, 4131, 5286, 5340, 6336, 7347, 8115, 9276.

A number is not divisible by 9 if the sum of its digits is not divisible by 9. Such numbers are 2343, 5286, 5340, 7347, 8115, 9276.

20. (b) Given number $= 893n + 193$

$= 47 \times 19 \times n + 47 \times 4 + 5$

$= 47(19n + 4) + 5$

$\therefore$ Required remainder = 5.

21. (d) Required number = LCM of (1, 2, 3, 4, 5, 6, 7, 8, 9) = 2520.

23. (d) Let integer = 4 and Rational number = $\frac{2}{1}$.

Then, product $= 4 \times \frac{2}{1} = 8$ (a natural number)

and Quotient $= 4 \div \frac{2}{1} = 4 \times \frac{1}{2} = 2$ (an integer)

24. (d) The rational number lying between $\frac{5}{6}$ and $\frac{6}{7}$ is

$\frac{1}{2}\left(\frac{5}{6} + \frac{6}{7}\right) = \frac{1}{2}\left(\frac{35+36}{42}\right) = \frac{71}{84}$

25. (b) $-\frac{18}{5} = -3.6$, which lies between – 3 and –4

26. (a) $-\frac{24}{20} = -\frac{6}{5} = -\frac{30}{25}$

27. (b) The rational numbers $\frac{1}{4}$ and $\frac{3}{4}$ can be written as $\frac{250}{1000}$ and $\frac{750}{1000}$. Therefore, $\frac{63}{250} = \frac{252}{1000}$ and $\frac{187}{250} = \frac{748}{1000}$ satisfy this condition.

28. (b) $3.292929....... = 3.\overline{29} = 3\frac{29}{99} = \frac{326}{99}$, a rational number.

33. (c) $\frac{1}{5} + \frac{1}{-6} = \frac{1}{5} - \frac{1}{6} = \frac{6-5}{30} = \frac{1}{30}$

$\left|-\frac{7}{6} + 1\right| = \left|\frac{-7+6}{6}\right| = \left|-\frac{1}{6}\right| = \frac{1}{6}$

$$\left(-\frac{1}{2}\right)^3 = -\frac{1}{8}; \frac{5}{8} \div \frac{-15}{4} = \frac{5}{8} \times \frac{-4}{15} = -\frac{1}{6}$$

$\therefore$ Arranged in descending order, the given numbers are $\frac{1}{6}, \frac{1}{30}, -\frac{1}{8}, -\frac{1}{6}$

i.e., $\left|-\frac{7}{6}+1\right|, \frac{1}{5}+\frac{1}{-6}, \left(-\frac{1}{2}\right)^3, \frac{5}{8} \div \frac{-15}{4}$.

34. (c) Given exp. $= \frac{9 \times 2 - 5 \times 4 \div 10}{-3 \times 5 - 2 \times 2}$

$$= \frac{18-2}{-15-4} = -\frac{16}{19}$$

35. (c) $25 - 5\,[2 + 3\,\{2 - 2 \times 2 + 5\} - 10] \div 4$

$= 25 - 5\,[2 + 3\,(2 - 4 + 5) - 10] \div 4$

$= 25 - 5\,[2 + 3 \times 3 - 10] \div 4$

$= 25 - 5\,[2 + 9 - 10] \div 4$

$= 25 - 5\,(1) \div 4 = 25 - \frac{5}{4}$

$= \frac{95}{4} = 23.75.$

Self Assessment Sheet–1

1. If a and b are such numbers that $a > 0$ and $b < 0$, then which one of the following is always correct ?

(a) $a - b > 0$ (b) $a + b > 0$
(c) $a + b < 0$ (d) $a - b < 0$

2. The rational numbers lying between $\frac{1}{3}$ and $\frac{3}{4}$ are

(a) $\frac{97}{300}, \frac{299}{500}$ (b) $\frac{99}{300}, \frac{301}{400}$
(c) $\frac{95}{300}, \frac{301}{400}$ (d) $\frac{117}{300}, \frac{287}{400}$

3. Match List-I with List-II and select the correct answer using the codes given below the lists :

List I (Number)	List II Divisible by
(A) 4926549	(1) 11
(B) 54192039	(2) 5
(C) 394192045	(3) 4
(D) 19706196	(4) 3

Codes :	(A)	(B)	(C)	(D)
(a)	4	1	2	3
(b)	1	3	2	4
(c)	2	2	3	1
(d)	2	1	4	3

4. Of the numbers 29540, 53416 and 21543

(a) none is divisible by 12
(b) one is divisible by 12
(c) two are divisible by 12
(d) all are divisible by 12

5. On dividing 4996 by a certain number, the quotient is 62, and the remainder is 36. What is the divisor ?

(a) 80 (b) 85
(c) 90 (d) 95

6. Consider the following statements about natural numbers :

1. There exists a smallest natural number.
2. There exists a largest natural number.
3. Between two natural numbers, there is always a natural number.

Which of these is/are correct ?

(a) None (b) Only 1
(c) 1 and 2 (d) 2 and 3

7. The number 311 311 311 311 311 311 311 is

(a) Divisible by both 3 and 11
(b) Divisible by 3 but not by 11
(c) Divisible by 11 but not by 3
(d) Neither divisible by 3 nor by 11.

8. The least perfect square number which is divisible by 3, 4, 5, 6 and 8 is

(a) 900 (b) 1600
(c) 2500 (d) 3600

9. If we divide a positive integer by another positive integer, what is the resulting number ?

(a) It is always a natural number
(b) It is always an integer
(c) It is always a rational number
(d) It is an irrational number

10. If p is a number between 0 and 1, which one of the following is true ?

(a) $p > \sqrt{p}$ (b) $\frac{1}{p} < \sqrt{p}$
(c) $p < \frac{1}{p}$ (d) $p^3 > p^2$

Answers

1. (a)	2. (d)	3. (a)	4. (a)	5. (a)	6. (b)	7. (d)	8. (d)	9. (c)	10. (c)

Chapter 2

FRACTIONS

KEY FACTS

1. Numbers of the form $\frac{a}{b}$ where a and b are whole numbers and $b \neq 0$ are called **fractions**.
2. **Improper fractions** are those in which numerator is greater than or equal to the denominator.
3. A **mixed fraction** has a whole number part and a fractional part.
4. **Equivalent fractions** represent the same value.
5. Multiplying or dividing the numerator and denominator by the same number does not change the value of the fraction.
6. A fraction can be changed to its simplest form by dividing both the numerator and denominator by their HCF.
7. **Comparison of fractions** can be done in the following ways :
   (i) Changing them into equivalent fractions having same denominator by finding the LCM of the denominators and then comparing the numerators. Greater the numerator, greater the fraction.
   (ii) Changing them into equivalent fractions having same numerator by finding the LCM of the numerators and then comparing the denominators. Smaller the denominator, greater the fraction.
   (iii) Finding the cross-product of $\frac{a}{b}$ and $\frac{c}{d}$ (two unlike fractions)
      (a) If $ad > bc$, then $\frac{a}{b} > \frac{c}{d}$
      (b) If $ad < bc$, then $\frac{a}{b} < \frac{c}{d}$
      (c) If $ad = bc$, then $\frac{a}{b} = \frac{c}{d}$

$\frac{a}{b} \times \frac{c}{d}$ (cross-multiplication diagram)

8. To **add** or **subtract** find equivalent fractions that have the same denominator.
9. To **multiply** : $\frac{a}{b} \times \frac{c}{d} = \frac{a \times c}{b \times d}$, then simplify.
10. Reciprocal of a fraction $\frac{a}{b}$ $(a \neq 0, b \neq 0) = \frac{b}{a}$.
11. To divide $\frac{a}{b} \div \frac{c}{d} = \frac{a}{b} \times \frac{d}{c} = \frac{a \times d}{b \times c}$
12. If $\frac{a}{b}$ and $\frac{c}{d}$ $(a \in N, b \in N, c \in N, d \in N)$ are two fractions, then $\frac{a+b}{c+d}$ is a fraction between $\frac{a}{b}$ and $\frac{c}{d}$.
13. **BODMAS** is followed while simplifying a numerical expression having fractions.

Solved Examples

Ex. 1. *Find the value of* $\dfrac{2}{1+\dfrac{1}{1-\dfrac{1}{2}}} \times \dfrac{3}{\dfrac{5}{6} \text{ of } \dfrac{3}{2} \div 1\dfrac{1}{4}}$

Sol. Given exp. $= \dfrac{2}{1+\dfrac{1}{\dfrac{1}{2}}} \times \dfrac{3}{\dfrac{15}{12} \div \dfrac{5}{4}} = \dfrac{2}{1+2} \times \dfrac{3}{\dfrac{15}{12} \times \dfrac{4}{5}} = \dfrac{2}{3} \times \dfrac{3}{1} = \mathbf{2}.$

Ex. 2. ***Find x if*** $5\frac{1}{6} - \left[1\frac{1}{5} + \left\{2\frac{3}{4} \div 5\frac{1}{2} \div x - \left(\frac{5}{6} - \frac{2}{3}\right)\right\}\right] = 2\frac{61}{120}$

Sol. $\frac{31}{6} - \left[\frac{6}{5} + \left\{\frac{11}{4} \div \frac{11}{2} \div x - \left(\frac{5}{6} - \frac{2}{3}\right)\right\}\right] = \frac{301}{120} \Rightarrow \frac{31}{6} - \left[\frac{6}{5} + \left\{\frac{11}{4} \times \frac{2}{11} \times \frac{1}{x} - \left(\frac{5-4}{6}\right)\right\}\right] = \frac{301}{120}$

$\Rightarrow \frac{31}{6} - \left[\frac{6}{5} + \left\{\frac{1}{2x} - \frac{1}{6}\right\}\right] = \frac{301}{120} \Rightarrow \frac{31}{6} - \frac{6}{5} - \frac{1}{2x} + \frac{1}{6} = \frac{301}{120}$

$\Rightarrow \frac{32}{6} - \frac{6}{5} - \frac{1}{2x} = \frac{301}{120} \Rightarrow \frac{160-36}{30} - \frac{1}{2x} = \frac{301}{120}$

$\Rightarrow -\frac{1}{2x} = \frac{301}{120} - \frac{124}{30} \Rightarrow -\frac{1}{2x} = \frac{301-496}{120} = -\frac{195}{120}$

$\Rightarrow x = \frac{120}{2 \times 195} = \mathbf{\frac{4}{13}}.$

Ex. 3. ***Find the value of*** $\dfrac{1}{1+\dfrac{1}{3-\dfrac{4}{2+\dfrac{1}{3-\dfrac{1}{2}}}}} + \dfrac{3}{3-\dfrac{4}{3+\dfrac{1}{1-\dfrac{1}{2}}}}$

Sol. Given exp. $= \dfrac{1}{1+\dfrac{1}{3-\dfrac{4}{2+\dfrac{1}{\dfrac{5}{2}}}}} + \dfrac{3}{3-\dfrac{4}{3+\dfrac{1}{\dfrac{1}{2}}}} = \dfrac{1}{1+\dfrac{1}{3-\dfrac{4}{2+\dfrac{2}{5}}}} + \dfrac{3}{3-\dfrac{4}{3+2}}$

$= \dfrac{1}{1+\dfrac{1}{3-\dfrac{4}{12/5}}} + \dfrac{3}{3-\dfrac{4}{5}} = \dfrac{1}{1+\dfrac{1}{3-\dfrac{20}{12}}} + \dfrac{3}{\dfrac{11}{5}} = \dfrac{1}{1+\dfrac{1}{16/12}} + \dfrac{15}{11}$

$= \dfrac{1}{1+\dfrac{\cancel{12}^{3}}{\cancel{16}^{4}}} + \dfrac{15}{11} = \dfrac{1}{\dfrac{7}{4}} + \dfrac{15}{11} = \dfrac{4}{7} + \dfrac{15}{11} = \dfrac{44+105}{77} = \mathbf{\dfrac{149}{77}}.$

Ex. 4. ***If*** $\frac{1}{4} \times \frac{2}{6} \times \frac{3}{8} \times \frac{4}{10} \times \frac{5}{12} \times \dots\dots\dots\dots \times \frac{31}{64} = \frac{1}{2^x}$***, then what is the value of x ?***

Sol. $\frac{1}{2^x} = \frac{1}{\cancel{4}_2} \times \frac{\cancel{2}^1}{\cancel{6}_2} \times \frac{\cancel{3}^1}{\cancel{8}_2} \times \frac{\cancel{4}^1}{\cancel{10}_2} \times \frac{\cancel{5}^1}{\cancel{12}_2} \times \dots\dots\dots \times \frac{\cancel{30}^1}{\cancel{62}_2} \times \frac{\cancel{31}^1}{64} = \frac{1}{(2 \times 2 \times 2 \times \dots 30 \text{ times}) \times 64}$

$\Rightarrow \frac{1}{2^x} = \frac{1}{2^{30} \times 64} = \frac{1}{2^{30} \times 2^6} = \frac{1}{2^{36}} \Rightarrow x = \mathbf{36}.$

Ex. 5. ***If*** $N = \frac{1}{2}+\frac{1}{6}+\frac{1}{12}+\frac{1}{20}+\frac{1}{30}+ \ldots\ldots\ldots\ldots + \frac{1}{156}$, ***what is the value of N ?***

Sol. Given exp. $= \frac{1}{1.2}+\frac{1}{2.3}+\frac{1}{3.4}+\frac{1}{4.5}+\frac{1}{5.6}+\ldots\ldots+\frac{1}{11.12}+\frac{1}{12.13}$

$$= \left(1-\frac{1}{2}\right)+\left(\frac{1}{2}-\frac{1}{3}\right)+\left(\frac{1}{3}-\frac{1}{4}\right)+\left(\frac{1}{4}-\frac{1}{5}\right)+\left(\frac{1}{5}-\frac{1}{6}\right)+\ldots\ldots\ldots+\left(\frac{1}{11}-\frac{1}{12}\right)+\left(\frac{1}{12}-\frac{1}{13}\right) = 1-\frac{1}{13} = \mathbf{\frac{12}{13}}.$$

Ex. 6. ***What is the value of*** $\frac{3}{1^2.2^2}+\frac{5}{2^2.3^2}+\frac{7}{3^2.4^2}+\frac{9}{4^2.5^2}+\frac{11}{5^2.6^2}+\frac{13}{6^2.7^2}+\frac{15}{7^2.8^2}+\frac{17}{8^2.9^2}+\frac{19}{9^2.10^2}$?

Sol. Given exp. $= \left(1-\frac{1}{2^2}\right)+\left(\frac{1}{2^2}-\frac{1}{3^2}\right)+\left(\frac{1}{3^2}-\frac{1}{4^2}\right)+\left(\frac{1}{4^2}-\frac{1}{5^2}\right)+\left(\frac{1}{5^2}-\frac{1}{6^2}\right)+\left(\frac{1}{6^2}-\frac{1}{7^2}\right)+$

$$\left(\frac{1}{7^2}-\frac{1}{8^2}\right)+\left(\frac{1}{8^2}-\frac{1}{9^2}\right)+\left(\frac{1}{9^2}-\frac{1}{10^2}\right)$$

$$= 1-\frac{1}{10^2} = 1-\frac{1}{100} = \mathbf{\frac{99}{100}}.$$

Ex. 7. ***Production of wheat is*** $2\frac{1}{4}$ ***times that of rice, but the cost of rice is*** $1\frac{1}{4}$ ***times that of wheat. If a farmer produces wheat in place of rice, then what is his income in terms of the previous income ?***

Sol. Let production of rice = x quintals and cost of 1 quintal rice = ₹ y

Then, original income of the farmer = ₹ $(x \times y)$ = ₹ xy

Now as per question, production of wheat = $2\frac{1}{4}$ × production of rice = $\frac{9}{4}x$ quintals

Cost of wheat × $1\frac{1}{4}$ = Cost of rice = ₹ y

⇒ Cost of wheat = ₹ $\frac{4}{5}y$

∴ Present income of the farmer = ₹ $\left(\frac{9x}{4}\times\frac{4y}{5}\right)$ = ₹ $\frac{9}{5}xy = 1\frac{4}{5}$ times of xy

$= 1\frac{4}{5}$ times the previous income.

Ex. 8. ***If a man spends*** $\frac{5}{6}$***th part of money and then earns*** $\frac{1}{2}$ ***part of the remaining money, what part of the money is with him now ?***

Sol. Let the money with the man at first be ₹ 1.

∴ Money spent = $\frac{5}{6}$ of ₹ 1 = ₹ $\frac{5}{6}$

Remaining money = ₹ $\left(1-\frac{5}{6}\right)$ = ₹ $\frac{1}{6}$

Money earned = $\frac{1}{2}$ of ₹ $\frac{1}{6}$ = ₹ $\frac{1}{12}$

∴ Total money with the man = ₹ $\left(\frac{1}{6}+\frac{1}{12}\right)$ = ₹ $\frac{3}{12}$ = ₹ $\frac{1}{4}$

∴ The man now has $\frac{1}{4}$th part of the money.

Ex. 9. ***A woman sells to the first customer half her stock of apples and half an apple, to the second customer she sells half her remaining stock and half an apple, and so on to the third and to the fourth customer. She finds that she has now 15 apples left. How many apples did she have before she started selling ?***

Sol. Suppose she had x apples in the beginning.

Sold to the first customer $= \frac{x}{2} + \frac{1}{2} = \frac{x+1}{2}$

Remaining stock $= x - \frac{x+1}{2} = \frac{2x-x-1}{2} = \frac{x-1}{2}$

Sold to the second customer $= \frac{1}{2} \times \frac{x-1}{2} + \frac{1}{2} = \frac{x-1}{4} + \frac{1}{2} = \frac{x-1+2}{4} = \frac{x+1}{4}$

Remaining stock $= \left(\frac{x-1}{2}\right) - \left(\frac{x+1}{4}\right) = \frac{2x-2-x-1}{4} = \frac{x-3}{4}$

Sold to the third customer $= \frac{1}{2} \times \frac{x-3}{4} + \frac{1}{2} = \frac{x-3}{8} + \frac{1}{2} = \frac{x-3+4}{8} = \frac{x+1}{8}$

Remaining stock $= \left(\frac{x-3}{4}\right) - \left(\frac{x+1}{8}\right) = \frac{2x-6-x-1}{8} = \frac{x-7}{8}$

Sold to the fourth cutomer $= \frac{1}{2} \times \frac{x-7}{8} + \frac{1}{2} = \frac{x-7}{16} + \frac{1}{2} = \frac{x-7+8}{16} = \frac{x+1}{16}$

$\therefore\ x - \left[\frac{x+1}{2} + \frac{x+1}{4} + \frac{x+1}{8} + \frac{x+1}{16}\right] = 15 \Rightarrow x - \left[\frac{8x+8+4x+4+2x+2+x+1}{16}\right] = 15$

$\Rightarrow\ x - \left[\frac{15x+15}{16}\right] = 15 \Rightarrow \frac{16x-15x-15}{16} = 15$

$\Rightarrow x - 15 = 16 \times 15 = 240 \Rightarrow x = 240 + 15 =$ **255.**

Therefore, she had 255 apples before she started selling.

Ex. 10. ***Eight people are planning to share equally the cost of a rental car. If one person withdraws from the arrangement and the others share equally the entire cost of the car, then by how much is the share of each of the remaining persons is increased in terms of original share ?***

Sol. When there are 8 people, the share of each person is $\frac{1}{8}$ of the total cost.

When there are 7 people, each person's share is $\frac{1}{7}$ of the total cost.

$\therefore$ Increase in the share of each person $= \frac{1}{7} - \frac{1}{8} = \frac{1}{56}$, *i.e.*, $\frac{1}{7}$ of $\frac{1}{8}$, *i.e.*, $\frac{1}{7}$ of the original share of each person.

Question Bank–2

1. Which of the following fractions is the largest ?

(a) $\frac{13}{16}$ (b) $\frac{7}{8}$

(c) $\frac{31}{40}$ (d) $\frac{63}{80}$

2. Which of the following fractions is less than $\frac{7}{8}$ and greater than $\frac{1}{3}$?

(a) $\frac{1}{4}$ (b) $\frac{23}{24}$

(c) $\frac{11}{12}$ (d) $\frac{17}{24}$

3. Madan picks up three different digits from the set {1, 2, 3, 4, 5} and forms a mixed number by placing the digits in the spaces $\square\frac{\square}{\square}$. The fractional part of the mixed number should be less than 1 (for example $4\frac{2}{3}$). What is the difference between the largest and smallest possible mixed number that can be formed?

(a) $4\frac{7}{20}$ (b) $4\frac{3}{10}$

(c) $4\frac{9}{20}$ (d) $4\frac{3}{5}$

4. What fraction of $\frac{4}{7}$ must be added to itself to make the sum $1\frac{1}{14}$?

(a) $\frac{1}{2}$ (b) $\frac{4}{7}$

(c) $\frac{7}{8}$ (d) $\frac{15}{14}$

5. If $a=\left(\frac{1}{10}\right)^2$, $b=\frac{1}{5}$ and $c=\sqrt{\frac{1}{100}}$, then which of the following statements is correct ?

(a) $a<b<c$ (b) $a<c<b$

(c) $b<c<a$ (d) $c<a<b$

6. Mohan ate half a pizza on Monday. He ate half of what was left on Tuesday and so on. He followed this pattern for one week. How much of the pizza would he have eaten during the week ?

(a) 99.22% (b) 95%

(c) 98.22% (d) 100%

7. The least fraction that must be added to $1\frac{1}{3}\div1\frac{1}{2}\div1\frac{1}{9}$ to make the result an integer is

(a) $\frac{4}{5}$ (b) $\frac{3}{5}$

(c) $\frac{2}{5}$ (d) $\frac{1}{5}$

8. The expression $\dfrac{5\frac{5}{8}}{6\frac{3}{7}}$ of $\dfrac{6\frac{7}{11}}{9\frac{1}{8}}\div\frac{8}{9}\left(2\frac{3}{11}+\frac{13}{22}\right)of\frac{3}{5}$ equals

(a) 1 (b) $\frac{1}{2}$

(c) $\frac{5}{12}$ (d) $\frac{7}{9}$

9. Find the value of x in the following :

$$1\frac{2}{3}\div\frac{2}{7}\times\frac{x}{7}=1\frac{1}{4}\times\frac{2}{3}\div\frac{1}{6}$$

(a) 0.006 (b) $\frac{1}{6}$

(c) 0.6 (d) 6

10. At the first stop on his route, a driver unloaded $\frac{2}{5}$ of the packages in his van. After he unloaded another 3 packages at his next step, $\frac{1}{2}$ of the original number of packages remained. How many packages were in the van before the first delivery ?

(a) 25 (b) 10

(c) 30 (d) 36

11. A student was asked to solve the fraction $\dfrac{\frac{7}{3}+1\frac{1}{2}\text{of}\frac{5}{3}}{2+1\frac{2}{3}}$ and his answer was $\frac{1}{4}$. By how much was his answer wrong ?

(a) 1 (b) $\frac{1}{55}$

(c) $\frac{1}{220}$ (d) $1\frac{3}{44}$

12. Simplify : $\frac{1}{2}+\left[\frac{1}{2}\times\frac{1}{2}\div\left\{\frac{1}{2}\times\frac{1}{2}\div\frac{1}{2}+\left(\frac{1}{2}\div\frac{1}{2}\right)\right\}\right]$

(a) $\frac{2}{3}$ (b) $\frac{1}{2}$

(c) $\frac{1}{4}$ (d) $\frac{1}{5}$

13. If $\dfrac{2+\dfrac{1}{3\frac{4}{5}}}{2+\dfrac{1}{3+\dfrac{1}{1+\frac{1}{4}}}}=x$, then the value of x is

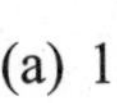

(a) 1 (b) $\frac{3}{7}$

(c) $\frac{1}{7}$ (d) $\frac{8}{7}$

14. If $x=1+\dfrac{1}{1+\dfrac{1}{1+\dfrac{1}{1+\frac{1}{2}}}}$, then the value of $2x+\frac{7}{4}$ is

(a) 3 (b) 4

(c) 5 (d) 6

15. Simplify :

$$\left[3\frac{1}{4}\div\left\{1\frac{1}{4}-\frac{1}{2}\left(2\frac{1}{2}-\overline{\frac{1}{4}-\frac{1}{6}}\right)\right\}\right]\div\left(\frac{1}{2}\text{ of }4\frac{1}{3}\right)$$

(a) 18 (b) 36

(c) 39 (d) 78

16. The expression $\left[\frac{1}{1.2}+\frac{1}{2.3}+\frac{1}{3.4}+\_\_\_\_+\frac{1}{n(n+1)}\right]$ for a natural number is

(a) always greater than 1
(b) always less than 1
(c) always equal to 0
(d) always a negative integer.

17. The value of $\left(1+\frac{1}{1\times2}+\frac{1}{1\times2\times4}+\frac{1}{1\times2\times4\times8}+\frac{1}{1\times2\times4\times8\times16}\right)$ up to 4 places of decimal is

(a) 1.6416 (b) 1.2937
(c) 1.6414 (d) 1.6415

18. Find the value of

$$\left(1-\frac{1}{2^2}\right)\left(1-\frac{1}{3^2}\right)\left(1-\frac{1}{4^2}\right)\left(1-\frac{1}{5^2}\right)\ldots\ldots\ldots\left(1-\frac{1}{9^2}\right)\left(1-\frac{1}{10^2}\right)$$

(a) $\frac{5}{12}$ (b) $\frac{1}{2}$
(c) $\frac{11}{20}$ (d) $\frac{7}{10}$

19. $\frac{1}{20}+\frac{1}{30}+\frac{1}{42}+\frac{1}{56}+\frac{1}{72}+\frac{1}{90}+\frac{1}{110}+\frac{1}{132}$ is equal to

(a) $\frac{1}{8}$ (b) $\frac{1}{7}$
(c) $\frac{1}{6}$ (d) $\frac{1}{10}$

20. A lamp post has half of its length in mud, $\frac{1}{3}$ of its length in water and $3\frac{1}{3}$m above the water, but in the mud. Find the total length of the post ?

(a) 20 m (b) 15 m
(c) 25 m (d) 30 m

21. If the difference between the reciprocal of a positive proper fraction and the fraction itself be $\frac{9}{20}$, then the fraction is

(a) $\frac{3}{5}$ (b) $\frac{3}{10}$
(c) $\frac{4}{5}$ (d) $\frac{5}{4}$

22. What would be the reciprocal of the sum of the reciprocals of the numbers $\frac{3}{5}$ and $\frac{7}{3}$?

(a) $\frac{1}{42}$ (b) $\frac{21}{44}$
(c) $\frac{4}{5}$ (d) $\frac{36}{55}$

23. The product of two fractions is $\frac{14}{15}$ and their quotient is $\frac{35}{24}$. The greater fraction is

(a) $\frac{7}{4}$ (b) $\frac{7}{6}$
(c) $\frac{7}{3}$ (d) $\frac{4}{5}$

24. Chandran gave one-fourth of his money to Suresh. Suresh in turn gave one-third of what he received to Jayesh. If the difference between the amount of Suresh and Jayesh is ₹ 100, how much did Chandran have ?

(a) ₹ 450 (b) ₹ 600
(c) ₹ 800 (d) ₹ 900

25. From a number of mangoes, a man sells half the number of existing mangoes plus 1 to the first customer, then sells $\frac{1}{3}$rd of the remaining number of mangoes plus 1 to the second customer, then $\frac{1}{4}$th of the remaining number of mangoes plus 1 to third customer and $\frac{1}{5}$th of the remaining number of mangoes plus 1 to the fourth customer. He then finds that he does not have any mango left. How many mangoes did he have originally ?

(a) 12 (b) 14
(c) 15 (d) 13

Answers

1. (b)	**2.** (d)	**3.** (a)	**4.** (c)	**5.** (b)	**6.** (a)	**7.** (d)	**8.** (c)	**9.** (d)	**10.** (c)
11. (d)	**12.** (a)	**13.** (a)	**14.** (c)	**15.** (b)	**16.** (b)	**17.** (a)	**18.** (c)	**19.** (c)	**20.** (a)
21. (c)	**22.** (b)	**23.** (b)	**24.** (b)	**25.** (b)					

Hints and Solutions

1. (b) $\frac{13}{16}=\frac{13\times5}{16\times5}=\frac{65}{80},\frac{7}{8}=\frac{7\times10}{8\times10}=\frac{70}{80},\frac{31}{40}=\frac{31\times2}{40\times2}$

$=\frac{62}{80}$ and last fraction is $\frac{63}{80}$

Out of these, the largest fraction $=\frac{70}{80}=\mathbf{\frac{7}{8}}$.

2. (d) $\frac{1}{3}=0.333000,\ \frac{7}{8}=0.875$

$\frac{1}{4}=0.25,\ \frac{23}{24}=0.9583000,\ \frac{11}{12}=0.9166000,$

$\frac{17}{24}=0.7083000$

Since $0.7083000\left(=\frac{17}{24}\right)$ is greater than $0.333000\left(=\frac{1}{3}\right)$ and less than $0.875\left(=\frac{7}{8}\right)$

therefore, $\frac{17}{24}$ lies between $\frac{1}{3}$ and $\frac{7}{8}$.

3. (a) The largest number $=5\frac{3}{4}$

The smallest number $=1\frac{2}{5}$

$\therefore$ Required difference $=5\frac{3}{4}-1\frac{2}{5}=\frac{23}{4}-\frac{7}{5}$

$=\frac{115-28}{20}=\frac{87}{20}=\mathbf{4\frac{7}{20}}$.

4. (c) Let the required fraction be x. Then,

x of $\frac{4}{7}+\frac{4}{7}=1\frac{1}{14}\Rightarrow\frac{4x}{7}+\frac{4}{7}=\frac{15}{14}$

$\Rightarrow\ \frac{4x}{7}=\frac{15}{14}-\frac{4}{7}=\frac{7}{14}=\frac{1}{2}$

$\Rightarrow\ x=\frac{1}{2}\times\frac{7}{4}=\mathbf{\frac{7}{8}}$.

5. (b) $a=\frac{1}{100},\ b=\frac{1}{5}=\frac{20}{100},\ c=\frac{1}{10}=\frac{10}{100}$

$\therefore\ a<c<b$

6. (a) Pizza he ate on Monday $=\frac{1}{2}$

Pizza left $=\frac{1}{2}$

Pizza he ate on Tuesday $=\frac{1}{2}\times\frac{1}{2}=\frac{1}{4}$

Pizza left $=\frac{1}{2}-\frac{1}{4}=\frac{1}{4}$

Pizza he ate on Wednesday $=\frac{1}{2}\times\frac{1}{4}=\frac{1}{8}$

Pizza left $=\frac{1}{4}-\frac{1}{8}=\frac{1}{8}$

Continuing this pattern, we see that

Pizza he ate on Thursday, Friday, Saturday and Sunday is $\frac{1}{16},\frac{1}{32},\frac{1}{64}$ and $\frac{1}{128}$ respectively.

$\therefore$ Quantity of pizza he ate during the week

$=\frac{1}{2}+\frac{1}{4}+\frac{1}{8}+\frac{1}{16}+\frac{1}{32}+\frac{1}{64}+\frac{1}{128}$

$=\frac{64+32+16+8+4+2+1}{128}=\frac{127}{128}$

$=\frac{127}{128}\times100\%=\mathbf{99.22\%}$.

7. (d) $1\frac{1}{3}\div1\frac{1}{2}\div1\frac{1}{9}=\frac{4}{3}\div\frac{3}{2}\div\frac{10}{9}$

$=\frac{4}{3}\times\frac{2}{3}\div\frac{10}{9}=\frac{8}{9}\times\frac{9}{10}=\frac{4}{5}$

$\therefore$ $\frac{1}{5}$ should be added to $\frac{4}{5}$ to make it an integer.

8. (c) Given exp. $=\dfrac{\frac{45}{8}}{\frac{45}{7}}$ of $\dfrac{\frac{73}{11}}{\frac{73}{8}}\div\frac{8}{9}\left(\frac{25}{11}+\frac{13}{22}\right)$ of $\frac{3}{5}$

$=\left[\left(\frac{45}{8}\times\frac{7}{45}\right)\times\left(\frac{73}{11}\times\frac{8}{73}\right)\right]\div\frac{8}{9}\left(\frac{50+13}{22}\right)\times\frac{3}{5}$

$=\left(\frac{7}{8}\times\frac{8}{11}\right)\div\left(\frac{8}{9}\times\frac{63}{22}\right)\times\frac{3}{5}$

$=\frac{7}{11}\div\frac{84}{55}=\frac{7}{11}\times\frac{55}{84}=\mathbf{\frac{5}{12}}$.

9. (d) $\frac{5}{3}\div\frac{2}{7}\times\frac{x}{7}=\frac{5}{4}\times\frac{2}{3}\div\frac{1}{6}$

$\Rightarrow\ \frac{5}{3}\times\frac{7}{2}\times\frac{x}{7}=\frac{5}{4}\times\frac{2}{3}\times\frac{6}{1}$

$\Rightarrow\ \frac{5}{6}\times x=5\ \Rightarrow\ x=\frac{5\times6}{5}=\mathbf{6}$.

10. (c) Let x be the number of packages originally in the van.

Then, $\frac{x}{2}-\frac{2x}{5}=3 \Rightarrow \frac{5x-4x}{10}=3$

$\Rightarrow \frac{x}{10}=3 \Rightarrow x=\mathbf{30}.$

11. (d) Given exp. $=\dfrac{\frac{7}{3}+\frac{3}{2}\times\frac{5}{3}}{2+\frac{5}{3}}=\dfrac{\frac{7}{3}+\frac{5}{2}}{\frac{6+5}{3}}$

$=\dfrac{\frac{14+15}{6}}{\frac{11}{3}}=\frac{29}{6}\times\frac{3}{11}=\frac{29}{22}$

$\therefore$ Required answer $=\frac{29}{22}-\frac{1}{4}=\frac{58-11}{44}$

$=\frac{47}{44}=\mathbf{1\frac{3}{44}}.$

12. (a) Given exp. $=\frac{1}{2}+\left[\frac{1}{2}\times\frac{1}{2}\div\left\{\frac{1}{2}\times\frac{1}{2}\div\frac{1}{2}+1\right\}\right]$

$=\frac{1}{2}+\left[\frac{1}{2}\times\frac{1}{2}\div\left\{\frac{1}{2}\times\frac{1}{2}\times\frac{2}{1}+1\right\}\right]$

$=\frac{1}{2}+\left[\frac{1}{2}\times\frac{1}{2}\div\left\{\frac{1}{2}\times1+1\right\}\right]$

$=\frac{1}{2}+\left[\frac{1}{2}\times\frac{1}{2}\div\left\{\frac{1}{2}+1\right\}\right]$

$=\frac{1}{2}+\left[\frac{1}{2}\times\frac{1}{2}\div\frac{3}{2}\right]=\frac{1}{2}+\left[\frac{1}{2}\times\frac{1}{2}\times\frac{2}{3}\right]$

$=\frac{1}{2}+\frac{1}{6}=\frac{3+1}{6}=\frac{4}{6}=\mathbf{\frac{2}{3}}.$

13. (a) $x=\dfrac{\dfrac{2+\frac{1}{19}}{5}}{2+\dfrac{1}{3+\frac{1}{\frac{5}{4}}}}=\dfrac{2+\frac{5}{19}}{2+\dfrac{1}{3+\frac{4}{5}}}$

$=\dfrac{\frac{43}{19}}{2+\dfrac{1}{\frac{19}{5}}}=\dfrac{\frac{43}{19}}{2+\frac{5}{19}}=\dfrac{\frac{43}{19}}{\frac{43}{19}}=\mathbf{1}.$

14. (c) $x=1+\dfrac{1}{1+\dfrac{1}{1+\frac{1}{\frac{3}{2}}}}=1+\dfrac{1}{1+\dfrac{1}{1+\frac{2}{3}}}$

$=1+\dfrac{1}{1+\frac{1}{\frac{5}{3}}}=1+\dfrac{1}{1+\frac{3}{5}}=1+\dfrac{1}{\frac{8}{5}}=1+\frac{5}{8}=\frac{13}{8}$

$\therefore\ 2x+\frac{7}{4}=2\times\frac{13}{8}+\frac{7}{4}=\frac{13}{4}+\frac{7}{4}=\frac{20}{4}=\mathbf{5}.$

15. (b) Given exp.

$=\left[\frac{13}{4}\div\left\{\frac{5}{4}-\frac{1}{2}\left(\frac{5}{2}-\frac{3-2}{12}\right)\right\}\right]\div\left(\frac{1}{2}\text{ of }\frac{13}{3}\right)$

$=\left[\frac{13}{4}\div\left\{\frac{5}{4}-\frac{1}{2}\left(\frac{5}{2}-\frac{1}{12}\right)\right\}\right]\div\frac{13}{6}$

$=\left[\frac{13}{4}\div\left\{\frac{5}{4}-\frac{1}{2}\times\frac{30-1}{12}\right\}\right]\div\frac{13}{6}$

$=\left[\frac{13}{4}\div\left\{\frac{5}{4}-\frac{29}{24}\right\}\right]\div\frac{13}{6}$

$=\left[\frac{13}{4}\div\frac{30-29}{24}\right]\div\frac{13}{6}$

$=\left(\frac{13}{4}\div\frac{1}{24}\right)\div\frac{13}{6}=\frac{13}{4}\times24\times\frac{6}{13}=\mathbf{36}.$

16. (b) Given exp.

$=\left[\left(1-\frac{1}{2}\right)+\left(\frac{1}{2}-\frac{1}{3}\right)+\left(\frac{1}{3}-\frac{1}{4}\right)+\dots\dots\dots+\left(\frac{1}{n}-\frac{1}{n+1}\right)\right]$

$=1-\frac{1}{2}+\frac{1}{2}-\frac{1}{3}+\frac{1}{3}-\frac{1}{4}+\dots\dots\dots+\frac{1}{n}-\frac{1}{n+1}$

$=1-\frac{1}{n+1}=\frac{n}{n+1}$, which is always less than 1.

17. (a) Given exp. $=1+\dfrac{4\times8\times16+8\times16+16+1}{1\times2\times4\times8\times16}$

$=1+\dfrac{512+128+16+1}{1024}=1+\dfrac{657}{1024}$

$=1+0.6416=\mathbf{1.6416}.$

18. (c) Given exp. $=\left(1-\frac{1}{4}\right)\left(1-\frac{1}{9}\right)\left(1-\frac{1}{16}\right)\left(1-\frac{1}{25}\right)\dots\dots\dots\left(1-\frac{1}{81}\right)\left(1-\frac{1}{100}\right)$

$=\frac{3}{4}\times\frac{8}{9}\times\frac{15}{16}\times\frac{24}{25}\times\frac{35}{36}\times\frac{48}{49}\times\frac{63}{64}\times\frac{80}{81}\times\frac{99}{100}=\mathbf{\frac{11}{20}}.$

19. (c) Given exp. $= \frac{1}{4.5}+\frac{1}{5.6}+\frac{1}{6.7}+\frac{1}{7.8}+\frac{1}{8.9}+\frac{1}{9.10}+\frac{1}{10.11}+\frac{1}{11.12}$

$=\left(\frac{1}{4}-\frac{1}{5}\right)+\left(\frac{1}{5}-\frac{1}{6}\right)+\left(\frac{1}{6}-\frac{1}{7}\right)+\left(\frac{1}{7}-\frac{1}{8}\right)+\left(\frac{1}{8}-\frac{1}{9}\right)+\left(\frac{1}{9}-\frac{1}{10}\right)+\left(\frac{1}{10}-\frac{1}{11}\right)+\left(\frac{1}{11}-\frac{1}{12}\right)$

$=\frac{1}{4}-\frac{1}{12}=\frac{3-1}{12}=\frac{2}{12}=\mathbf{\frac{1}{6}}.$

20. (a) Let the total length of the post be x m.

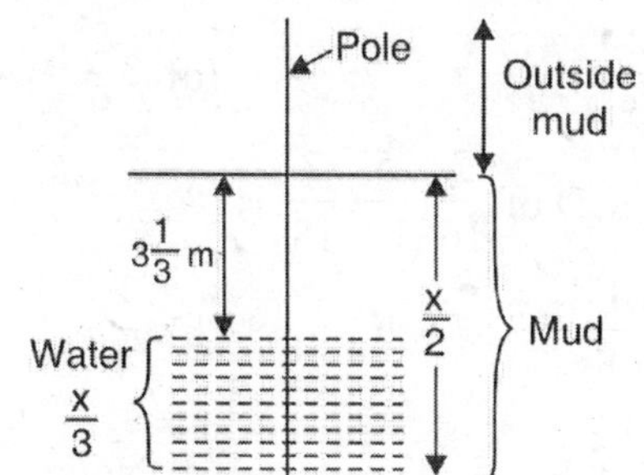

Then, length of post in mud $= \frac{x}{2}$

Length of post in water $= \frac{x}{3}$

Given, $\frac{x}{2}-\frac{x}{3}=3\frac{1}{3}=\frac{10}{3}$

$\Rightarrow \frac{x}{6}=\frac{10}{3} \Rightarrow x = \mathbf{20\ m}.$

21. (c) Let the fraction be $\frac{x}{y}$. Then, reciprocal $= \frac{y}{x}$

Given, $\frac{y}{x}-\frac{x}{y}=\frac{9}{20} \Rightarrow \frac{y^2-x^2}{xy}=\frac{9}{20}$

The only options where $xy = 20$ are (c) and (d), but in (d) y will be 4 and x will be 9.

$\therefore$ $y^2 - x^2 = 16 - 25 = -9$ (negative), hence the option satisfying the given condition is option (c), *i.e.*, $\frac{x}{y}=\frac{4}{5}$.

22. (b) Sum of the reciprocals of $\frac{3}{5}$ and $\frac{7}{3}$

$=\frac{5}{3}+\frac{3}{7}=\frac{35+9}{21}=\frac{44}{21}$

$\therefore$ Required number $= \mathbf{\frac{21}{44}}$.

23. (b) Let the two fractions be x and y.

Give $xy = \frac{14}{15}$ and $\frac{x}{y}=\frac{35}{24}$

$\Rightarrow xy \times \frac{x}{y}=\frac{14}{15}\times\frac{35}{24}$

$\Rightarrow x^2=\frac{49}{36} \quad \Rightarrow \quad x=\frac{7}{6}$

$\therefore \quad y=\frac{14}{15}\div\frac{7}{6}=\frac{4}{5}$

$\therefore$ Greater fraction $= \mathbf{\frac{7}{6}}$.

24. (b) Suppose Chandran has ₹ x. Then,

Suresh has ₹ $\frac{x}{4}$ and Jayesh has $\frac{1}{3}\times$ ₹ $\frac{x}{4}$ $=$ ₹ $\frac{x}{12}$

Given, $\frac{x}{4}-\frac{x}{12}=100 \Rightarrow \frac{3x-x}{12}=100$

$\Rightarrow \frac{2x}{12}=100 \Rightarrow x =$ **₹ 600.**

25. (b) Let the number of mangoes that the man has originally $= x$

Number of mangoes sold to	**Balance**
1st customer $= \frac{x}{2}+1$	$\frac{x-2}{2}$
2nd customer $= \frac{x-2}{6}+1$	$\frac{x-5}{3}$
3rd customer $= \frac{x-5}{12}+1$	$\frac{x-9}{4}$
4th customer $= \frac{x-9}{20}+1$	0

$\frac{x-9}{20}+1=\frac{x-9}{4} \Rightarrow \frac{x+11}{20}=\frac{x-9}{4}$

$\Rightarrow 4x + 44 = 20x - 180$

$\Rightarrow 16x = 224 \Rightarrow x = \mathbf{14}.$

Self Assessment Sheet-2

1. Which one of the following sets of fractions is in the correct sequence of ascending order of their values ?

(a) $-\frac{1}{2}, \frac{5}{6}, \frac{-4}{9}$ (b) $-\frac{3}{7}, \frac{-5}{6}, \frac{3}{5}$

(c) $-\frac{1}{2}, -\frac{4}{9}, \frac{5}{6}$ (d) $-\frac{4}{9}, \frac{5}{6}, \frac{1}{6}$

2. If $\frac{3}{4}$ of an estate is worth ₹ 90,000, then the value of $\frac{2}{3}$ of the same will be

(a) ₹ 60, 000 (b) ₹ 65,000
(c) ₹ 70, 000 (d) ₹ 80,000

3. What is the value of x in $1+\frac{1}{1+\frac{1}{1+\frac{1}{x}}}=\frac{11}{7}$?

(a) 1 (b) 3
(c) $\frac{1}{2}$ (d) $\frac{7}{11}$

4. The fraction $\frac{3}{5}$ is found between which pair of fractions on a number line ?

(a) $\frac{7}{10}$ and $\frac{3}{4}$ (b) $\frac{2}{5}$ and $\frac{1}{2}$
(c) $\frac{1}{3}$ and $\frac{5}{13}$ (d) $\frac{2}{7}$ and $\frac{8}{11}$

5. $\frac{4}{15}$ of $\frac{5}{7}$ of a number is greater than $\frac{4}{9}$ of $\frac{2}{5}$ of the same number by 8. What is half of the number ?

(a) 630 (b) 315
(c) 210 (d) 105

6. If two-third of three-fourth of a number added to three-fourth of the fourth-fifth of the number is x times the number, the value of x is

(a) $\frac{11}{10}$ (b) $1\frac{1}{11}$
(c) $\frac{10}{11}$ (d) $\frac{9}{11}$

7. What is the missing figure in the expression given below ?

$$\frac{16}{7}\times\frac{16}{7}-\frac{\otimes}{7}\times\frac{9}{7}+\frac{9}{7}\times\frac{9}{7}=1$$

(a) 1 (b) 7
(c) 4.57 (d) 32

8. Simplify : $\frac{2\frac{3}{4}}{1\frac{5}{6}}\div\frac{7}{8}\times\left(\frac{1}{3}+\frac{1}{4}\right)+\frac{5}{7}\div\frac{3}{4}$ of $\frac{3}{7}$

(a) $\frac{56}{77}$ (b) $\frac{49}{80}$
(c) $\frac{2}{3}$ (d) $3\frac{2}{9}$

9. The GCD of $\frac{3}{16},\frac{5}{12},\frac{7}{18}$ is

(a) $\frac{105}{48}$ (b) $\frac{1}{4}$
(c) $\frac{1}{48}$ (d) None

10. Suppose $a=\frac{2}{3}b$, $b=\frac{2}{3}c$, and $c=\frac{2}{3}d$, what would be the value of b as a fraction of d ?

(a) $\frac{2}{3}$ (b) $\frac{4}{3}$
(c) $\frac{4}{9}$ (d) $\frac{8}{27}$

Answers

1. (c)	2. (d)	3. (b)	4. (d)	5. (b)	6. (a)	7. (d)	8. (d)	9. (d)	10. (c)

Chapter 3 DECIMALS

KEY FACTS

1. A fraction whose denominator is 10 or a power of 10 like 10^2, 10^3, etc., i.e., 10, 100, 1000 etc., is called a **decimal fraction.**
2. **A decimal number has two parts, whole number part** to the left of decimal point and *decimal part* to the right of the decimal point.
3. **To add or subtract decimals**

   (i) *Line up the decimal points* (ii) *Add or subtract* (iii) *Put the point in the answer.*
4. To multiply decimals by 10^n where n is a positive integer move the decimal point n places to the right, whereas to divide decimals by 10^n where n is a positive integer, move the decimal point n places to the left.
5. To multiply decimal with whole numbers or decimals multiply the number without taking the decimal point. Then the product has the same number of decimal places as there are in the multiplier and the multiplicand put together.
6. **When dividing a decimal by a whole number** put the decimal point in the answer above the decimal point in the number being divided.
7. **To round a decimal number to given decimal places**

   (i) *Take one digit extra than the required decimal places.*

   (ii) *If it is 5 or more, then add 1 to the previous digit and if it is less than 5, leave the previous digit as such omit the extra digit.*
8. **To change fractions to decimals** divide the numerator by the denominator.
9. A fraction $\frac{p}{q}$ is **terminating decimal** if in its lowest form the denominator has factors as 2 or 5 or both 2 and 5.
10. A fraction $\frac{p}{q}$ is a **non-terminating repeating** or **recurring decimal** if a digit or a block of digits are repeated in the decimal part.

    $\frac{2}{3} = 0.66.................. = 0.\bar{6}$, $\frac{1}{11} = 0.0909............. = 0.\overline{09}$, $\frac{4}{7} = 0.571428571428 = 0.\overline{571428}$

    (a) **Pure recurring decimals :** A decimal number in which all the figures after the decimal point are repeated is known as a pure recurring decimal, e.g., $0.\bar{8}, 0.\overline{13}, 0.\overline{462}$.

    To convert a pure recurring decimal into a vulgar fraction, write the repeated block only once in the numerator and write as many nines in the denominator as is the number of repeated digits.

    $0.\bar{6} = \frac{6}{9} = \frac{2}{3}$, $0.\overline{89} = \frac{89}{99}$, $0.\overline{6705} = \frac{6705}{9999}$, $1.\overline{37} = 1 + \frac{37}{99} = \frac{136}{99}$.

(b) **Mixed recurring decimals** : A decimal number in which at least one figure after the decimal point is not repeated and then some figure or figures are repeated is known as a mixed recurring decimal, *e.g.*, $0.4\overline{7}$, $0.31\overline{67}$, $14.06\overline{23}$ etc.

To convert a mixed recurring decimal into a vulgar fraction, consider the decimal part. Subtract the non-repeating part from the whole and divide it by as many nines as there are in the repeating block X as many tens as there are non-repeating digits.

$$0.5\overline{76} = \frac{576-5}{990} = \frac{571}{990}$$

$$1.415\overline{5} = 1 + \frac{4155-415}{9000} = 1 + \frac{3740}{9000} = 1 + \frac{187}{450} = \frac{637}{450}.$$

Solved Examples

Ex. 1. ***What is the value of (4.7 × 13.26 + 4.7 × 9.43 + 4.7 × 77.31) ?***

Sol. $4.7 \times 13.26 + 4.7 \times 9.43 + 4.7 \times 77.31 = 4.7 \times (13.26 + 9.43 + 77.31) = 4.7 \times 100 = \mathbf{470}$.

Ex. 2. ***Which pair of operations will make the equation below true when inserted into the blank spaces in the order shown ?***

$$2\frac{3}{10} \_\_\_ 1.5 \_\_\_ 2 = 1.8$$

(a) – and + *(b) × and +* *(c) + and –* *(d) × and –*

Sol. Trying all the given options we get,

(a) $2.3 - 1.5 + 2 = 0.8 + 2 = 2.8$ (b) $2.3 \times 1.5 + 2 = 3.45 + 2 = 5.45$

(c) $2.3 + 1.5 - 2 = 3.8 - 2 = 1.8$ (d) $2.3 \times 1.5 - 2 = 3.45 - 2 = 1.45$

Hence option (c) makes the given equation true.

Ex. 3. ***Simplify :*** $\dfrac{0.2 \times 0.2 + 0.2 \times 0.02}{0.044}$

Sol. $\dfrac{0.2 \times 0.2 + 0.2 \times 0.02}{0.044} = \dfrac{0.04 + 0.004}{0.044} = \dfrac{0.044}{0.044} = \mathbf{1}.$

Ex. 4. ***Evaluate:*** $\dfrac{0.0203 \times 2.92}{0.0073 \times 14.5 \times 0.7}$

Sol. Since the number of decimal places in the numerator and denominator are equal.

Given expression $= \dfrac{\cancel{203}^{\cancel{29}^{1}} \times \cancel{292}^{4}}{\cancel{73}_{1} \times \cancel{145}_{5} \times \cancel{7}_{1}} = \dfrac{4}{5} = \mathbf{0.8}.$

Ex. 5. ***Simplify :*** $8.7 - [7.6 - \{6.5 - (5.4 - \overline{4.3 - 2})\}]$

Sol. Given exp. $= 8.7 - [7.6 - \{6.5 - (5.4 - 2.3)\}] = 8.7 - [7.6 - \{6.5 - 3.1\}]$

$= 8.7 - [7.6 - 3.4] = 8.7 - 4.2 = \mathbf{4.5}.$

Ex. 6. ***In the expression***

$2.5 + 0.05 - [1.6 - \{3.2 - (3.2 + 2.1 \div x)\}] = 0.65$, ***find the value of x.***

Sol. Given, $2.5 + 0.05 - \left[1.6 - \left\{3.2 - \left(3.2 + \dfrac{2.1}{x}\right)\right\}\right] = 0.65$

$$\Rightarrow 2.5 + 0.05 - \left[1.6 - \left\{3.2 - 3.2 - \frac{2.1}{x}\right\}\right] = 0.65 \Rightarrow 2.55 - \left[1.6 + \frac{2.1}{x}\right] = 0.65$$

$$\Rightarrow 2.55 - 1.6 - \frac{2.1}{x} = 0.65 \Rightarrow 0.95 - \frac{2.1}{x} = 0.65$$

$$\Rightarrow \frac{2.1}{x} = 0.95 - 0.65 = 0.30 \Rightarrow x = \frac{2.10}{0.30} = \frac{210}{30} = 7.$$

Ex. 7. ***Convert (i) $0.\overline{89}$ (ii) $4.\overline{173}$ (iii) $6.27\overline{97}$ to vulgar fractions***

Sol. (i) $0.\overline{89} = \mathbf{\frac{89}{99}}$

(ii) $4.\overline{173} = 4 + 0.\overline{173} = 4 + \frac{173-1}{990} = 4 + \frac{172}{990} = 4 + \frac{86}{495} = \mathbf{4\frac{86}{495}}$

(iii) $6.27\overline{97} = 6 + \frac{2797-27}{9900} = 6 + \frac{2770}{9900} = 6 + \frac{277}{990} = \mathbf{6\frac{277}{990}}$

Ex. 8. ***Simplify : $0.\bar{4} + 0.\overline{61} + 0.\overline{11} - 0.\overline{36}$***

Sol. $0.\bar{4} + 0.\overline{61} + 0.\overline{11} - 0.\overline{36} = \frac{4}{9} + \frac{61}{99} + \frac{11}{99} - \frac{36}{99} = \frac{4}{9} + \frac{72}{99} - \frac{36}{99} = \frac{4}{9} + \frac{36}{99}$

$$= \frac{44}{99} + \frac{36}{99} = \frac{80}{99} = \mathbf{0.\overline{80}}.$$

Ex. 9. ***Which of the following numbers 0.1, 0.11, $(0.11)^2$, $\sqrt{0.0001}$ is the greatest ?***

Sol. $(0.11)^2 = 0.0121$

$\sqrt{0.0001} = 0.01$

∴ Arranging the numbers, 0.1, 0.11, 0.0121 and 0.01 in ascending order we get $0.01 < 0.0121 < 0.1 < 0.11$.

∴ 0.11 is the greatest number.

Ex. 10. ***What is $0.\overline{09} \times 7.\bar{3}$ equal to ?***

Sol. $0.\overline{09} \times 7.\bar{3} = \frac{9}{99} \times 7\frac{3}{9} = \frac{9}{99} \times \frac{66}{9} = \frac{6}{9} = \mathbf{0.\bar{6}}.$

Question Bank–3

1. If the numbers $\frac{4}{5}$, 81% and 0.801 are arranged from smallest to largest, the correct order is :
   (a) $\frac{4}{5}$, 81%, 0.801 (b) 81%, 0.801, $\frac{4}{5}$
   (c) 0.801, $\frac{4}{5}$, 81% (d) $\frac{4}{5}$, 0.801, 81%

2. If all the fractions $\frac{3}{5}, \frac{1}{8}, \frac{8}{11}, \frac{4}{9}, \frac{2}{7}, \frac{5}{7}$ and $\frac{5}{12}$ are arranged in descending order of their values, which one will be the third ?
   (a) $\frac{1}{8}$ (b) $\frac{3}{5}$
   (c) $\frac{5}{12}$ (d) $\frac{8}{11}$

3. Evaluate the expression $6\frac{1}{4} \times 0.25 + 0.75 - 0.3125$.
   (a) 5.9375 (b) 4.2968
   (c) 2.1250 (d) 2

4. Find the values of $\frac{0.34 - 0.034}{0.0034 \div 34}$.
   (a) 0.306 (b) 306
   (c) 3060 (d) 0.0306

5. $\frac{42.31 - 26.43}{42.31 + 26.43} \div \frac{423.1 - 264.3}{4.231 + 2.643}$ is equal to
   (a) 10^{-2} (b) 10^{-1}
   (c) 10 (d) 10^{2}

6. $2.8\overline{768}$ expressed as a rational number is

(a) $2\frac{878}{999}$ (b) $2\frac{9}{10}$

(c) $2\frac{292}{333}$ (d) $2\frac{4394}{4995}$

7. When 0.125125....... is converted into a fraction the result is

(a) $\frac{63}{487}$ (b) $\frac{250}{990}$

(c) $\frac{125}{999}$ (d) $\frac{125}{990}$

8. $0.29\overline{56}$ when expressed as a vulgar fraction is

(a) $\frac{2956}{1000}$ (b) $\frac{2956}{10000}$

(c) $\frac{2927}{9900}$ (d) $\frac{2900}{9999}$

9. The value of $0.\overline{2}+0.\overline{3}+0.\overline{4}+0.\overline{9}+0.\overline{39}$ is

(a) $0.\overline{57}$ (b) $1\frac{20}{33}$

(c) $2\frac{1}{3}$ (d) $2\frac{13}{33}$

10. $(0.34\overline{67}+0.13\overline{33})$ is equal to

(a) $0.\overline{48}$ (b) 0.4803

(c) $0.48\overline{01}$ (d) $0.4\overline{8}$

11. $[(5.\overline{88}-4.\overline{58})-(0.\overline{64}+0.\overline{36})$ is equal to

(a) $1.\overline{01}$ (b) $1.\overline{30}$

(c) $1.\overline{19}$ (d) $0.\overline{29}$

12. The simplification of $3.\overline{36}-2.\overline{05}+1.\overline{33}$ is equal to

(a) 2.6 (b) 2.64

(c) $2.\overline{61}$ (d) $2.\overline{64}$

13. Simplify : $(0.\bar{1})^2\{1-9(0.1\bar{6})^2\}$

(a) $-\frac{1}{162}$ (b) $\frac{1}{108}$

(c) $\frac{7696}{10^6}$ (d) $\frac{1}{106}$

14. If $\frac{547.527}{0.0082}=x$, then the value of $\frac{547527}{82}$ is

(a) $\frac{x}{10}$ (b) $10x$

(c) $100\,x$ (d) $\frac{x}{100}$

15. $\frac{24.23\times 1.423\times 34.21}{521.3\times 413.32\times 2.53}$ is same as

(a) $\frac{2423\times 1423\times 3421}{5213\times 41332\times 253}$

(b) $\frac{242.3\times 142.3\times 3421}{5213\times 4133.2\times 2.53}$

(c) $\frac{2.423\times 14.23\times 342.1}{521.3\times 4133.2\times 2.53}$

(d) $\frac{24.23\times 14.23\times 3.421}{5.213\times 41332\times 0.253}$

16. Which of the following is greater than 1000.01 ?

(a) 0.00010001×10^7 (b) 0.00001×10^8

(c) 1.1×10^2 (d) 1.00001×10^3

17. Simplify : $\frac{3.6\times 0.48\times 2.50}{0.12\times 0.09\times 0.5}$

(a) 80 (b) 800

(c) 8000 (d) 80,000

18. Simplify : $[0.9-\{2.3-3.2-(7.1-5.4-3.5)\}]$

(a) 0.18 (b) 1.8

(c) 0 (d) 2.6

19. In the expression $24-[2.4-\{0.24\times 2-(0.024-x)\}]=22.0589$, the value of x is

(a) 0.0024 (b) 0.024

(c) 0.24 (d) 2.4

20. $1+\frac{1}{4\times 3}+\frac{1}{4\times 3^2}+\frac{1}{4\times 3^3}$ is equal to

(a) 1.120 (b) 1.250

(c) 1.140 (d) 1.160

Answers

1. (d)	2. (b)	3. (d)	4. (c)	5. (a)	6. (c)	7. (c)	8. (c)	9. (d)	10. (c)
11. (d)	12. (d)	13. (b)	14. (a)	15. (c)	16. (a)	17. (b)	18. (c)	19. (a)	20. (a)

Hints and Solutions

1. (d) $\frac{4}{5} = 0.80$, $81\% = \frac{81}{100} = 0.81$.

Arranged in ascending order, the given numbers are 0.80, 0.801, 0.81, *i.e.*, $\frac{4}{5}$, 0.801, 81%.

2. (b) $\frac{3}{5} = 0.6, \frac{1}{8} = 0.125, \frac{8}{11} = 0.73$ (approx),

$\frac{4}{9} = 0.44$ (approx.), $\frac{2}{7} = 0.286$ (approx),

$\frac{5}{7} = 0.714$ (approx), $\frac{5}{12} = 0.42$ (approx)

Therefore, arranged in descending order, the fractions are $\frac{8}{11}, \frac{5}{7}, \frac{3}{5}, \frac{4}{9}, \frac{5}{12}, \frac{2}{7}, \frac{1}{8}$.

The third fraction is $\frac{3}{5}$.

3. (d) $6\frac{1}{4} \times 0.25 + 0.75 - 0.3125$

$= 6.25 \times 0.25 + 0.75 - 0.3125$

$= 1.5625 + 0.75 - 0.3125$

$= 2.3125 - 0.3125 = \mathbf{2}$.

4. (c) $\frac{0.34 - 0.034}{0.0034 \div 34} = \frac{0.306}{0.0001} = \frac{0.3060}{0.0001}$

$= \frac{3060}{1} = \mathbf{3060}$.

5. (a) $\frac{42.31 - 26.43}{42.31 + 26.43} \div \frac{423.1 - 264.3}{4.231 + 2.643}$

$= \frac{15.88}{68.74} \div \frac{158.8}{6.874}$

$= \frac{15.88}{68.74} \times \frac{6.874}{158.8} = \frac{1588}{6874} \times \frac{68.74}{1588}$

$= \frac{68.74}{6874} = \frac{6874}{687400} = \frac{1}{100} = \mathbf{0.01}$.

6. (c) $2.8\overline{768} = 2 + 0.8\overline{768}$

$= 2 + \frac{8768 - 8}{9990} = 2 + \frac{8760}{9990}$

$= 2 + \frac{292}{333} = \mathbf{2\frac{292}{333}}$.

7. (c) $0.125125 \ldots = 0.\overline{125} = \mathbf{\frac{125}{999}}$.

8. (c) $0.29\overline{56} = \frac{2956 - 29}{9900} = \mathbf{\frac{2927}{9900}}$.

9. (d) $0.\bar{2} + 0.\bar{3} + 0.\bar{4} + 0.\bar{9} + 0.\overline{39}$

$= \frac{2}{9} + \frac{3}{9} + \frac{4}{9} + \frac{9}{9} + \frac{39}{99}$

$= \frac{22 + 33 + 44 + 99 + 39}{99}$

$= \frac{237}{99} = \mathbf{2\frac{13}{33}}$.

10. (c) $0.34\overline{67} + 0.13\overline{33} = \frac{3467 - 34}{9900} + \frac{1333 - 13}{9900}$

$= \frac{3433}{9900} + \frac{1320}{9900} = \frac{4753}{9900}$

$= \frac{4801 - 48}{9900} = \mathbf{0.48\overline{01}}$.

11. (d) $(5.\overline{88} - 4.\overline{58}) - (0.\overline{64} + 0.\overline{36})$

$= \left[\left(5 + 0.\overline{88}\right) - \left(4 + 0.\overline{58}\right)\right] - \left(\frac{64}{99} + \frac{36}{99}\right)$

$= \left(5 + \frac{88}{99} - 4 - \frac{58}{99}\right) - \left(\frac{100}{99}\right)$

$= \left(1 + \frac{30}{99} - \frac{100}{99}\right) = \left(1 - \frac{70}{99}\right) = \frac{29}{99} = \mathbf{0.\overline{29}}$.

12. (d) $3.\overline{36} - 2.\overline{05} + 1.\overline{33}$

$= 3 + 0.\overline{36} - (2 + 0.\overline{05}) + 1 + 0.\overline{33}$

$= 3 + \frac{36}{99} - 2 - \frac{5}{99} + 1 + \frac{33}{99}$

$= 2 + \frac{64}{99} = 2 + 0.\overline{64} = \mathbf{2.\overline{64}}$.

13. (b) $(0.\bar{1})^2\{1 - 9(0.1\bar{6})^2\}$

$= \left(\frac{1}{9}\right)^2 \left\{1 - 9\left(\frac{16-1}{90}\right)^2\right\} = \frac{1}{81}\left\{1 - 9 \times \left(\frac{15}{90}\right)^2\right\}$

$= \frac{1}{81}\left\{1 - 9 \times \frac{1}{36}\right\} = \frac{1}{81} \times \frac{3}{4} = \mathbf{\frac{1}{108}}$.

14. (a) Given, $\frac{547.527}{0.0082} = x \Rightarrow \frac{5475270}{82} = x$

$\Rightarrow \frac{547527}{82} = \frac{x}{10}$.

15. (c) In the given expression the number of decimal places in the numerator = 7 and number of

decimal places in the denominator = 5.
Checking the number of decimal places in the numerator and denominator of all the options given, we see that in option (c) No. of decimal places in numerator = 7 and No. of decimal places in denominator = 5.

16. (a) $0.00010001 \times 10^7 = 1000.1$
$0.00001 \times 10^8 = 1000$
$1.1 \times 10^2 = 110$
$1.00001 \times 10^3 = 1000.01$
We observe that 1000.1 > 1000.01
i.e., $0.00010001 \times 10^7 > 1000.01$.

17. (b) $\dfrac{3.6 \times 0.48 \times 2.50}{0.12 \times 0.09 \times 0.5} = \dfrac{36 \times 48 \times 250}{12 \times 9 \times 5} = 800$
No. of decimal places in num. and den. being equal.

18. (c) $[0.9 - \{2.3 - 3.2 - (7.1 - 5.4 - 3.5)\}]$
$= [0.9 - \{2.3 - 3.2 - (7.1 - 8.9)\}]$
$= [0.9 - \{2.3 - 3.2 - (-1.8)\}]$
$= [0.9 - \{2.3 - 3.2 + 1.8\}]$
$= [0.9 - \{4.1 - 3.2\}]$
$= [0.9 - 0.9] = \mathbf{0}$.

19. (a) $24 - [2.4 - \{0.24 \times 2 - (0.024 - x)\}] = 22.0584$
$\Rightarrow 24 - [2.4 - \{0.48 - 0.024 + x\}] = 22.0584$
$\Rightarrow 24 - [2.4 - \{0.456 + x\}] = 22.0584$
$\Rightarrow 24 - [2.4 - 0.456 - x] = 22.0584$
$\Rightarrow 24 - 1.944 + x = 22.0584$
$\Rightarrow 22.056 + x = 22.0584$
$\Rightarrow x = 22.0584 - 22.056 = \mathbf{0.0024}$.

20. (a) Given expression $= 1 + \dfrac{1}{12} + \dfrac{1}{36} + \dfrac{1}{108}$
$= \dfrac{108 + 9 + 3 + 1}{108} = \dfrac{121}{108} = \mathbf{1.120}$ (approx.)

Self Assessment Sheet–3

1. 5.63 divided by 0.01 is equal to
(a) 563 (b) 56.3
(c) 0.563 (d) 5630

2. Consider the following statements :
1. $\dfrac{1}{22}$ cannot be written as a terminating decimal
2. $\dfrac{2}{15}$ can be written as a terminating decimal.
3. $\dfrac{1}{16}$ can be written as a terminating decimal.

Which of the statements given above is/are correct ?
(a) 1 only (b) 2 only
(c) 1 and 3 (d) 2 and 3

3. If $2.5252525 \ldots = \dfrac{p}{q}$ (in the lowest form), then what is the value of $\dfrac{q}{p}$?
(a) 0.4 (b) 0.42525
(c) 0.0396 (d) 0.396

4. Which one of the following is not a correct statement ?
(a) $0.\overline{01} = \dfrac{1}{90}$ (b) $0.\bar{1} = \dfrac{1}{9}$
(c) $0.\bar{2} = \dfrac{2}{9}$ (d) $0.\bar{3} = \dfrac{1}{3}$

5. If $\dfrac{1}{3.718} = 0.2689$, then the value of $\dfrac{1}{0.0003718}$ is
(a) 2689 (b) 2.689
(c) 26890 (d) 0.2689

6. If $47.2506 = 4A + \dfrac{7}{B} + 2C + \dfrac{5}{D} + 6E$, then the value of $5A + 3B + 6C + D + 3E$ is :
(a) 53.6003 (b) 53.603
(c) 153.6003 (d) 213.0003

7. The value of $(0.4\overline{467} + 0.1\overline{444})$ is :
(a) 0.59 (b) $0.59\overline{12}$
(c) $0.\overline{59}$ (d) $0.5\bar{9}$

8. The value of $\dfrac{20 \times (0.3)^2}{0.018} \div 0.5$ of 0.2 is
(a) 10 (b) 100
(c) 20 (d) 1000

9. If 178 × 34 = 6052, then 6.052 ÷ 17.8 = ?
(a) 34 (b) 0.34
(c) 0.034 (d) None of these

10. 58.326 × 463.9 × 0.0081 is the same as
(a) 5.8326 × 4.639 × 8.1
(b) 5.8326 × 4.639 × 0.81
(c) 58326 × 4639 × 0.0000081
(d) None of these

Answers

1. (a) **2.** (c) **3.** (d) **4.** (a) **5.** (a) **6.** (c) **7.** (b) **8.** (d) **9.** (b)
10. (a) [**Hint.** The number of decimal places in the given problem and the answer should be the same.]

HCF AND LCM

KEY FACTS

1. Exact divisors of a number are called its **factors.** A factor of a number is either less than or equal to the number.
2. Number obtained on multiplying a given number by any counting number is called its **multiple**.
   The multiple of a number is either equal to or greater than the number.
3. The highest common factor of two or more numbers that exactly divides them is called their **Highest Common Factor (HCF).**
   Methods of finding HCF.
   (i) **Prime factorization.**
   Express each of the given numbers as the product of their prime factors. The product of the least powers of the common factors gives the HCF of the given numbers.
   (ii) **Continued division method.**
   Divide the larger number by the smaller number and get a remainder. Divide the previous divisor by the remainder last obtained. Repeat this until the remainder becomes zero.
   (iii) To find the HCF of more than two numbers, first find the HCF of any two numbers and then find the HCF of the result and the third number and so on. The final HCF is the required HCF.
4. The **Least Common Multiple (LCM)** of two or more numbers is the smallest number into which each of the numbers divides without a remainder.
   Methods of finding LCM
   (i) **Prime factorisation method.**
   Express each number as a product of prime factors. The product of the greatest powers of these prime factors is the LCM of the given numbers.
   (ii) **Common division method.**
   Arrange the given numbers in a row in any order. Now divide by a prime number which divides exactly at least two of the given numbers and carry forward the numbers which are not divisible. Repeat this process till no numbers have a common factor other than 1. The product of the divisors and the remaining numbers is the LCM of the given numbers.
5. The product of the HCF and LCM of two numbers is equal to the product of the numbers.
6. Two numbers are co-prime of their HCF = 1 and then, product of the two given numbers = Their LCM.
7. The HCF always divides the LCM for a set of numbers.
8. HCF of given fractions $= \dfrac{\text{HCF of numerators}}{\text{LCM of denominators}}$

   LCM of given fractions $= \dfrac{\text{LCM of numerators}}{\text{HCF of denominators}}$

9. **HCF and LCM of decimals.** To find the HCF and LCM of decimals we convert the given decimals to like decimals, adding zeroes in the numbers wherever necessary. Now we find the HCF and LCM of the numbers without taking the decimal point into consideration (*i.e.*, treating them as whole numbers). The number of decimal places in the answer are equal to the number of decimal places in the given numbers. Accordingly, put the decimal point in the answer.

SOME IMPORTANT RESULTS

10. The greatest number that will divide x, y and z leaving remainders a, b and c respectively is given by [HCF of $(x - a)$, $(y - b)$, $(z - c)$].
11. The greatest number that will divide x, y and z leaving the same remainder in each case is given by [HCF of $(x - y)$, $(y - z)$, $(z - x)$].
12. The least number which when divided by x, y and z leaves the same remainder R in each case is given by [LCM of $(x, y, z) + R$]
13. The least number which when divided by x, y and z leaves the remainders a, b and c respectively is given by [LCM of $(x, y, z) - p$]
where $p = x - a = y - b = z - c$.

Solved Examples

Ex. 1. ***What is the product of the digits of the greatest number that divides 690 and 875 leaving remainders 10 and 25 respectively ?***

Sol. Required no. = HCF of (690 – 10, 875 – 25)
= HCF of 680 and 850
= 170

```
        1
680)850
   -680
    170)680(4
       -680
          0
```

∴ Product of the digits of 170 = 1 × 7 × 0 = **0**.

Ex. 2. ***Find the largest number which divides 55, 127 and 175 so as to leave the same remainder in each case.***

Sol. Since the remainders are the same, the difference of every pair of given numbers would be exactly divisible by the required number.

127 – 55 = 72, 175 – 127 = 48, 175 – 55 = 120

∴ Required number = HCF of 72, 48 and 120

$72 = 2^3 \times 3^2$, $48 = 2^4 \times 3$, $120 = 2^3 \times 3 \times 5$

∴ Required number = $2^3 \times 3 = 8 \times 3 =$ **24**.

Ex. 3. ***Find the smallest number which when decreased by 10 is exactly divisible by 16, 21, 24 and 42 ?***

Sol. Required no. = (LCM of 16, 21, 24, 42) + 10
= (2 × 2 × 2 × 3 × 7 × 2) + 10
= 336 + 10
= **346.**

2	16, 21, 24, 42
2	8, 21, 12, 21
2	4, 21, 6, 21
3	2, 21, 3, 21
7	2, 7, 1, 7
	2, 1, 1, 1

Ex. 4. ***What is the least number which when divided by 36, 38, 57, 114 and 19 leaves the remainder 26, 28, 47, 104 and 9 respectively ?***

Sol. Since the difference between any divisor and the corresponding remainder is same *i.e.*, 10 (36 – 26 = 10, 38 – 28 = 10, 57 – 47 = 10, 114 – 104 = 10 and 19 – 9 = 10) therefore, the required number will be obtained by subtracting 10 from the LCM of divisors.

19	36, 38, 57, 114, 19
2	36, 2, 3, 6, 1
3	18, 1, 3, 3, 1
	6, 1, 1, 1, 1

$\therefore$ Reqd. no. $= (\text{LCM of } 36, 38, 57, 114, 19) - 10$
$= (19 \times 2 \times 3 \times 6) - 10$
$= 684 - 10$
$= \mathbf{674}$.

Ex. 5. ***Find the smallest number of six digits that is exactly divisible by 40, 35, 85, 119 and 136.***

Sol. LCM of 40, 35, 85, 119 and 136 $= 2 \times 2 \times 2 \times 17 \times 5 \times 7 = 4760$
Least number of six digits = 100000
Dividing 100000 by 4760, we get remainder = 40
$\therefore$ Least number of 6-digits exactly divisible by 4760
$= 100000 + (4760 - 40)$
$= 100000 + 4720 = \mathbf{104720}$.

$$\begin{array}{r} 21 \\ 4760\overline{)100000} \\ -9520 \\ \hline 4800 \\ 4760 \\ \hline 40 \end{array}$$

2	40, 35, 85, 119, 136
2	20, 35, 85, 119, 68
2	10, 35, 85, 119, 34
17	5, 35, 85, 119, 17
5	5, 35, 5, 7, 1
7	1, 7, 1, 7, 1
	1, 1, 1, 1, 1

Ex. 6. ***The sum of two numbers is 204 and their HCF is 17. Find all the possible pairs of such numbers.***

Sol. Since the HCF is 17, the number must be of the form $17a$ and $17b$, where a and b are prime to each other.
Given, $17a + 17b = 204 \Rightarrow a + b = 12$
The possible pairs of numbers whose sum is 12 are 1, 11; 2, 10; 3, 9; 4, 8; 5, 7; 6, 6
The only pairs of numbers which are prime to each other are 1, 11 and 5, 7.
$\therefore$ The required pairs are $17 \times 1 = \mathbf{17}$ and $17 \times 11 = \mathbf{187}$ and $17 \times 5 = \mathbf{85}$ and $17 \times 7 = \mathbf{119}$.

Ex. 7. ***The product of two numbers is 6300 and their HCF is 15. How many pairs of such numbers are there ?***

Sol. Since the HCF is 15, the numbers must be of the form $15a$ and $15b$ where a and b are prime to each other.
Given, $15a \times 15b = 6300 \Rightarrow ab = \frac{6300}{225} = 28$
The possible pairs of numbers whose product is 28 are 1, 28; 2, 14; 4, 7; of these, the only pairs of numbers which are prime to each other are (1, 28) and (4, 7), *i.e.*, 2 pairs.

Ex. 8. ***HCF and LCM of two numbers are 7 and 140 respectively. If the numbers are between 20 and 45, then what is the sum of the numbers ?***

Sol. The numbers are between 20 and 45. Taking into consideration that HCF of the numbers is 7, the possible numbers are 21, 28, 35 and 42. Also we find that the LCM of the numbers does not have 3 as a factor. But 21 and 42 are numbers with 3 as a factor. So the required numbers between 20 and 45 are 28 and 35.
Sum of these numbers = 28 + 35 = **63**.

Ex. 9. ***Find the HCF and LCM of 3.6, 0.24 and 1.2.***

Sol. Converting them to like decimals we have 3.60, 0.24 and 1.20. Treating them as whole numbers, the numbers are 360, 24 and 120.
$360 = 2^3 \times 3^2 \times 5$, $24 = 2^3 \times 3$, $120 = 2^3 \times 3 \times 5$
$\therefore$ HCF of 360, 24 and 120 $= 2^3 \times 3 = 24$
LCM of 360, 24 and 120 $= 2^3 \times 3^2 \times 5 = 360$
$\Rightarrow$ HCF of 3.6, 0.24 and 1.2 = 0.24
LCM of 3.6, 0.24 and 1.2 = 3.6
(Since all of them have two places of decimal).

Ex. 10. ***A number lying between 1000 and 2000 is such that on division by 2, 3, 4, 5, 6, 7 and 8 leaves remainders 1, 2, 3, 4, 5, 6 and 7 respectively. What is the number ?***

Sol. Note that the difference between the divisors and the corresponding remainders is 1.
$(2 - 1 = 3 - 2 = 4 - 3 = 5 - 4 = 6 - 5 = 7 - 6 = 8 - 7 = 1)$

$\therefore$ LCM of 2, 3, 4, 5, 6, 7 and 8
$= 2 \times 2 \times 3 \times 5 \times 7 \times 2 = 840$

$\therefore$ Required number = $840K - 1$

Since the number lies between 1000 and 2000, $K = 2$. The number = $840 \times 2 - 1 =$ **1679.**

2	2, 3, 4, 5, 6, 7, 8
2	1, 3, 2, 5, 3, 7, 4
3	1, 3, 1, 5, 3, 7, 2
	1, 1, 1, 5, 1, 7, 2

Question Bank–4

1. The sides of a triangular piece of ground measure 15547, 17647, 3521 metres respectively. Find the length of the largest hurdle that can be used to fence it exactly without bending or cutting a hurdle.
(a) 6 m (b) 6.5 m
(c) 7 m (d) 7.5 m

2. The greatest number which divides 261, 933 and 1381 leaving remainders 5 in each case is
(a) 128 (b) 64
(c) 32 (d) 16

3. If 60, 82 and 126 are each divided by a number, then the remainder is the same in each case. The greatest possible value of the divisor is
(a) 16 (b) 8
(c) 22 (d) 11

4. Let N be the greatest number that will divide 1305, 4665 and 6905 leaving the same remainder in each case. The sum of the digits of N is
(a) 4 (b) 5
(c) 6 (d) 8

5. Three sets of English, Mathematics and Science books containing 336, 240 and 96 books respectively have to be stacked in such a way that all the books are stored subject-wise and the height of each stack is the same. Total number of stacks will be
(a) 14 (b) 21
(c) 22 (d) 48

6. Swapnil, Aakash and Vinay begin to jog around a circular stadium. They complete their revolutions in 36 seconds, 48 seconds and 42 seconds respectively. After how many seconds will they be together at the starting point ?
(a) 504 seconds (b) 940 seconds
(c) 1008 seconds (d) 470 seconds

7. The largest number of five digits which when divided by 16, 24, 30 or 36 leaves the same remainder 10 in each case is
(a) 99279 (b) 99370
(c) 99269 (d) 99350

8. What least number x must be subtracted from 797 so that $(797 - x)$ on being divided by 8, 9 and 11 leaves in each case the same remainder 4 ?
(a) 0 (b) 1
(c) 2 (d) 3

9. The least number which when increased by 4 is divisible by each of the numbers 10, 15, 20 and 25 is
(a) 296 (b) 300
(c) 304 (d) 308

10. A certain type of wooden board is sold only in lengths of multiples of 25 cm from 2 to 10 metres. A carpenter needs a large quantity of this type of board in 1.65 m length. For the minimum waste, the length to be purchased should be
(a) 3.30 m (b) 6.60 m
(c) 8.25 m (d) 9.95 m

11. Find the least number which when divided by 35 leaves a remainder 25, when divided by 45 leaves a remainder 35, and when divided by 55 leaves a remainder 45.
(a) 3465 (b) 4575
(c) 3455 (d) 3670

12. A number which when divided by 10 leaves a remainder of 9, when divided by 9 leaves remainder of 8 and when divided by 8 leaves a remainder of 7 is
(a) 1539 (b) 539
(c) 359 (d) 1359

13. The sum of two 2-digit numbers is 132. If their HCF is 11, the numbers are
(a) 55, 77 (b) 44, 88
(c) 33, 99 (d) 22, 110

14. The sum of two numbers is 684 and their H.C.F is 57. The number of possible pairs of such numbers is
(a) 2 (b) 3
(c) 4 (d) none

15. The product of two numbers is 2028 and their HCF is 13. The number of such pairs is
(a) 1 (b) 2
(c) 3 (d) 4

16. The LCM and HCF of two numbers are 84 and 21 respectively. If the ratio of the two numbers is 1 : 4, then the larger of the two numbers is

(a) 12 (b) 48
(c) 84 (d) 108

17. HCF of 3240, 3600 and a third number is 36 and their LCM is $2^4 \times 3^5 \times 5^2 \times 7^2$. The third number is :

(a) $2^4 \times 5^3 \times 7^2$ (b) $2^2 \times 3^5 \times 7^2$
(c) $2^3 \times 3^5 \times 7^2$ (d) $2^5 \times 5^2 \times 7^2$

18. The HCF of two numbers is 12 and their difference is also 12. The numbers are

(a) 12, 84 (b) 84, 96
(c) 64, 76 (d) 100, 112

19. If two numbers are in the ratio 2 : 3 and the product of their HCF and LCM is 33750, then the sum of the numbers is

(a) 250 (b) 425
(c) 325 (d) 375

20. Three numbers which are co-prime to each other are such that the product of the first two is 551 and that of the last two is 1073. The sum of the three numbers is

(a) 75 (b) 81
(c) 85 (d) 89

21. The HCF and LCM of two numbers are 21 and 4641 respectively. If one of the numbers lies between 200 and 300, then the two numbers are :

(a) 273, 363 (b) 273, 359
(c) 273, 361 (d) 273, 357

22. The LCM and HCF of two numbers are 1530 and 51. Find how many such pairs are possible.

(a) 2 (b) 3
(c) 4 (d) 1

23. Three wheels can complete 60, 36, 24 revolutions per minute respectively. There is a red spot on each wheel that touches the ground at time zero. After how much time, all these spots will simultaneously touch the ground again.

(a) $\frac{5}{2}$s (b) $\frac{5}{3}$s
(c) 6s (d) 7.5s

24. Two numbers have 16 as their HCF and 146 as their LCM. How many such pairs of numbers are there ?

(a) Zero (b) Only 1
(c) Only 2 (d) Many

25. If the LCM of three numbers is 9570, then their HCF is

(a) 11 (b) 12
(c) 19 (d) 21

26. Find the least number of five digits which when divided by 16, 24, 30 and 32 leaves a remainder 2 in each case.

(a) 10084 (b) 10071
(c) 10082 (d) 10002

27. The HCF of 1.08, 0.36 and 0.9 is

(a) 0.03 (b) 0.9
(c) 0.18 (d) 0.108

28. Product of three natural numbers is 24000 and their HCF is 10. How many such triplets of numbers are there ?

(a) 5 (b) 4
(c) 6 (d) 7

29. The HCF and LCM of two numbers are 13 and 455 respectively. If one of the numbers lies between 75 and 125, then the number is

(a) 78 (b) 91
(c) 104 (d) 117

30. The least perfect square, which is divisible by each of 21, 36 and 66 is

(a) 214344 (b) 214434
(c) 213444 (d) 231444

Answers

1. (c)	**2.** (c)	**3.** (c)	**4.** (a)	**5.** (a)	**6.** (c)	**7.** (b)	**8.** (b)	**9.** (a)	**10.** (c)
11. (c)	**12.** (c)	**13.** (a)	**14.** (a)	**15.** (b)	**16.** (c)	**17.** (b)	**18.** (b)	**19.** (d)	**20.** (c)
21. (d)	**22.** (c)	**23.** (c)	**24.** (a)	**25.** (a)	**26.** (c)	**27.** (c)	**28.** (d)	**29.** (b)	**30.** (c)

Hints and Solutions

1. (c) Length of the largest hurdle required = HCF of 15547, 17647 and 3521. For that first find the HCF of two numbers, say 15547 and 17647 by long division and then HCF of the above HCF and 3521 by long division. The final HCF will be obtained as 7 which is the required answer.

2. (c) When each of the numbers 261, 933 and 1381 is divided by the required number, the remainder is 5.

Therefore, the required number

= HCF of (261 – 5, 933 – 5, 1381 – 5)

= HCF of 256, 928, 1376

= 32

Find the HCF yourself.

3. (c) Since the remainder is same in each case, the difference of every pair of the given numbers would be exactly divisible by the required number.

Now, 82 – 60 = 22, 126 – 82 = 44, 126 – 60 = 66

∴ Required number = HCF of 22,44 and 66 = **22.**

4. (a) Similar to Q. No. 3.

4665 – 1305 = 3360, 6905 – 4665 = 2240,

6905 – 1305 = 5600

Required number (N) = HCF of 3360, 2240 and 5600

```
       1                    5
2240 )3360        1120 )5600
     - 2240    2        - 5600
      1120 )2240            0
            - 2240
                 0
```

⇒ N=1120

⇒ Sum of digits of N = 1 + 1 + 2 + 0 = **4.**

5. (a) Height of each stack = HCF of (336, 240, 96) books

	336		240		96
2	336	2	240	2	96
2	168	2	120	2	48
2	84	2	60	2	24
2	42	2	30	2	12
3	21	3	15	2	6
	7		5		3

∴ $336 = 2^4 \times 3 \times 7$, $240 = 2^4 \times 3 \times 5$, $96 = 2^5 \times 3$ $\Rightarrow$ HCF $= 2^4 \times 3 = 48$

Height of each stack = 48 books

$\Rightarrow$ Number of stacks $= \frac{336}{48} + \frac{240}{48} + \frac{96}{48}$

$= 7 + 5 + 2 =$ **14.**

6. (c) Required time (in seconds)

= LCM of 36, 48 and 42

= (2 × 2 × 3 × 3 × 4 × 7) seconds

= **1008 seconds.**

2	36, 48, 42
2	18, 24, 21
3	9, 12, 21
	3, 4, 7

7. (b)

2	16, 24, 30, 36
2	8, 12, 15, 18
2	4, 6, 15, 9
3	2, 3, 15, 9
	2, 1, 5, 3

∴ LCM = 2 × 2 × 2 × 3 × 2 × 5 × 3 = 720

Largest number of 5-digits = 99999

```
      138
720)99999
   -720
    2799
   -2160
     6399
    -5760
      639
```

∴ Required number = 720 × 138 + 10 = **99370.**

8. (b) LCM of 8, 9 and 11 = 8 × 9 × 11 = 792

∴ 797 – x = 792 + 4

⇒ x = 797 – 796 = **1.**

9. (a) Required number = (LCM of 10, 15, 20, 25) – 4 (Note : Find the LCM yourself)

10. (c) Length to be purchased = LCM of (25, 165) cm

11. (c) Since 35 – 25 = 10, 45 – 35 = 10, 55 – 45 = 10, *i.e.*, the difference between any divisor and the corresponding remainder is the same, the required number will be obtained by subtracting 10 from the LCM of 35, 45 and 55.

∴ Required number = (LCM of 35, 45, 55) – 10

= 3465 – 10 = **3455.**

12. (c) Since the common difference between any divisor and corresponding remainder is 1, then Required number = (LCM of 10, 9, 8) – 1

13. (a) Since the HCF is 11, the numbers must be of the form $11a$ and $11b$, where a and b are prime to each other.

$11a + 11b = 132 \quad \Rightarrow \quad a + b = 12$

The possible pairs of numbers whose sum is 12 are 1, 11; 2, 10; 3, 9; 4, 8; 5, 7; 6, 6.

Out of these pairs, the pairs of numbers that are prime to each other are 1, 11 and 5,7. The pairs which satisfies the given condition is 5, 7. Hence the numbers are 11 × 5 and 11 × 7, *i.e.*, 55 and 77.

14. (a) Similar to Q. No. 13.

15. (b) Since the HCF is 13, the numbers must be of the form $13a$ and $13b$, where a and b are prime to each other.

$\therefore$ $13a \times 13b = 2028 \Rightarrow 169ab = 2028$

$\Rightarrow ab = 12$

The possible pairs of numbers whose product is 12 are 1, 12; 2, 6; 3, 4. Of these, the only pairs of numbers which are prime to each other are 1, 12 and 3, 4, *i.e.*, 2 in number.

16. (c) Let the two numbers be x and $4x$.

$\therefore$ Product of the two numbers = HCF × LCM

$\Rightarrow x \times 4x = 84 \times 21 \Rightarrow 4x^2 = 84 \times 21$

$\Rightarrow x^2 = \frac{84 \times 21}{4} = 21 \times 21 \Rightarrow x = 21$

$\therefore$ Larger number = 4 × 21 = **84.**

17. (b) $3240 = 2^3 \times 3^4 \times 5$; $3600 = 2^4 \times 3^2 \times 5^2$;

$\text{HCF} = 2^2 \times 3^2$

$\text{LCM} = 2^4 \times 3^5 \times 5^2 \times 7^2$

Since HCF is the product of lowest powers of common factors, third number should have $2^2 \times 3^2$ as factor.

LCM is the product of the highest powers of all factors, therefore third number should have $3^5 \times 7^2$ as its factors.

$\therefore$ Third number = $\mathbf{2^2 \times 3^5 \times 7^2}$.

18. (b) Since the HCF of the two numbers is 12, the numbers are $12a$ and $12b$ where a and b are prime to each other.

Given, $12a - 12b = 12 \Rightarrow a - b = 1$

$\Rightarrow$ a and b are consecutive numbers that are prime to each other. By inspection we can see that 84, 96, *i.e.*, 12 × 7 and 12 × 8 are the two numbers satisfying this condition. (8 – 7 = 1)

19. (d) Let the numbers be $2x$ and $3x$. Then,

$2x \times 3x = 33750 \Rightarrow 6x^2 = 33750$

$\Rightarrow x^2 = 5625 \Rightarrow x^2 = 5^4 \times 3^2$

$\Rightarrow x = 5^2 \times 3 = 75$

$\therefore$ The two numbers are 2 × 75 and 3 × 75, *i.e.*, 150 and 225.

$\therefore$ Required sum = 150 + 225 = **375.**

20. (c) Let the three numbers be a, b and c.

Then, $ab = 551$ and $bc = 1073$

Middle number = HCF of 551 and 1073 = 29

First number $= \frac{551}{29} = 19$

Third number $= \frac{1073}{29} = 37$

Required sum = 19 + 29+ 37 = **85.**

21. (d) Let the numbers be $21a$ and $21b$, where a and b are co-primes.

Then, $21a \times 21b = 21 \times 4641 \Rightarrow ab = 221$

Two co-primes with product 221 are 13 and 17.

Numbers are (21 × 13, 21 × 17), *i.e.*, (273, 357)

22. (c) Since the HCF is 51, let the two numbers be $51a$ and $51b$, where a and b are prime to each other.

Given $51a \times 51b = 51 \times 1530$

$\Rightarrow ab = 30$

$\therefore$ The possible pairs of numbers whose product is 30 are 1, 30; 2, 15; 3, 10 and 5, 6. All these pairs of numbers are prime to each other, *i.e.*, 4 in number.

23. (c) Time taken to complete one revolution by

First wheel $= \frac{60}{60}\text{s} = 1\text{s}$

Second wheel $= \frac{36}{60}\text{s} = \frac{3}{5}\text{s}$

Third wheel $= \frac{24}{60}\text{s} = \frac{2}{5}\text{s}$

$\therefore$ Required time = LCM of $\left(1, \frac{3}{5}, \frac{2}{5}\right)$s

$= \frac{\text{LCM of } 1, 3, 2}{\text{LCM of } 1, 5, 5} = \frac{6}{1}\text{s} = \mathbf{6s}.$

24. (a) There are zero pairs of numbers whose HCF is 16 and LCM is 146, as HCF should divide LCM completely. Here, 146 is not divided completely.

25. (a) By inspection, we see that 9570 is completely divisible by 11, so the HCF is 11.

26. (c) LCM of 16, 24, 30 and 32 = 480

(Note : Find LCM by division method)

```
         2
480) 10000
     -960
     ----
      400
```

Least number of 5 digits = 10000

$\therefore$ Required number = 10000 + (480 – 400) + 2

= 10000 + 82 = **10082.**

27. (c) Converting them to like decimals we have, 1.08, 0.36 and 0.90.

Treating them as whole numbers, the numbers are 108, 36 and 90

$108 = 2^2 \times 3^3$, $36 = 2^2 \times 3^2$, $90 = 2 \times 3^2 \times 5$

$\therefore$ HCF of 108, 36, 90 = $2 \times 3^2 = 18$

$\Rightarrow$ HCF of 1.08, 0.36, 0.90 = **0.18.**

($\because$ All of them have two places of decimal).

28. (d) Since the HCF of the three numbers is 10, the numbers are of the form $10a$, $10b$, $10c$.

Given, $10a \times 10b \times 10c = 24000$

$\Rightarrow a \times b \times c = 24$

$\therefore$ The triplets (a, b, c) can be (1, 1, 24), (1, 2, 12), (1, 3, 8), (1, 4, 6), (2, 3, 4), (2, 6, 2) and (3, 3, 4), *i.e.*, 7 triplets.

29. (b) We know that LCM × HCF = $a \times b$, where a and b are the two numbers.

$\therefore$ $13 \times 455 = a \times b \Rightarrow 13 \times 5 \times 13 \times 7 = a \times b$

$\Rightarrow 65 \times 91 = a \times b$.

$\Rightarrow$ 91 is the required number as it lies between **75** and **125**.

30. (c) $21 = 3 \times 7$

$36 = 2 \times 2 \times 3 \times 3$

$66 = 2 \times 3 \times 11$

$\therefore$ LCM of 21, 36, 66 $= 2 \times 2 \times 3 \times 3 \times 7 \times 11$

$\therefore$ Least perfect square number divisible by 21, 36 and 66 $= 2 \times 2 \times 3 \times 3 \times 7 \times 7 \times 11 \times 11$

$=$ **213444.**

Self Assessment Sheet–4

1. Match List - I with List - II and select the correct answer using the codes given below the lists :

	List - I (Numbers)		List - II (Their LCM)
A.	12, 18, 20	1.	48
B.	12,16, 24	2.	720
C.	5, 18, 80	3.	180
D.	18, 24, 56	4.	504

Codes:

	A.	B.	C.	D.
(a)	3	2	4	1
(b)	2	1	3	4
(c)	3	4	2	1
(d)	3	1	2	4

2. Let p, q, r be natural numbers. If m is their LCM and n is their HCF, consider the following:

1. $mn = pqr$ if each one of p, q, r is prime
2. $mn = pqr$, if p, q, r are relatively prime in pairs.

(a) 1 only (b) 2 only
(c) both 1 and 2 (d) neither 1 nor 2

3. The LCM of three different numbers is 150. Which of the following cannot be their HCF ?

(a) 15 (b) 25
(c) 50 (d) 55

4. 5 bells start tolling together and toll at intervals of 2, 4, 6, 8 and 10 seconds, respectively. How many times do the five bells toll together in 20 minutes ?

(a) 10 (b) 11
(c) 12 (d) 15

5. If HCF of m and n is 1, then what are the HCF of $m+n$, m and HCF of $m-n$, n respectively ? $(m > n)$.

(a) 1 and 2
(b) 2 and 1
(c) 1 and 1
(d) cannot be determined

6. If HCF of p and q is x and $q = xy$, then the LCM of p and q is

(a) xy (b) py
(c) qy (d) pq

7. The LCM and HCF of two given numbers are 960 and 8 respectively. If one of them is 64, then the other number is

(a) 60 (b) 120
(c) 240 (d) 480

8. What is the least number which who divided by 7, 9 and 12 leaves the same remainder 1 in each case ?

(a) 253 (b) 352
(c) 505 (d) 523

9. There are 576 boys and 448 girls in a school that are to be divided into equal sections of either boys or girls alone. Find the total number of sections thus formed.

(a) 24 (b) 32
(c) 16 (d) None of these

10. What is the largest number which when divides 1475, 3155 and 5255 leaves the same remainder in each case ?

(a) 320 (b) 420
(c) 350 (d) 410

Answers

1. (d) **2.** (c) **3.** (d) [**Hint.** HCF should divide LCM exactly] **4.** (a) **5.** (c) **6.** (b) **7.** (b)
8. (a) **9.** (c) **10.** (b) [**Hint.** Find HCF of relative differences of the given numbers.]

POWERS AND ROOTS

KEY FACTS

1. Any rational number a when multiplied by itself n times can be written in exponential form as a^n.
   Thus, $a \times a \times a \times$ ______ n times $= a^n$, where n is an integer.
2. **Laws of exponents**

(a) $a^m \times a^n = a^{m+n}$	(b) $\frac{a^m}{a^n} = a^{m-n}$	(c) $(a^m)^n = a^{mn}$	(d) $(ab)^n = a^n \times b^n$
(e) $\left(\frac{a}{b}\right)^n = \frac{a^n}{b^n}$	(f) $a^0 = 1$	(g) $a^{-n} = \frac{1}{a^n} (a \neq 0)$	

3. **Square root** : The square root of a positive number x is a number whose square is x. It is denoted by $\sqrt{x}$.

   For example: $\sqrt{49} = 7$, $\sqrt{\frac{225}{289}} = \frac{\sqrt{225}}{\sqrt{289}} = \frac{15}{17}$

   To find the square root of a given number :
   (a) *Write the number as the product of its prime factors.*
   (b) *Form pairs of like factors.*
   (c) *From each pair, pick one factor.*
   (d) *The product of the factors so picked gives the square root of the given number.*

   For example : $676 = \underline{2 \times 2} \times \underline{13 \times 13}$

   $\sqrt{676} = 2 \times 13 = 26$

4. **Cube root** : Cube root of a number x is that number which when multiplied by itself three times, gives the original number x. Cube root is denoted by $\sqrt[3]{x}$. Cube root of a negative number is always negative.

   For example : $\sqrt[3]{125} = 5$, $\sqrt[3]{-\frac{8}{27}} = -\frac{\sqrt[3]{8}}{\sqrt[3]{27}} = -\frac{2}{3}$

   To find the cube root of a give number :
   (a) *Write the number as the product of its prime factors.*
   (b) *Form triplets of like factors.*
   (c) *From each triplet, pick one factor.*
   (d) *The product of the factors so picked gives the cube root of the given number.*

   For example : $\sqrt[3]{-1728} = -\sqrt[3]{\underline{2 \times 2 \times 2} \times \underline{2 \times 2 \times 2} \times \underline{3 \times 3 \times 3}}$

   $= -(2 \times 2 \times 3) = -12$

Section-A
EXPONENTS

Solved Examples

Ex. 1. *Find the value of the expression* $\dfrac{10^{-1}\times 5^{x-3}\times 4^{x-1}}{10\times 5^{x-5}\times 4^{x-2}}$

Sol. $\dfrac{10^{-1}\times 5^{x-3}\times 4^{x-1}}{10\times 5^{x-5}\times 4^{x-2}}=10^{-1-1}\times 5^{(x-3)-(x-5)}\times 4^{(x-1)-(x-2)}$

$=10^{-2}\times 5^{(x-3-x+5)}\times 4^{(x-1-x+2)}=\dfrac{1}{10^2}\times 5^2\times 4^1=\dfrac{100}{100}=\mathbf{1}.$

Ex. 2. *Evaluate* $\left[(x^y)^{1-\frac{1}{y}}\right]^{\frac{1}{y-1}}$

Sol. $\left[(x^y)^{1-\frac{1}{y}}\right]^{\frac{1}{y-1}}=\left[(x^y)^{\frac{y-1}{y}}\right]^{\frac{1}{y-1}}=\left[x^{y\times\frac{y-1}{y}}\right]^{\frac{1}{y-1}}=x^{y-1\times\frac{1}{y-1}}=x^1=\mathbf{x}.$

Ex. 3. *Find the value of* $(1296)^{0.75}(36)^{-1}$.

Sol. $(1296)^{0.75}\times(36)^{-1}=(6^4)^{0.75}\times(6^2)^{-1}=6^{4\times0.75}\times6^{2\times(-1)}=6^3\times6^{-2}=6^{3+(-2)}=6^1=\mathbf{6}.$

Ex. 4. *If* $a^m.a^n=a^{mn}$*, then find the value of* $m(n-2)+n(m-2)$.

Sol. Given, $a^m.a^n=a^{mn}$

$\Rightarrow a^{m+n}=a^{mn}\Rightarrow m+n=mn$

Now, $m(n-2)+n(m-2)=mn-2m+nm-2n$

$=2mn-2(m+n)=2mn-2mn=\mathbf{0}.$

Ex. 5. *If* $\dfrac{6^6+6^6+6^6+6^6+6^6+6^6}{3^6+3^6+3^6}\div\dfrac{4^6+4^6+4^6+4^6}{2^6+2^6}=2^n$*, then find the value of n.*

Sol. Given exp. $=\dfrac{6\times6^6}{3\times3^6}\div\dfrac{4\times4^6}{2\times2^6}=2^n$

$\Rightarrow\dfrac{6^7}{3^7}\div\dfrac{4^7}{2^7}=2^n\Rightarrow\left(\dfrac{6}{3}\right)^7\div\left(\dfrac{4}{2}\right)^7=2^n\Rightarrow 2^7\div2^7=2^n\Rightarrow1=2^n\Rightarrow2^0=2^n\Rightarrow n=\mathbf{0}.$

Ex. 6. *If* $\left(\dfrac{a}{b}\right)^{x-1}=\left(\dfrac{b}{a}\right)^{x-3}$*, then find x.*

Sol. $\left(\dfrac{a}{b}\right)^{x-1}=\left(\dfrac{b}{a}\right)^{x-3}\Rightarrow\left(\dfrac{a}{b}\right)^{x-1}=\left(\dfrac{a}{b}\right)^{-(x-3)}\Rightarrow\left(\dfrac{a}{b}\right)^{x-1}=\left(\dfrac{a}{b}\right)^{3-x}\Rightarrow x-1=3-x\Rightarrow2x=4\Rightarrow x=\mathbf{2}.$

Ex. 7. *If $5^{x+3} = 25^{3x-4}$, then find x.*

Sol. $5^{x+3} = 25^{3x-4} \Rightarrow 5^{x+3} = (5^2)^{3x-4} \Rightarrow 5^{x+3} = 5^{6x-8} \Rightarrow x+3 = 6x-8 \Rightarrow 5x = 11 \Rightarrow x = \mathbf{\frac{11}{5}}$.

Ex. 8. *If $2^x = 4^y = 8^z$ and $\left(\frac{1}{2x}+\frac{1}{4y}+\frac{1}{6z}\right) = \frac{24}{7}$, then find the value of z.*

Sol. $2^x = 4^y = 8^z \Rightarrow 2^x = (2^2)^y = (2^3)^z$

$\Rightarrow 2^x = 2^{2y} = 2^{3z} \Rightarrow x = 2y = 3z$

$\therefore \frac{1}{2x}+\frac{1}{4y}+\frac{1}{6z} = \frac{1}{2\times 3z}+\frac{1}{4\times\frac{3z}{2}}+\frac{1}{6z} = \frac{1}{6z}+\frac{1}{6z}+\frac{1}{6z} = \frac{24}{7} \Rightarrow \frac{3}{6z} = \frac{24}{7}$

$\Rightarrow \frac{1}{2z} = \frac{24}{7} \Rightarrow z = \frac{7}{2\times 24} \Rightarrow z = \mathbf{\frac{7}{48}}$.

Ex. 9. *Simplify :* $\frac{1}{1+a^{n-m}}+\frac{1}{1+a^{m-n}}$.

Sol. $\frac{1}{1+a^{n-m}}+\frac{1}{1+a^{m-n}} = \frac{1}{1+\frac{a^n}{a^m}}+\frac{1}{1+\frac{a^m}{a^n}} = \frac{1}{\frac{a^m+a^n}{a^m}}+\frac{1}{\frac{a^n+a^m}{a^n}}$

$= \frac{a^m}{a^m+a^n}+\frac{a^n}{a^n+a^m} = \frac{a^m+a^n}{a^m+a^n} = \mathbf{1}.$

Ex. 10. *Find the value of* $\frac{3^{12+n}\times 9^{2n-7}}{3^{5n}}$.

Sol. $\frac{3^{12+n}\times 9^{2n-7}}{3^{5n}} = \frac{3^{12+n}\times(3^2)^{2n-7}}{3^{5n}} = \frac{3^{12+n}\times 3^{4n-14}}{3^{5n}} = \frac{3^{12+n+4n-14}}{3^{5n}}$

$= \frac{3^{5n-2}}{3^{5n}} = 3^{5n-2-5n} = 3^{-2} = \frac{1}{3^2} = \mathbf{\frac{1}{9}}$.

Question Bank–5(a)

1. If m is a positive integer, which of the following is not equal to $(2^4)^m$?

(a) 2^{4m} (b) 4^{2m}

(c) $2^m(2^{3m})$ (d) $4^m(2^m)$

2. Which of the following is not the reciprocal of $\left(\frac{2}{3}\right)^4$?

(a) $\left(\frac{3}{2}\right)^4$ (b) $\left(\frac{2}{3}\right)^{-4}$

(c) $\left(\frac{3}{2}\right)^{-4}$ (d) $\frac{3^4}{4^2}$

3. $[3^3+3^2+3^{-2}+3^{-3}]$ is equal to

(a) 0 (b) $36+\frac{1}{36}$

(c) $\frac{976}{27}$ (d) 3^5+3^{-5}

4. If $9^{8.6}\times 8^{3.9}\times 72^{4.4}\times 9^{3.9}\times 8^{8.6} = 72^x$, then value of x is

(a) 15.1 (b) 17.9

(c) 20.9 (d) 16.9

5. $\left(\dfrac{a^{-1}b^2}{a^2b^{-4}}\right)^7 \div \left(\dfrac{a^3b^{-5}}{a^{-2}b^3}\right)^{-5}$ is equal to

(a) a^4b^2 (b) a^2b^4
(c) a^3b^2 (d) a^2b^3

6. The expression $(p^{-2x}q^{3y})^6 \div (p^3q^{-1})^{-4x}$ after simplification becomes
(a) independent of p, but not of q
(b) independent of q, but not of p
(c) independent of both p and q
(d) dependent on both p and q but independent of x and y.

7. The expression $a^{\frac{2}{3}}\left\{a^{\frac{1}{3}}\left(a^{\frac{1}{4}}\right)^4\right\}^{\frac{1}{4}}$ is equal to

(a) $a^{\frac{1}{2}}$ (b) $a^{\frac{1}{6}}$
(c) a (d) 1

8. Evaluate : $\dfrac{(3^4)^4 \times 9^6}{(27)^7 \times 3^9}$

(a) 3 (b) 9
(c) $\dfrac{1}{3}$ (d) $\dfrac{1}{9}$

9. Evaluate : $\dfrac{2^{n+4} - 2.2^n}{2.2^{n+3}} + 2^{-3}$

(a) 2^{n+1} (b) 2^3
(c) 2^{-3} (d) 1

10. The expression $\dfrac{\left(x+\frac{1}{y}\right)^a \times \left(x-\frac{1}{y}\right)^b}{\left(y+\frac{1}{x}\right)^a \times \left(y-\frac{1}{x}\right)^b}$ reduces to

(a) $\left(\dfrac{y}{x}\right)^{a+b}$ (b) $\left(\dfrac{x}{y}\right)^{a+b}$
(c) $\left(\dfrac{y}{x}\right)^{a-b}$ (d) $\left(\dfrac{x}{y}\right)^{a-b}$

11. If $\left(\dfrac{p}{q}\right)^{rx-s} = \left(\dfrac{q}{p}\right)^{px-q}$, then the value of x is

(a) 1 (b) $\dfrac{q+s}{p+r}$
(c) $\dfrac{q+r}{q+s}$ (d) $\dfrac{q+r}{p+s}$

12. If $(ab^{-1})^{2x-1} = (ba^{-1})^{x-2}$, then what is the value of x ?
(a) 1 (b) 2
(c) 3 (d) 4

13. $\left(\dfrac{2^m}{2^n}\right)^t \times \left(\dfrac{2^n}{2^t}\right)^m \times \left(\dfrac{2^t}{2^m}\right)^n$ is equal to

(a) 1 (b) 2
(c) $\dfrac{1}{2}$ (d) 0

14. Find the value of x when $4^{2x} = \dfrac{1}{32}$

(a) $-\dfrac{5}{4}$ (b) $\dfrac{4}{5}$
(c) $\dfrac{3}{5}$ (d) $\dfrac{5}{3}$

15. If $2^{x+4} - 2^{x+2} = 3$, then x is equal to
(a) 0 (b) 2
(c) –1 (d) –2

16. $[1 - 2(1-2)^{-1}]^{-1}$ equals

(a) $\dfrac{1}{3}$ (b) $-\dfrac{1}{3}$
(c) –1 (d) $\dfrac{1}{2}$

17. In the expression $\dfrac{2^x+1}{(7)^{-1}+(2)^{-1}} = 14$, the value of x is
(a) 3 (b) 5
(c) 15 (d) 7

18. If $m^n . n^m = 800$, then the value of $\dfrac{n}{m}$ is

(a) $\dfrac{1}{2}$ (b) $\dfrac{1}{5}$
(c) $\dfrac{4}{5}$ (d) $\dfrac{5}{2}$

19. If $a^{2x+2} = 1$, where a is a positive real number other than 1, then the value of x is
(a) –2 (b) –1
(c) 0 (d) 1

20. If $3^x - 3^{x-1} = 18$, then the value of x^x is
(a) 3 (b) 8
(c) 27 (d) 216

21. If $x^{11} = y^0$ and $x = 2y$, then y is equal to

(a) $\frac{1}{2}$ (b) 1

(c) -1 (d) -2

22. The value of $\frac{(5)^{0.25} \times (125)^{0.25}}{(256)^{0.10} \times (256)^{0.15}}$ is

(a) $\frac{\sqrt{5}}{2}$ (b) $\frac{5}{4}$

(c) $\frac{25}{2}$ (d) $\frac{25}{16}$

23. $\left[3^{2^3} - (3^2)^3\right]$ is equal to

(a) 8532 (b) 5832

(c) 3852 (d) 5238

24. If $x^y = y^x$, then $\left(\frac{x}{y}\right)^{x/y}$ is equal to

(a) $x^{x/y}$ (b) $x^{\frac{x}{y}-1}$

(c) $x^{y/x}$ (d) $x^{\frac{y}{x-1}}$

25. If a and b are positive integers such that $a^b = 125$, then $(a - b)^{a+b-4}$ is equal to

(a) 16 (b) 25

(c) 28 (d) 30

26. The value of $\frac{(243)^{0.13} \times (243)^{0.07}}{(7)^{0.25} \times (49)^{0.075} \times (343)^{0.2}}$ is

(a) $\frac{3}{7}$ (b) $\frac{7}{3}$

(c) $1\frac{3}{7}$ (d) $2\frac{2}{7}$

27. Solve for x : $\left(\frac{9}{4}\right)^x \times \left(\frac{8}{27}\right)^{x-1} = \frac{2}{3}$

(a) 1 (b) 2

(c) 3 (d) 4

28. If $\frac{9^n \times 3^5 \times (27)^3}{3 \times (81)^4} = 27$, then n is equal to

(a) 0 (b) 2

(c) 3 (d) 4

29. If $pqr = 1$, then

$\left(\frac{1}{1+p+q^{-1}} + \frac{1}{1+q+r^{-1}} + \frac{1}{1+r+p^{-1}}\right)$ is equal to

(a) 1 (b) pq

(c) qr (d) $\frac{1}{pq}$

30. If a, b, c are real numbers, then the value of $\sqrt{a^{-1}b}.\sqrt{b^{-1}c}.\sqrt{c^{-1}a}$ is

(a) abc (b) $\sqrt{abc}$

(c) $\frac{1}{abc}$ (d) 1

Answers

1. (d)	**2.** (c)	**3.** (c)	**4.** (d)	**5.** (a)	**6.** (a)	**7.** (c)	**8.** (d)	**9.** (d)	**10.** (b)
11. (b)	**12.** (a)	**13.** (a)	**14.** (a)	**15.** (d)	**16.** (a)	**17.** (a)	**18.** (d)	**19.** (b)	**20.** (c)
21. (a)	**22.** (b)	**23.** (b)	**24.** (b)	**25.** (a)	**26.** (a)	**27.** (d)	**28.** (c)	**29.** (a)	**30.** (d)

Hints and Solutions

1. (d) $(2^4)^m = 2^{4m}$;

$4^{2m} = (2^2)^{2m} = 2^{4m}$;

$2^m (2^{3m}) = 2^{m+3m} = 2^{4m}$;

$4^m(2^m) = (2^2)^m.2^m = 2^{2m}.2^m = 2^{3m} \neq 2^{4m}$.

3. (c) $[3^3 + 3^2 + 3^{-2} + 3^{-3}] = 27 + 9 + \frac{1}{9} + \frac{1}{27}$

$= \frac{729 + 243 + 3 + 1}{27} = \mathbf{\frac{976}{27}}$.

4. (d) $9^{8.6} \times 8^{3.9} \times 72^{4.4} \times 9^{3.9} \times 8^{8.6}$

$= 9^{8.6+3.9} \times 8^{3.9+8.6} \times 72^{4.4}$

$= 9^{12.5} \times 8^{12.5} \times 72^{4.4} = (9 \times 8)^{12.5} \times 72^{4.4}$

$= 72^{12.5} \times 72^{4.4} = 72^{16.9}$

$\therefore$ $x = \mathbf{16.9}$.

5. (a) $\left(\frac{a^{-1}b^2}{a^2b^{-4}}\right)^7 \div \left(\frac{a^3b^{-5}}{a^{-2}b^3}\right)^{-5} = \frac{a^{-7}b^{14}}{a^{14}b^{-28}} \div \frac{a^{-15}b^{25}}{a^{10}b^{-15}}$

$$=\left[a^{-7-14}\times b^{14-(-28)}\right]\div\left[a^{-15-10}\times b^{25-(-15)}\right]$$

$$=\frac{a^{-21}b^{42}}{a^{-25}b^{40}}=a^{-21-(-25)}b^{42-40}=\boldsymbol{a^4b^2}.$$

6. (a) $\dfrac{\left(p^{-2x}q^{3y}\right)^6}{\left(p^3q^{-1}\right)^{-4x}}=\dfrac{p^{-12x}q^{18y}}{p^{-12x}q^{4x}}=\boldsymbol{q^{18y-4x}}.$

7. (c) $a^{\frac{2}{3}}\left\{a^{\frac{1}{3}}\left(a^{\frac{1}{4}}\right)^4\right\}^{\frac{1}{4}}=a^{\frac{2}{3}}\left\{a^{\frac{1}{3}}\times a\right\}^{\frac{1}{4}}$

$$=a^{\frac{2}{3}}\left(a^{\frac{4}{3}}\right)^{\frac{1}{4}}=a^{\frac{2}{3}}\times a^{\frac{1}{3}}=a^{\frac{2}{3}+\frac{1}{3}}=a^1=\boldsymbol{a}.$$

8. (d) $\dfrac{(3^4)^4\times 9^6}{(27)^7\times 3^9}=\dfrac{3^{16}\times(3^2)^6}{(3^3)^7\times 3^9}=\dfrac{3^{16}\times 3^{12}}{3^{21}\times 3^9}$

$$=\frac{3^{28}}{3^{30}}=3^{28-30}=3^{-2}=\frac{1}{3^2}=\boldsymbol{\frac{1}{9}}.$$

9. (d) $\dfrac{2^{n+4}-2.2^n}{2.2^{n+3}}+2^{-3}=\dfrac{2^4.2^n-2.2^n}{2.2^n.2^3}+\dfrac{1}{2^3}$

$$=\frac{2^n(16-2)}{2^n\times 16}+\frac{1}{8}=\frac{\cancel{14}^{\,7}}{\cancel{16}_{\,8}}+\frac{1}{8}=\frac{8}{8}=\boldsymbol{1}.$$

10. (b) $\dfrac{\left(x+\frac{1}{y}\right)^a\times\left(x-\frac{1}{y}\right)^b}{\left(y+\frac{1}{x}\right)^a\times\left(y-\frac{1}{x}\right)^b}$

$$=\frac{\left(\frac{xy+1}{y}\right)^a\times\left(\frac{xy-1}{y}\right)^b}{\left(\frac{xy+1}{x}\right)^a\times\left(\frac{xy-1}{x}\right)^b}=\frac{\frac{(xy+1)^a}{y^a}\times\frac{(xy-1)^b}{y^b}}{\frac{(xy+1)^a}{x^a}\times\frac{(xy-1)^b}{x^b}}$$

$$=\frac{x^a}{y^a}\times\frac{x^b}{y^b}=\frac{x^{a+b}}{y^{a+b}}=\boldsymbol{\left(\frac{x}{y}\right)^{a+b}}$$

11. (b) $\left(\dfrac{p}{q}\right)^{rx-s}=\left(\dfrac{q}{p}\right)^{px-q}$

$$\Rightarrow\left(\frac{p}{q}\right)^{rx-s}=\left(\frac{p}{q}\right)^{-(px-q)}$$

$$\Rightarrow rx-s=-px+q\Rightarrow rx+px=q+s$$

$$\Rightarrow x(r+p)=q+s\Rightarrow x=\boldsymbol{\frac{q+s}{r+p}}$$

12. (a) $(a\,b^{-1})^{2x-1}=(ba^{-1})^{x-2}$

$$\Rightarrow\left(\frac{a}{b}\right)^{2x-1}=\left(\frac{b}{a}\right)^{x-2}\Rightarrow\left(\frac{a}{b}\right)^{2x-1}=\left(\frac{a}{b}\right)^{-(x-2)}$$

$$\Rightarrow\left(\frac{a}{b}\right)^{2x-1}=\left(\frac{a}{b}\right)^{2-x}\Rightarrow 2x-1=2-x$$

$$\Rightarrow 3x=3\Rightarrow x=\boldsymbol{1}.$$

13. (a) $\left(\dfrac{2^m}{2^n}\right)^t\times\left(\dfrac{2^n}{2^t}\right)^m\times\left(\dfrac{2^t}{2^m}\right)^n$

$$=(2^{m-n})^t\times(2^{n-t})^m\times(2^{t-m})^n$$

$$=2^{mt-nt}\times 2^{nm-tm}\times 2^{tn-mn}$$

$$=2^{mt-nt+nm-tm+tn-mn}$$

$$=2^0=\boldsymbol{1}.$$

14. (a) $4^{2x}=\dfrac{1}{32}\Rightarrow(2^2)^{2x}=\dfrac{1}{2^5}$

$$\Rightarrow 2^{4x}=2^{-5}\Rightarrow 4x=-5\Rightarrow x=\boldsymbol{-\frac{5}{4}}.$$

15. (d) $2^{x+4}-2^{x+2}=3\Rightarrow 2^4.\,2^x-2^2.2^x=3$

$$\Rightarrow 16.2^x-4.2^x=3\Rightarrow 2^x(16-4)=3$$

$$\Rightarrow 2^x=\frac{3}{12}\Rightarrow 2^x=\frac{1}{4}\Rightarrow 2^x=2^{-2}\Rightarrow x=\boldsymbol{-2}.$$

16. (a) $[1-2(1-2)^{-1}]^{-1}=[1-2(-1)^{-1}]^{-1}$

$$=\left[1-2\times\frac{1}{-1}\right]^{-1}=[1+2]^{-1}=3^{-1}=\boldsymbol{\frac{1}{3}}.$$

17. (a) $\dfrac{2^x+1}{\frac{1}{7}+\frac{1}{2}}=14\Rightarrow\dfrac{2^x+1}{\frac{9}{14}}=14$

$$\Rightarrow(2^x+1)\times\frac{14}{9}=14\Rightarrow 2^x+1=9$$

$$\Rightarrow 2^x=8\Rightarrow 2^x=2^3\Rightarrow x=\boldsymbol{3}.$$

18. (d) $800=2\times2\times2\times2\times2\times5\times5$

$$=2^5\times5^2$$

$$\therefore\ m^n.\,n^m=800=2^5.5^2$$

$$\Rightarrow m=2,\ n=5\quad\therefore\ \frac{n}{m}=\boldsymbol{\frac{5}{2}}.$$

19. (b) $a^{2x+2}=1\Rightarrow a^{2x+2}=a^0$

$$\Rightarrow 2x+2=0\Rightarrow x=\boldsymbol{-1}.$$

20. (c) $3^x-3^{x-1}=18\Rightarrow 3^x-3^x.\,3^{-1}=18$

$$\Rightarrow 3^x\left(1-\frac{1}{3}\right)=18\Rightarrow 3^x\times\frac{2}{3}=18$$

$\Rightarrow\ 3^x = \frac{18\times 3}{2} = 27 = 3^3 \Rightarrow x = 3$

$\therefore\ x^x = 3^3 = \mathbf{27}.$

21. (a) $x^{11} = y^0 \Rightarrow x^{11} = 1 \Rightarrow x = 1$

Given, $x = 2y \Rightarrow 1 = 2y \Rightarrow y = \frac{\mathbf{1}}{\mathbf{2}}.$

22. (b) $\frac{(5)^{0.25}\times(125)^{0.25}}{(256)^{0.10}\times(256)^{0.15}} = \frac{(5)^{0.25}\times(5^3)^{0.25}}{(4^4)^{0.10}\times(4^4)^{0.15}}$

$= \frac{5^{0.25}\times 5^{0.75}}{4^{0.40}\times 4^{0.60}} = \frac{5^{0.25+0.75}}{4^{0.40+0.60}} = \frac{5^1}{4^1} = \frac{\mathbf{5}}{\mathbf{4}}.$

23. (b) $3^{2^3} - (3^2)^3 = 3^8 - 3^6 = 3^6(3^2 - 1)$

$= 729 \times 8 = \mathbf{5832}.$

24. (b) Given, $x^y = y^x$

$\left(\frac{x}{y}\right)^{x/y} = \frac{x^{x/y}}{y^{x/y}} = \frac{x^{x/y}}{(y^x)^{1/y}} = \frac{x^{x/y}}{(x^y)^{1/y}}$

$= \frac{x^{x/y}}{x} = x^{\frac{x}{y}-1}.$

25. (a) $a^b = 125 \Rightarrow a^b = 5^3 \Rightarrow a = 5,\ b = 3$

$\therefore\ (a-b)^{a+b-4} = (5-3)^{5+3-4} = 2^4 = \mathbf{16}.$

26. (a) $\frac{(243)^{0.13}\times(243)^{0.07}}{(7)^{0.25}\times(49)^{0.075}\times(343)^{0.2}}$

$= \frac{(243)^{0.13+0.07}}{(7)^{0.25}\times(7^2)^{0.075}\times(7^3)^{0.2}}$

$= \frac{(3^5)^{0.20}}{7^{0.25+0.15+0.60}} = \frac{3^1}{7^1} = \frac{\mathbf{3}}{\mathbf{7}}.$

27. (d) $\left(\frac{9}{4}\right)^x \times \left(\frac{8}{27}\right)^{x-1} = \frac{2}{3}$

$\Rightarrow \left(\frac{3}{2}\right)^{2x} \times \left(\frac{2}{3}\right)^{3(x-1)} = \frac{2}{3}$

$\Rightarrow \left(\frac{2}{3}\right)^{-2x} \times \left(\frac{2}{3}\right)^{3x-3} = \frac{2}{3}$

$\Rightarrow -2x + 3x - 3 = 1 \Rightarrow x - 3 = 1 \Rightarrow x = \mathbf{4}.$

28. (c) $\frac{9^n \times 3^5 \times (27)^3}{3\times(81)^4} = 27$

$\Rightarrow \frac{3^{2n}\times 3^5\times(3^3)^3}{3\times(3^4)^4} = 3^3 \quad \Rightarrow \frac{3^{2n}\times 3^5\times 3^9}{3\times 3^{16}} = 3^3$

$\Rightarrow 3^{2n+14-17} = 3^3 \ \Rightarrow\ 3^{2n-3} = 3^3$

$\Rightarrow 2n - 3 = 3 \Rightarrow 2n = 6 \Rightarrow n = \mathbf{3}.$

29. (a) $\frac{1}{1+p+q^{-1}} + \frac{1}{1+q+r^{-1}} + \frac{1}{1+r+p^{-1}}$

$= \frac{1}{1+p+q^{-1}} + \frac{q^{-1}}{q^{-1}+1+q^{-1}r^{-1}} + \frac{p}{p+pr+1}$

$= \frac{1}{1+p+q^{-1}} + \frac{q^{-1}}{1+q^{-1}+p} + \frac{p}{p+q^{-1}+1}$

($\because pqr = 1 \Rightarrow (qr)^{-1} = p$
$\Rightarrow q^{-1}r^{-1} = p$ and $pr = q^{-1}$)

$= \frac{1+q^{-1}+p}{1+q^{-1}+p} = \mathbf{1}.$

30. (d) $\sqrt{a^{-1}b}\cdot\sqrt{b^{-1}c}\cdot\sqrt{c^{-1}a}$

$= (a^{-1}b)^{1/2}\cdot(b^{-1}c)^{1/2}\cdot(c^{-1}a)^{1/2}$

$= a^{-1/2}b^{1/2}\cdot b^{-1/2}c^{1/2}\cdot c^{-1/2}a^{1/2}$

$= a^{-\frac{1}{2}+\frac{1}{2}}\cdot b^{\frac{1}{2}-\frac{1}{2}}\cdot c^{\frac{1}{2}-\frac{1}{2}} = a^0\cdot b^0\cdot c^0$

$= 1\times 1\times 1 = \mathbf{1}.$

Self Assessment Sheet–5(a)

1. The value of x, if $2^x + 2^x + 2^x = 192$ is

(a) 5 (b) $\frac{1}{6}$
(c) 6 (d) None of these

2. The value of $\sqrt{\frac{1}{4}} + (0.0001)^{1/2} - (1000)^{-2/3}$ is

(a) $\frac{1}{2}$ (b) $\frac{1}{4}$
(c) $\frac{1}{8}$ (d) 0

3. If $2^x = 4^y = 8^z$ and $xyz = 288$, then $\frac{1}{2x} + \frac{1}{4y} + \frac{1}{8z}$ is equal to

(a) $\frac{11}{8}$ (b) $\frac{11}{24}$
(c) $\frac{11}{48}$ (d) $\frac{11}{96}$

4. $\left(\sqrt[5]{\sqrt[5]{a^5}}\right)^{10}$ is equal to

(a) a^2 (b) 1

(c) $a^{1/5}$ (d) a^5

5. If $(4)^{x+y} = 1$ and $(4)^{x-y} = 4$, then the value of x and y will be respectively

(a) $\frac{1}{2}$ and $-\frac{1}{2}$ (b) $\frac{1}{2}$ and $\frac{1}{2}$

(c) $-\frac{1}{2}$ and $-\frac{1}{2}$ (d) $-\frac{1}{2}$ and $\frac{1}{2}$

6. If $4^{2x} = \frac{1}{32}$, then x is

(a) $\frac{5}{4}$ (b) $\frac{4}{5}$

(c) $\frac{3}{5}$ (d) $-\frac{5}{4}$

7. $16^5 + 2^{15}$ is divisible by

(a) 31 (b) 13

(c) 27 (d) 33

8. The value of $\left[\left[(2401)^{-1/2}\right]^{-1/4}\right]^2$ is

(a) 8 (b) 7

(c) $\frac{1}{7}$ (d) 16

9. The number of prime factors in the expression $(6)^4 \times (8)^6 \times (10)^8 \times (12)^{10}$ is

(a) 48 (b) 64

(c) 72 (d) 80

10. The value of $\frac{3^{(12+n)} \times 9^{(2n-7)}}{3^{5n}}$ is

(a) $\frac{1}{3}$ (b) $\frac{9}{13}$

(c) $\frac{1}{9}$ (d) $\frac{2}{3}$

Answers

1. (c) **2.** (a) **3.** (d) **4.** (a) **5.** (a) **6.** (d) **7.** (d) **8.** (b)
9. (c) [**Hint.** Given exp. = $2^{50} \times 3^{14} \times 5^8$] **10.** (c)

Section-B

SQUARE ROOTS AND CUBE ROOTS

Solved Examples

Ex. 1. ***Find the smallest number by which 5808 should be multiplied so that the product becomes a perfect square?***

Sol. $5808 = \underline{2 \times 2} \times \underline{2 \times 2} \times 3 \times \underline{11 \times 11}$

Forming pairs of like factors, we see that 3 is the only unpaired factor. So to make 5808 a perfect square, we have to multiply it by **3**.

2	5808
2	2904
2	1452
2	726
3	363
11	121
	11

Ex. 2. ***One - third of the square root of which number is 0.001 ?***

Sol. Let the required number be x. Then,

$\frac{1}{3} \times \sqrt{x} = 0.001 \Rightarrow \sqrt{x} = 0.001 \times 3 = 0.003$

$\therefore\ x = (0.003)^2 = \mathbf{0.000009}$.

Note that there are 3 decimal places in the number so there will be 6 decimal places in the square of the number.

Ex. 3. *What is the value of x in the equation* $\sqrt{1+\sqrt{1-\frac{2176}{2401}}} = 1+\frac{x}{7}$?

Sol. $\sqrt{1+\sqrt{1-\frac{2176}{2401}}}=1+\frac{x}{7} \Rightarrow \sqrt{1+\sqrt{\frac{2401-2176}{2401}}}=1+\frac{x}{7} \Rightarrow \sqrt{1+\sqrt{\frac{225}{2401}}}=1+\frac{x}{7}$

$\Rightarrow \sqrt{1+\frac{\sqrt{225}}{\sqrt{2401}}}=1+\frac{x}{7}$

$225 = \underline{5\times5}\times\underline{3\times3}$
$2401 = \underline{7\times7}\times\underline{7\times7}$

$\Rightarrow \sqrt{1+\frac{15}{49}}=1+\frac{x}{7} \Rightarrow \sqrt{\frac{64}{49}}=1+\frac{x}{7} \Rightarrow \frac{8}{7}=1+\frac{x}{7} \Rightarrow \frac{x}{7}=\frac{8}{7}-1 \Rightarrow \frac{x}{7}=\frac{1}{7} \Rightarrow x=\frac{1}{7}\times7 \Rightarrow x=\mathbf{1}.$

Ex. 4. *Evaluate the square root of* $\frac{0.342\times0.684}{0.000342\times0.000171}$.

Sol. $\sqrt{\frac{0.342\times0.684}{0.000342\times0.000171}}=\sqrt{\frac{342000\times684000}{342\times171}}=\sqrt{1000\times4000}=\sqrt{4000000}=\mathbf{2000}$

Note. The number of zeros in the square root of a number is half of the number of zeros in the given number.

Ex. 5. *Arrange the following numbers in ascending order : 3.5 ÷ 4,* $\sqrt{0.64}$ *, 0.204 × 4,* $(0.89)^2$.

Sol. $\frac{3.5}{4}=0.875$, $\sqrt{0.64}=0.8$, $0.204\times4=0.816$, $(0.89)^2=0.7921$

$\therefore$ $0.7921 < 0.8 < 0.816 < 0.875$

$\therefore$ The given numbers arranged in ascending order are $(0.89)^2$, $\sqrt{0.64}$, 0.204 × 4, 3.5 ÷ 4.

Ex. 6. *The product of two numbers is 37. What is the square root of their difference ?*

Sol. Since 37 is a prime number, 37 = 1 × 37

$\therefore$ Required square root = $\sqrt{(37-1)}=\sqrt{36}=\mathbf{6}$.

Ex. 7. *Find the smallest number by which 9000 should be divided so that the quotient becomes a perfect cube ?*

Sol. Resolving into prime factors, we have,

$9000 = \underline{2\times2\times2}\times3\times3\times\underline{5\times5\times5}$

Forming triplets of like factors we see that the prime factor 3 does not form a triplet, so 9000 has to be divided by 3 × 3 = 9, so as to make the quotient a perfect cube.

2	9000
2	4500
2	2250
3	1125
3	375
5	125
5	25
	5

Ex. 8. *If the cube root of 132651 is 51, then what is the value of*

$$\sqrt[3]{132.651}+\sqrt[3]{0.132651}+\sqrt[3]{0.000132651}$$

Sol. $\sqrt[3]{132.651}+\sqrt[3]{0.132651}+\sqrt[3]{0.000132651}=\sqrt[3]{\frac{132651}{1000}}+\sqrt[3]{\frac{132651}{1000000}}+\sqrt[3]{\frac{132651}{1000000000}}$

$=\frac{\sqrt[3]{132651}}{\sqrt[3]{1000}}+\frac{\sqrt[3]{132651}}{\sqrt[3]{1000000}}+\frac{\sqrt[3]{132651}}{\sqrt[3]{1000000000}}=\frac{51}{10}+\frac{51}{100}+\frac{51}{1000}=5.1+0.51+0.051=\mathbf{5.661}$

Note. The number of zeros in the cube root is one-third of the number of zeros in the given number.

Ex. 9. *If* $\frac{\sqrt{x}}{\sqrt{0.0064}} = \sqrt[3]{0.008}$ *, then find the value of x.*

Sol. $\sqrt{x} = \sqrt[3]{0.008} \times \sqrt{0.0064} = \sqrt[3]{\frac{8}{1000}} \times \sqrt{\frac{64}{10000}} = \frac{\sqrt[3]{8}}{\sqrt[3]{1000}} \times \frac{\sqrt{64}}{\sqrt{10000}}$

$= \frac{2}{10} \times \frac{8}{100} = \frac{16}{1000} = 0.016$

$\therefore$ $x = (0.016)^2 = \mathbf{0.000256}$.

Ex. 10. *Evaluate :* $\sqrt[3]{\sqrt{0.000729}} + \sqrt[3]{0.008}$

Sol. $\sqrt[3]{\sqrt{0.000729}} + \sqrt[3]{0.008} = \sqrt[3]{\sqrt{\frac{729}{1000000}}} + \sqrt[3]{\frac{8}{1000}} = \sqrt[3]{\frac{27}{1000}} + \frac{2}{10}$

$= \frac{3}{10} + \frac{2}{10} = 0.3 + 0.2 = \mathbf{0.5}$.

Question Bank–5(b)

1. The value of $\sqrt{10+\sqrt{25+\sqrt{108+\sqrt{154+\sqrt{225}}}}}$ is

(a) 4 (b) 6
(c) 8 (d) 10

2. If $\sqrt{x} \div \sqrt{441} = 0.02$, then the value of x is

(a) 1.64 (b) 2.64
(c) 1.764 (d) 0.1764

3. $\frac{\sqrt{0.49}}{0.25} \times \frac{\sqrt{0.81}}{0.36}$ is equal to

(a) 7 (b) $2\frac{9}{10}$
(c) $7\frac{9}{10}$ (d) $9\frac{9}{10}$

4. Simplify : $\sqrt{0.0025} \times \sqrt{2.25} \times \sqrt{0.0001}$

(a) 0.00075 (b) 0.0075
(c) 0.075 (d) 0.75

5. Given that $\sqrt{1225} = 35$, find the value of

$\sqrt{12.25} + \sqrt{0.1225} + \sqrt{0.001225}$

(a) 0.3885 (b) 388.5
(c) 38.85 (d) 3.885

6. The square root of $0.\bar{4}$ is

(a) $0.\bar{8}$ (b) $0.\bar{6}$
(c) $0.\bar{7}$ (d) $0.\bar{9}$

7. If $\sqrt{1+\frac{25}{144}} = \frac{x}{12}$, then x equals

(a) 1 (b) 11
(c) 13 (d) 7

8. Of the numbers 0.16, $\sqrt{0.16}$, $(0.16)^2$ and $0.1\bar{6}$, the least number is

(a) $(0.16)^2$ (b) $\sqrt{0.16}$
(c) 0.16 (d) $0.1\bar{6}$

9. 1008 divided by which single digit number gives a perfect square ?

(a) 9 (b) 4
(c) 8 (d) 7

10. Find the smallest natural number by which 980 should be multiplied to make it a perfect square.

(a) 7 (b) 5
(c) 3 (d) 6

11. Simplify $\sqrt[3]{\frac{1}{8} \times \frac{125}{64}}$

(a) $\frac{5}{8}$ (b) $\frac{375}{512}$
(c) $2\frac{1}{2}$ (d) $15\frac{5}{8}$

12. Find the value of $\sqrt[3]{\sqrt{441+\sqrt{16}+\sqrt{4}}}$

(a) 3 (b) 5
(c) 7 (d) 9

13. Simplify $\sqrt{\sqrt[3]{0.000729}}$

(a) 3 (b) 0.9
(c) 0.3 (d) 0.09

14. Simplify $\frac{\sqrt[3]{8}}{\sqrt{16}} \div \frac{\sqrt{100}}{\sqrt{49}} \times \sqrt[3]{125}$

(a) 7 (b) $1\frac{3}{4}$
(c) $\frac{7}{100}$ (d) $\frac{4}{7}$

15. If $\sqrt{24} = 4.899$, then the value of $\sqrt{\frac{8}{3}}$ is

(a) 0.544 (b) 2.666
(c) 1.633 (d) 1.333

16. If $\sqrt{6} = 2.55$, then the value of $\sqrt{\frac{2}{3}} + 3\sqrt{\frac{3}{2}}$ is

(a) 4.48 (b) 4.49
(c) 3.27 (d) None of these

17. What should come in place of both the x in the equation $\frac{x}{\sqrt{128}} = \frac{\sqrt{162}}{x}$

(a) 12 (b) 14
(c) 144 (d) 196

18. What number should be divided by $\sqrt{0.25}$ to give the result 25 ?

(a) 25 (b) 50
(c) 12.5 (d) 125

19. $\sqrt[3]{333 + \sqrt[3]{987 + \sqrt[3]{2197}}}$ is equal to

(a) 21 (b) 18
(c) 7 (d) 3

20. If $\sqrt{\frac{x}{y}} + \sqrt{\frac{y}{x}} = \frac{10}{3}$ and $x + y = 10$, then the value of xy is

(a) 36 (b) 24
(c) 16 (d) 9

21. The sum of the squares of 2 numbers is 146 and the square root of one of them is $\sqrt{5}$. The cube of the other number is

(a) 1111 (b) 1221
(c) 1331 (d) 1441

22. If $(28)^2$ is added to the square of a number, the answer so obtained is 1808. What is the number ?

(a) 34 (b) 26
(c) 36 (d) 32

Answers

1. (a)	**2.** (d)	**3.** (a)	**4.** (a)	**5.** (d)	**6.** (b)	**7.** (c)	**8.** (a)	**9.** (d)	**10.** (b)
11. (a)	**12.** (a)	**13.** (c)	**14.** (b)	**15.** (c)	**16.** (c)	**17.** (a)	**18.** (c)	**19.** (c)	**20.** (d)
21. (c)	**22.** (d)								

Hints and Solutions

1. (a) $\sqrt{10 + \sqrt{25 + \sqrt{108 + \sqrt{154 + \sqrt{225}}}}}$

$= \sqrt{10 + \sqrt{25 + \sqrt{108 + \sqrt{154 + 15}}}}$

$= \sqrt{10 + \sqrt{25 + \sqrt{108 + \sqrt{169}}}}$

$= \sqrt{10 + \sqrt{25 + \sqrt{108 + 13}}}$

$= \sqrt{10 + \sqrt{25 + \sqrt{121}}}$

$= \sqrt{10 + \sqrt{25 + 11}} = \sqrt{10 + \sqrt{36}}$

$= \sqrt{10 + 6} = \sqrt{16} = \mathbf{4}.$

2. (d) $\frac{\sqrt{x}}{\sqrt{441}} = 0.02 \Rightarrow \sqrt{x} = 0.02 \times 21$

$\Rightarrow \sqrt{x} = 0.42 \Rightarrow x = (0.42)^2 = \mathbf{0.1764}.$

3. (a) $\frac{\sqrt{0.49}}{0.25} \times \frac{\sqrt{0.81}}{0.36} = \frac{0.7}{0.25} \times \frac{0.9}{0.36}$

$= \frac{0.7 \times 0.9}{\frac{25}{100} \times 0.36} = \frac{0.63}{\frac{1}{4} \times 0.36} = \frac{0.63}{0.09} = \frac{63}{9} = 7.$

4. (a) Given exp. = 0.05 × 1.5 × 0.01 = **0.00075**.

5. (d) Given exp. $= \sqrt{\frac{1225}{100}} + \sqrt{\frac{1225}{10000}} + \sqrt{\frac{1225}{1000000}}$

$= \frac{\sqrt{1225}}{\sqrt{100}} + \frac{\sqrt{1225}}{\sqrt{10000}} + \frac{\sqrt{1225}}{\sqrt{1000000}}$

$= \frac{35}{10} + \frac{35}{100} + \frac{35}{1000}$

$= 3.5 + 0.35 + 0.035 = \mathbf{3.885}.$

6. (b) $0.\bar{4} = \frac{4}{9} \therefore \sqrt{0.\bar{4}} = \sqrt{\frac{4}{9}} = \frac{2}{3} = \mathbf{0.\bar{6}}.$

7. (c) $\sqrt{1 + \frac{25}{144}} = \frac{x}{12} \Rightarrow \sqrt{\frac{169}{144}} = \frac{x}{12}$

$\Rightarrow \frac{x}{12} = \frac{13}{12} \Rightarrow x = \frac{13}{12} \times 12 \Rightarrow x = \mathbf{13}.$

8. (a) $0.16 = 0.16$

$\left.\begin{array}{l}\sqrt{0.16} = 0.4 \\ (0.16)^2 = 0.0256 \\ 0.1\bar{6} = 0.1666.....\end{array}\right\} \Rightarrow 0.0256 = (0.16)^2$ is the least.

9. (d) $1008 = \underline{2 \times 2} \times \underline{2 \times 2} \times 7 \times \underline{3 \times 3}$

Prime factorising 1008 and forming pairs of like factors we see that 7 is the only unpaired prime factor. So we divide 1008 by 7 to get a perfect square.

2	1008
2	504
2	252
2	126
7	63
3	9
	3

10. (b) The prime factorisation of 980 can be written as $980 = \underline{2 \times 2} \times 5 \times \underline{7 \times 7}$.

Forming pairs of like factors, we see that 5 is the only unpaired prime factor. So, to make 980 a perfect square. We have to multiply it by 5 to complete the pair.

11. (a) $\sqrt[3]{\frac{1}{8} \times \frac{125}{64}} = \sqrt[3]{\frac{1}{8}} \times \sqrt[3]{\frac{125}{64}} = \frac{\sqrt[3]{1}}{\sqrt[3]{8}} \times \frac{\sqrt[3]{125}}{\sqrt[3]{64}} = \frac{1}{2} \times \frac{5}{4} = \mathbf{\frac{5}{8}}.$

12. (a) $\sqrt[3]{\sqrt{441} + \sqrt{16} + \sqrt{4}} = \sqrt[3]{21 + 4 + 2} = \sqrt[3]{27} = \mathbf{3}.$

13. (c) $\sqrt{\sqrt[3]{0.000729}} = \sqrt{\sqrt[3]{\frac{729}{1000000}}}$

$= \sqrt{\frac{\sqrt[3]{729}}{\sqrt[3]{1000000}}} = \sqrt{\frac{9}{100}} = \frac{\sqrt{9}}{\sqrt{100}} = \frac{3}{10} = \mathbf{0.3}.$

14. (b) Given exp. $= \frac{2}{4} \div \frac{10}{7} \times 5 = \frac{2}{4} \times \frac{7}{10} \times 5 = \frac{7}{4} = \mathbf{1\frac{3}{4}}.$

15. (c) $\sqrt{\frac{8}{3}} = \sqrt{\frac{8 \times 3}{3 \times 3}} = \frac{\sqrt{24}}{\sqrt{9}} = \frac{4.899}{3} = \mathbf{1.633}.$

16. (c) $\sqrt{\frac{2}{3}} + 3\sqrt{\frac{3}{2}} = \sqrt{\frac{2 \times 3}{3 \times 3}} + 3\sqrt{\frac{3 \times 2}{2 \times 2}}$

$= \sqrt{\frac{6}{9}} + 3\sqrt{\frac{6}{4}} = \frac{\sqrt{6}}{3} + \frac{3\sqrt{6}}{2} = \frac{2.45}{3} + \frac{3 \times 2.45}{3}$

$= 0.82 + 2.45 = \mathbf{3.27}.$

17. (a) $\frac{x}{\sqrt{128}} = \frac{\sqrt{162}}{x}$

$\Rightarrow x^2 = \sqrt{128} \times \sqrt{162}$

$\Rightarrow x^2 = \sqrt{128 \times 162}$

$= \sqrt{\underline{2 \times 2} \times \underline{2 \times 2} \times \underline{2 \times 2} \times \underline{2 \times 2} \times \underline{3 \times 3} \times \underline{3 \times 3}}$

$\Rightarrow x^2 = 2 \times 2 \times 2 \times 2 \times 3 \times 3$

$\Rightarrow x = \sqrt{\underline{2 \times 2} \times \underline{2 \times 2} \times \underline{3 \times 3}} \Rightarrow x = 2 \times 2 \times 3 = \mathbf{12}.$

18. (c) Given, $\frac{x}{\sqrt{0.25}} = 25 \Rightarrow \frac{x}{0.5} = 25$

$\Rightarrow x = 25 \times 0.5 = \mathbf{12.5}.$

19. (c) $\sqrt[3]{333 + \sqrt[3]{987 + \sqrt[3]{2197}}} = \sqrt[3]{333 + \sqrt[3]{987 + 13}}$

$= \sqrt[3]{333 + \sqrt[3]{1000}} = \sqrt[3]{333 + 10} = \sqrt[3]{343} = \mathbf{7}.$

20. (d) $\sqrt{\frac{x}{y}} + \sqrt{\frac{y}{x}} = \frac{10}{3} \Rightarrow \frac{\sqrt{x}}{\sqrt{y}} + \frac{\sqrt{y}}{\sqrt{x}} = \frac{10}{3}$

$\Rightarrow \frac{(\sqrt{x})^2 + (\sqrt{y})^2}{\sqrt{y} \cdot \sqrt{x}} = \frac{10}{3} \Rightarrow \frac{x + y}{\sqrt{xy}} = \frac{10}{3}$

$\Rightarrow \frac{10}{\sqrt{xy}} = \frac{10}{3} \Rightarrow \sqrt{xy} = 3 \Rightarrow xy = \mathbf{9}.$

21. (c) Let the two numbers be x and y.

$x^2 + y^2 = 146$ and $\sqrt{x} = \sqrt{5} \Rightarrow x = 5$

$\therefore \; 5^2 + y^2 = 146 \Rightarrow 25 + y^2 = 146 \Rightarrow y^2 = 121$

$\Rightarrow y = \sqrt{121} = 11 \therefore (y)^3 = (11)^3 = \mathbf{1331}.$

22. (d) $(28)^2 + x^2 = 1808 \Rightarrow x^2 = 1808 - (28)^2$

$\Rightarrow x^2 = 1808 - 784 = 1024 \Rightarrow x = \sqrt{1024} = \mathbf{32}.$

Self Assessment Sheet–5(b)

1. The value of $\sqrt{\frac{16}{36} + \frac{1}{4}}$ is

(a) $\frac{2}{5}$ (b) $\frac{1}{3}$

(c) $\frac{5}{6}$ (d) $\frac{7}{6}$

2. A decimal number has 16 decimal places. The number of decimal places in the square root of this number will be

(a) 2 (b) 4

(c) 8 (d) 16

3. The value of $\dfrac{5}{\sqrt{0.0025}}$ is
(a) $\dfrac{1}{5}$ (b) 5
(c) 100 (d) 50

4. What is the value of
$\sqrt{7.84}+\sqrt{0.0784}+\sqrt{0.000784}+\sqrt{0.00000784}$?
(a) 3.08 (b) 3.108
(c) 3.1008 (d) 3.1108

5. The ratio of three numbers is 3 : 4 : 5 and the sum of their squares is 1250. The sum of three numbers is
(a) 60 (b) 90
(c) 30 (d) 50

6. $\dfrac{(225)^{0.2}\times(225)^{0.3}}{(225)^{0.8}\times(225)^{0.2}}$ is equal to :
(a) 15 (b) 1/15
(c) 1/25 (d) 1.5

7. Given that $\sqrt{24025}=155$, then
$\sqrt{240.25}+\sqrt{2.4025}+\sqrt{0.024025}+\sqrt{0.00024025}$ is equal to
(a) 16.2205 (b) 16.2402
(c) 17.2205 (d) 155.2205

8. The number of integral values of x if the following statement is valid ?
$0\le x^2\le 100$
(a) 19 (b) 20
(c) 22 (d) 21

9. Sum of digits of the smallest number by which 1440 should be multiplied so that it becomes a perfect cube, is
(a) 4 (b) 6
(c) 7 (d) 8

10. First find the number in place of P in the following number series and then find the value of the expression given after the series.
188 186 P 174 158 126
The value of $\sqrt{P-13}$ is
(a) 14.03 (b) 14.10
(c) 13.00 (d) 13.67

Answers

1. (c) **2.** (c) **3.** (c) **4.** (d) **5.** (a) **6.** (b) **7.** (c)
8. (d) [**Hint.** The possible values are – 10 to – 1, 0, 1 to 10] **9.** (b)
10. (c) [**Hint.** 188 186 182 174 158 126; differences – 2 – 4 – 8 – 16 – 32]

Unit Test–1

1. The expression
$$\left(\frac{x^a}{x^b}\right)^{a^2+ab+b^2}\times\left(\frac{x^b}{x^c}\right)^{b^2+bc+c^2}\times\left(\frac{x^c}{x^a}\right)^{c^2+ca+a^2}$$
simplifies to
(a) 1 (b) –1
(c) 0 (d) None of these

2. Which one of the following is correct ? the number 222222 is :
(a) divisible by 3, but not divisible by 7
(b) divisible by 3 and 7, but not divisible by 11
(c) divisible by 2 and 7, but not divisible by 11
(d) divisible by 3, 7 and 11

3. In a division operation, the divisor is 5 times the quotient and twice the remainder. If the remainder is 15, then what is the dividend ?
(a) 175 (b) 185
(c) 195 (d) 205

4. LCM of two numbers is 16 times their HCF. The sum of LCM and HCF is 850. If one number is 50, then what is the number ?
(a) 800 (b) 1200
(c) 1600 (d) 2400

5. The value of $\sqrt{\dfrac{0.289}{0.00121}}$ is equal to
(a) $\dfrac{1.7}{11}$ (b) $\dfrac{0.17}{11}$
(c) $\dfrac{17}{110}$ (d) $\dfrac{170}{11}$

6. Consider the following statements :
A number $a_1a_2a_3a_4a_5$ is divisible by 9 if
1. $a_1+a_2+a_3+a_4+a_5$ is divisible by 9
2. $a_1-a_2+a_3-a_4+a_5$ is divisible by 9

Which of the above statements is/are correct ?

(a) 1 only (b) 2 only

(c) Both 1 and 2 (d) Neither 1 nor 2

7. A bell rings every 5 seconds. A second bell rings every 6 seconds and a third one rings every 8 seconds. If all the three rings at the same time at 8.00 a.m., at what time will they all ring together next ?

(a) 1 minute past 8.00 a.m.

(b) 2 minutes past 8.00 a.m.

(c) 3 minutes past 8.00 a.m.

(d) 4 minutes past 8.00 a.m.

8. If $4^x - 4^{x-1} = 24$, then the value of $(2x)^x$ is

(a) $(5)^{3/2}$ (b) 4

(c) $(5)^{5/2}$ (d) $\frac{1}{5}$

9. The first twenty natural numbers from 1 to 20 are written next to each other to form a 31 digit number N= 1 2 3 4 5 6 7 8 9 10 11 12 13 14 15 16 17 18 19 20. What is the remainder when this number is divided by 16 ?

(a) 0 (b) 4

(c) 7 (d) 9

10. x, y and z are the natural numbers. Which of the following statements is true ?

I. If x is divisible by y and y is divisible by z then x must be divisible by z.

II. If x is a factor of y and z, then x must be a factor of $y + z$.

III. If x is a factor of y and z, then x must be a factor of $\frac{y}{z}$.

(a) I, II and III (b) I only

(c) I and II (d) II only

11. The number of prime factors in

$\left(\frac{1}{6}\right)^{12} \times (8)^{25} \times \left(\frac{3}{4}\right)^{15}$ is

(a) 33 (b) 37

(c) 52 (d) None of these

12. The value of $\frac{2}{3} \times \frac{3}{\frac{5}{6} \div \frac{2}{3} \text{ of } 1\frac{1}{4}}$ is :

(a) 2 (b) 1

(c) $\frac{1}{2}$ (d) $\frac{2}{3}$

13. If $\left(\frac{1}{5}\right)^{3y} = 0.008$, then the value of $(0.25)^{y/2}$ will be

(a) 1.00 (b) 0.5

(c) 0.25 (d) 0.125

14. Simplify: $\left[\sqrt[3]{\sqrt[6]{5^9}}\right]^4 \left[\sqrt[6]{\sqrt[3]{5^9}}\right]^4$:

(a) 5^2 (b) 5^4

(c) 5^8 (d) 5^{12}

15. The value of $2.6\overline{34} + 0.1\overline{7}$ is

(a) $2\frac{731}{999}$ (b) $2\frac{731}{990}$

(c) $2\frac{731}{900}$ (d) $2\frac{731}{9900}$

16. The least values of x and y so that $7x342y$ is divisible by 88 are

(a) 4, 4 (b) 4, 3

(c) 5, 6 (d) 6, 7

17. The total number of 8 digit numbers is

(a) 9,000 (b) 9,00,000

(c) 9,00,00,000 (d) None of these

18. If X, Y are positive real numbers such that $X > Y$ and A is any positive real number, then

(a) $\frac{X}{Y} \geq \frac{X+A}{Y+A}$ (b) $\frac{X}{Y} > \frac{X+A}{Y+A}$

(c) $\frac{X}{Y} \leq \frac{X+A}{Y+A}$ (d) $\frac{X}{Y} < \frac{X+A}{Y+A}$

19. **Assertion (A)** : The number 90356294 is divisible by 4.

Reason (R) : A number with an even digit in the units place is always divisible by 2.

(a) both A and R are true and R is the correct explanation of A.

(b) both A and R are true but R is not the correct explanation of A.

(c) A is true but R is false.

(d) A is false but R is true.

20. Given two different prime numbers P and Q, find the number of divisors of the following choose the correct answer from the given options.

Dividend	**No. of divisors**			
(1) PQ	(a) 2	(b) 4	(c) 6	(d) 8
(2) P^2Q	(a) 2	(b) 4	(c) 6	(d) 8
(3) P^3Q^2	(a) 2	(b) 4	(c) 6	(d) 12

Answers

1. (a) **2.** (d) **3.** (c) **4.** (a) **5.** (d) **6.** (a) **7.** (b) **8.** (c)

9. (a) [**Hint.** Apply test for divisibility by 16, *i.e.,* number formed by last 4 digits should be divisible by 16.]

10. (c) **11.** (d) The answer is 36. **12.** (a) **13.** (b) **14.** (b) **15.** (c) **16.** (a)

17. (c) [**Hint.** Total number = $9 \times 10^{n-1}$. Here $n = 8$]

18. (a) [**Hint.** If XYY then $\frac{X}{Y}$ increases if both the num. and den. are increased by a +ve number] **19.** (d)

20. (1) → (b), (2) → (c), (3) → (d).

[**Hint.** Divisors of PQ are 1, P, Q, PQ. Divisors of P^2Q are 1, P, P^2, Q, PQ, P^2Q

$\Rightarrow$ Number of divisors of $P^2Q = 6 = (2 + 1)(1 + 1)$].

UNIT-2

ALGEBRA

- *Algebraic Expressions*
- *Algebraic Identities*
- *Factorisation*
- *Linear Equations in One Variable*

Chapter 6 ALGEBRAIC EXPRESSIONS

KEY FACTS

1. **Constants and variables:** A symbol such as 5, $\frac{-6}{7}$, 0.216 etc., in algebra having a fixed value is called a **constant** whereas a symbol as x, y, a, p etc., which can be given or assigned a varied number of values is called a **variable.**
2. **Algebraic expressions:** A combination of constants and variables connected by basic mathematical operators, *i.e.*, +, –, ×, ÷, is called an **algebraic expression,** *e.g.*, $7xy$, $8a + 7b - 4c$, $\frac{x}{z}$ etc.
3. **Term:** The various parts of an algebraic expression connected by + or – sign are called the **terms** of the expression. E.g., $6a - 4b^2 + 3c^3$ has three terms namely $6a$, $-4b^2$, $3c^3$.
4. **Like terms and unlike terms:** The terms having the *same literal factors* are called **like terms** and these having *different literal factors* are called **unlike terms.**
   E.g., $3a^2b$ and $-4ba^2$ are like terms, whereas $2xy$, $-6xy^2$, $3x^2y^2$ are unlike terms.
5. **Coefficient:** The numerical part of a term is known as **numerical coefficient** and the variable part is called the **literal coefficient.**
   E.g., in the term $4x^2y$, 4 is the numerical coefficient and x^2y is the literal coefficient.
6. **Types of algebraic expressions :**
   (i) **Monomials:** Expressions with a single term are called monomials.
   E.g., $4x^2, 7, -8abc$
   (ii) **Binomials:** Expressions with two terms are called binomials.
   E.g., $7x+9y, 2a^2b-8c, x+\frac{1}{x}$
   (iii) **Trinomials:** Expressions with three terms are called trinomials.
   E.g., $8ab+9a^2b-7c, 4x+7y+2z^2$
   (iv) **Multinomials:** Expressions with two or more terms are called multinomials.
7. **Polynomials:**
   (i) **Polynomial in one variable:** An algebraic expression of the form $a+bx+cx^2+dx^3$, where a, b, c and d are constants and x is a variable is called a polynomial. The powers of the variable involved are non-negative integers. The degree of the polynomial is the greatest power of the variable present in the polynomial.
   E.g., $-7+5x$ is a polynomial of degree 1,
   $5x^3+3x^2+2x-8$ is a polynomial of degree 3.

(ii) **Polynomial in two or more variables:** It is an algebraic expression involving two or more variables with non-negative integral powers.
In such a polynomial, the degree of any term is the greatest sum of the powers of the variables of that term.
E.g., $6x^2y^2 - 4x^2y^3 + 7$ is a polynomial in x and y of degree 5.

Note: Terms where the powers of the variable are negative or fractional, *i.e.*, x^{-2} or $\frac{1}{x^2}, \frac{1}{y}, \frac{z}{x^2}, \frac{x^4}{y}, x^{-\frac{1}{2}}$ etc., do not form a polynomial.

8. For **addition or subtraction of algebraic expressions,** first we collect the like terms and then find the sum or difference of the numerical coefficients of these terms.

E.g., $8x + 9y - 15x + 6y = (8x - 15x) + (9y + 6y) = -7x + 15y$

9. Multiplication: Recall the following results.

$(+x)\times(+y) = (+xy)$	$(+x)\times(-y) = (-xy)$
$(-x)\times(-y) = (+xy)$	$(-x)\times(+y) = (-xy)$

$x^m \times x^n = x^{m+n}$

(a) **To multiply monomials**

(i) *Multiply numerical coefficients*	(ii) *Multiply literal coefficients*	(iii) *Multiply the results.*

E.g., $\frac{-2}{5}a^5b^2c \times \frac{25}{8}a^3xc^2 = \left(\frac{-2}{5}\times\frac{25}{8}\right)\times\left(a^5b^2c \times a^3xc^2\right) = \frac{-5}{4}a^8b^2xc^3.$

(b) **To multiply a polynomial by a monomial,** multiply each term of the polynomial by the monomial and then add the products to get the result.

E.g., $-7a^2b^2(a^3 - b^3 + 3abc) = (-7a^2b^2 \times a^3) + (-7a^2b^2 \times -b^3) + (-7a^2b^2 \times 3abc)$

$= -7a^5b^2 + 7a^2b^5 - 21a^3b^3c$

(c) **To multiply any two polynomials,** multiply each term of one polynomial by each term of the other. Finally add the products combining like terms together.

E.g., $(a - 2b)(a^2 + 4ab - b^2)$

$= a(a^2 + 4ab - b^2) - 2b(a^2 + 4ab - b^2) = a^3 + 4a^2b - ab^2 - 2ba^2 - 8ab^2 + 2b^3$

$= a^3 + 2a^2b - 9ab^2 + 2b^3$

10. Division of algebraic expressions

Recall, $x^m \div x^n = x^{m-n}$, where m, n are rational numbers and $m > n$. Also, $x^0 = 1$ and $x^{-n} = \frac{1}{x^n}$

(a) **To divide a monomial by a monomial:**

(i) *Divide the numerical coefficients*	(ii) *Divide the literal coefficients*	(iii) *Multiply the result*

E.g., $(-63x^5y^2z^4) \div 9xyz^5 = \frac{-63x^5y^2z^4}{9xyz^5} = \left(\frac{-63}{9}\right)\times\left(\frac{x^5y^2z^4}{xyz^5}\right)$

$= -7x^{5-1}y^{2-1}z^{4-5} = -7x^4yz^{-1} = \frac{-7x^4y}{z}$

(b) **To divide a polynomial by a monomial,** divide each term of the polynomial by a monomial.

E.g., $(-24x^3y + 6x^2y^2 - 7xy + 10xy^3) \div (-4x) = \frac{-24x^3y}{-4x} + \frac{6x^2y^2}{-4x} - \frac{7xy}{-4x} + \frac{10xy^3}{-4x}$

$= 6x^2y - \frac{3}{2}xy^2 + \frac{7}{4} - \frac{5}{2}y^3.$

(c) **To divide a polynomial by a polynomial :**

Step 1: *Set up as a form of long division in which the polynomials are arranged in descending order, leaving space for missing terms.*

Step 2: *The first term of the quotient is obtained by dividing the first term of the dividend by the first term of the divisor.*

Step 3: *Then, multiply each term of the divisor by the first term of the quotient and subtract the result from the dividend.*

Step 4: *Now use the remainder as the new dividend and repeat steps 2 and 3.*

Step 5: *Repeat the above process till the remainder becomes zero or a polynomial of degree less than that of the divisor.*

E.g.,

$$\begin{array}{r|l} & 2x - 3 \\ \hline x-4 & 2x^2 - 11x + 12 \\ & 2x^2 - 8x \\ & - \quad + \\ \hline & -3x + 12 \\ & -3x + 12 \\ & + \quad - \\ \hline & 0 \end{array}$$

Note: You can use the result **Dividend = Divisor × Quotient + Remainder** to check the result of division.

Question Bank–6

1. Which of the following expressions is not a polynomial ?

(a) $6y^3 + 5y^2 - 2y - 9$

(b) $-\frac{2}{9}x^2y + \frac{4}{13}x^2y^2 + 6y^3$

(c) $(a^3 - 8a)(x^4 + 6)$

(d) $\frac{5x^4 + 7x^2y^2 - 8}{y}$

2. Which of the following expression is a polynomial?

(a) $y^2 + \sqrt{2}y(x-4) + x$ (b) $\sqrt[3]{9x} + x^4 - x$

(c) $a^{-\frac{1}{2}} + \sqrt{5}a + 6$ (d) $4\sqrt{x} + xy - 1$

3. What is the degree of the polynomial $2a^2 + 4b^8$?

(a) 2 (b) 10
(c) 8 (d) 0

4. Degree of a constant term is

(a) 1 (b) 0
(c) 2 (d) not defined

5. Degree of the polynomial $(a^2+1)(a+2)(a^3+3)$ is

(a) 3 (b) 6
(c) 2 (d) 7

6. If the degree of the polynomial $\left(p^6 + \frac{3}{7}\right)(p^n + 3p)$ is 9, then the value of n is

(a) 1 (b) 3
(c) 6 (d) 18

7. The sum of three expressions is $x^2 + y^2 + z^2$. If two of them are $4x^2 - 5y^2 + 3z^2$ and $-3x^2 + 4y^2 + 2z^2$, the third expression is

(a) $2x^2 + 2z^2$ (b) $2y^2$
(c) $2x^2 + 2y^2 - z^2$ (d) $2y^2 + 2z^2$

8. If $P = 3x - 4y - 8z$, $Q = -10y + 7x + 11z$ and $R = 19z - 6y + 4x$, then $P - Q + R$ is equal to

(a) $13x - 20y + 16z$ (b) 0
(c) $x + y + z$ (d) $2x - 4y + 3z$

9. The product of $4a^2, -6b^2$ and $3a^2b^2$ is

(a) a^2b^2 (b) $13a^4b^4$
(c) $-72a^4b^4$ (d) a^4b^4

10. $(14x^2yz - 28x^2y^2z^3 + 32y^2z^2) \div (-4xy)$ is equal to

(a) $\frac{7}{2}yz + 7xyz^2 + 8xyz$

(b) $-\frac{7}{2}xz + 7xyz^3 - \frac{8yz^2}{x}$

(c) $-\frac{7}{2}xz - 7xyz^3 + \frac{8yz^2}{x}$

(d) $\frac{7}{2}xz - 7xyz^2 - \frac{8yz^2}{x}$

11. The product of $\left(\frac{1}{5}x^2 - \frac{1}{6}y^2\right)$ and $(5x^2 + 6y^2)$ is

(a) 1

(b) $x^4 + \frac{11}{60}x^2y^2 + y^4$

(c) $x^4 + \frac{11}{30}x^2y^2 - y^4$

(d) $x^4 - \frac{11}{30}x^2y^2 - y^4$

12. The product $(x+2)(x^2 - 2x + 4)$ is equal to

(a) $x^3 + 8$ (b) $x^3 - 8$

(c) $x^3 - 4x^2 + 4x - 8$ (d) $x^3 + 4x^2 + 2x + 8$

13. $(x+4)(x+3) - (x-4)(x-3)$ is equal to

(a) $2x^2 - 14x + 24$ (b) $2x^2 + 14x - 24$

(c) $14x$ (d) 24

14. If $(x^2 + 4x - 21)$ is divided by $x + 7$, then the quotient is

(a) $x + 3$ (b) $x - 3$

(c) $x^2 - 2$ (d) $x - 4$

15. What is the remainder when $13x^2 + 22x - 10$ is divided by $x + 2$?

(a) 2 (b) –2

(c) 0 (d) – 4

16. A polynomial when divided by $(x - 6)$, gives a quotient $x^2 + 2x - 13$ and leaves a remainder –8. The polynomial is

(a) $x^3 + 4x^2 + 25x - 78$

(b) $x^3 - 4x^2 - 25x + 70$

(c) $x^3 - 4x^2 - 25x - 70$

(d) $x^3 + 4x^2 - 25x + 78$

17. For a polynomial, dividend is $x^4 + 4x - 2x^2 + x^3 - 10$, quotient is $x^3 + 3x^2 + 4x + 12$ and remainder is 14, then divisor is equal to

(a) $x^2 + 2$ (b) $x^2 - 2$

(c) $x + 2$ (d) $x - 2$

18. What is the quotient when $10a^2 + 3a - 27$ is divided by $2a - 3$

(a) $5a - 9$ (b) $(-5a - 9)$

(c) $(-5a + 9)$ (d) $5a + 9$

19. If $(14x^2 + 13x - 15)$ is divided by $(7x - 4)$, the degree of the remainder is

(a) 1 (b) 2

(c) 0 (d) 3

20. $x + y - (z - x - [y + z - (x + y - \{z + x - (y + z + x)\})])$ is equal to

(a) $3x$ (b) $2y$

(3) c (d) 0

21. The remainder when $x^3 - 2x^2 + 4x$ is divided by x^2 is

(a) 1 (b) $x - 2 + \frac{4}{x}$

(c) 0 (d) $4x - 2x^2$

22. What should be added to $\frac{1}{x}$, to make it equal to x?

(a) $\frac{x^2 - x}{x^2}$ (b) $\frac{x}{x^2 - 1}$

(c) $\frac{x^2 + 1}{x}$ (d) $\frac{x^2 - 1}{x}$

23. Using the formula : $W = np + \frac{1}{2}NX^2$, frame a formula for X.

(a) $\sqrt{\frac{Wn - P}{2N}}$ (b) $\sqrt{\frac{2(W - np)}{N}}$

(c) $\sqrt{\frac{np - W}{2N}}$ (d) $\sqrt{\frac{W + np}{2N}}$

24. If $A = \pi(R^2 - r^2)$, then R is equal to

(a) $\sqrt{\frac{A - \pi r^2}{\pi}}$ (b) $\sqrt{\frac{A + \pi r^2}{\pi}}$

(c) $\sqrt{\frac{r^2\pi - A}{\pi}}$ (d) $\sqrt{\frac{r^2\pi - A}{r}}$

25. $\frac{3}{4}(a+y) - \left[y + a - \frac{1}{3}\left(y + a - \frac{1}{4}(a+y)\right)\right]$ is equal to

(a) $a + y$ (b) $3a$

(c) $-4y$ (d) 0

Answers

1. (d)	2. (a)	3. (c)	4. (b)	5. (b)	6. (b)	7. (b)	8. (b)	9. (c)	10. (b)
11. (c)	12. (a)	13. (c)	14. (b)	15. (b)	16. (b)	17. (d)	18. (d)	19. (c)	20. (a)
21. (c)	22. (d)	23. (b)	24. (b)	25. (d)					

Hints and Solutions

1. (d) Here the variable is in the denominator.

2. (a) In all the other algebraic expressions, the variables have fractional powers.

4. (b) In a constant term, the power of the variable is zero, so its degree is 0, *e.g.*, $4 = 4 \times 1 = 4 \times x^0$.

5. (b) $(a^2+1)(a+2)(a^3+3)$

$= \{a^2(a+2)+1(a+2)\}(a^3+3)$

$= \{a^3+2a^2+a+2\}(a^3+3)$

$= a^3 \times a^3 + 2a^2 \times a^3 + \ldots$

$\therefore$ Highest power of $a = 3 + 3 = 6$.

5. (b) $\left(p^6+\frac{3}{7}\right)\left(p^n+3p\right)$

$= p^6 \times p^n + p^6 \times 3p + \frac{3}{7}p^n + \frac{3}{7} \times 3p$

$\therefore$ Degree $= 9 \Rightarrow$ Highest power of $p = 9$

Term containing highest power

$= p^6 \times p^n = p^{6+n}$

$\Rightarrow 6+n = p \Rightarrow 6+n = 9 \Rightarrow n = 3$.

7. (b) Third exp. $= x^2+y^2+z^2-\{(4x^2-5y^2+3z^2)$

$+(-3x^2+4y^2-2z^2)\}$

8. (b) $P-Q+R = (3x-4y-8z)-(-10y+7x+11z)$

$+(19z-6y+4x)$

$= 3x-4y-8z+10y-7x-11z+19z-6y+4x$

$= 0$.

9. (c) $(4a^2)(-6b^2)(3a^2b^2)$

$= 4 \times (-6) \times 3 \times a^{2+2} . b^{2+2}$

$= -72\, a^4b^4$.

10. (b) $(14x^2yz - 28x^2y^2z^3 + 32y^2z^2) \div (-4xy)$

$= \frac{14x^2yz}{-4xy} - \frac{28x^2y^2z^3}{-4xy} + \frac{32y^2z^2}{-4xy}$

$= -\frac{7}{2}xz + 7xyz^3 - \frac{8yz^2}{x}$.

11. (c) $\left(\frac{1}{5}x^2-\frac{1}{6}y^2\right)\left(5x^2+6y^2\right)$

$= \frac{1}{5}x^2(5x^2+6y^2) - \frac{1}{6}y^2(5x^2+6y^2)$

$= x^4 + \frac{6}{5}x^2y^2 - \frac{5}{6}x^2y^2 - y^4$

$= x^4 + \frac{36x^2y^2 - 25x^2y^2}{30} - y^4$

$= x^4 + \frac{11x^2y^2}{30} - y^4$.

12. (a) $(x+2)(x^2-2x+4)$

$= x(x^2-2x+4) + 2(x^2-2x+4)$

Now solve.

13. (c) $(x+4)(x+3)-(x-4)(x-3)$

$= (x^2+3x+4x+12)-(x^2-3x-4x+12)$

$= (x^2+7x+12)-(x^2-7x+12)$

$= x^2+7x+12-x^2+7x-12$

$= 7x+7x = 14x$.

14. (b)

```
          x - 3
x + 7 ) x² + 4x - 21
        x² + 7x
        -    -
        ----------
           -3x - 21
           -3x - 21
            +    +
        ----------
                 0
```

15. (b)

```
         13x - 4
x + 2 ) 13x² + 22x - 10
        13x² + 26x
        -      -
        ------------
              -4x - 10
              -4x - 8
               +    +
        ------------
                   -2
```

16. (b) Use the division algorithm:

Dividend = Divisor × Quotient + Remainder

17. (d) Similar to Q. 16.

18. (d) Use long division method.

19. (c)

$$\begin{array}{r|l} & 2x + 3 \\ 7x-4 & 14x^2 + 13x - 15 \\ & 14x^2 - 8x \\ & - \quad + \\ \hline & 21x - 15 \\ & 21x - 12 \\ & - \quad + \\ \hline & -3 \end{array}$$

Remainder being a constant term, its degree = 0.

20. (a) $x+y-(z-x-[y+z-(x+y-\{z+x -(y+z-x)\})]$

$= x+y-(z-x-[y+z-(x+y -\{\not{z}+x-y-\not{z}+x\})]$

$= x+y-(z-x-[y+z-(x+y-2x+y)]$

$= x+y-(z-x-[y+z-(-x+2y)]$

$= x+y-(z-x-[y+z+x-2y]$

$= x+y-(z-x-[-y+z+x]$

$= x+y-(z-x+y-z-x)$

$= x+y-(-2x+y)$

$= x+y+2x-y = 3x.$

21. (c) $(x^3-2x^2+4x)\div x^2 = x-2+\frac{4}{x}$

Hence, remainder = 0

22. (d) Required addend $= x-\frac{1}{x}$

$= \frac{x^2-1}{x}.$

23. (b) $W = np + \frac{1}{2}NX^2$

$\Rightarrow W - np = \frac{1}{2}NX^2$

$\Rightarrow \frac{2(W-np)}{N} = X^2$

$\Rightarrow X = \sqrt{\frac{2(W-np)}{N}}.$

24. (b) $A = \pi R^2 - \pi r^2 \Rightarrow A + \pi r^2 = \pi R^2$

$\Rightarrow R^2 = \frac{A+\pi r^2}{\pi}$

$\Rightarrow R = \sqrt{\frac{A+\pi r^2}{\pi}}$

25. (d) $\frac{3}{4}(a+y) - \left[y+a-\frac{1}{3}\left(y+a-\frac{1}{4}(a+y)\right)\right]$

$= \frac{3}{4}a+\frac{3}{4}y-\left[y+a-\frac{1}{3}\left(y+a-\frac{1}{4}a-\frac{1}{4}y\right)\right]$

$= \frac{3}{4}a+\frac{3}{4}y-\left[y+a-\frac{1}{3}\left(\frac{3}{4}a+\frac{3}{4}y\right)\right]$

$= \frac{3}{4}a+\frac{3}{4}y-\left[y+a-\frac{1}{4}a-\frac{1}{4}y\right]$

$= \frac{3}{4}a+\frac{3}{4}y-\left[\frac{3}{4}a+\frac{3}{4}y\right] = 0.$

Self Assessment Sheet–6

1. (i) Figureshows a number of equal steps. If the 'rise' of each step is *h* cm, and there are *n* steps, make a formula for the height (*H*) in centimeters of the steps.

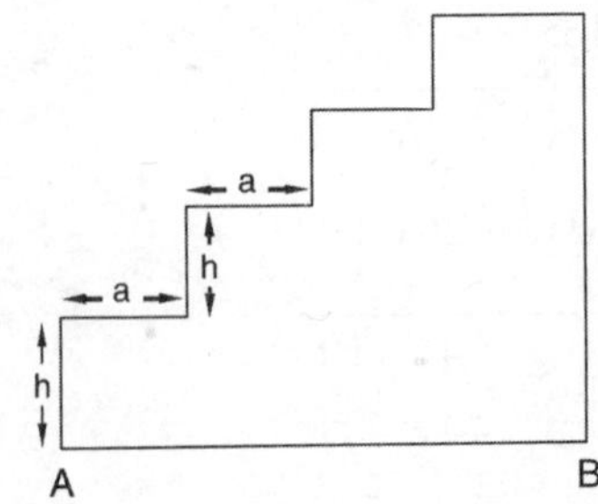

(a) $H = \frac{1}{2}nH$ (b) $H = 2nh$

(c) $H = nh$ (d) $H = n^2h$

(ii) If the 'tread' of each step is *a* cm, and there are *n* steps, make a formula for the length (*d* cm) of *AB*.

(a) $d = \frac{1}{2}na$ (b) $d = na$

(c) $d = 3na$ (d) $d = na^2$

(iii) If stair carpet is laid, starting at *A*, how many centimeters will be required for *n* steps ?

(a) $\frac{1}{2}nH + na$ (b) $nH + \frac{1}{2}na$

(c) $2(nH + na)$ (d) $nh + na$

2. The formula for the area, *A* sq cm of the white cross is

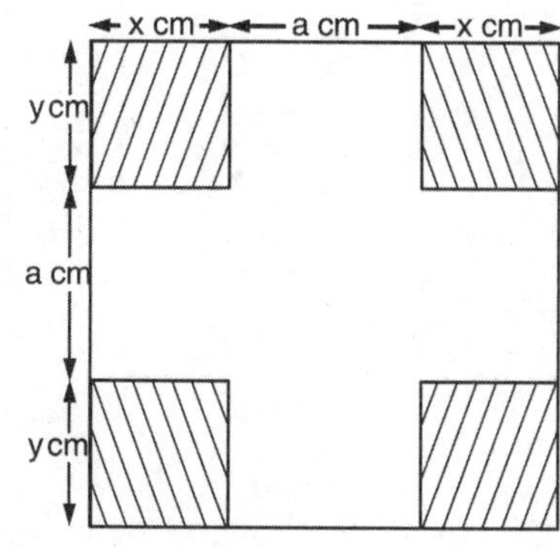

(a) $A = 4ax + 4ay + a^2$

(b) $A = 2ax + 4ay + a^2$

(c) $A = 2ax + 2ay + a^2$

(d) None of these

3. If $P = \frac{W}{2g}(v^2 - u^2)$, then the value of P when $W = 40$, $g = 32$, $u = 4$, $v = 12$ is

(a) 80 (b) 82
(c) 90 (d) 78

4. If $\frac{1}{f} = \frac{1}{p} + \frac{1}{q}$ and $p = 2$, $q = 3$, then f is

(a) $2\frac{1}{5}$ (b) $1\frac{1}{5}$

(c) $3\frac{1}{2}$ (d) $3\frac{2}{5}$

5. The remainder when $3x^2 + 5x - 7$ is divided by $x + 3$ is

(a) –5 (b) 4
(c) 2 (d) 5

6. When $a^2b(a^3 - a + 1) - ab(a^4 - 2a^2 + 2a) - b(a^3 - a^2 - 1)$ is simplified, the answer is

(a) $-a^2b$ (b) ab
(c) b (d) 0

7. The value of the product $(4a^2 + 3b)(9b^2 + 4a)$ at $a = 1$, $b = -2$ is

(a) 60 (b) – 80
(c) 70 (d) – 50

8. The expression that should be subtracted from $4x^4 - 2x^3 - 6x^2 + x - 5$ so that it may be exactly divisible by $2x^2 + x - 2$ is

(a) $3x + 5$ (b) $-3x - 5$
(c) $-3x + 5$ (d) $3x - 5$

9. The area of a rectangular courtyard is $(10x^3 - 11x^2 + 19x + 10)$ sq units. If one of its sides is $(2x^2 - 3x + 5)$ units, then the other side is

(a) $(5x + 2)$ units (b) $-5x + 2$ units
(c) $-(5x + 2)$ units (d) $5x - 2$ units

10. What must be added to the sum of $2a^2 - 3a + 7$, $-5a^2 - 2a - 11$ and $3a^2 + 5a - 8$ to get 0 ?

(a) – 12 (b) 12
(c) $a^2 + a$ (d) $a - 1$

Answers

1. (i) (c) (ii) (b) (iii) (d) **2.** (c) **3.** (a) **4.** (b) **5.** (d) **6.** (c) **7.** (b) **8.** (b)
9. (a) **10.** (b)

Chapter

7

ALGEBRAIC IDENTITIES

KEY FACTS

1. **Identities** or **Special products** are equations or formulae consisting of variable/variables which are true for all real values of the variable/variables.
   They can be thought of as shortcuts for multiplying binomials.
   Identity 1: $(x + a)(x + b) = x^2 + (a + b)x + ab$
   E.g., $(x+5)(x-2) = x^2 + (5+(-2))\,x + (5\times -2) = x^2 + 3x - 10$
   Identity 2: $(a + b)^2 = a^2 + 2ab + b^2$
   E.g., $(4x+5y)^2 = (4x)^2 + 2\times 4x \times 5y + (5y)^2 = 16x^2 + 40xy + 25y^2$
   Identity 3: $(a - b)^2 = a^2 - 2ab + b^2$
   E.g., $\left(3a - \frac{1}{2}b\right)^2 = (3a)^2 - 2\times 3a \times \frac{1}{2}b + \left(\frac{1}{2}b\right)^2 = 9a^2 - 3ab + \frac{1}{4}b^2$
   Identity 4: $(a - b)(a + b) = a^2 - b^2$
   E.g., $(xy - \frac{1}{3}a^2)(xy + \frac{1}{3}a^2) = (xy)^2 - \left(\frac{1}{3}a^2\right)^2 = x^2y^2 - \frac{1}{9}a^4$
   Identity 5: $(a + b + c)^2 = a^2 + b^2 + c^2 + 2ab + 2bc + 2ca$
   E.g., $(2x+4y-3z)^2 = (2x)^2 + (4y)^2 + (-3z)^2 + 2[2x\times 4y + 4y\times(-3z) + (-3z)\times 2x]$
   $= 4x^2 + 16y^2 + 9z^2 + 2(8xy - 12yz - 6zx) = 4x^2 + 16y^2 + 9z^2 + 16xy - 24yz - 12zx.$
   Identity 6: $(a + b)^3 = a^3 + 3a^2b + 3ab^2 + b^3 = a^3 + b^3 + 3ab(a + b)$
   E.g., $(a+3b)^3 = a^3 + 3\times a^2 \times 3b + 3\times a \times (3b)^2 + (3b)^3 = a^3 + 9a^2b + 27ab^2 + 27b^3.$
   Identity 7: $(a - b)^3 = a^3 - 3a^2b + 3ab^2 - b^3 = a^3 - b^3 - 3ab(a - b)$
   E.g., $\left(\frac{p}{3} - \frac{q}{2}\right)^3 = \left(\frac{p}{3}\right)^3 - 3\times\left(\frac{p}{3}\right)^2 \times \frac{q}{2} + 3\times\frac{p}{3}\times\left(\frac{q}{2}\right)^2 - \left(\frac{q}{2}\right)^3$
   $= \frac{p^3}{27} - 3\times\frac{p^2}{9}\times\frac{q}{2} + 3\times\frac{p}{3}\times\frac{q^2}{4} - \frac{q^3}{8} = \frac{p^3}{27} - \frac{p^2q}{6} + \frac{pq^2}{4} - \frac{q^3}{8}.$
   Identity 8: $(x + a)(x + b)(x + c) = x^3 + (a + b + c)x^2 + (ab + bc + ca)x + abc$
   E.g., $(y+4)(y+5)(y-2) = y^3 + [4+5+(-2)]y^2 + [4\times 5 + 5\times(-2) + 4\times(-2)]y + 4\times 5\times(-2)$
   $= y^3 + 7y^2 + (20-10-8)y - 40 = y^3 + 7y^2 + 2y - 40.$
   Identity 9(a): $(a + b)(a^2 - ab + b^2) = a^3 + b^3$
   E.g., $(x+2)(x^2 - 2x + 4) = (x+2)(x^2 - 2\times x + 2^2) = x^3 + 2^3 = x^3 + 8.$

Identity 9(b): $(a-b)(a^2+ab+b^2)=a^3-b^3$

E.g., $(2x-4y)(4x^2+8xy+16y^2)=(2x-4y)[(2x)^2+2x\times 4y+(4y)^2]=(2x)^3-(4y)^3=8x^3-64y^3$.

Identity 10: $(a+b+c)(a^2+b^2+c^2-ab-bc-ca)=a^3+b^3+c^3-3abc$

E.g., $(x+2y+3z)(x^2+4y^2+9z^2-2xy-6yz-3xz)$

$= (x+2y+3z)[(x)^2+(2y)^2+(3z)^2-x\times 2y-2y\times 3z-3z\times x]$

$= (x)^3+(2y)^3+(3z)^3-3\times x\times 2y\times 3z.$

$= x^3+8y^3+27z^3-18xyz.$

2. Some Important Formulae :

1. $(a+b)^2+(a-b)^2=2(a^2+b^2)$
2. $(a+b)^2-(a-b)^2=4ab$
3. $\left(a+\frac{1}{a}\right)^2=a^2+\frac{1}{a^2}+2$
4. $\left(a-\frac{1}{a}\right)^2=a^2+\frac{1}{a^2}-2$

Solved Examples

Ex. 1. ***Find the product :*** $(a^2+b^2)(a^4+b^4)(a+b)(a-b)$.

Sol. $(a^2+b^2)(a^4+b^4)(a+b)(a-b)=(a^2+b^2)(a^4+b^4)(a^2-b^2)$ $\quad [\because (x+y)(x-y)=x^2-y^2]$

$=(a^4+b^4)(a^4-b^4)=\mathbf{a^8-b^8}$.

Ex. 2. ***Find the value of*** $c+d$ ***if*** $c-d=2,\ cd=63$.

Sol. $(c+d)^2=(c-d)^2+4cd=2^2+4\times 63=4+252=256$

$\Rightarrow (c+d)=\sqrt{256}=\mathbf{16}$.

Ex. 3. ***Find the value of*** a^2+b^2 ***if*** $a+b=10$ ***and*** $a-b=2$.

Sol. $2(a^2+b^2)=(a+b)^2+(a-b)^2=10^2+2^2=100+4=104$

$\Rightarrow a^2+b^2=\frac{104}{2}=\mathbf{52}$.

Ex. 4. ***If*** $a^2+\frac{1}{a^2}=7$, ***find the values of***

(a) $\left(a+\frac{1}{a}\right)$ ***(b)*** $\left(a-\frac{1}{a}\right)$ ***(c)*** $\left(a^2-\frac{1}{a^2}\right)$

Sol. (a) $\left(a+\frac{1}{a}\right)^2=a^2+\frac{1}{a^2}+2=7+2=9 \Rightarrow a+\frac{1}{a}=\sqrt{9}=\mathbf{\pm 3}$

(b) $\left(a-\frac{1}{a}\right)^2=a^2+\frac{1}{a^2}-2=7-2=5 \Rightarrow \left(a-\frac{1}{a}\right)=\mathbf{\pm\sqrt{5}}$

(c) $\left(a^2-\frac{1}{a^2}\right)=\left(a+\frac{1}{a}\right)\left(a-\frac{1}{a}\right)=(\pm 3)(\pm\sqrt{5})=\mathbf{\pm 3\sqrt{5}}$.

Ex. 5. ***If*** $p-\frac{1}{p}=4$, ***find the value of*** $p^4+\frac{1}{p^4}$.

Sol. $p^2+\frac{1}{p^2}=\left(p-\frac{1}{p}\right)^2+2=4^2+2=18$

$p^4+\frac{1}{p^4}=\left(p^2+\frac{1}{p^2}\right)^2-2=18^2-2=324-2=\mathbf{322}$.

Ex. 6. ***Find the value of pq if $p^3 - q^3 = 68$ and $p - q = -4$.***

Sol. $(p-q)^3 = p^3 - q^3 - 3pq(p-q)$

$\Rightarrow 3pq(p-q) = p^3 - q^3 - (p-q)^3 \Rightarrow 3pq\,(-4) = 68 - (-4)^3$

$\Rightarrow -12pq = 68 - (-64) = 68 + 64 = 132 \Rightarrow pq = \dfrac{132}{-12} = \mathbf{-11}.$

Ex. 7. ***If $\left(x+\dfrac{1}{x}\right)^2 = 3$, show that $x^3 + \dfrac{1}{x^3} = 0$.***

Sol. $\left(x+\dfrac{1}{x}\right)^2 = 3 \Rightarrow x + \dfrac{1}{x} = \pm\sqrt{3}$

Now $\left(x+\dfrac{1}{x}\right)^3 = x^3 + \dfrac{1}{x^3} + 3\left(x+\dfrac{1}{x}\right)$

$\Rightarrow x^3 + \dfrac{1}{x^3} = \left(x+\dfrac{1}{x}\right)^3 - 3\left(x+\dfrac{1}{x}\right) = (\pm\sqrt{3})^3 - 3\times(\pm\sqrt{3})$ $\quad\left[\begin{array}{l}\therefore (-)\times(+) = (-)\\ (-)\times(-) = (+)\end{array}\right]$

$= \pm 3\sqrt{3} \mp 3\sqrt{3} = (3\sqrt{3} - 3\sqrt{3})$ or $(-3\sqrt{3} + 3\sqrt{3}) = \mathbf{0}.$

Ex. 8. ***If $a^2 + b^2 + c^2 = 50$ and $ab + bc + ca = 47$, find the value of $a + b + c$.***

Sol. $(a+b+c)^2 = a^2 + b^2 + c^2 + 2(ab+bc+ca) = 50 + 2\times 47 = 50 + 94 = 144$

$\Rightarrow (a+b+c) = \sqrt{144} = \mathbf{\pm 12}.$

Ex. 9. ***Find the continued product of $x + y$, $x - y$, $x^2 + xy + y^2$, $x^2 - xy + y^2$***

Sol. Required product $= [(x+y)(x^2 - xy + y^2)][(x-y)(x^2 + xy + y^2)]$

$= (x^3 + y^3)(x^3 - y^3) = \mathbf{x^6 - y^6}.$

Ex. 10. ***Multiply by using the correct identity***

$(x + y - z)\,(x^2 + y^2 + z^2 - xy + xz + yz)$

Sol. $(x+y-z)(x^2 + y^2 + z^2 - xy + xz + yz) = (x)^3 + (y)^3 + (-z)^3 - 3\times(x)\times(y)\times(-z)$

$= x^3 + y^3 - z^3 + 3xyz$ $\quad\boxed{\because (a+b+c)\,(a^2 + b^2 + c^2 - ab - bc - ca) = a^3 + b^3 + c^3 - 3abc}$

Ex. 11. ***Find the value of $a^3 + b^3 + c^3 - 3abc$ if $a + b + c = 8$ and $ab + bc + ca = 19$.***

Sol. First we find the value of $a^2 + b^2 + c^2$

$a^2 + b^2 + c^2 = (a+b+c)^2 - 2(ab+bc+ca) = 8^2 - 2\times 19 = 64 - 38 = 26$

$\therefore\ a^3 + b^3 + c^3 - 3abc = (a+b+c)\,(a^2 + b^2 + c^2 - ab - bc - ca) = (8)(26-19) = 8\times 7 = \mathbf{56}.$

Question Bank-7

1. $\left(3m + \dfrac{1}{5}\right)^2 =$ ____

(a) $3m^2 + \dfrac{6m}{5} + \dfrac{1}{25}$ (b) $9m^2 + \dfrac{3m}{5} - \dfrac{1}{25}$

(c) $9m^2 + \dfrac{6m}{5} + \dfrac{1}{25}$ (d) $3m^2 + \dfrac{3m}{5} + \dfrac{1}{25}$

2. $(mx - ny)(mx - ny) =$ ____

(a) $m^2x^2 + 2mxny - n^2y^2$

(b) $m^2x^2 - 2mxny - n^2y^2$

(c) $m^2x^2 - 2mxny + n^2y^2$

(d) $m^2x^2 + 2mxny + n^2y^2$

3. $\left(5z^6+\frac{1}{z^6}\right)^2 =$ _____

(a) $5z^8+10z^6+\frac{1}{z^8}$

(b) $25z^{12}+10z^6+\frac{1}{z^{12}}$

(c) $25z^{12}+10+\frac{1}{z^{12}}$

(d) $5z^8+10+\frac{1}{z^8}$

4. $(z^2+13)(z^2-5) =$ _____

(a) $2z^4+18z^2-8$ (b) z^4+8z^2-65

(c) z^4-8z^2-65 (d) z^4+8z^2+65

5. $(x-y)^2+2xy =$ _____

(a) $x^2-4xy-y^2$ (b) x^2+y^2

(c) x^2-y^2 (d) $x^2-4xy+y^2$

6. $(p-q)^2+4pq$

(a) p^2-q^2 (b) $(p+q)^2$

(c) $(2p-q)^2$ (d) $(2p-2q)^2$

7. $x^2+\frac{1}{x^2} =$ _____

(a) $\left(x+\frac{1}{x}\right)^2-2$ (b) $\left(x+\frac{1}{2}\right)^2+2$

(c) $\left(x-\frac{1}{x}\right)^2-2$ (d) $\left(x-\frac{1}{x}\right)^2-1$

8. If $x+\frac{1}{x}=m$, then find the value of $x^2+\frac{1}{x^2}$:

(a) $\frac{m^2}{4}$ (b) m^2-2

(c) $2m^2+1$ (d) $2m^2-1$

9. If $x-\frac{1}{x}=8$, the value of $x^2+\frac{1}{x^2}$ is

(a) 10 (b) 62

(c) 6 (d) 66

10. If $x^2+\frac{1}{x^2}=83$, the value of $x-\frac{1}{x}$ is

(a) 9 (b) $\sqrt{85}$

(c) 81 (d) 85

11. If $x+\frac{1}{x}=7$, the value of $x^4+\frac{1}{x^4}$ is

(a) 2401 (b) 2023

(c) 2209 (d) 2207

12. $\left(\frac{2}{5}ab+c\right)\left(\frac{2}{5}ab-c\right) =$ _____

(a) $\frac{4}{25}a^2b^2-\frac{4}{5}abc+c^2$

(b) $\frac{4}{25}a^2b^2+\frac{4}{5}abc+c^2$

(c) $\frac{4}{25}a^2b^2-c^2$

(d) $\frac{4}{25}a^2b^2+c^2$

13. $(x+4)(x-4)(x^2+16) =$ _____

(a) x^2-64 (b) x^4-64

(c) x^4-256 (d) x^2-256

14. $25^2-15^2 =$ _____

(a) $(25+15)^2$ (b) $(25-15)^2$

(c) $(25+15)(25-15)$ (d) 25×15

15. $\left(3x+\left(\frac{-1}{4}\right)+2y\right)^2 =$ _____

(a) $9x^2-\frac{1}{16}+4y^2-\frac{6x}{4}-y+12xy$

(b) $9x^2+\frac{1}{16}+4y^2+\frac{6x}{4}+y-12xy$

(c) $9x^2+\frac{1}{16}+4y^2-\frac{3x}{2}-y+12xy$

(d) $9x^2-\frac{1}{16}+4y^2+\frac{3x}{2}+y-12xy$

16. $(x-y-z)^2-(x+y+z)^2$ is equal to

(a) $4xy+4yz$ (b) $-4xy-4xz$

(c) $4xy+4xz$ (d) $-4yz$

17. If $x+y+z=0$, then x^2+xy+y^2 equals

(a) y^2+yz+z^2 (b) y^2-yz+z^2

(c) z^2-zx+x^2 (d) z^2+zx+x^2

18. If $a^2+b^2=6,\ a+b=4$, then $ab =$ _____

(a) 10 (b) 5

(c) 20 (d) 7

19. If $x^4+\frac{1}{x^4}=727$, then $x-\frac{1}{x}$ is equal to

(a) 5 (b) $\sqrt{29}$

(c) 25 (d) 27

20. If $a+b+c=9$ and $a^2+b^2+c^2=21$, then $ab+bc+ca$ is equal to

(a) 30 (b) 15
(c) 51 (d) 60

21. If $a^2+b^2+c^2=31$ and $ab+bc+ca=25$, then $a+b+c$ is equal to

(a) 81 (b) 56
(c) 9 (d) 6

22. If $a+b+c=0$, then the value of $\frac{a^2+b^2+c^2}{bc+ca+ab}$ is

(a) 2 (b) –2
(c) 4 (d) –4

23. $\left(\frac{4}{3}x-\frac{3}{4}y\right)^3$ is equal to

(a) $\frac{64}{27}x^3+\frac{27}{64}y^3+4x^2y+\frac{9}{4}xy^2$

(b) $\frac{64}{27}x^3+\frac{27}{64}y^3-4x^2y-\frac{9}{4}xy^2$

(c) $\frac{64}{27}x^3-\frac{27}{64}y^3-4x^2y+\frac{9}{4}xy^2$

(d) $\frac{64}{27}x^3-\frac{27}{64}y^3+4x^2y-\frac{9}{4}xy^2$

24. $\left(x^2-\frac{1}{x^2}\right)^3=x^6-\frac{1}{x^6}+\ldots\ldots$. The missing part is

(a) $3x^2-\frac{3}{x^2}$ (b) $-3x^4+\frac{3}{x^4}$
(c) $-3x^2+\frac{3}{x^2}$ (d) $3x^4-\frac{3}{x^4}$

25. $(p-q)^3-(p+q)^3$ is equal to

(a) $2p^3-2q^3$ (b) $2p^3+2q^3$
(c) $2p^3+6pq^2$ (d) $-6p^2q-2q^3$

26. The value of a^3-b^3, when $a-b=4$ and $ab=-2$ is

(a) 88 (b) 40
(c) 72 (d) 64

27. The value of x^3+y^3, when $x+y=5$ and $xy=6$ is

(a) 125 (b) 35
(c) 215 (d) 107

28. If $x+\frac{1}{x}=5$, then $x^3+\frac{1}{x^3}$ is equal to

(a) 125 (b) 140
(c) 110 (d) 100

29. If $x^2+\frac{1}{x^2}=51$, then the value of $x^3-\frac{1}{x^3}$ is

(a) 364 (b) 343
(c) 153 (d) 103

30. $(0.8x+0.3y)(0.64x^2-0.24xy+0.09y^2)$ equals

(a) $(0.8x+0.3y)^2$ (b) $(0.8x+0.3y)^3$
(c) $(0.8x)^3+(0.3y)^3$ (d) $(0.8x)^3-(0.3y)^3$

31. $(2x-5)(4x^2+10x+25)$ is equal to

(a) $(2x-5)^3$ (b) $8x^3+125$
(c) $(2x+5)^3$ (d) $8x^3-125$

32. The value of $\frac{4.359\times4.359-1.641\times1.641}{4.359-1.641}$ is

(a) 6.3 (b) 6
(c) 3.2 (d) 4.6

33. $\frac{(6.4)^2-(5.4)^2}{(8.9)^2+8.9\times2.2+(1.1)^2}=$ ____

(a) 0.118 (b) 0.92
(c) 1.5 (d) 0.61

34. $\frac{3.7\times3.7+2.3\times2.3+2\times3.7\times2.3}{4.6\times4.6-3.4\times3.4}=$ ____

(a) $3\frac{3}{4}$ (b) $4\frac{3}{4}$
(c) $3\frac{1}{4}$ (d) $3\frac{1}{2}$

35. $\frac{8.73\times8.73\times8.73+4.27\times4.27\times4.27}{8.73\times8.73-8.73\times4.27+4.27\times4.27}=$ ____

(a) 11 (b) 12
(c) 13 (d) 10

36. $\frac{0.06\times0.06\times0.06-0.05\times0.05\times0.05}{0.06\times0.06+0.06\times0.05+0.05\times0.05}$ gives

(a) 0.01 (b) 0.001
(c) 0.1 (d) 0.02

37. $(1-a)(1+a+a^2)+(1+a)(1-a+a^2)$ is equal to

(a) $2a^3$ (b) 2
(c) $-2a^3$ (d) 0

38. The product $(2x-3y+5z)(4x^2+9y^2+25z^2+6xy+15yz-10xz)$ is

(a) $8x^3+27y^3+125z^3-90xyz$
(b) $8x^3+27y^3+125z^3+90xyz$
(c) $8x^3-27y^3+125z^3-90xyz$
(d) $8x^3-27y^3+125z^3+90xyz$

39. If $x+y+z=7$, $xy+yz+zx=12$, find the value of $x^3+y^3+z^3-3xyz$.

(a) 25 (b) 91

(c) 105 (d) 59

40. The product $(x+y+z)\left[(x-y)^2+(y-z)^2+(z-x)^2\right]$ is equal to

(a) $x^3+y^3+z^3-3xyz$

(b) $x^3+y^3+z^2$

(c) $3xyz$

(d) $2\left[(x^3+y^3+z^3)-3xyz\right]$

Answers

1. (c)	**2.** (c)	**3.** (c)	**4.** (b)	**5.** (b)	**6.** (b)	**7.** (a)	**8.** (b)	**9.** (d)	**10.** (a)
11. (d)	**12.** (c)	**13.** (c)	**14.** (c)	**15.** (c)	**16.** (b)	**17.** (d)	**18.** (b)	**19.** (a)	**20.** (a)
21. (c)	**22.** (b)	**23.** (c)	**24.** (c)	**25.** (d)	**26.** (b)	**27.** (b)	**28.** (c)	**29.** (a)	**30.** (c)
31. (d)	**32.** (b)	**33.** (a)	**34.** (a)	**35.** (c)	**36.** (a)	**37.** (b)	**38.** (d)	**39.** (b)	**40.** (d)

Hints and Solutions

1. (c) $\left(3m+\frac{1}{5}\right)^2=(3m)^2+2\times 3m\times\frac{1}{5}+\left(\frac{1}{5}\right)^2$

$=9m^2+\frac{6m}{5}+\frac{1}{25}.$

2. (c) $(mx-ny)(mx-ny)=(mx-ny)^2$

$=(mx)^2-2mx\times ny+(ny)^2$

$=m^2x^2-2mxny+n^2y^2.$

3. (c) $\left(5z^6+\frac{1}{z^6}\right)^2=(5z^6)^2+2\times 5z^6\times\frac{1}{z^6}+\left(\frac{1}{z^6}\right)^2$

$=25z^{12}+10+\frac{1}{z^{12}}.$

4. (b) $(z^2+13)(z^2-5)$

$=(z^2)^2+(13+(-5))z^2+(13\times -5)$

$=z^4+8z^2-65.$

5. (b) $(x-y)^2+2xy=x^2-2xy+y^2+2xy$

$=x^2+y^2.$

6. (b) $(p-q)^2+4pq=(p^2-2pq+q^2)+4pq$

$=p^2+2pq+q^2=(p+q)^2.$

7. (a) $\left(x+\frac{1}{x}\right)^2=x^2+\frac{1}{x^2}+2$

$\Rightarrow x^2+\frac{1}{x^2}=\left(x+\frac{1}{x}\right)^2-2.$

8. (b) Similar to Q. 7.

9. (d) Use : $\left(x-\frac{1}{x}\right)^2=x^2+\frac{1}{x^2}-2$

10. (a) Similar to Q. 9.

11. (d) Type Solved Example 5.

12. (c) $\left(\frac{2}{5}ab+c\right)\left(\frac{2}{5}ab-c\right)=\left(\frac{2}{5}ab\right)^2-c^2.$

13. (c) $(x+4)(x-4)(x^2+16)$

$=(x^2-4^2)(x^2+16)$

$=(x^2-16)(x^2+16)=(x^2)^2-(16)^2$

$=x^4-256.$

15. (c) $\left(3x+\left(\frac{-1}{4}\right)+2y\right)^2=(3x)^2+\left(\frac{-1}{4}\right)^2+(2y)^2$

$+2\left(3x\times\frac{-1}{4}+\frac{-1}{4}\times 2y+3x\times 2y\right)$

16. (b) $(x-y-z)^2-(x+y+z)^2$

$=(x^2+y^2+z^2-2xy+2yz-2xz)$

$-(x^2+y^2+z^2+2xy+2yz+2xz)$

$=-4xy-4xz.$

17. (d) $x+y+z=0\Rightarrow x+y=-z$ and $y=-z-x$

Now, $x^2+xy+y^2=x^2+2xy+y^2-xy$

$=(x+y)^2-xy=(-z)^2-x(-z-x)$

$=z^2+xz+x^2.$

18. (b) $(a+b)=4 \Rightarrow (a+b)^2 = 4^2$

$\Rightarrow a^2+2ab+b^2=16 \Rightarrow 6+2ab=16$

$\Rightarrow 2ab=10 \Rightarrow ab=5.$

19. (a) $x^4+\frac{1}{x^4}=727 \Rightarrow x^4+\frac{1}{x^4}+2=727+2$

$\Rightarrow \left(x^2+\frac{1}{x^2}\right)^2=729$

$\Rightarrow x^2+\frac{1}{x^2}=\sqrt{729}=27$

$\Rightarrow x^2+\frac{1}{x^2}-2=27-2 \Rightarrow \left(x-\frac{1}{x}\right)^2=25$

$\Rightarrow x-\frac{1}{x}=5.$

20. (a) $(a+b+c)^2=a^2+b^2+c^2+2(ab+bc+ca)$

$\Rightarrow 9^2=21+2(ab+bc+ca)$

$\Rightarrow ab+bc+ca=\frac{81-21}{2}=\frac{60}{2}=30.$

21. (c) Similar to Q. 20.

22. (b) $(a+b+c)^2=a^2+b^2+c^2+2(ab+bc+ca)$

$a+b+c=0$

$\Rightarrow a^2+b^2+c^2+2(ab+bc+ca)=0$

$\Rightarrow a^2+b^2+c^2=-2(ab+bc+ca)$

$\Rightarrow \frac{a^2+b^2+c^2}{ab+bc+ca}=-2.$

23. (c) $\left(\frac{4}{3}x-\frac{3}{4}y\right)^3=\left(\frac{4}{3}x\right)^3-3\times\left(\frac{4}{3}x\right)^2\times\frac{3}{4}y$

$+3\times\left(\frac{4}{3}x\right)\times\left(\frac{3}{4}y\right)^2-\left(\frac{3}{4}y\right)^3$

Now solve.

24. (c) $\left(x^2-\frac{1}{x^2}\right)^3=(x^2)^3-3\times(x^2)^2\times\frac{1}{x^2}$

$+3\times x^2\times\left(\frac{1}{x^2}\right)^2-\left(\frac{1}{x^2}\right)^3$

$=x^6-3\times x^4\times\frac{1}{x^2}+3\times x^2\times\frac{1}{x^4}-\frac{1}{x^6}$

$=x^6-3x^2+\frac{3}{x^2}-\frac{1}{x^6}$

$\therefore$ The missing part is $-3x^2+\frac{3}{x^2}$.

25. (d) $(p-q)^3-(p+q)^3=(p^3-3p^2q+3pq^2-q^3)$

$-(p^3+3p^2q+3pq^2+q^3)$

$=-6p^2q-2q^3.$

26. (b) $(a-b)^3=a^3-b^3-3ab(a-b)$

$\therefore\ 4^3=a^3-b^3-3\times(-2)\times 4$

$\Rightarrow 64=a^3-b^3+24 \Rightarrow a^3-b^3=40.$

27. (b) $(a+b)^3=a^3+b^3+3ab(a+b)$

28. (c) Similar to Q. 27.

29. (a) $x^2+\frac{1}{x^2}=51 \Rightarrow x^2+\frac{1}{x^2}-2=51-2=49$

$\Rightarrow \left(x-\frac{1}{x}\right)^2=49 \Rightarrow x-\frac{1}{x}=7$

Now, use the formula

$\left(x-\frac{1}{x}\right)^3=x^3-\frac{1}{x^3}-3\times x\times\frac{1}{x}\left(x-\frac{1}{x}\right)$

32. (b) Given exp. $=\frac{(4.359)^2-(1.641)^2}{4.359-1.641}$

$(\because a^2-b^2=(a+b)(a-b))$

$=\frac{(4.359+1.641)\cancel{(4.359-1.641)}}{\cancel{(4.359-1.641)}}$

$=4.359+1.641=6.$

33. (a) Given exp. $=\frac{(6.4+5.4)(6.4-5.4)}{(8.9+1.1)^2}$

$\left[\begin{array}{l} a^2-b^2=(a+b)(a-b) \\ a^2+2ab+b^2=(a+b)^2 \end{array}\right]$

$=\frac{11.8\times 1}{100}=0.118.$

34. (a) Given exp. $=\frac{(3.7+2.3)^2}{(4.6+3.4)(4.6-3.4)}$

$=\frac{6^2}{8\times 1.2}=\frac{36}{9.6}=\frac{360}{96}=\frac{15}{4}=3\frac{3}{4}.$

35. (c) Given exp. $=\frac{(8.73)^3+(4.27)^3}{(8.73)^2-8.73\times 4.27+(4.27)^2}$

$\left[\because a^3+b^3=(a+b)(a^2+ab+b^2)\right]$

$$= \frac{(8.73+4.27)\{(8.73)^2 - 8.73\times4.27 + (4.27)^2\}}{\{(8.73)^2 - 8.73\times4.27+(4.27)^2\}}$$

$= (8.73+4.27) = 13.$

36. (a) Use $a^3 - b^3 = (a-b)(a^2+ab+b^2)$

37. (b) $(1-a)(1+a+a^2)+(1+a)(1-a+a^2)$

$= (1-a^3)+(1+a^3) = 2.$

38. (d) Use the identity

$(x+y+z)(x^2+y^2+z^2-xy-yz-zx)$

$= x^3+y^3+z^3-3xyz.$

39. (b) Type Solved Example 11.

40. (d) $(x+y+z)\left[(x-y)^2+(y-z)^2+(z-x)^2\right]$

$= (x+y+z)\left[x^2-2xy+y^2+y^2-2yz+z^2 + z^2-2zx+x^2\right]$

$= (x+y+z)(2x^2+2y^2+2z^2-2xy-2yz-2zx)$

$= 2(x+y+z)(x^2+y^2+z^2-xy-yz-zx)$

$= 2\left[(x^3+y^3+z^3)-3xyz\right]$

Self Assessment Sheet–7

1. $(-a-b)(b-a)$ is

(a) a^2+b^2 (b) b^2-a^2

(c) a^2-b^2 (d) $-(b^2+a^2)$

2. $\frac{0.25\times0.25-0.24\times0.24}{0.49} = ?$

(a) 0.0006 (b) 0.49

(c) 0.01 (d) 0.1

3. The value of $(5x-3y)^2-(5x+3y)^2$ when $x=-1,\ y=\sqrt{\frac{1}{25}}$ is

(a) 12 (b) $\frac{1}{15}$

(c) 10 (d) –30

4. $8x^3-27$ divided by $4x^2+6x+9$ is

(a) $2x+3$ (b) $-2x+3$

(c) $2x-3$ (d) $-(2x+3)$

5. $\frac{2.3\times2.3\times2.3-1}{2.3\times2.3+2.3+1} = ?$

(a) 0.3 (b) 1.3

(c) 2.2 (d) 3.3

6. $a^3+3a^2b+3ab^2+b^3$ divided by $a^2+2ab+b^2$ is

(a) a^2+b^2 (b) $a+2b$

(c) $2a^2+b^2$ (d) $a+b$

7. $(25.732)^2-(15.732)^2 = ?$

(a) 4.1464 (b) 41.464

(c) 414.64 (d) 4164.4

8. If $x+\frac{1}{x}=8$, then the value of $\left(x-\frac{1}{x}\right)^2$ is

(a) 64 (b) 60

(c) 16 (d) 62

9. The continued product of $(x+3)(x-3)(x^2+9)$ is

(a) $-x^4+81$ (b) x^2-81

(c) x^4-18 (d) x^4-81

10. If $x^2+\frac{1}{x^2}=7$, then the vaue of $x^3+\frac{1}{x^3}$ is

(a) 9 (b) 18

(c) 27 (d) 14

Answers

1. (c)	2. (c)	3. (a)	4. (c)	5. (b)	6. (d)	7. (c)	8. (b)	9. (d)	10. (b)

Chapter

8 FACTORISATION

KEY FACTS

1. When an algebraic expression or polynomial is the product of two or more expressions, then each of these expressions is called a **factor** of the given polynomial.
E.g. (i) 2, a, b, $2a$, $2b$, ab are all factors of $2ab$
(ii) y and $(c^2 + d^2)$ are factors of $y\ (c^2 + d^2)$
2. The process of writing a given polynomial as a product of two or more factors is called **factorisation.**
E.g., factorising $6x^2 + 12x$ means writing it in the form $6x(x + 2)$
3. The greatest common factor of two or more monomials is the product of the greatest common factors of the numerical coefficients and the common letters with the smallest powers.
E.g., Greatest common factor of $25a^3y^5$ and $35a^2y^6$ is $5a^2y^5$.
4. **Types of Factorisation**

Type I: Factorising by taking out the common factor:

A binomial may be factorised by taking out the greatest common factor of the two terms of the binomial.
E.g., $25a^3y^5 + 35\ a^2y^6 = 5a^2y^5\ (5a + 7y)$

Type II: Factorising by grouping the terms:

Here we group the terms and take out common factors from each group
E.g., $ax - by + bx - ay = ax - ay + bx - by = a\ (x - y) + b(x - y) = (a + b)(x - y)$

Type III: Factorising perfect square trinomials:

A perfect square trinomial is the square of a binomial. Thus,

$$a^2 + 2ab + b^2 = (a + b)^2 = (a + b)\ (a + b)$$
$$a^2 - 2ab + b^2 = (a - b)^2 = (a - b)\ (a - b)$$

E.g. (i) $x^2 + 10x + 25 = (x)^2 + 2 \times x \times 5 + (5)^2$
$= (x + 5)^2 = (x + 5)(x + 5)$

(ii) $1 - 14x + 49x^2 = (1)^2 - 2 \times 1 \times 7x + (7x)^2$
$= (1 - 7x)^2 = (1 - 7x)(1 - 7x)$

Type IV: Factorising the difference of two squares:

When a polynomial is expressible as a difference of two squares, then its two factors are

(i) Sum of the two square roots (ii) Difference of the two square roots

E.g. (i) $9x^2 - 25y^2 = (3x)^2 - (5y)^2 = (3x+5y)(3x-5y)$

(ii) $49a^2 - 4(b-a)^2 = (7a)^2 - (2(b-a))^2 = [7a+2(b-a)][7a-2(b-a)]$

$= [7a+2b-2a][7a-2b+2a] = (5a+2b)(9a-2b).$

Type V: $a^2 + b^2 + c^2 + 2ab + 2bc + 2ca = (a+b+c)^2$

E.g., $4a^2 + b^2 + 9c^2 + 4ab + 6bc + 12ac = (2a)^2 + b^2 + (3c)^2 + 2\times 2a\times b + 2\times b\times 3c + 2\times 3c\times 2a$

$= (2a+b+3c)^2.$

Type VI: Factorising a non-perfect square quadratic polynomial $x^2 + px + q$ by splitting the middle term.

$x^2 + px + q$ can be expressed as the product of two linear factors $(x + a)$ and $(x + b)$ as,

$$x^2 + px + q = (x+a)(x+b) = x^2 + (a+b)\,x + ab$$

$\Rightarrow \quad p = a+b$ and $q = ab$

Thus, $x^2 + px + q$ can be factorised by finding two factors of the constant term whose sum is the numerical coefficient of the middle term, *i.e., the numerical coefficient of the middle term can be split into two numbers whose product is equal to the constant term.*

E.g. $x^2 + 8x + 15 = x^2 + 5x + 3x + (5\times 3) = (x^2 + 5x) + (3x + 15)$

$= x(x+5) + 3(x+5) = (x+5)(x+3)$

The set of factors of 15 is {1, 3, 5, 15}. Thus, the constant term could be 1×15 or 3×5. The pair that gives the sum as 8 (coefficient of middle term) is 3×5, therefore the middle term $8x = 3x + 5x$.

Study the following examples and the explanation given besides them to understand all types of problems of the type $x^2 + px + q$.

(i) $x^2 + 17x + 60 = x^2 + 12x + 5x + 60$

$= x\,(x+12) + 5\,(x+12) = (x+12)\,(x+5)$

Since the trinomial is of the form +, +, + find mentally two numbers whose product is + 60 and sum + 17. The numbers are +12 and + 5.

(ii) $x^2 - 5x + 6 = x^2 - 3x - 2x + 6$

$= x(x-3) - 2(x-3) = (x-3)(x-2).$

Since the trinomial is of the form +, –, +, find mentally two numbers whose product is +6 and sum –5. Both the numbers will be negative. The numbers are –3 and –2.

(iii) $x^2 - 4x - 12 = x^2 - 6x + 2x - 12$

$= x(x-6) + 2(x-6) = (x-6)\,(x+2).$

Since the trinomial is of the form +, –, –, i.e., product (–12) is negative, one number will be +ve and the other –ve. Since the sum is to be –4, the greater number will be –ve and smaller + ve. Such numbers are –6 and +2.

(iv) $x^2 + x - 42 = x^2 + 7x - 6x - 42$

$= x(x+7) - 6(x+7) = (x+7)\,(x-6).$

Since the trinomial is of the form, +, +, –, i.e., the product (–42) is negative, one number will be +ve and the other –ve. Since the sum is to be +1, the greater number will be +ve and the smaller –ve. Such numbers are +7 and –6.

Type VII: Factorising $ax^2 + bx + c$

The factorisation is done by splitting the middle term as in type VI. Here, we first multiply the coefficient of x^2 and the constant term and get *ac*. Now we find two numbers whose product is *ac* and whose algebraic sum is equal to the coefficient of the middle terms, *i.e.*, *b*.

(i) $2x^2+3x+1 = 2x^2+2x+x+1$

$= 2x\,(x+1)+1(x+1) = (2x+1)\,(x+1)$

Here we find two numbers whose product is (2 × 1) i.e., +2 and sum is (+3). Such numbers are +2 and +1.

(ii) $3a^2-10a+8 = 3a^2-6a-4a+8$

$= 3a(a-2)-4(a-2) = (3a-4)(a-2).$

Here we find two numbers whose product is (3 × 8), i.e., +24 and sum is (–10). Such numbers are –6 and –4.

(iii) $2x^2+9x-5 = 2x^2+10x-x-5$

$= 2x(x+5)-1(x+5) = (2x-1)\,(x+5)$

Here find two numbers whose product is (2 × –5), i.e., –10 and sum is (+9). The numbers are +10 and –1.

(iv) $6t^2-13t-5 = 6t^2-15t+2t-5 = 3t(2t-5)+1(2t-5) = (3t+1)(2t-5)$

Here find two numbers whose product is (6 × –5), i.e., –30 and sum is –13. Such numbers are –15 and +2.

Type VIII: Factorising sum or difference of cubes :

$$a^3+b^3 = (a+b)(a^2-ab+b^2)$$

$$a^3-b^3 = (a-b)(a^2+ab+b^2)$$

E.g. (i) $x^3+64 = (x)^3+(4)^3$

$= (x+4)((x)^2-x\times4+(4)^2) = (x+4)(x^2-4x+16)$

(ii) $250x^3-2 = 2(125x^3-1) = 2((5x)^3-1^3)$

$= 2(5x-1)((5x)^2+5x\times1+1) = 2(5x-1)(25x^2+5x+1).$

Solved Examples

Ex. 1. ***Factorise : $4m^3n^2 + 12\,m^2n^2 + 18\,m^4\,n^3$***

Sol. $4m^3n^2+12m^2n^2+18m^4n^3 = 2m^2n^2(2m+6+9m^2n).$

Ex. 2. ***Factorise : $32\,(x+y)^2 - 2x - 2y$***

Sol. $32(x+y)^2-2x-2y = 32(x+y)^2-2(x+y)$

$= 2(x+y)\{16(x+y)-1\} = 2(x+y)(16x+16y-1).$

Ex. 3. ***Factorise : $a^2 - ac + xc - xa + 6a - 6c$***

Sol. $a^2-ac+xc-xa+6a-6c = a(a-c)+x(c-a)+6(a-c)$ **(Note:** $c-a=-(a-c)$**)**

$= a(a-c)-x(a-c)+6(a-c)$

$= (a-c)(a-x+6).$

Ex. 4. ***Factorise : $\dfrac{x^2}{9}+\dfrac{xy^2}{3}+\dfrac{y^4}{4}$***

Sol. $\dfrac{x^2}{9}+\dfrac{xy^2}{3}+\dfrac{y^4}{4} = \left(\dfrac{x}{3}\right)^2+2\times\dfrac{x}{3}\times\dfrac{y^2}{2}+\left(\dfrac{y^2}{2}\right)^2 = \left(\dfrac{x}{3}+\dfrac{y^2}{2}\right)^2 = \left(\dfrac{x}{3}+\dfrac{y^2}{2}\right)\left(\dfrac{x}{3}+\dfrac{y^2}{2}\right).$

Ex. 5. *Factorise : $36x^4 - 84x^2y^2 + 49y^4$*

Sol. $36x^4 - 84x^2y^2 + 49y^4 = (6x^2)^2 - 2\times 6x^2 \times 7y^2 + (7y^2)^2$

$= (6x^2 - 7y^2)^2 = (6x^2 - 7y^2)(6x^2 - 7y^2).$

Ex. 6. *Factorise : $81 - (x+y)^2$*

Sol. $81 - (x+y)^2 = (9)^2 - (x+y)^2$

$= (9+x+y)\{9-(x+y)\} = (9+x+y)(9-x-y).$

Ex. 7. *Factorise : $16x^4 - y^4$*

Sol. $16x^4 - y^4 = (4x^2)^2 - (y^2)^2 = (4x^2 - y^2)(4x^2 + y^2)$

$= \{(2x)^2 - y^2\}(4x^2 + y^2) = (2x+y)(2x-y)(4x^2+y^2).$

Ex. 8. *Factorise the following:*

(i) $x^2 + 7x + 10$ (ii) $x^2 - 16x + 39$ (iii) $x^2 - 9x - 36$ (iv) $x^2 + 2x - 48$

Sol. (i) $x^2 + 7x + 10 = x^2 + 5x + 2x + 10$

$= x(x+5) + 2(x+5) = (x+5)(x+2).$

(ii) $x^2 - 16x + 39 = x^2 - 13x - 3x + 39$

$= x(x-13) - 3(x-13) = (x-3)(x-13).$

(iii) $x^2 - 9x - 36 = x^2 - 12x + 3x - 36$

$= x(x-12) + 3(x-12) = (x-12)(x+3).$

(iv) $x^2 + 2x - 48 = x^2 + 8x - 6x - 48$

$= x(x+8) - 6(x+8) = (x-6)(x+8).$

Ex. 9. *Factorise the following:*

(i) $6x^2 + 11x + 3$ (ii) $2x^2 - 15x + 7$ (iii) $5m^2 - 8m - 4$ (iv) $3a^2 + 7ab - 6b^2$

Sol. (i) $6x^2 + 11x + 3$

$= 6x^2 + 9x + 2x + 3$ $\begin{pmatrix}\text{Sum} = 11\\ \text{Prod.} = 18\end{pmatrix}$

$= 3x(2x+3) + 1(2x+3) = (2x+3)(3x+1).$

(ii) $2x^2 - 15x + 7$

$= 2x^2 - 14x - x + 7$ $\begin{pmatrix}\text{Sum} = -15\\ \text{Prod.} = 14\end{pmatrix}$

$= 2x(x-7) - 1(x-7) = (x-7)(2x-1).$

(iii) $5m^2 - 8m - 4$

$= 5m^2 - 10m + 2m - 4$ $\begin{pmatrix}\text{Sum} = -8\\ \text{Prod.} = -20\end{pmatrix}$

$= 5m(m-2) + 2(m-2) = (5m+2)(m-2).$

(iv) $3a^2 + 7ab - 6b^2$

$= 3a^2 + 9ab - 2ab - 6b^2$ $\begin{pmatrix}\text{Sum} = +7\\ \text{Prod.} = -18\end{pmatrix}$

$= 3a(a+3b) - 2b(a+3b) = (a+3b)(3a-2b).$

Ex. 10. *Factorise : (i)* $a^3 + \frac{8}{27}b^3$ *(ii)* $a^3 - b^3 + 4(a-b)$

Sol. (i) $a^3 + \frac{8}{27}b^3 = (a)^3 + \left(\frac{2}{3}b\right)^3$

$= \left(a + \frac{2}{3}b\right)\left((a)^2 - a \times \frac{2}{3}b + \left(\frac{2}{3}b\right)^2\right) = \left(a + \frac{2}{3}b\right)\left(a^2 - \frac{2}{3}ab + \frac{4}{9}b^2\right)$

(ii) $a^3 - b^3 + 4(a-b) = (a-b)(a^2 + ab + b^2) + 4(a-b)$

$= (a-b)(a^2 + ab + b^2 + 4).$

Ex. 11. *Factorise:* $9x^2 - 6xy + y^2 - z^2$

Sol. $9x^2 - 6xy + y^2 - z^2 = (9x^2 - 6xy + y^2) - z^2$

$= \left\{(3x)^2 - 2 \times 3x \times y + y^2\right\} - z^2 = (3x - y^2) - z^2 = (3x - y + z)(3x - y - z).$

Ex. 12. *Evaluate the following products without directly multiplying:*
(i) 102 × 105 *(ii) 196 × 193* *(iii) 97 × 103*

Sol. (i) $102 \times 105 = (100 + 2)(100 + 5)$

$= (100)^2 + (2 + 5) \times 100 + (2 \times 5)$

$= 10000 + 700 + 10 = 10710.$

(ii) $196 \times 193 = (200 - 4)(200 - 7)$

$= (200)^2 - (4 + 7) \times 200 + 4 \times 7$

$= 40000 - 5600 + 28 = 34428.$

(iii) $97 \times 103 = (100 - 3)(100 + 3)$

$= (100)^2 - 3^2 = 10000 - 9$

$= 9991.$

Question Bank–8

1. $28yz - 21y^3z^4 + 35y^2z^2 =$ ____
(a) $7y^2z(4z - 3yz^3 + 5z)$
(b) $7yz(4 - 3y^2z^3 + 5yz)$
(c) $7yz(4yz - 3y^2z^2 + 5yz^2)$
(d) $7y^2z^2(4 - 3z^3 + 5)$

2. $(3a-1)^2 - 6a + 2 =$ ____
(a) $(3a - 1)(3a - 2)$ (b) $(3a - 1)(3a + 2)$
(c) $3(3a - 1)(a - 1)$ (d) $(3a - 1)(3a + 3)$

3. $2xya^2 + 10y + 3xa^2 + 15 =$ ____
(a) $(2a^2 + 5)(2y + 3x)$ (b) $(xa^2 + 5)(2y + 3)$
(c) $(ya^2 + 3)(2x + 5)$ (d) $(xy + 5)(2a^2 + 3)$

4. $xy^2 - yz^2 - xy + z^2 =$ ____
(a) $(z - 1)(xy - z)$ (b) $(y - 1)(xy - z^2)$
(c) $(xy - 1)(z^2 - y)$ (d) $(y + 1)(xy - z^2)$

5. $x^2 + (a + b + c)x + ab + bc =$ ____
(a) $(x + a)(x + b + c)$ (b) $(x + c)(x + a + b)$
(c) $(x + b)(x + a + c)$ (c) $(x + a)(x + b - c)$

6. $9x^2 + 30x + 25 =$ ____
(a) $(3x + 5)(3x - 10)$ (b) $(3x + 5)(3x + 5)$
(c) $(3x + 12)(3x + 13)$ (d) $(9x + 5)(x + 5)$

7. $16a^2 - 56ab + 49b^2 =$ ____
(a) $(4a + 7b)(4b - 8b)$ (b) $(2a + 7b)(2a + 8b)$
(c) $(4a - 7b)(4a - 7b)$ (d) $(4a + 7b)(4a + 7b)$

8. $4x^4 - 6x^2 + \frac{9}{4} =$ ____
(a) $\left(2x^3 - \frac{3}{2}\right)\left(2x - \frac{3}{2}\right)$
(b) $\left(2x^2 + \frac{3}{2}\right)\left(2x^2 - \frac{3}{2}\right)$

(c) $\left(2x^2-\frac{3}{2}\right)\left(2x^2-\frac{3}{2}\right)$

(d) $\left(4x^3-\frac{9}{4}\right)(x-1)$

9. $4y-y^2-4=$ _____

(a) $(y-2)(y-2)$ (b) $(y-2)(2-y)$

(c) $(y-2)(y+2)$ (d) $(y+2)(y+2)$

10. $121-16a^2b^2=$ _____

(a) $(11+16ab)(11-ab)$

(b) $(11+16ab)(11+ab)$

(c) $(11-4ab)(11+4ab)$

(d) $(11-4ab)(11-4ab)$

11. $\frac{49a^2}{25b^2}-\frac{x^2}{100y^2}=$ _____

(a) $\left(\frac{7a}{5b}+\frac{x}{10y}\right)\left(\frac{7a}{5b}-\frac{x}{y}\right)$

(b) $\left(\frac{7a}{5b}+\frac{x}{10y}\right)\left(\frac{7a}{5b}+\frac{x}{y}\right)$

(c) $\left(\frac{7a}{5b}+\frac{x}{10y}\right)\left(\frac{7a}{5b}-\frac{x}{10y}\right)$

(d) $\left(\frac{7a}{5b}+\frac{x}{10y}\right)\left(\frac{7a}{5b}-\frac{x}{10y}\right)$

12. $x^4-81=$ _____

(a) $(x^2+81)(x^2-1)$ (b) $(x^2-9)(x^2-9)$

(c) $(x+3)(x-3)(x^2+9)$ (d) $(x^2-9)(x+3)$

13. $27a^2x^2-48b^2=$ _____

(a) $(9ax-16b)(3ax-3b)$

(b) $(3ax+16b)(9ax-3b)$

(c) $3(3ax+4b)(3ax-4b)$

(d) $3(3ax+6b)(3ax-4b)$

14. $16(x+y)^2-4y^2=$ _____

(a) $(4x-2y)(4x-2y)$

(b) $(4x+2y)(4x-2y)$

(c) $(4x-6y)(4x+2y)$

(d) $(4x+2y)(4x+6y)$

15. $9a^2-(2b-c)^2=$ _____

(a) $(3a-2b-c)(3a+2b-c)$

(b) $(3a+2b-c)(3a-2b+c)$

(c) $(-3a+2b+c)(3a+2b-c)$

(d) $(3a+2b+c)(3a+2b-c)$

16. $9a^2-24ab+16b^2-4x^2=$ _____

(a) $(3a+4b-2x)(3a+4b+2x)$

(b) $(3a-4b+2x)(3a-4b-2x)$

(c) $(3a-4b-2x)(3a+4b+2x)$

(d) $(3a-4b-2x)(3a-4b-2x)$

17. $x^2+6xy-z^2+9y^2=$

(a) $(x+3y-z)(x+3y-z)$

(b) $(x-3y+z)(x-3y+z)$

(c) $(x+3y+z)(x+3y-z)$

(d) $(x-3y-z)(x-3y-z)$

18. $25a^2-4b^2+5a+2b=$ _____

(a) $(5a+2b-1)(5a-2b)$

(b) $(5a+2b+1)(5a-2b)$

(c) $(5a+2b)(5a-2b-1)$

(d) $(5a+2b)(5a-2b+1)$

19. $x^2+y^2+2xy+yz+zx=$ _____

(a) $(x+y+1)(x+z)$ (b) $(x+y)(x+y+z)$

(c) $(x+y-z)(x+1)$ (d) $(x+z+1)(x+y)$

20. $(a+b)^2-(b-a)^2=$ _____

(a) $(2a+2b)$ (b) $(2a-2b)$

(c) $4ab$ (d) $-4ab$

21. $x^2+11x+24=$ _____

(a) $(x+6)(x+4)$ (b) $(x+12)(x-1)$

(c) $(x+12)(x+2)$ (d) $(x+8)(x+3)$

22. $x^2-22x+40=$ _____

(a) $(x-10)(x+4)$ (b) $(x+2)(x-20)$

(c) $(x+10)(x-4)$ (d) $(x-20)(x-2)$

23. $m^2-5m-36=$ _____

(a) $(m+9)(m-4)$ (b) $(m-9)(m-4)$

(c) $(m+9)(m+4)$ (d) $(m-9)(m+4)$

24. $a^2+15a-54=$ _____

(a) $(a+9)(x-6)$ (b) $(a-18)(a+3)$

(c) $(a+18)(a-3)$ (d) $(a-27)(a+2)$

25. $2x^2+11x+14=$ _____

(a) $(2x+2)(x+7)$ (b) $(2x+7)(x+4)$

(c) $(2x+7)(x+2)$ (d) $(2x+4)(x+7)$

26. $2z^2-13z+15=$ _____

(a) $(z-5)(2z-3)$ (b) $(z+5)(2z-3)$

(c) $(z+5)(2z+3)$ (d) $(z-5)(2z+3)$

27. $3x^2 + 11x - 4 =$ _____

(a) $(3x-1)(x+4)$ (b) $(3x-1)(x-4)$
(c) $(3x+1)(x-4)$ (d) $(3x+1)(x+4)$

28. $6y^2 - 7y - 5 =$ _____

(a) $(3y+5)(2y-1)$ (b) $(3y-5)(2y+1)$
(c) $(3y-5)(2y-1)$ (d) $(3y+5)(2y+1)$

29. $8x^3 + 1 =$ ____

(a) $(2x+1)(4x^2-2x+1)$
(b) $(2x-1)(4x^2-2x+1)$
(c) $(2x-1)(4x^2+2x+1)$
(d) $(2x+1)(4x^2+2x+1)$

30. $a^3 - 0.216 =$ ____

(a) $(a-0.6)(a^2+0.6a+0.36)$
(b) $(a^2-0.36)(a-0.6a+0.6)$
(c) $(a-0.6)(a^2-0.6a+0.36)$
(d) $(a^2+0.36)(a+0.6a+0.6)$

31. $a^3 + b^3 + a + b =$ ____

(a) $(a+b+1)(a^2-ab+b^2)$
(b) $(a+b)(a^2-ab+b^2+1)$
(c) $(a^2+b^2)(a+b-ab+1)$
(d) $(a-b)(a^2+ab+b^2+1)$

32. $25x^2 - (x^2-36)^2 =$ _____

(a) $(x-4)(x+4)(x+9)(x-9)$
(b) $(x-4)(4+x)(x+9)(9-x)$
(c) $(x+4)(x+4)(x-9)(x-9)$
(d) $(x-4)(4-x)(x+9)(9+x)$

33. $(6-x)^2 - 3x =$ ____

(a) $(x-9)(x-4)$ (b) $(x-12)(x-3)$
(c) $(x+12)(x-3)$ (d) $(x-12)(x+3)$

34. $x^4 - x^2y^2 - 72y^4 =$ _____

(a) $(x^2+9y^2)(x^2-8y^2)$
(b) $(x+3y)(x+3y)(x^2-8y^2)$
(c) $(x+3y)(x-3y)(x^2+8y^2)$
(d) $(x^2-9y^2)(x^2-8y^2)$

35. $x^6 - 9^3 =$ _____

(a) $(x^2+9)(x^4-9x^2-81)$
(b) $(x^2-9)(x^4-9x^2+81)$
(c) $(x+3)(x-3)(x^4+9x^2+81)$
(d) $(x^2+9)(x^4+9x^2+81)$

Answers

1. (b)	**2.** (c)	**3.** (b)	**4.** (b)	**5.** (c)	**6.** (b)	**7.** (c)	**8.** (c)	**9.** (b)	**10.** (c)
11. (d)	**12.** (c)	**13.** (c)	**14.** (d)	**15.** (b)	**16.** (b)	**17.** (c)	**18.** (c)	**19.** (b)	**20.** (c)
21. (d)	**22.** (c)	**23.** (d)	**24.** (c)	**25.** (c)	**26.** (a)	**27.** (a)	**28.** (b)	**29.** (a)	**30.** (a)
31. (b)	**32.** (b)	**33.** (b)	**34.** (c)	**35.** (c)					

Hints and Solutions

2. (c) $(3a-1)^2 - 6a + 2$
$= (3a-1)^2 - 2(3a-1)$
$= (3a-1)(3a-1-2)$
$= (3a-1)(3a-3)$
$= 3(3a-1)(a-1).$

3. (b) $2xya^2 + 10y + 3xa^2 + 15$
$= 2y(xa^2+5) + 3(xa^2+5)$
$= (xa^2+5)(2y+3).$

4. (b) $xy^2 - yz^2 - xy + z^2$
$= xy^2 - xy - yz^2 + z^2$
$= xy(y-1) - z^2(y-1)$
$= (xy-z^2)(y-1).$

5. (c) $x^2 + (a+b+c)x + ab + bc$
$= x^2 + ax + bx + cx + ab + bc$
$= (x^2+bx) + (ax+ab) + (cx+bc)$
$= x(x+b) + a(x+b) + c(x+b)$
$= (x+b)(x+a+c).$

6. (b) $9x^2 + 30x + 25 = (3x)^2 + 2\times 3x \times 5 + (5)^2$
$= (3x+5)^2.$

7. (c) A perfect square trinomial

8. (c) $4x^4 - 6x^2 + \frac{9}{4} = (2x^2)^2 - 2 \times 2x^2 \times \frac{3}{2} + \left(\frac{3}{2}\right)^2$

$= \left(2x^2 - \frac{3}{2}\right)^2$

9. (b) $4y - y^2 - 4$

$= -(y^2 - 4y + 4)$

$= -(y^2 - 2 \times 2 \times y + 2^2)$

$= -(y-2)(y-2) = (y-2)(2-y).$

10. (c) $121 - 16a^2b^2$

$= (11)^2 - (4ab)^2 = (11 + 4ab)(11 - 4ab).$

11. (d) $\frac{49a^2}{25b^2} - \frac{x^2}{100y^2} = \left(\frac{7a}{5b}\right)^2 - \left(\frac{x}{10y}\right)^2$

Now factorise.

12. (c) $x^4 - 81 = (x^2)^2 - (9)^2 = (x^2 + 9)(x^2 - 9)$

$= (x^2 + 9)(x - 3)(x + 3).$

13. (c) $27a^2x^2 - 48b^2$

$= 3(9a^2x^2 - 16b^2) = 3\{(3ax)^2 - (4b)^2\}$

$= 3(3ax + 4b)(3ax - 4b).$

14. (d) $16(x+y)^2 - 4y^2 = \{4(x+y)\}^2 - (2y)^2$

$= (4x + 4y + 2y)(4x + 4y - 2y)$

$= (4x + 6y)(4x + 2y).$

15. (b) $9a^2 - (2b - c)^2$

$= (3a)^2 - (2b - c)^2$

$= \{3a + (2b - c)\}\{3a - (2b - c)\}$

$= (3a + 2b - c)(3a - 2b + c).$

16. (b) $9a^2 - 24ab + 16b^2 - 4x^2 = (3a - 4b)^2 - (2x)^2$

$= (3a - 4b + 2x)(3a - 4b - 2x).$

17. (c) $x^2 + 6xy - z^2 + 9y^2$

$= x^2 + 6xy + 9y^2 - z^2$

$= (x + 3y)^2 - z^2$

$= (x + 3y + z)(x + 3y - z).$

18. (c) $25a^2 - 4b^2 + 5a + 2b$

$= (5a)^2 - (2b)^2 + (5a + 2b)$

$= (5a + 2b)(5a - 2b) + (5a + 2b)$

$= (5a + 2b)(5a - 2b + 1).$

19. (b) $x^2 + y^2 + 2(xy + yz + zx)$

$= (x^2 + y^2 + 2xy) + yz + zx$

$= (x + y)^2 + z(x + y)$

$= (x + y)(x + y + z).$

20. (c) $(a + b)^2 - (b - a)^2$

$= \{a + b + (b - a)\}\{a + b - (b - a)\}$

$= 2b \times 2a = 4ab$

21. (d) $x^2 + 11x + 24 = x^2 + 8x + 3x + 24$

Now factorise.

22. (c) $x^2 - 22x + 40 = x^2 - 20x - 2x + 40$

Now factorise.

23. (d) $m^2 - 5m - 36 = m^2 - 9m + 4m - 36$

Now factorise.

24. (c) $a^2 + 15a - 54 = a^2 + 18a - 3a - 54$

Now factorise.

25. (c) $2x^2 + 11x + 14 = 2x^2 + 7x + 4x + 14$

Now factorise.

26. (a) $2z^2 - 13z + 15 = 2z^2 - 10z - 3z + 15$

Now factorise.

27. (a) $3x^2 + 11x - 4 = 3x^2 + 12x - x - 4$

Now factorise.

28. (b) $6y^2 - 7y - 5 = 6y^2 - 10y + 3y - 5$

Now factorise.

29. (a) $8x^3 + 1 = (2x)^3 + 1^3$

$= (2x + 1)((2x)^2 - 2 \times x \times 1 + 1^2).$

30. (a) $a^3 - 0.216 = (a)^3 - (0.6)^3$

$= (a - 0.6)((a)^2 + a \times 0.6 + (0.6)^2).$

31. (b) $a^3 + b^3 + a + b = (a + b)(a^2 - ab + b^2) + (a + b)$

$= (a + b)(a^2 - ab + b^2 + 1).$

32. (b) $25x^2 - (x^2 - 36)^2$

$= (5x)^2 - (x^2 - 36)^2$

$= (5x + x^2 - 36)(5x - x^2 + 36)$

$= (x^2 + 5x - 36)\{-(x^2 - 5x - 36)\}$

$= (x^2 + 9x - 4x - 36)\{-(x^2 - 9x + 4x - 36)\}$

$= \{x(x + 9) - 4(x + 9)\}\{-(x(x - 9) + 4(x - 9)\}$

$= (x+9)(x-4)\{-(x-9)(x+4)\}$
$= (x+9)(x-4)(9-x)(4+x).$

33. (b) $(6-x)^2 - 3x = 36 - 12x + x^2 - 3x$
$= x^2 - 15x + 36$
$= x^2 - 12x - 3x + 36$
$= x(x-12) - 3(x-12)$
$= (x-3)(x-12).$

34. (c) $x^4 - x^2y^2 - 72y^4$
$= x^4 - 9x^2y^2 + 8x^2y^2 - 72y^4$
$= x^2(x^2 - 9y^2) + 8y^2(x^2 - 9y^2)$
$= (x^2 - 9y^2)(x^2 + 8y^2)$
$= (x+3y)(x-3y)(x^2 + 8y^2).$

35. (c) $x^6 - 9^3 = (x^2)^3 - 9^3 = (x^2 - 9)\ (x^4 + 9x^2 + 81)$
$= (x+3)(x-3)(x^4 + 9x^2 + 81).$

Self Assessment Sheet–8

1. Which one of the following statements is not correct?
(a) $a^2 + b^2$ cannot be factorized
(b) $(a-2c)$ is a factor of $a^4 - 4a^2c^2$
(c) $2x + 7$ is a factor of $8x^3 - 27$
(d) $\dfrac{3a^2 - 13a - 10}{9a^2 - 4} = \dfrac{a-5}{3a-2}$

2. $p(q^2 + r^2) - q(r^2 + p^2)$ can be factorised as
(a) $(p+q)(r^2 - pq)$ (b) $(p-q)(r^2 - pq)$
(c) $(p-q)(pq - r^2)$ (d) $(p - qr)\ (p+q)$

3. $\dfrac{a^3 - 64b^3}{a^2 + 3ab - 28b^2} = ?$
(a) $\dfrac{a-4b}{a+3b}$ (b) $\dfrac{a^2 + 4ab + 16b^2}{a-4b}$
(c) $\dfrac{a^2 - 2ab + 3b^2}{a+2b}$ (d) $\dfrac{a^2 + 4ab + 16b^2}{a+7b}$

4. Factorise : $16(a-b)^3 - 24(a-b)^2$. The factors are:
(a) $4(a-b)(2a-b-3)$
(b) $8(a-b)^2(2a-2b-3)$
(c) $8(a-b)(2a-b-4)$
(d) None of these

5. Factorise and simplify : $\dfrac{2x^5 - 2x}{2x^3 + 2x}$. The answer is
(a) $x^2 - 1$ (b) $x^2 + 1$
(c) $x - 1$ (d) $x + 1$

6. Factorise : $16 - x^2 - 2xy - y^2$. The factors are
(a) $(4 + x - y)\ (4 + x + y)$
(b) $(4 - x - y)\ (4 - x + y)$
(c) $(2 + x + y)\ (2 - x - y)$
(d) $(4 + x + y)(4 - x - y)$

7. The area of a rectangle is $12y^4 + 28y^3 - 5y^2$. If its length is $6y^3 - y^2$, then its width is
(a) $y + 5$ (b) $-2y + 5$
(c) $-2y^2 + 5$ (d) $2y + 5$

8. Factorise : $n^3 - 3n - 2$, given that $n + 1$ is a factor. The factors are
(a) $(n+1)^2(n-2)$ (b) $(n+1)\ (n-2)^2$
(c) $(n-1)^2(n+2)$ (d) None of these

9. What are the possible expressions for the dimensions of the cuboid whose volume is $12kx^2 + 8kx - 20k$?
(a) $k, (x+1)$ and $(3x-5)$ units
(b) $4k, (x-1)$ and $(3x+5)$ units
(c) $k, 4(x-1)$ and $(2y-5)$ units
(d) $4k, 4(x+1)$ and $(3x+5)$ units

10. Factorise : $8x^3 - 27y^3 - 2x + 3y$: The factors are
(a) $(2x-3y)(4x^2 + xy + 9y^2 - 1)$
(b) $(2x-3y)(4x^2 - 6xy + 9y^2)$
(c) $(2x-3y)(4x^2 + 3xy + 9y^2 - 1)$
(d) $(2x-3y)(4x^2 + 6xy + 9y^2 - 1)$

Answers

1. (d) **2.** (b) **3.** (d) **4.** (b) **5.** (a) **6.** (d) **7.** (d) **8.** (a) **9.** (b) **10.** (d)

Chapter 9

LINEAR EQUATIONS IN ONE VARIABLE

KEY FACTS

1. A statement of equality containing one or more unknowns (variables) is called an algebraic equation.
   E.g., $6x + 4 = 16$, $3x + 2y = 6$, $4x^2 + 8x + 5 = 17$, etc.
2. An equation is called a **simple equation** or **linear equation in one variable** if the variable is raised to power 1 only.

   E.g., $7x + 4 = 5$, $5(x - 3) = 7 + 2(x + 5)$, $\frac{4}{x+4} = \frac{6}{x+5}$ etc.
3. All equations consists of a **left side**, an equal sign ('=') and a **right side**. Thus, in the equation $x + 4 = 10$, the **left side** is $(x + 4)$ and the **right side** is 10.
4. **A number which satisfies an equation or the value of the unknown is called the solution or root of the equation.**
   E.g., $6 + 2x = 8$ is satisfied when $x = 1$. Thus 1 is the solution of the given equation.
5. An equation remains unchanged if :
   (a) same number is added to both the sides.
   (b) same number is subtracted from both the sides.
   (c) both sides are multiplied by the same number.
   (d) both sides are divided by the same number.
6. **Rule of Transposition**: A term may be transposed from one side of the equation to the other side, but its sign will change.
   E.g., $x + 4 = 10 \Rightarrow x = 10 - 4 = 6$
   (+4 when transferred to right side becomes –4)
7. When moved from one side to the other (*i.e.*, left to right or vice-versa) inverse operations come into play
   (a) addition changes to subtraction, subtraction changes to additon.
   (b) divison changes to multiplication, multiplication to division.

   E.g. (a) $4x + 5 = 17 \Rightarrow 4x = 17 - 5$

   + to –

   (b) $3x - 4 = 7x - 3 \Rightarrow 3x - 7x = -3 + 4$

   + to –

   – to +

(c) $\frac{4x}{5}=12 \Rightarrow 4x=12\times5$

÷ to ×

(d) $7x=49 \Rightarrow x=49\div7$

× to ÷

Solved Examples

Ex. 1. ***Solve :*** $\frac{3}{5}(4x-9)-\frac{5}{4}(3x-8)=5-\frac{7}{10}(2x-1).$

Sol. $\frac{3}{5}(4x-9)-\frac{5}{4}(3x-8)=5-\frac{7}{10}(2x-1)$

Multiply both sides by the LCM of the denominators 5, 4 and 10 to clear the fractions
LCM of 5, 4, 10 = 20

∴ Multiplying both sides by 20 we have $20\times\frac{3}{5}(4x-9)-20\times\frac{5}{4}(3x-8) = 20\times5-20\times\frac{7}{10}(2x-1)$

$\Rightarrow 12(4x-9)-25(3x-8)=100-14(2x-1) \Rightarrow 48x-108-75x+200=100-28x+14$

$\Rightarrow 48x-75x+28x=100+14+108-200$

$\Rightarrow x=222-200 \Rightarrow \mathbf{x=22.}$

Ex. 2. ***Solve :*** $(x+6)(x-6)-(x-5)^2=40-17(x-2)$

Sol. Step 1. Simplify both

$(x+6)(x-6)-(x-5)^2=40-17(x-2)$

$\Rightarrow x^2-36-(x^2-10x+25)=40-17x+34 \Rightarrow x^2-36-x^2+10x-25=40-17x+34$

$\Rightarrow 10x+17x=40+34+36+25 \Rightarrow 27x=135 \Rightarrow x=\frac{135}{27} \Rightarrow \mathbf{x=5.}$

Check : Putting $x=5$,

LHS $=(5+6)(5-6)-(5-5)^2=11\times-1-0=-11.$

RHS $=40-17(5-2)=40-17\times3=40-51=-11=$ LHS.

Ex. 3. ***The sum of three numbers is 264. If the first number be twice the second and third number be one-third of the first, then find the second number?***

Sol. Let the second number be x. Then,

First number = $2x$ and third number = $\frac{2x}{3}$

Given, $2x+x+\frac{2x}{3}=264 \Rightarrow \frac{6x+3x+2x}{3}=264 \Rightarrow 11x=264\times3 \Rightarrow x=\frac{264\times3}{11}=\mathbf{72.}$

Ex. 4. ***If 50 is subtracted from two-third of a number, the result is equal to sum of 40 and one-fourth of that number. What is the number?***

Sol. Let the number be x.

Given, $\frac{2x}{3}-50=\frac{x}{4}+40 \Rightarrow \frac{2x}{3}-\frac{x}{4}=50+40$

$\Rightarrow \frac{8x-3x}{12}=90 \Rightarrow \frac{5x}{12}=90 \Rightarrow x=\frac{90\times12}{5}=\mathbf{216.}$

Ex. 5. ***The difference between two numbers is 16. If one-third of the smaller number is greater than one-seventh of the larger number by 4, then what are the two numbers?***

Sol. Let the smaller number be x. Then,

larger number = $x + 16$

Given , $\frac{x}{3} - \frac{(x+16)}{7} = 4 \Rightarrow \frac{7x - 3x - 48}{21} = 4 \Rightarrow 4x - 48 = 84$

$\Rightarrow 4x = 84 + 48 = 132 \Rightarrow x = \frac{132}{4} = 33$

∴ The two numbers are 33 and (33 + 16), *i.e.*, 49.

Ex. 6. ***The sides of a triangle are in the ratio*** $\frac{1}{2}:\frac{1}{3}:\frac{1}{4}$***. If the perimeter of the triangle is 52 cm, then what is the length of the smallest side?***

Sol. Let the three sides of the triangle be $\frac{1}{2}x, \frac{1}{3}x, \frac{1}{4}x$

Given, $\frac{x}{2} + \frac{x}{3} + \frac{x}{4} = 52 \Rightarrow \frac{6x + 4x + 3x}{12} = 52 \Rightarrow 13x = 52 \times 12 \Rightarrow x = \frac{52 \times 12}{13} = 48$

∴ The smallest side = $\frac{1}{4} \times 48 = \mathbf{12\,cm}$.

Ex. 7. ***A labourer was engaged for 20 days on the condition that he will receive ₹ 60 for each day he works and he will be fined ₹ 5 for each day he is absent. If he received ₹ 745 in all, then what is the number of days he was absent ?***

Sol. Let the number of days for which the labourer worked be x.

Then, the number of days he was absent are $(20 - x)$. According to given condition,

$60 \times x + (-5)(20 - x) = 745 \Rightarrow 60x - 100 + 5x = 745 \Rightarrow 65x = 845 \Rightarrow x = 13$

∴ The labourer was absent for (20 – 13) = **7 days.**

Ex. 8. ***The numerator and denominator of a fraction are in the ratio 2:3. If 6 is subtracted from the numerator, the result is a fraction that has a value*** $\frac{2}{3}$ ***of the original fraction. What is the numerator of the original fraction?***

Sol. Let the numerator and denominator of the fraction be $2x$ and $3x$.

∴ The original fraction is $\frac{2x}{3x}$.

Given , $\frac{2x - 6}{3x} = \frac{2}{3} \times \frac{2x}{3x} \Rightarrow \frac{2x - 6}{3x} = \frac{4}{9} \Rightarrow 18x - 54 = 12x \Rightarrow 6x = 54 \Rightarrow x = 9$

∴ The numerator of the fraction is $2 \times 9 = \mathbf{18}$.

Ex. 9. ***A number consists of two digits such that the digit in the ten's place is less by 2 than the digit in the unit's place. Three times the number added to*** $\frac{6}{7}$ ***times the number obtained by reversing the digits equals 108 . What is the sum of the digits in the number?***

Sol. Let the unit's digit be x. Then ten's digit = $x - 2$

T	U
$x - 2$	x

$\therefore$ Original number $= 10(x-2) + x = 11x - 20$

After reversing the digits, the number is

T	U
x	$x-2$

i.e., $10x + (x-2) = 11x - 2$

Given , $3(11x-20) + \frac{6}{7}(11x-2) = 108 \Rightarrow \quad 33x - 60 + \frac{66x-12}{7} = 108$

$\Rightarrow 231x - 420 + 66x - 12 = 108 \times 7 = 756 \Rightarrow 297x = 1188 \Rightarrow x = 4$

$\therefore$ Sum of digits $= x + x - 2 = 4 + 4 - 2 =$ **6.**

Ex. 10. ***The present ages of three persons are in the ratio 4:7:9. Eight years ago, the sum of their ages was 56. Find their present ages (in years).***

Sol. Let the present ages of the three persons be $4x$, $7x$ and $9x$. Their ages, 8 years ago were $(4x-8)$, $(7x-8)$ and $(9x-8)$ respectively.

Given, $4x - 8 + 7x - 8 + 9x - 8 = 56 \Rightarrow 20x = 56 + 24 = 80 \Rightarrow x = 4$

$\therefore$ Their present ages are 4×4, 7×4, 9×4 *i.e.*, 16 years, 28 years, 36 years.

Ex. 11. ***The ratio of the present ages of two brothers is 1:2 and 5 years back, the ratio was 1:3. What will be the ratio of their ages after 5 years?***

Sol. Let the present ages of the two brothers be x and $2x$ years 5 years back, $\frac{x-5}{2x-5} = \frac{1}{3}$

$\Rightarrow \; 3x - 15 = 2x - 5 \Rightarrow x = 10$

$\therefore$ The present ages are 10 years and 20 years

Required ratio $= \frac{10+5}{20+5} = \frac{15}{25} = \frac{3}{5} = \mathbf{3:5}.$

Ex. 12. ***The total value of a collection of coins of denominations ₹ 1.00, 50 paise, 25 paise, 10 paise and 5 paise is ₹ 380. If the number of coins of each denomination is the same, then what is the number of one-rupee coins?***

Sol. Let the number of coins of each denomination be x. Then,

$$x \times 1 + x \times \frac{1}{2} + x \times \frac{1}{4} + x \times \frac{1}{10} + x \times \frac{1}{20} = 380 \Rightarrow \quad \frac{20x + 10x + 5x + 2x + x}{20} = 380$$

$$\Rightarrow \frac{38x}{20} = 380 \Rightarrow x = \frac{380 \times 20}{38} = \mathbf{200}.$$

Ex. 13. ***A person travelled $\frac{5}{8}$th of the distance by train, $\frac{1}{4}$th by bus and the remaining 15 km by boat. What was the total distance travelled by him?***

Sol. Let the total distance travelled by him be x km.

Then , $\frac{5x}{8} + \frac{x}{4} + 15 = x$

$$\Rightarrow \quad x - \frac{5x}{8} - \frac{x}{4} = 15 \Rightarrow \quad \frac{8x - 5x - 2x}{8} = 15 \Rightarrow \frac{x}{8} = 15 \Rightarrow x = 120 \text{ km}$$

Total distance travelled by him = **120 km.**

Question Bank–9

1. The value of m that satisfies the equation $\frac{7}{4m-2}=\frac{5}{3m-4}$ is:

(a) 9 (b) $3\frac{1}{2}$
(c) 18 (d) 38

2. Given, $5(x+4)-8(4-x)=25-3(7-x)+144$, then the value of x is

(a) 8 (b) 16
(c) 4 (d) 11

3. If $(x-2)(x+3)=x^2-4$, the value of x is

(a) 3 (b) 2
(c) $\frac{1}{2}$ (d) –2

4. The solution of $0.2(2x-1)-0.5(3x-1)=0.4$ is

(a) $\frac{1}{11}$ (b) $-\frac{1}{11}$
(c) $\frac{3}{11}$ (d) $-\frac{3}{11}$

5. Match the equations with their solutions.

(1) $14-\frac{x-1}{10}=\frac{x+5}{6}-3$ (a) 4
(2) $(x-5)^2-(x+3)^2=48$ (b) $-11\frac{1}{2}$
(3) $6(x-4)=4(x-3)-(3x-8)$ (c) 61
(4) $(2x-1)(2x+3)=(2x-7)(2x+7)$ (d) –2

6. The sum of three consecutive multiples of 3 is 72. What is the largest number ?

(a) 21 (b) 24
(c) 27 (d) 36

7. A number is doubled and 9 is added. If the result is trebled it becomes 75. What is that number?

(a) 3.5 (b) 6
(c) 8 (d) 7

8. If the sum of one-half and one-fifth of a number exceeds one-third of that number by $7\frac{1}{3}$, the number is

(a) 15 (b) 18
(c) 20 (d) 30

9. $\frac{1}{2}$ is subtracted from a number and the difference is multiplied by 4. If 25 is added to the product and the sum is divided by 3, the result is equal to 10. Find the number

(a) $\frac{3}{5}$ (b) $\frac{7}{4}$
(c) $\frac{6}{7}$ (d) $\frac{2}{3}$

10. The denominator of a fraction is 1 more than its numerator. If 1 is deducted from both the numerator and the denominator, the fraction becomes equivalent to 0.5. The fraction is

(a) $\frac{3}{4}$ (b) $\frac{4}{5}$
(c) $\frac{2}{3}$ (d) $\frac{7}{8}$

11. Divide 224 into three parts so that the second will be twice the first and third will be twice the second

(a) 26, 52, 104 (b) 24, 48, 96
(c) 18, 36, 72 (d) 32, 64,128

12. A rectangle is 8 cm long and 5 cm wide. Its perimeter is doubled when each of its sides is increased by x cm.

(a) 15 cm (b) 14.5 cm
(c) 10.5 cm (d) 9 cm

13. In an isosceles triangle, each of the two equal sides is 3 cm more than twice the base. If the perimeter of the triangle is 31 cm, find the sides of the triangle

(a) 7 cm, 12 cm, 12 cm
(b) 5 cm, 13 cm, 13 cm
(c) 10 cm, 10.5 cm, 10.5 cm
(d) 9 cm, 11 cm, 11 cm

14. In a two digit number, the digit at the unit's place is four times the digit in the ten's place and the sum of the digits is equal to 10. What is the number?

(a) 14 (b) 44
(c) 82 (d) 28

15. In a two digit number, unit's digit is 3 more than ten's digit. The number formed by interchanging the digits and the original number are in the ratio 7 : 4. Find the number.

(a) 63 (b) 36
(c) 57 (d) 75

16. One third of a pole is painted yellow, one-fifth is painted white and the remaining 7 metres is painted black. The length of the pole is :

(a) 15 m (b) 30 m
(c) $10\frac{7}{15}$ m (d) $7\frac{1}{15}$ m

17. The number that should be added to both the numerator and denominator of $\frac{4^2}{9^2}$, so that the fraction becomes $\frac{4}{9}$ is

(a) 0 (b) 16
(c) 36 (d) 81

18. The sum of two numbers is 18 and the difference of their squares is 108. The difference between the numbers is

(a) 2 (b) 12
(c) 6 (d) 8

19. X is 36 years old and Y is 16 years old. In how many years will X be twice as old as Y?

(a) 1 year (b) 2 years
(c) 3 years (d) 4 years

20. A is 2 years older than B who is twice as old as C. If the total of the ages of A, B and C be 27 years, then how old is B?

(a) 7 years (b) 8 years
(c) 9 years (d) 10 years

21. Present ages of Amit and his father are in the ratio 2 : 5 respectively. Four years hence the ratio of their ages becomes 5 : 11 respectively. What was the father's age five years ago?

(a) 40 years (b) 45 years
(c) 30 years (d) 35 years

22. The sum of the ages of 5 children born at intervals of 3 years each is 50 years. What is the age of the youngest child?

(a) 4 years (b) 8 years
(c) 10 years (d) 6 years

23. A train starts will full number of passengers. At the first station, the train drops one-third of the passengers and takes in 96 more. At the next station, one half of the passengers on board get down while 12 new passengers get on board. If the passengers on board now are 240, the number of passengers in the beginning was:

(a) 540 (b) 600
(c) 444 (d) 430

24. Shuba got three fourth of what Alka had. Alka gave half of what remained with her to Mohini. If Mohini got ₹ 625, how much did Alka have in the beginning?

(a) ₹ 3750 (b) ₹ 7000
(c) ₹ 5000 (d) ₹ 5625

25. A person was asked to state his age in years. His reply was, "Take my age three years hence, multiply it by 3 and then subtract three times my age three years ago and you will know how old I am." What was the age of the person?

(a) 24 years (b) 20 years
(c) 32 years (d) 18 years

26. In an examination, a student attempted 15 questions correctly and secured 40 marks. If there were two types of questions (2 marks and 4 marks questions), how many questions of 2 marks did he attempt correctly?

(a) 5 (b) 10
(c) 20 (d) 40

27. A sum of ₹ 36.90 is made up of 180 coins which are either 10 paise coins or 25 paise coins. Determine the number of each type of coins.

(a) 126 of 10 p coins and 54 of 25 p coins
(b) 54 of 10 p coins and 126 of 25 p coins
(c) 90 of 10 p coins and 90 of 25 p coins
(d) 54 of 10 p coins and 90 of 25 p coins

28. ₹ 770 have to be divided among A, B and C such that A receives 2/9th of what B and C together receive. Then A's share is:

(a) ₹ 140 (b) ₹ 154
(c) ₹ 165 (d) ₹ 170

29. In a zoo, there are rabbits and pigeons. If their heads are counted these are 90 while their legs are 224. The number of pigeons in the zoo are

(a) 70 (b) 68
(c) 72 (d) 22

30. The ratio of three numbers is 3 : 4 : 5 and the sum of their squares is 1250. The sum of the three numbers is:

(a) 60 (b) 90
(c) 30 (d) 50

31. Find a number such that if 6,12 and 20 are added to it, the product of the first and third sums may be equal to the square of the second.

(a) 10 (b) 8
(c) 12 (d) 9

32. The sum of the digits of a three digit number is 16. If the ten's digit of the number is 3 times the unit's digit and the unit's digit is one-fourth of the hundredth digit, then what is the number?

(a) 446 (b) 561
(c) 682 (d) 862

33. The ratio between the present ages of M and N is 5 : 3 respectively. The ratio between M's age 4 years ago and N's age after 4 years is 1 : 1. What is the ratio between M's age after 4 years and N's age 4 years ago?

(a) 2 : 1 (b) 1 : 3
(c) 4 : 1 (d) 3 : 1

34. A man engaged a servant on the condition that he would pay him ₹ 90 and a turban after a service of one year. He served only for nine months and received the turban and ₹ 65. The price of the turban is :

(a) ₹ 25 (b) ₹ 18.75
(c) ₹ 10 (d) ₹ 2.50

35. Ashok gave 40 per cent of the amount he had to Jayant. Jayant in turn gave one-fourth of what he received from Ashok to Prakash. After paying ₹ 200 to the taxi driver out of the amount he got from Jayant, Prakash how has ₹ 600 left with him. How much amount did Ashok have?

(a) ₹ 1200 (b) ₹ 4000
(c) ₹ 8000 (d) ₹ 6000

Answers

1. (c)	**2.** (b)	**3.** (b)	**4.** (b)	**5.** (1) → c, (2) → d, (3) → a, (4) → b				**6.** (c)	**7.** (c)
8. (c)	**9.** (b)	**10.** (c)	**11.** (d)	**12.** (b)	**13.** (b)	**14.** (d)	**15.** (b)	**16.** (a)	**17.** (c)
18. (c)	**19.** (d)	**20.** (d)	**21.** (d)	**22.** (a)	**23.** (a)	**24.** (c)	**25.** (d)	**26.** (b)	**27.** (b)
28. (a)	**29.** (b)	**30.** (a)	**31.** (c)	**32.** (d)	**33.** (d)	**34.** (c)	**35.** (c)		

Hints and Solutions

1. (c) $\frac{7}{4m-2}=\frac{5}{3m-4} \Rightarrow 7(3m-4)=5(4m-2)$

$\Rightarrow 21m - 28 = 20m - 10$

$\Rightarrow m = \mathbf{18}.$

2. (b) $5(x+4) - 8(4-x) = 25 - 3(7-x) + 144$

$\Rightarrow 5x + 20 - 32 + 8x = 25 - 21 + 3x + 144$

$\Rightarrow 13x - 12 = 3x + 148$

$\Rightarrow 13x - 3x = 148 + 12$

$\Rightarrow 10x = 160 \Rightarrow x = \frac{160}{10} = \mathbf{16}.$

3. (b) $(x-2)(x+3) = x^2 - 4$

$x^2 + [(-2)+3]x + (-2\times 3) = x^2 - 4$

$\Rightarrow x^2 + x - 6 = x^2 - 4$

$\Rightarrow x^2 + x - x^2 = -4 + 6$

$\Rightarrow x = \mathbf{2}.$

4. (b) $0.2(2x-1) - 0.5(3x-1) = 0.4$

$\Rightarrow 0.4x - 0.2 - 1.5x + 0.5 = 0.4$

$\Rightarrow -1.1x = 0.4 + 0.2 - 0.5 = 0.1$

$\Rightarrow x = -\frac{0.1}{1.1} = \mathbf{\frac{-1}{11}}.$

6. (c) Let the three consecutive multiples of 3 be $3x$, $3(x+1)$ and $3(x+2)$, *i.e.*, $3x$, $3x+3$ and $3x+6$.

Given, $3x + 3x + 3 + 3x + 6 = 72$

$\Rightarrow 9x + 9 = 72 \Rightarrow 9x = 63 \Rightarrow x = 7$

$\therefore$ Largest multiple of 3 $= 3(x+2) = 3(7+2) = 3\times 9 = \mathbf{27}.$

7. (c) Let the number be x.

According to the question

$3(2x+9) = 75 \Rightarrow 6x + 27 = 75$

$\Rightarrow 6x = 75 - 27 = 48 \Rightarrow x = \mathbf{8}.$

8. (c) Let the number be x.

Given, $\left(\frac{x}{2}+\frac{x}{5}\right) - \frac{x}{3} = 7\frac{1}{3} = \frac{22}{3}$

$\Rightarrow \frac{15x+6x-10x}{30} = \frac{22}{3}$

$\Rightarrow \frac{11x}{30} = \frac{22}{3} \Rightarrow x = \frac{22}{3}\times\frac{30}{11} = \mathbf{20}.$

9. (b) Let the number be x.

Given, $\left\{\left(x-\frac{1}{2}\right)\times 4 + 25\right\} \div 3 = 10$

$\Rightarrow \{4x - 2 + 25\}\times\frac{1}{3} = 10$

$\Rightarrow \left(\frac{4x+23}{3}\right) = 10 \Rightarrow 4x + 23 = 30$

$\Rightarrow 4x = 7 \Rightarrow x = \mathbf{\frac{7}{4}}.$

10. (c) Let the numerator of the fraction be x.

Then, the denominator $= x + 1$

$\Rightarrow$ Given fraction $= \frac{N}{D} = \frac{x}{x+1}$

According to question, $\frac{x-1}{(x+1)-1} = 0.5 = \frac{1}{2}$

$\Rightarrow \frac{x-1}{x} = \frac{1}{2} \Rightarrow 2(x-1) = x \Rightarrow 2x - 2 = x$

$\Rightarrow x = 2.$

$\therefore$ Given fraction $= \frac{2}{2+1} = \mathbf{\frac{2}{3}}.$

11. (d) Let the first part be x.

Then, second part = $2x$

Third part = $2 \times 2x = 4x$

Given, $x + 2x + 4x = 224$

$\Rightarrow 7x = 224 \Rightarrow x = \frac{224}{7} = 32$

$\therefore$ The three parts are $32, 2\times32, 4\times32$, *i.e.*, **32, 64, 128.**

12. (b) Perimeter of a rectangle = $2\,(l + b)$

$= 2\,(8 + 5)$ cm = 26 cm

New length = $(8 + x)$ cm,

New breadth = $(5 + x)$ cm

Given, $2\,(8 + x + 5 + x) = 26 \times 2$

$\Rightarrow 2(13 + 2x) = 52 \Rightarrow 13 + 2x = \frac{52}{2} = 26$

$\Rightarrow 2x = 13 \Rightarrow x = \frac{13}{2} = 6.5$ cm

$\therefore$ New length = $(8 + 6.5)$ cm = **14.5 cm.**

13. (b) Let the base = x cm

Then, each of equal sides = $2x + 3$

Given, perimeter of Δ = 31 cm

$\Rightarrow x + 2x + 3 + 2x + 3 = 31$

$\Rightarrow 5x + 6 = 31 \Rightarrow 5x = 25 \Rightarrow x = 5$

$\therefore$ The sides are 5 cm, $(2 \times 5 + 3)$ cm, $(2 \times 5 + 3)$ cm, *i.e.*, 5 cm, 13 cm, 13 cm.

14. (d) Let the digit at ten's place = x

Then, digit at unit's place = $4x$

Sum of digits =10

$\Rightarrow x + 4x = 10 \Rightarrow 5x = 10 \Rightarrow x = 2$

$\therefore$ The number = $10 \times x + 4x$

$= 10 \times 2 + 4 \times 2 = 20 + 8 =$ **28.**

15. (b) Let the ten's digit = x

Then, unit's digit = $x + 3$

$\therefore$ Original number = $10 \times x + (x + 3)$

$= 11x + 3$

T	U
x	$x + 3$

Number formed after reversing digits

$= 10\,(x + 3) + x$

$= 10x + 30 + x = 11x + 30$

T	U
$x + 3$	x

According to given condition

$\frac{11x + 30}{11x + 3} = \frac{7}{4} \Rightarrow 44x + 120 = 77x + 21$

$\Rightarrow 77x - 44x = 120 - 21 \Rightarrow 33x = 99 \Rightarrow x = 3$

The number = $11 \times 3 + 3 = 33 + 3 =$ **36.**

16. (a) Let the total length of the pole be x m.

$\frac{x}{3} + \frac{x}{5} + 7 = x$

$\Rightarrow \frac{5x + 3x + 105}{15} = x$

$\Rightarrow 8x + 105 = 15x \Rightarrow 7x = 105$

$\Rightarrow x = \frac{105}{7} =$ **15.**

17. (c) Let the number to be added be x. Then,

$\frac{4^2 + x}{9^2 + x} = \frac{4}{9} \Rightarrow \frac{16 + x}{81 + x} = \frac{4}{9}$

$\Rightarrow 144 + 9x = 324 + 4x$

$\Rightarrow 9x - 4x = 324 - 144 \Rightarrow 5x = 180$

$\Rightarrow x = \frac{180}{5} =$ **36.**

18. (c) Let one number be x.

Then, other number = $18 - x$

Given, $(18 - x)^2 - x^2 = 108$

$\Rightarrow 324 - 36x + \cancel{x^2} - \cancel{x^2} = 108$

$\Rightarrow 36x = 324 - 108 = 216$

$\Rightarrow x = \frac{216}{36} = 6$

$\therefore$ The difference between the two numbers

$= 18 - x - x$

$= 18 - 2x = 18 - 2 \times 6$

$= 18 - 12 =$ **6.**

19. (d) Let X be twice as old as Y after x years. Then, after x years.

X's age = $(36 + x)$ yrs., Y's age = $(16 + x)$ yrs.

Given, $36 + x = 2(16 + x)$

$\Rightarrow 36 + x = 32 + 2x \Rightarrow x =$ **4 years.**

20. (d) Let C's age = x years. Then,

B's age = $2x$ years

A's age $(2x + 2)$ years

Given, $(2x + 2) + 2x + x = 27$

$\Rightarrow 5x + 2 = 27 \Rightarrow 5x = 25 \Rightarrow x = 5$

$\therefore$ B's age = (2×5) years = **10 years.**

21. (d) Let the present ages of Amit and his father be $2x$ and $5x$ years respectively.

Four years hence,

Amit's age = $(2x + 4)$ years

Father's age = $(5x + 4)$ years

Given, $\frac{2x + 4}{5x + 4} = \frac{5}{11} \Rightarrow 11(2x + 4) = 5(5x + 4)$

$\Rightarrow$ $22x + 44 = 25x + 20 \Rightarrow 3x = 24 \Rightarrow x = 8$

$\therefore$ Father's present age = (5×8) years = 40 years

Father's age 5 years ago = $(40 - 5)$ = **35 years.**

22. (a) Let the age of the youngest child be x years.

Then, $x + x + 3 + x + 6 + x + 9 + x + 12 = 50$

$\Rightarrow 5x + 30 = 50 \Rightarrow 5x = 20 \Rightarrow x = 4$

$\therefore$ Age of the youngest child = **4 years.**

23. (a) Let the full number of passengers be x.

Number of passengers at the first station

$= \left(x - \frac{1}{3}x\right) + 96 = \frac{2x}{3} + 96$

Number of passengers at the second station

$= \frac{1}{2}\left(\frac{2x}{3} + 96\right) + 12 = \frac{x}{3} + 48 + 12 = \frac{x}{3} + 60$

Given, $\frac{x}{3} + 60 = 240 \Rightarrow \frac{x}{3} = 240 - 60 = 180$

$\Rightarrow x = 180 \times 3 =$ **540.**

24. (c) Let Alka have ₹ x in the beginning.

Amount that Shubha got = ₹ $\frac{3x}{4}$

Amount remaining with Alka $= x - \frac{3x}{4} =$ ₹ $\frac{x}{4}$

Amount that Mohini got $= \frac{1}{2} \times \frac{x}{4} =$ ₹ $\frac{x}{8}$

Given, $\frac{x}{8} = 625 \Rightarrow x = 625 \times 8 =$ **₹ 5000.**

25. (d) Let the present age of the person be x years.

His age 3 years hence = $(x + 3)$ years

His age 3 years ago = $(x - 3)$ years

Given, his present age $= (x + 3)\,3 - 3(x - 3)$

$= (3x + 9 - 3x + 9)$ years = **18 years.**

26. (b) Let the number of 2 marks questions attempted correctly be x.

Then, the number of 4 marks questions attempted correctly = $(15 - x)$

Given, $x \times 2 + (15 - x) \times 4 = 40$

$\Rightarrow 2x + 60 - 4x = 40 \Rightarrow -2x = -20 \Rightarrow x =$ **10.**

27. (b) Let the number of 10 paise coins be x.

Then, the number of 25 paise coins is $(180 - x)$.

Given, $x \times 10 + (180 - x) \times 25 = 3690$

($\because$ ₹ 36.90 = 3690 paise)

Now solve for x to get the required answer.

28. (a) Let B and C together receive ₹ x.

Then, A's share = ₹ $\frac{2x}{9}$

Given, $\frac{2x}{9} + x = 770 \Rightarrow \frac{2x + 9x}{9} = 770$

$\Rightarrow \frac{11x}{9} = 770 \Rightarrow x = \frac{770 \times 9}{11} =$ ₹ 630

$\therefore$ A's share $= \frac{2}{9} \times$ ₹ 630 = **₹ 140.**

29. (b) Let the number of rabbits be x.

No. of heads = 90 $\Rightarrow$ No. of pigeons = $(90 - x)$

Since, a pigeon has 2 legs and a rabbit has 4 legs.

$x \times 4 + (90 - x) \times 2 = 224$

$\Rightarrow 4x + 180 - 2x = 224 \Rightarrow 2x = 224 - 180 = 44$

$\Rightarrow x = 22$

$\therefore$ No. of pigeons = 90 – 22 = **68.**

30. (a) Let the three numbers be $3x$, $4x$ and $5x$.

Given, $(3x)^2 + (4x)^2 + (5x)^2 = 1250$

$9x^2 + 16x^2 + 25x^2 = 1250$

$\Rightarrow 50x^2 = 1250 \Rightarrow x^2 = 25 \Rightarrow x = 5$

$\therefore$ The three numbers are 3×5, 4×5 and 5×5, *i.e.*, 15, 20 and 25.

Required sum = 15 + 20 + 25 = **60.**

31. (c) Let the number be x. Then,

$(x + 6)(x + 20) = (x + 12)^2$

$\Rightarrow x^2 + (6 + 20)x + 120 = x^2 + 24x + 144$

$\Rightarrow 26x + 120 = 24x + 144$

$\Rightarrow 2x = 24 \Rightarrow x =$ **12.**

32. (d) Let the hundreth's digit be x.

Then, unit's digit = $x/4$, ten's digit $= \frac{3x}{4}$

Given, $x + \frac{x}{4} + \frac{3x}{4} = 16 \Rightarrow \frac{4x + x + 3x}{4} = 16$

$\Rightarrow 2x = 16 \Rightarrow x = 8$

$\therefore$ Ten's digit $= 3 \times \frac{8}{4} = 6$, unit's digit $= \frac{8}{4} = 2$

$\therefore$ Required number = $100 \times 8 + 10 \times 6 + 2$ = 800 + 60 + 2 = **862.**

33. (d) Let the present ages of M and N be $5x$ years and $3x$ years respectively.

Given, $(5x - 4) : (3x + 4) = 1:1 \Rightarrow 5x - 4 = 3x + 4$

$\Rightarrow 2x = 8 \Rightarrow x = 4$

$\therefore$ Present ages of M and N respectively are (5×4) year and (3×4) years, *i.e.*, 20 years and 12 years.

$\therefore$ Required ratio $= \dfrac{20+4}{12-4} = \dfrac{24}{8} = \dfrac{3}{1} = \mathbf{3:1.}$

34. (c) Let the price of the turban be ₹ x.

Then, total income for 12 months = ₹ $(90 + x)$

$\Rightarrow$ Income for 9 months $= \left(\dfrac{90+x}{12}\right)\times 9 = \dfrac{3x+270}{4}$

Given, $\dfrac{3x+270}{4} = 65 + x$

$\Rightarrow$ $3x + 270 = 4x + 260$

$\Rightarrow$ $x =$ **₹ 10.**

35. (c) Let Ashok have ₹ x.

Amount given to Jayant = 40% of ₹ x = ₹ $\dfrac{2x}{5}$

Amount given to Prakash $= \dfrac{1}{4}\times$ ₹ $\dfrac{2x}{5}$ = ₹ $\dfrac{x}{10}$

Given, $\dfrac{x}{10} - 200 = 600$

$\Rightarrow$ $\dfrac{x}{10} = 800 \Rightarrow x =$ **₹ 8000.**

Self Assessment Sheet–9

1. For which equation(s) is $x = 3$ a solution?

(I) $2x - 5 + 3x = 10$ (II) $\dfrac{-x+7}{2} = 2$

(III) $4x - 11 = 17$ (IV) $9 = -(x-1) + 11$

(a) I only (b) I and II

(c) I, II and III (d) I, II and IV

2. Solve : $\dfrac{n+3}{\frac{1}{3}} - \dfrac{n+2}{\frac{1}{2}} = \dfrac{n-4}{\frac{1}{10}}$. The value of n is obtained as

(a) –5 (b) 5

(c) 6 (d) 4

3. Solve : $(x-7)^2 - (x+8)^2 = 75$. The value of x is:

(a) 3 (b) 1

(c) –3 (d) –1

4. Solve : $(t-4)(t+4) = 54 + (t-5)(t-10)$. The value of t is :

(a) 4 (b) 8

(c) – 8 (d) 2

5. An old rhyme, and problem:

"If to my age, there added be,
One half of it, and three times three,
Four score and seven my age will be.
How old am I, pray tell me ?"

(a) 52 years (b) 50 years

(c) 48 years (d) 54 years

6. Solve : $\dfrac{x+5}{4} - \dfrac{3x-1}{8} + \dfrac{4-x}{6} = 2\dfrac{1}{24}$. The value of x is

(a) 1 (b) –1

(c) 0 (d) 2

7. Mr. Joshi has 430 cabbage - plants which he wants to plant out, some 25 to a row, the rest 20 to a row. If there are to be 18 rows in all, how many rows of 25 will there be ?

(a) 10 (b) 14

(c) 8 (d) 12

8. A choir is singing at a festival. On the first night, 12 choir members were absent , so the choir stood in 5 equal rows. On the second night, only 1 member was absent, so the choir stood in 6 equal rows. The same member of people stood in each row each night. How many members are in the choir ?

(a) 65 (b) 67

(c) 70 (d) 64

9. Solve : $\dfrac{2}{3}(n+6) - \dfrac{1}{5}(n-4) = \dfrac{3}{7}(n+12)$

(a) –9 (b) 8

(c) $3\dfrac{1}{9}$ (d) 9

10. There are 90 multiple choice questions in a test. Suppose you get two marks for every correct answer and for every question you leave unattempted or answer wrongly, one mark is deducted from your total score of correct answers. If you get 60 marks in the test, then how many questions did your answer correctly ?

(a) 40 (b) 60

(c) 50 (d) 48

Answers

1. (d)	2. (b)	3. (c)	4. (b)	5. (a)	6. (c)	7. (b)	8. (b)	9. (d)	10. (c)

Unit Test–2

1. If $\frac{b}{a}=0.25$, then $\frac{2a-b}{2a+b}+\frac{2}{9}=?$

(a) $\frac{4}{9}$ (b) $\frac{5}{9}$

(c) 1 (d) 2

2. The value of $[(0.98)^3+(0.02)^3+3\times 0.98\times 0.02-1]$ is

(a) 0 (b) 1

(c) 1.09 (d) 1.98

3. $a^2b(a^3-a+1)-ab(a^4-2a^2+2a)-b(a^3-a^2-1)$ is

(a) ab (b) ab^2

(c) a (d) b

4. Match the problem with their answers :

Column I

Problem

1. $(3x-2)(2x-3)+(5x-3)(x+1)$
2. $3x^3+4x^2+5x+18$ divided by $x+2$
3. Sum of $2x-x^2+5$ and $-4x-3+7x^2$ subtracted from 5
4. $(3x-2)^3-(3x^2+2)(9x-7)$

Column II

Answer

(a) $3+2x-6x^2$

(b) $-33x^2+18x+6$

(c) $11x^2-11x+3$

(d) $3x^2-2x+9$

5. $\frac{(0.35)^2-(0.03)^2}{0.19}=?$

(a) 0.32 (b) 0.48

(c) 0.76 (d) 0.64

6. $\frac{(2.3)^3-0.027}{(2.3)^2+0.69+0.09}=?$

(a) 2.6 (b) 2

(c) 2.33 (d) 2.8

7. $2n$ is an even number. What are the odd numbers on each side of it ? The sum of two consecutive odd numbers is 96. What are they ?

(a) 41, 43 (b) 57, 59

(c) 47, 49 (d) 31, 33

8. The factors of a^4-4a^2 are

(a) $a^2(a-2)(a+2)$ (b) $a(a-2)(a+2)$

(c) $a(a+2)(a+2)$ (d) $a^2(a-2)^2$

9. If $a+\frac{1}{a}=4$, then the value of $a^2-\frac{1}{a^2}$ is

(a) $8\sqrt{3}$ (b) $2\sqrt{3}$

(c) 12 (d) $4\sqrt{2}$

10. The value of

$$\frac{(1.5)^3+(4.7)^3+(3.8)^3-3\times1.5\times4.7\times3.8}{(1.5)^2+(4.7)^2+(3.8)^2-1.5\times4.7-4.7\times3.8-1.5\times3.8}$$

is

(a) 0 (b) 1

(c) 10 (d) 30

11. Solve: $(x-4)^2-(x+4)^2=48$. The value of x is

(a) 3 (b) -3

(c) 4 (d) 2

12. Divide $3x^4-5x^3y+6x^2y^2-3xy^3+y^4$ by x^2-xy+y^2. What should be subtracted from the quotient to make it a perfect square ?

(a) $2x^2$ (b) x^2

(c) y^2 (d) $-xy$

13. Simplify, giving your answer in factors : $3(x-1)^2+5(x-1)(x+4)-2(x+4)^2$. The answer is

(a) $(2x+7)(3x-7)$ (b) $(x-7)(2x+7)$

(c) $(2x-3)(x+7)$ (d) $(2x-7)(3x+7)$

14. Factorize $2x^2-5xy+3y^2$ and use your result to factorize $2(3a-b)^2-5(3a-b)(2a-b)+3(2a-b)^2$. The factors are

(a) a^2b (b) $(a+b)(a-b)$

(c) ab (d) $(2a+b)(a-b)$

15. $\frac{a^2-b^2-2bc-c^2}{a^2+b^2+2ab-c^2}$ is equivalent to

(a) $\frac{a+b+c}{a-b+c}$ (b) $\frac{a-b-c}{a+b-c}$

(c) $\frac{a-b-c}{a-b+c}$ (d) $\frac{a-b+c}{a+b+c}$

16. One of the factors of $a^3-b^3-a^2b+ab^2+a^2-b^2$ is

(a) $a+b$ (b) $b-a$

(c) $a-b$ (d) a^2+b^2

17. If $x + y - 1 = 0$, then $x^3 + y^3 - 1$ is equal to
(a) $x^2 + y^2 - 1$
(b) $x^2 - xy + y^2$
(c) $x^2 + xy + y^2$
(d) $-3xy$

18. If $x + y = a$ and $xy = b$, then the value of $\frac{1}{x^3} + \frac{1}{y^3}$ is
(a) $a^3 - 3ab$
(b) $\frac{a^3 + 3ab}{b^3}$
(c) $\frac{a^3 - 3ab}{b^3}$
(d) $a^3 + 3ab$

19. If $2x + ky + z$ is a factor of $9y^2 - z^2 - 2xz + 6xy$, then the value of k is equal to
(a) –3
(b) –1
(c) 1
(d) 3

20. In a lottery, a total of 200 prizes are to be given. A prize is either ₹ 500 or ₹ 100. If the total prize money is ₹ 50,000, then the number of ₹ 500 and ₹ 100 prizes are
(a) 70, 130
(b) 75, 125
(c) 60, 140
(d) 80, 120

Answers

1. (c) **2.** (a) [**Hint.** $a^3 + b^3 + 3ab(a + b) - 1 = (a + b)^3 - 1$, Here $a = 0.98$, $b = 0.2$]
3. (d) **4.** (1)→ (*c*), (2)→ (*d*), (3)→ (*a*), (4) → (*b*) **5.** (d) **6.** (*b*) [**Hint.** 0.69 = 2.3 × 0.3]
7. (c) **8.** (a) **9.** (a) **10.** (c) **11.** (b) **12.** (a) **13.** (d) **14.** (c) **15.** (b) **16.** (c)
17. (d) **18.** (c) **19.** (d) **20.** (b)

UNIT–3

COMMERCIAL MATHEMATICS

- *Percentage and its Applications*
- *Profit and Loss and Discount*
- *Simple Interest*
- *Time and Work*
- *Distance, Time and Speed*
- *Ratio and Proportion*
- *Average*

Chapter 10 PERCENTAGE AND ITS APPLICATIONS

KEY FACTS

1. (i) Per cent means per hundred or for every hundred. The symbol % is used to denote per cent. 45 out of 100 students are girls means 45% of students are girls.

   (ii) A fraction with denominator 100, such as $\frac{19}{100}$ is called nineteen per cent. $\frac{63}{100}$ means 63%.

2. **To convert**

   (i) **A fraction into a per cent,** *multiply the fraction by 100, e.g.,* $\frac{4}{5}=\left(\frac{4}{5}\times 100\right)\%=80\%$.

   (ii) **A ratio into a per cent,** *write it as a fraction and multiply by 100, e.g.,* $19:25=\left(\frac{19}{25}\times 100\right)\%=76\%$.

   (iii) **A decimal into a per cent,** *multiply by 100, i.e.,* shift the decimal point two places to the right, *e.g.,* 0.045 = (0.045 × 100) % = 4.5%.

   (iv) **A per cent into a fraction,** *drop the per cent symbol (%) and divide by 100, e.g.,* $46\% = \frac{46}{100}=\frac{23}{50}$

   (v) **A per cent into a ratio,** *drop the per cent symbol (%) and form a ratio by taking the given number as the 1st term and 100 as the 2nd term, e.g.,* 52% = 52 : 100 = 13 : 25.

   (vi) **A per cent into a decimal,** *drop the per cent symbol and shift the decimal two places to the left, e.g.,* 99% = 0.99, 4.5% = 0.045

3. **Increase %** $=\frac{\text{Increase}}{\text{Original value}}\times 100$; **Decrease%** $=\frac{\text{Decrease}}{\text{Original Value}}\times 100$; **Error%** $=\frac{\text{Error}}{\text{Actual value}}\times 100$.

Solved Examples

Ex. 1. ***Evaluate 15.2% of 726 × 12.8% of 643.***

Sol. 15.2% of 726 × 12.8% of 643 $=\frac{15.2}{100}\times 726\times\frac{12.8}{100}\times 643$

= 9082.411 = **9082** (approx.)

Ex. 2. ***If S is 150 per cent of T, then T is what per cent of S + T ?***

Sol. $S = 150\%$ of $T \Rightarrow S=\frac{150}{100}\times T=1.5\,T$

Required per cent $= \left(\frac{T}{S+T}\times 100\right)\% = \left(\frac{T}{1.5T+T}\times 100\right)\%$

$= \left(\frac{T}{2.5T}\times 100\right)\% = \frac{1000}{25}\% = \mathbf{40\%}.$

Ex. 3. ***30% of 28% of 480 is same as***

(a) 60% of 56% of 240 *(b) 15% of 56% of 240* *(c) 60% of 28% of 240* *(d) 15% of 28% of 480*

Sol. 30% of 28% of 480 = (30 × 2)% of 28% of $\frac{480}{2}$ = 60% of 28% of 240.

Ex. 4. ***A number is 68% of the other number. If the sum of these two numbers is 84% of 500, what is the bigger number?***

Sol. Let the greater number = x.

Then, other number = 68% of $x = \frac{68}{100}x$

Given, $x + \frac{68}{100}x = \frac{84}{100}\times 500 \Rightarrow \frac{168x}{100} = \frac{84}{100}\times 500$

$\Rightarrow \quad x = \frac{84\times 500}{168} \Rightarrow x = \mathbf{250}.$

Ex. 5. ***A sugar solution of 3 litres contains 40% sugar. One litre of water is added to this solution. What is the percentage of sugar in the new solution?***

Sol. In 3 litres sugar solution, sugar = 40% of 3 litres = 1.2 litres

As 1 litre water is added to the sugar solution, now 4 litres of sugar solution contains 1.2 litres of sugar, *i.e.,*

% of sugar $= \left(\frac{1.2}{4}\times 100\right)\% = \mathbf{30\%}.$

Ex. 6. ***Out of the total candidates appeared for a competitive examination, 20% qualified and 10% of the qualified candidates got finally selected. If 290 candidates were finally selected, then how many candidates appeared?***

Sol. Let the number of candidates appeared be x.

Then, number of candidates who qualified = 20% of $x = \frac{x}{5}$

Number of candidates selected = 10% of $\frac{x}{5} = \frac{x}{50}$

Given, $\frac{x}{50} = 290 \quad \Rightarrow \quad x = \mathbf{14500}.$

Ex. 7. ***A report consists of 20 sheets of 56 lines and each such line consists of 65 characters. The report is reduced into sheets, each of 65 lines such that each line consists of 70 characters. What is the percentage reduction in the number of sheets?***

Sol. Let the new number of sheets be x. Then,

$20 \times 56 \times 65 = x \times 65 \times 70 \Rightarrow x = \frac{20\times 56\times 65}{65\times 70} = 16$

$\therefore$ Percentage reduction in number of sheets $= \frac{(20-16)}{20}\times 100\% = \mathbf{20\%}.$

Ex. 8. ***If a^2% of b = b^3% of c and c^4% of a = b% of b, then the relation between a and b is***

(a) $a = b$ *(b) $a = b^2$* *(c) $a^9 = b^{10}$* *(d) $a = b^{10}$*

Sol. Given, $a^2\%$ of $b = b^3\%$ of $c \Rightarrow \frac{a^2}{100}\times b = \frac{b^3}{100}\times c \Rightarrow a^2b = b^3c$ ...(*i*) and $c^4\%$ of $a = b\%$ of b $\Rightarrow \frac{c^4}{100} \times a = \frac{b}{100} \times b$ $\Rightarrow c^4 a = b^2$ ...(ii)

From (i) $c = \frac{a^2}{b^2}$

Putting the value of c in (ii), we get $\left(\frac{a^2}{b^2}\right)^4 \times a = b^2$ $\Rightarrow \frac{a^8 \times a}{b^8} = b^2$ $\Rightarrow$ $\boldsymbol{a^9 = b^{10}}$.

Ex. 9. ***A quantity of 30 ml of 20% alcohol is mixed with 20 ml of 25% alcohol. What is the strength of alcohol in the mixture?***

Sol. Quantity of alcohol in 30 ml solution = 20% of 30 ml = 6 ml

Quantity of alcohol in 20 ml solution = 25% of 20 ml = 5 ml

$\therefore$ Percentage of alcohol in the mixture $= \frac{6+5}{(30+20)} \times 100 = \frac{11}{50} \times 100 = \mathbf{22\%}$.

Ex. 10. ***A man spends 80% of his income with the increase in the cost of living, his expenditure increases by $37\frac{1}{2}\%$ and his income increases by $16\frac{2}{3}\%$. Find his present per cent saving.***

Sol. If the income was ₹ 100, then the expenditure was ₹ 80.

Income increases by $16\frac{2}{3}\%$ $\Rightarrow$ Income becomes ₹ $116\frac{2}{3}$

Expenditure increases by $37\frac{1}{2}\%$ $\Rightarrow$ Expenditure $= 80 + 37\frac{1}{2}\%$ of $80 = 80 + \frac{37.5}{100} \times 80 = 80 + 30 =$ ₹ 110.

$\therefore$ Present saving = ₹$116\frac{2}{3}$ − ₹ 110 = ₹ $6\frac{2}{3}$

Present per cent saving $= \frac{6\frac{2}{3}}{116\frac{2}{3}} \times 100 = \frac{\frac{20}{3}}{\frac{350}{3}} \times 100 = \frac{2000}{350} = \mathbf{5\frac{5}{7}\%}$.

Question Bank–10

Which of the given values should replace the question mark in Questions 1 to 4?

1. (0.9% of 450) ÷ (0.2% of 250) = ?
(a) 5.04 (b) 7.5
(c) 8.1 (d) 9.1

2. (180% of ?) ÷ 2 = 504
(a) 400 (b) 480
(c) 560 (d) 600

3. 12% of 980 – ?% of 450 = 30% of 227
(a) 14 (b) 17
(c) 11 (d) 8

4. 80% of 50% of 250% of 34 = ?
(a) 34 (b) 40
(c) 42.5 (d) 43

5. 15% of 45% of a number is 105.3. What is 24% of that number?
(a) 385.5 (b) 374.4
(c) 390 (d) 375

6. 40% of 60% of $\frac{3}{5}$th of a number is 504. What is 25% of $\frac{2}{5}$th of that number?
(a) 180 (b) 175
(c) 360 (d) 350

7. $x\%$ of x is the same as 10% of
(a) $\frac{x}{10}$ (b) $\frac{x^2}{10}$
(c) $\frac{x^3}{10}$ (d) $\frac{x}{100}$

8. A period of 4 hrs 30 min is what per cent of a day?
(a) $18\frac{3}{4}\%$ (b) 20%
(c) $16\frac{2}{3}\%$ (d) 19%

9. Two-third of one-seventh of a number is 87.5% of 240. What is the number?
(a) 2670 (b) 2450
(c) 2205 (d) 1470

10. In a shipment of 120 machine parts, 5 per cent were defective. In a shipment of 80 machine parts, 10 per cent were defective. For the two shipments combined, what per cent of the machine parts were defective?
(a) 6.5% (b) 7.0%
(c) 7.5% (d) 8.0%

11. Which of the following multipliers will cause a number to be increased by 25.3%?
(a) 12.53 (b) 125.3
(c) 1.253 (d) 1253

12. A earns $12\frac{1}{2}\%$ more than B. If they jointly earn ₹ 35,700 in a year, then the amount that A earns annually is
(a) ₹ 16,800 (b) ₹ 17,875
(c) ₹ 18,225 (d) ₹ 18,900

13. A litre of water is evaporated from 6 litres of sugar solution containing 4% of sugar. Find the percentage of sugar in the remaining solution?
(a) 4% (b) 5%
(c) 4.2% (d) 4.8%

14. A mixture of water and milk is 40 litres. There is 10% water in it. How much water should now be added in this mixture so that the new mixture contains 20% water?
(a) 4 litres (b) 5 litres
(c) 6.5 litres (d) 7.5 litres

15. 8% of the voters in an election did not cast their votes. In the election there were only two candidates. The winner by obtaining 48% of the total votes defeated his contestant by 1100 votes. The total number of the voters in the election was
(a) 21000 (b) 23500
(c) 22000 (d) 27500

16. p is six times as large as q. The per cent that q is less than p is
(a) $83\frac{1}{3}$ (b) 70
(c) $63\frac{1}{3}$ (d) 50

17. The population of a village was 9800. In a year, with the increase in population of males by 8% and that of females by 5%, the population of the village became 10458. What was the number of males in the village before increase?
(a) 4200 (b) 4410
(c) 5600 (d) 6048

18. Prasad's monthly income is 25% more than that of Anam. Anam's monthly income is 75% less than that of Bhushan. If the difference between the monthly incomes of Prasad and Bhushan is ₹ 26,125, then what is Anam's monthly income?
(a) ₹ 8000 (b) ₹ 16000
(c) ₹ 9500 (d) ₹ 10,000

19. If A exceeds B by 40% and B is less than C by 20%, then $A : C$ is
(a) 28 : 25 (b) 26 : 25
(c) 3 : 2 (d) 3 : 1

20. If x is 90% of y, then what per cent of x is y?
(a) 90% (b) 190%
(c) $101\frac{1}{9}\%$ (d) $111\frac{1}{9}\%$

21. Ramesh credits 15% of his salary in his fixed deposit plan and spends 30% of the remaining amount on groceries. If the cash in hand is ₹ 2380, what is his salary?
(a) ₹ 5000 (b) ₹ 4500
(c) ₹ 4000 (d) ₹ 3500

22. $\frac{2}{5}$ of the voters promise to vote for P and the rest promised to vote for Q. Of these, on the last day 25% of the voters went back on their promise to vote for P and 15% of the voters went back on their promise to vote for Q and reversed their votes. Q lost by 44 votes. Then the total number of votes is
(a) 200 (b) 210
(c) 220 (d) 110

23. A store sells boxes of 6 pens for ₹ 30 each and boxes of 12 pens for ₹ 48 each. The price per pen is what per cent less, when purchased in a box of 12 than in a box of 6?
(a) 80% (b) 75%
(c) 50% (d) 20%

24. A shopkeeper has certain number of eggs of which 5% are found to be broken. He sells 93% of the remainder and still has 266 eggs left. How many eggs did he originally have?
(a) 3800 (b) 4000
(c) 4200 (d) 4800

25. A spider climbed $62\frac{1}{2}\%$ of the height of the pole in one hour and in the next hour covered $12\frac{1}{2}\%$ of the remaining height. If the height of the pole is 192 m, then the distance climbed in the second hour is

(a) 3 m (b) 5 m
(c) 7 m (d) 9 m

26. There are two candidates Bhiku and Mhatre for an election. Bhiku gets 65% of the total valid votes. If the total votes were 6000, what is the number of valid votes that the other candidate Mhatre gets, if 25% of the total votes were declared invalid?

(a) 1575 (b) 1625
(c) 1675 (d) 1525

27. A person saves every year 20% of his income. If his income increases every year by 10%, then his savings increase every year by

(a) 10% (b) 5%
(c) 2% (d) 1%

28. A rainy day occurs once in every 10 days. Half of the rainy days produce rainbows. What per cent of all the days do not produce rainbow?

(a) 95% (b) 10%
(c) 50% (d) 5%

29. A cube of white chalk is painted red, and then cut parallel to the sides to form two rectangular solids of equal volume. What per cent of the surface area of each of the new solids not painted red?

(a) 50% (b) 25%
(c) 20% (d) 40%

30. A large water-melon weighs 20 kg with 96% of its weight being water. It is allowed to stand in the sun and some of the water evaporates so that now only 95% of its weight is water. Its reduced weight will be

(a) 18 kg (b) 17 kg
(c) 16.5 kg (d) 16 kg

31. The ratio of the prices of two houses A and B was 4 : 5 last year. This year the price of A increased by 25% and that of B by ₹ 50,000. If their prices are now in the ratio 9 : 10, the price of A last year was

(a) ₹ 3,60,000 (b) ₹ 4,50,000
(c) ₹ 4,80,000 (d) ₹ 5,00,000

32. When a train started from station A, there were certain number of passengers in it. On the next station, $\frac{1}{11}$ of these passengers got down and 20% of these got down passengers got into the train. Now if the number of passengers in the train is 510, how many passengers were there at station A?

(a) 600 (b) 580
(c) 560 (d) 550

33. In a test containing of 80 questions carrying one mark each Arpita answered 65% of the first 40 questions correctly. What per cent of the other 40 questions does she need to answer correctly for her grade in the entire exam to be 75%?

(a) 60 (b) 80
(c) 85 (d) 75

34. If a number x is 10% less than another number y and y is 10% more than 125, then x is equal to

(a) 150 (b) 143
(c) 140.55 (d) 123.75

35. In measuring the sides of a rectangle, errors of 5% and 3% in excess are made. The error per cent in the calculated area is

(a) 8.35% (b) 7.15%
(c) 8.15% (d) 6.25%

Answers

1. (c)	**2.** (c)	**3.** (c)	**4.** (a)	**5.** (b)	**6.** (d)	**7.** (b)	**8.** (a)	**9.** (c)	**10.** (b)
11. (c)	**12.** (d)	**13.** (d)	**14.** (b)	**15.** (d)	**16.** (a)	**17.** (c)	**18.** (c)	**19.** (a)	**20.** (d)
21. (c)	**22.** (a)	**23.** (d)	**24.** (b)	**25.** (d)	**26.** (a)	**27.** (a)	**28.** (a)	**29.** (b)	**30.** (d)
31. (a)	**32.** (d)	**33.** (c)	**34.** (d)	**35.** (c)					

Hints and Solutions

1. (c) (0.9% of 450) ÷ (0.2% of 250)

$$= \left(\frac{0.9}{100} \times 450\right) \div \left(\frac{0.2}{100} \times 250\right)$$

$= 4.05 \div 0.5 = \mathbf{8.1}$.

2. (c) (180% of ?) ÷ 2 = 504 $\Rightarrow$ 180% of ? = 504 × 2

$$\Rightarrow ? = \frac{504 \times 2 \times 100}{180} = \mathbf{560}.$$

3. (c) 12% of 980 – x% of 450 = 30% of 227

$$\Rightarrow \frac{12}{100}\times 980-\frac{x}{100}\times 450=\frac{30}{100}\times 227$$

$$\Rightarrow \frac{x}{100}\times 450=117.6-68.1=49.5$$

$$\Rightarrow x=\frac{49.5\times 100}{450}=\mathbf{11}.$$

4. (a) 80% of 50% of 250% of 34

$$=\frac{80}{100}\times\frac{50}{100}\times\frac{250}{100}\times 34=\mathbf{34}.$$

5. (b) Let the number be x. Then, 15% of 45% of x = 105.3

$$\Rightarrow \frac{15}{100}\times\frac{45}{100}\times x=105.3$$

$$\Rightarrow x=\frac{105.3\times 100\times 100}{15\times 45}=1560$$

$$\therefore\ 24\% \text{ of } 1560=\frac{24}{100}\times 1560=\mathbf{374.4}.$$

6. (d) Given $\frac{40}{100}\times\frac{60}{100}\times\frac{3}{5}\times$ Number = 504

$$\Rightarrow \text{Number}=\frac{504\times 100\times 100\times 5}{40\times 60\times 3}=3500$$

$$\therefore\ \frac{25}{100}\times\frac{2}{5}\times 3500=\mathbf{350}.$$

7. (b) $x\%\text{ of }x=\frac{x}{100}\times x=\frac{x^2}{100}=\frac{10}{100}\times\frac{x^2}{10}$

$$=10\%\text{ of }\frac{\mathbf{x^2}}{\mathbf{10}}.$$

8. (a) 4 hrs 30 min = $4\frac{1}{2}$ hrs = $\frac{9}{2}$ hrs

$$\therefore\ \text{Reqd. }\%=\left(\frac{\frac{9}{2}}{24}\times 100\right)\%=\left(\frac{9}{2\times 24}\times 100\right)\%$$

$$=\mathbf{18\frac{3}{4}\%}.$$

9. (c) $\frac{2}{3}\times\frac{1}{7}\times$ Number $=\frac{87.5}{100}\times 240$

$$\Rightarrow \text{Number}=\frac{87.5\times 240\times 3\times 7}{2\times 100}=\mathbf{2205}.$$

10. (b) Defective parts in the shipment of 120

$$=5\%\text{ of }120=\frac{5}{100}\times 120=6$$

Defective part in the shipment of 80 = 10% of 80

$$=\frac{10}{100}\times 80=8$$

$\therefore$ Total machine parts = 120 + 80 = 200

Total defective parts = 6 + 8 = 14

$$\%\text{ of defective parts}=\left(\frac{14}{200}\times 100\right)\%=\mathbf{7\%}.$$

11. (c) Let the number be x. Then, an increase by 25.3% means

$$x+25.3\%\text{ of }x=x+\frac{25.3}{100}\times x=\frac{125.3}{100}x=1.253$$

$\therefore$ Required multiplier = **1.253**.

12. (d) If B earns ₹ x, then A earns ₹ (x + 12.5% of x)

$$=\frac{112.5}{100}x=1.125x$$

$$\Rightarrow x+1.125x=35700$$

$$\Rightarrow 2.125x=35700\ \Rightarrow\ x=\frac{35700}{2.125}=₹\,16800$$

$\therefore$ A's earning = ₹ 16800 × 1.125 = **₹ 18900**.

13. (d) 6 litres of sugar solution contains 4% of sugar, *i.e.*, $\frac{4}{100}\times 6$ litres of sugar or 0.24 litres of sugar and 5.76 litres of water.

If one litre of water is evaporated, then there is 0.24 litre of sugar in 5 litres of water.

$\therefore$ Percentage of sugar in the solution

$$=\left(\frac{0.24}{5}\times 100\right)\%=\mathbf{4.8\%}.$$

14. (b) Quantity of water in the mixture = 4 litres

Suppose x litres of water is added to the mixture. Then, x + 4 = 20% of (40 + x)

$\Rightarrow 5x + 20 = 40 + x \Rightarrow 4x = 20 \Rightarrow x = 5$

$\therefore$ Water to be added = **5 litres**.

15. (d) Per cent of votes cast = 100% – 8% = 92%

Winner got 48% of the total votes

$\Rightarrow$ Loser got (92 – 48)%, *i.e.*, 44% of the total votes.

Difference = 4%

Given, 4% of total votes cast = 1100

$$\Rightarrow \text{Total votes cast}=\frac{1100\times 100}{4}=\mathbf{27500}.$$

16. (a) $p = 6q$

$\therefore$ Required per cent

$$= \left(\frac{p-q}{p}\right)\times 100\% = \left(\frac{6q-q}{6q}\right)\times 100\%$$

$$= \left(\frac{5q}{6q}\times 100\right)\% = \frac{250}{3}\% = \mathbf{83\frac{1}{3}\%}.$$

17. (c) Suppose the number of males in the village before increase was x.

Then, the number of females = $(9800 - x)$

Given, $(x + 8\% \text{ of } x) + (9800 - x) + 5\%$ of $(9800 - x) = 10458$

$\Rightarrow 1.08x + 1.05 \times (9800 - x) = 10458$

$\Rightarrow 1.08x + 1.05 \times 9800 - 1.05x = 10458$

$\Rightarrow 0.03x = 10458 - 10290 = 168$

$$\Rightarrow x = \frac{168\times 100}{3} = \mathbf{5600}.$$

18. (c) Let Bhushan's monthly income be ₹ x. Then,

$$\text{Anam's monthly income} = x - \frac{75}{100}x = x - \frac{3}{4}x = \frac{1}{4}x$$

$$\text{Prasad's monthly income} = \frac{1}{4}x + \frac{25}{100}\times\frac{1}{4}x$$

$$= \frac{1}{4}x + \frac{1}{16}x = \frac{5}{16}x$$

$$\text{Given, } x - \frac{5}{16}x = 26125 \Rightarrow \frac{11x}{16} = 26125$$

$$\Rightarrow x = \frac{26125\times 16}{11} = 38000$$

$$\therefore \text{ Anam's monthly income} = \frac{1}{4}\times ₹\ 38000$$

$$= \mathbf{₹\ 9500}.$$

19. (a) $A = B + 40\% \text{ of } B = B + \frac{40}{100} \text{ of } B$

$$= B + \frac{2}{5}B = \frac{7}{5}B$$

$$B = C - 20\% \text{ of } C = C - \frac{20}{100} \text{ of } C = C - \frac{1}{5}C$$

$$= \frac{4}{5}C$$

$$\therefore A = \frac{7B}{5} = \frac{7}{5}\times\frac{4}{5}C = \frac{28}{25}C$$

$$\Rightarrow \frac{A}{C} = \frac{28}{25} \Rightarrow A : C = \mathbf{28 : 25}.$$

20. (d) $x = \frac{90}{100}y \Rightarrow y = \frac{100}{90}x$

$$\Rightarrow \frac{1}{100}\times\frac{10000}{90}x = \frac{1000}{9}\% \text{ of } x = \mathbf{111\frac{1}{9}\%} \text{ of } x.$$

21. (c) Let Ramesh's salary = ₹ x. Then,

Salary credited in fixed deposit plan

$$= 15\% \text{ of } ₹\ x = \frac{3}{20}x$$

$$\text{Balance} = x - \frac{3}{20}x = \frac{17}{20}x$$

$$\text{Expenditure on groceries} = 30\% \text{ of } \frac{17}{20}x = \frac{51}{200}x$$

$$\text{Balance} = \frac{17x}{20} - \frac{51x}{200} = \frac{119x}{200}$$

$$\text{Given, } \frac{119x}{200} = 2380$$

$$\Rightarrow x = \frac{2380\times 200}{119} = \mathbf{₹\ 4000}.$$

22. (a) Let the total number of voters = x

Promised voters for $P = \frac{2}{5}$ of $x = 0.4\,x$

Promised voters for $Q = 0.6\,x$

Numbers of voters who voted for

$P = 0.75 \times 0.4x + 0.15 \times 0.6x$

$= 0.3x + 0.09x = 0.39x$

Number of voters who voted for

$Q = 0.85 \times 0.6x + 0.25 \times 0.4x$

$= 0.51x + 0.1x = 0.61x$

Given, $0.61x - 0.39x = 44$

$$\Rightarrow 0.22x = 44 \Rightarrow x = \frac{44}{0.22} = \frac{4400}{22} = \mathbf{200}.$$

23. (d) Price per pen (in a box of 6) = ₹ $\frac{30}{6}$ = ₹ 5

Price per pen (in a box of 12) = ₹ $\frac{48}{12}$ = ₹ 4

$$\therefore \text{ Required per cent} = \left(\frac{5-4}{5}\right)\times 100\%$$

$$= \mathbf{20\%}.$$

24. (b) Let the original number of eggs be x.

Then, broken eggs = 5% of x

$$\Rightarrow \text{Eggs not broken} = \frac{95x}{100} = 0.95x$$

$$\text{Eggs sold} = 93\% \text{ of } \frac{95x}{100} = \frac{93}{100}\times\frac{95}{100}x = 0.8835x$$

$\therefore$ Remainder $= 0.95x - 0.8835x = 0.0665x$

Given, $0.0665x = 266 \Rightarrow x = \dfrac{266}{0.0665} = \mathbf{4000}$.

25. (d) Distance climbed by the spider in one hour

$= 62.5\%$ of 192 m $= \dfrac{62.5}{100} \times 192$ m $= 120$ m

$\therefore$ Remaining distance = 192 m – 120 m = 72 m

$\therefore$ Distance climbed in the second hour

$= 12.5\%$ of 72 m $= \dfrac{12.5}{100} \times 72$ m

$= \dfrac{125}{1000} \times 72$ m $=$ **9 m.**

26. (a) Total votes = 6000

Invalid votes = 25% of 6000 = 0.25×6000 = 1500

$\therefore$ Valid votes = 6000 – 1500 = 4500

Bhiku gets 65% of the valid votes

$\Rightarrow$ Mhatre gets 35% of the valid votes.

$\therefore$ Votes got by Mhatre = 35% of 4500

$= \dfrac{35}{100} \times 4500 = \mathbf{1575}.$

27. (a) Let the income of the person be ₹ x.

Then, his savings = 20% of $x = \dfrac{20}{100} \times x = 0.2x$

Income next year $= x + 10\%$ of $x = \dfrac{110}{100}x = 1.1x$

$\therefore$ Savings next year $= 0.2 \times 1.1\,x = 0.22\,x$

Increase in savings $= 0.22x - 0.2x = 0.02x$

Per cent increase in savings $= \dfrac{0.02x}{0.2x} \times 100$

$= \dfrac{2}{20} \times 100 = \mathbf{10\%}.$

28. (a) Let the total number of days $= x$

$\therefore$ Number of rainy days $= \dfrac{x}{10}$

Number of days with rainbows $= \dfrac{1}{2} \times \dfrac{x}{10} = \dfrac{x}{20}$

Number of days without rainbows $= x - \dfrac{x}{20} = \dfrac{19x}{20}$

$\therefore$ Per cent of all days with no rainbows $= \dfrac{\frac{19x}{20}}{x} \times 100\% = \mathbf{95\%}.$

29. (b) Let each side of the cube be 2 cm.

It is cut from the middle. Then, we have two cuboids of dimensions $2 \times 2 \times 1$. One face in each is not coloured.

Area of this face $= 2 \times 2 = 4$

Total surface area of one cuboid

$= 2(2 \times 2 + 2 \times 1 + 2 \times 1)$

$= 16$

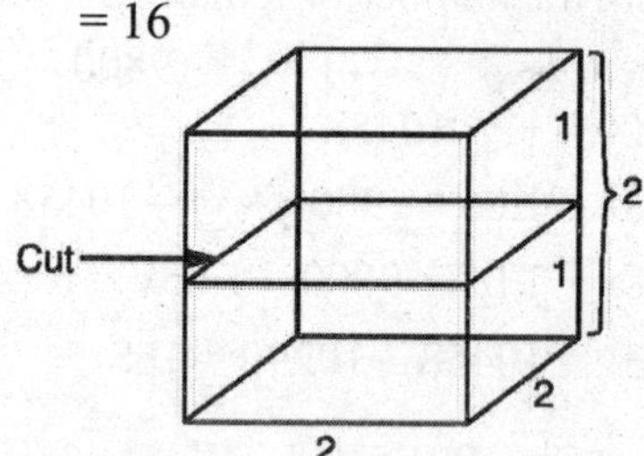

$\therefore$ Per cent of area not painted $= \dfrac{4}{16} \times 100 = \mathbf{25\%}.$

30. (d) Weight of water in the 20 kg water-melon

$= \dfrac{96 \times 20}{100} = 19.2$ kg

Weight of water-melon pulp = 20 kg – 19.2 kg = 0.8 kg

Let the reduced weight of the water-melon be x kg. Then,

Weight of water in the water-melon after evaporation = 95%

$\Rightarrow$ Weight of pulp = 5%

$\therefore$ 5% of $x = 0.8$ kg $\Rightarrow x = \dfrac{0.8 \text{ kg} \times 100}{5} = 16$ kg.

31. (a) Let the cost of the two houses last year be ₹ $4x$ and ₹ $5x$. Then,

$\dfrac{4x + 25\% \text{ of } 4x}{5x + 50{,}000} = \dfrac{9}{10} \Rightarrow \dfrac{1.25 \times 4x}{5x + 50000} = \dfrac{9}{10}$

$\Rightarrow 50x = 45x + 9 \times 50000 \Rightarrow 5x = 9 \times 50000$

$\Rightarrow x = 90{,}000$

$\therefore$ A's price last year $= 4 \times$ ₹ 90,000 = **₹ 3,60,000.**

32. (d) Suppose there were x passengers at station A. Then,

$x - \dfrac{x}{11} + 20\%$ of $\dfrac{x}{11} = 510$

$\Rightarrow \dfrac{10x}{11} + \dfrac{x}{55} = 510 \Rightarrow \dfrac{51x}{55} = 510 \Rightarrow x = \mathbf{550}.$

33. (c) Let the required per cent of questions be x. Then,

$\Rightarrow$ 65% of 40 + x% of 40 = 75% of 80

$$\Rightarrow \frac{65}{100}\times 40+\frac{x}{100}\times 40=\frac{75}{100}\times 80$$

$$\Rightarrow x=\left(\frac{75\times 80}{100}-\frac{65\times 40}{100}\right)\times\frac{100}{40}\%$$

$$=(60-26)\times\frac{100}{40}\%=\mathbf{85\%}.$$

34. (d) $y=125+10\%\text{ of }125=\frac{110}{100}\times 125=137.5$

$$x=137.5-10\%\text{ of }137.5=\frac{90}{100}\times 137.5=\mathbf{123.75}$$

35. (c) Let the original dimensions of the rectangle be x and y units.

Original area = xy sq. units.

Length shown = (x + 5% of x) units

$$=\frac{105}{100}x\text{ units}=\frac{21}{20}x\text{ units}$$

Breadth shown = (y + 3% of y) units = $\frac{103}{100}y$ units

$$\therefore\ \text{Calculated area}=\left(\frac{21x}{20}\times\frac{103}{100}y\right)\text{ sq units}$$

$$=\frac{2163}{2000}xy\text{ sq units}$$

$$\text{Error}=\frac{2163xy}{2000}-xy=\frac{163xy}{2000}$$

$$\%\text{ error}=\frac{\frac{163xy}{2000}}{xy}\times 100=\frac{163}{20}\%=\mathbf{8.15\%}.$$

Self Assessment Sheet–10

1. A number is mistakely divided by 5 instead of being multiplied by 5. Find the percentage change in the result due to this mistake

(a) 96% (b) 95%
(c) 2400% (d) 200%
(e) None of these

2. What is 20% of 50% of 75% of 80?

(a) 5.5 (b) 6
(c) 8 (d) 6.5

3. 25% of a number is less than 18% of 650 by 19. The number is

(a) 380.8 (b) 392
(c) 450 (d) 544

4. In an examination, A got 10% marks less than B. B got 25% marks more than C. C got 20% marks less than D. If A got 360 marks out of 500, the percentage of marks obtained by D was

(a) 70 (b) 75
(c) 80 (d) 85

5. If A's salary is 30% more than that of B, then how much per cent is B's salary less than that of A's?

(a) 30% (b) 25%
(c) $23\frac{1}{3}\%$ (d) $33\frac{1}{3}\%$

6. If a exceeds b by x %, then which one of these equations is correct ?

(a) $a-b=\frac{x}{100}$ (b) $b=a+100x$
(c) $a=\frac{bx}{100+x}$ (d) $a=b+\frac{bx}{100}$

7. In an examination, 35% of the total students failed in Hindi, 45% failed in English and 20% in both. The percentage of those who passed in both the subjects is :

(a) 20 (b) 30
(c) 40 (d) 80

8. Half of 1 per cent written as a decimal is :

(a) 0.2 (b) 0.02
(c) 0.05 (d) 0.005

9. Salary of a person is first increased by 20%, then it is decreased by 20%. Change in his salary is :

(a) 4% decrease (b) 4% increase
(c) 8 % decrease (d) 20 increase

10. In an examination, there are three papers and a candidate has to get 35% of the total to pass. In one paper, he gets 62 out of 150 and in the second 35 out of 150. How much must he get out of 180 in the third paper to just qualify to pass ?

(a) 65 (b) 62.5
(c) 72 (d) None of these

Answers

1. (c)	2. (b)	3. (b)	4. (c)	5. (c)	6. (d)	7. (c)	8. (d)	9. (a)	10. (d)

Chapter 11

PROFIT AND LOSS AND DISCOUNT

Section-A

PROFIT AND LOSS

KEY FACTS

1. (i) Profit = S.P. – C.P. (if S.P. > C.P.) (ii) Loss = C.P. – S.P. (if S.P. < C.P.)
   (iii) Overhead charges such as transportation, toll tax, repairs etc., are included in the C.P.
2. (i) Profit % = $\frac{\text{Profit}}{\text{C.P.}} \times 100$ (ii) Loss % = $\frac{\text{Loss}}{\text{C.P.}} \times 100$ (iii) S.P. = $\frac{\text{C.P.} \times (100 + \text{Profit\%})}{100}$

   or S.P. = $\frac{\text{C.P.} \times (100 - \text{Loss\%})}{100}$ (iv) C.P. = $\frac{\text{S.P.} \times 100}{(100 + \text{Profit\%})}$ or C.P. = $\frac{\text{S.P.} \times 100}{(100 - \text{Loss\%})}$

Solved Examples

Ex. 1. ***A person incurs 5% loss by selling a watch for ₹ 1140. At what price should the watch be sold to earn 5% profit?***

Sol. S.P. = ₹ 1140, Loss = 5%

$\therefore$ C.P. = ₹ $\frac{1140 \times 100}{(100 - 5)}$ = ₹ $\frac{1140 \times 100}{95}$ = ₹ 1200

New, C.P. = ₹ 1200, Profit = 5%

$\therefore$ New S.P. = ₹ $\frac{1200 \times 105}{100}$ = **₹ 1260.**

Ex. 2. ***A vendor bought lemons at the rate of 6 for a rupee. How many for a rupee must he sell to gain 20%?***

Sol. C.P. of 6 lemons = ₹ 1, Gain = 20%

$\therefore$ S.P. of 6 lemons = $\frac{1 \times 120}{100}$ = ₹ 1.20, ₹ 1.20 is the S.P. of 6 lemons

$\therefore$ ₹ 1 is the S.P. of $\frac{6}{1.2}$ = **5 lemons.**

Ex. 3. ***If 3 articles are sold for the cost price of 4, then what is the profit percentage?***

Sol. Let the C.P. of 1 article = ₹ 1. Then, S.P. of 3 articles = C.P. of 4 articles = ₹ 4

C.P. of 3 articles = ₹ 3

Profit = ₹ 4 – ₹ 3 = ₹ 1

Profit % = $\frac{1}{3} \times 100 = \mathbf{33\frac{1}{3}\%}$.

Ex. 4. ***A shopkeeper gains 20% while buying the goods and 30% while selling them. Find his total gain percentage.***

Sol. Let the C.P. of the goods be ₹ 100. As the shopkeeper gains 20% at the time of buying, the shopkeeper buys the articles worth ₹ 120.

$$\therefore \quad \text{S.P.} = ₹\ \frac{120 \times (100+30)}{100} = ₹\ \frac{120 \times 130}{100} = ₹\ 156$$

$$\therefore \ \% \text{ gain} = \frac{(156-100)}{100} \times 100 = \mathbf{56\%}.$$

Ex. 5. ***A man sells an article at a gain of 10%. If he had bought it at 10% less and sold it for ₹ 132 less, he would have still gained 10%. What is the cost price of the article?***

Sol. Let the C.P. = ₹ x. Gain = 10%

$$\text{S.P.} = ₹\ \frac{110x}{100}, \qquad \text{New C.P.} = ₹\ \frac{90x}{100} = ₹\ \frac{9x}{10}$$

$$\text{New S.P.} = 110\% \text{ of } ₹\ \frac{9x}{10} = ₹\ \frac{99x}{100}, \quad \frac{11x}{10} - \frac{99x}{100} = 132 \Rightarrow \frac{11x}{100} = 132$$

$$\Rightarrow \quad x = \frac{132 \times 100}{11} = ₹\ \mathbf{1200}.$$

Ex. 6. ***A shopkeeper sells two T.V. sets at the same price. There is a gain of 20% on one T.V. and loss of 20% on the other. State which of the following statements is correct.***

(a) The shopkeeper neither gains nor loses.
(b) The shopkeeper loses by 2%.
(c) The shopkeeper gains by 4%.
(d) The shopkeeper loses by 4%.

Sol. Let the S.P. of both the T.V. sets be ₹ x. Then,

$$\text{C.P. of the T.V. set at a gain of 20\%} = ₹\ \frac{100 \times x}{120} = ₹\ \frac{5x}{6}$$

$$\text{C.P. of the T.V. set at a loss of 20\%} = ₹\ \frac{100 \times x}{80} = ₹\ \frac{5x}{4}$$

$$\text{Total C.P.} = ₹\left(\frac{5x}{6} + \frac{5x}{4}\right) = ₹\ \frac{10x+15x}{12} = ₹\ \frac{25x}{12}$$

$$\text{Total S.P.} = ₹\ 2x$$

$$\text{Loss} = ₹\left(\frac{25x}{12} - 2x\right) = ₹\ \frac{x}{12}$$

$$\text{Loss \%} = \frac{x/12}{25x/12} \times 100 = \mathbf{4\%}$$

∴ Statement (*d*) is correct.

Ex. 7. ***A book seller bought 200 textbooks for ₹ 12000. He wanted to sell them at a profit so that he gets 20 books free. At what profit per cent should he sell them?***

Sol. C.P. of 200 textbooks = ₹ 12000

Profit = C.P. of 20 textbooks

∴ S.P. of 200 textbooks = C.P. of 200 textbooks + C.P. of 20 textbooks = C.P. of 220 textbooks

$$\text{C.P. of 1 textbook} = ₹\ \frac{12000}{200} = ₹\ 60$$

∴ S.P. of 200 textbooks = 220 × ₹ 60 = ₹ 13200

$$\therefore \ \text{Profit \%} = \frac{(13200-12000)}{12000} \times 100 = \frac{1200}{12000} \times 100 = \mathbf{10\%}.$$

Ex. 8. ***A owned an article worth ₹ 10,000. He sold it to B at a profit of 10% based on the worth of the article. B sold the article back to A at a loss of 10%. How much did A make in these transactions?***

Sol. Selling price of the article by A to B = ₹ $\frac{110}{100}\times 10000$ = ₹ 11000

Selling price of the article by B to A = ₹ $\frac{90}{100}\times 11000$ = ₹ 9900

∴ Profit earned by A = ₹ (11000 – 9000) = **₹ 1100.**

Ex. 9. ***I bought two tables for ₹ 5400. I sold one at a loss of 5% and the other at a gain of 7%. In the whole transaction, I neither gained nor lost. Find the costs of the tables separately.***

Sol. Let the C.P. of one table = ₹ x Then,

C.P. of the other table = ₹ $(5400 - x)$

S.P. of the first table at a loss of 5% = ₹ $\frac{95}{100}\times x$

S.P. of the other table at a gain of 7% = ₹ $\frac{107}{100}\times(5400 - x)$

Since there is no gain no loss.

Total S.P. = Total C.P. $\Rightarrow \frac{95x}{100}+\frac{107\times(5400-x)}{100}=5400 \Rightarrow \frac{95x}{100}-\frac{107x}{100}+\frac{107\times 5400}{100}=5400$

$\Rightarrow \frac{-12x}{100}=\frac{5400\times 100-5400\times 107}{100} = 5400\times -7 \Rightarrow -12x = 5400\times -7$

$\Rightarrow x=\frac{5400\times 7}{12}$ = ₹ 3150

∴ Cost of the two tables are ₹ 3150 and ₹ 2250.

Ex. 10. ***Ram mixes 15 kg of sugar purchased at the rate of ₹ 8.00 per kg with 25 kg of sugar purchased at the rate of ₹ 10.00 per kg. At what rate per kg should Ram sell the mixture to get a profit of 3 per kg?***

Sol. C.P. of 40 kg of sugar = ₹ (15 × 8 + 25 × 10) = ₹ 120 + ₹ 250 = ₹ 370

∴ C.P. of 1 kg of sugar = ₹ $\frac{370}{40}$ = ₹ 9.25

∴ S.P. of 1 kg of sugar = ₹ 9.25 + ₹ 3 = **₹ 12.25.**

Question Bank–11(a)

1. 100 oranges are bought for ₹ 350 and sold at ₹ 48 per dozen. The percentage of profit or loss is

(a) 15% loss (b) 15% gain

(c) $14\frac{2}{7}$% loss (d) $14\frac{2}{7}$% profit

2. If Ashok makes a profit of 25% on the selling price, what will be his profit on C.P.?

(a) 40% (b) 20%

(c) $33\frac{1}{3}$% (d) 25%

3. A person sells 320 mangoes at the cost price of 400 mangoes. What is his profit per cent?

(a) 10% (b) 15%

(c) 20% (d) 25%

4. A shopkeeper sells an article at a loss of $12\frac{1}{2}$%. Had he sold it for ₹ 51.80 more than he would have earned a profit of 6%. The cost price of the article is

(a) ₹ 280 (b) ₹ 300

(c) ₹ 380 (d) ₹ 400

5. A dealer buys an old cooler listed at ₹ 950 and gets a discount of 10%. He spends ₹ 45 for its repair. If he sells the cooler at a profit of 25%, then the selling price of the cooler is

(a) ₹ 1125 (b) ₹ 1215

(c) ₹ 1251 (d) ₹ 1512

6. A merchant has 1000 kg of sugar, part of which he sells at 8% profit and the rest at 18% profit. He gains 14% on the whole. The quantity of sugar sold at 18% profit is
(a) 560 kg (b) 600 kg
(c) 400 kg (d) 640 kg

7. The cost price of 20 articles is the same as selling price of x articles. If the profit is 25%, then the value of x is
(a) 15 (b) 16
(c) 18 (d) 25

8. An article passing through two hands is sold at a profit of 38% at the original cost price. If the dealer makes a profit of 20%, then the profit per cent made by the second is
(a) 15 (b) 12
(c) 10 (d) 5

9. A person bought some articles at the rate of 5 per rupee and the some number at the rate of 4 per rupee. He mixed both the types and sold at the rate of ₹ 9 for 2. In this business he suffered a loss of ₹ 3. The total number of articles bought by him was
(a) 1090 (b) 1080
(c) 540 (d) 545

10. A person sells a table at a profit of 20%. If he had bought it at 10% less cost and sold for ₹ 105 more, he would have gained 35%. The cost price of the table is
(a) ₹ 7000 (b) ₹ 7350
(c) ₹ 9200 (d) ₹ 8140

11. The percentage of loss when an article is sold at ₹ 50 is the same as that of the profit when it is sold at ₹ 70. The above mentioned percentage of profit or loss on the article is
(a) 10% (b) 12%
(c) $16\frac{2}{3}\%$ (d) $8\frac{1}{3}\%$

12. In what ratio must a grocer mix teas at ₹ 60 a kg and ₹ 65 a kg, so that by selling the mixture at ₹ 68.20 a kg, he may gain 10%.
(a) 3 : 2 (b) 3 : 4
(c) 3 : 5 (d) 4 : 5

13. A tradesman sold an article at a loss of 20%. If the selling price had been increased by ₹ 100, there would have been a gain of 5%. The cost price of the article was
(a) ₹ 200 (b) ₹ 225
(c) ₹ 400 (d) ₹ 250

14. A man bought two goats for ₹ 1008. He sold one at a loss of 20% and the other a profit of 44%. If each goat was sold for the same price, the cost price of the goat which was sold at a loss was
(a) ₹ 648 (b) ₹ 360
(c) ₹ 568 (d) ₹ 440

15. Two-thirds of a consignment was sold at a profit of 6% and the rest at a loss of 3%. If there was an overall profit of ₹ 540, the value of the consignment was
(a) ₹ 15000 (b) ₹ 16000
(c) ₹ 18000 (d) ₹ 16500

16. A man sells two horses for ₹ 1485. The cost price of the first is equal to the selling price of the second. If the first is sold at 20% loss and the second at 25% gain, what is his total gain or loss (in rupees)?
(a) ₹ 80 gain
(b) ₹ 60 gain
(c) ₹ 60 loss
(d) Neither gain nor loss

17. A trader purchases a watch and a wall clock for ₹ 390. He sells them making a profit of 10% on the watch and 15% on the wall clock. He earns a profit of ₹ 51.50. The difference between the original prices of the wall clock and the watch is equal to
(a) ₹ 110 (b) ₹ 100
(c) ₹ 80 (d) ₹ 120

18. The profit earned by selling an article for ₹ 900 is double the loss incurred when the same article is sold for ₹ 450. At what price should the article be sold to make 25% profit?
(a) ₹ 750 (b) ₹ 800
(c) ₹ 600 (d) ₹ 900

19. A person *A* sells a table costing ₹ 2000 to a person *B* and earns a profit of 6%. The person *B* sells it to another person *C* at a loss of 5%. At what price did *B* sell the table?
(a) ₹ 2054 (b) ₹ 2050
(c) ₹ 2024 (d) ₹ 2014

20. Sanjay sold a bicycle to Salman at 46% profit. Salman spent ₹ 40 on repairs and sold it to Sunil for ₹ 1500. In this deal, Salman made neither profit nor loss. What is the cost price for Sanjay?
(a) ₹ 900 (b) ₹ 960
(c) ₹ 1000 (d) ₹ 1060

21. In a certain store, the profit is 320% of the cost. If the cost increases by 25%, but the selling price remains constant, approximately what percentage of the selling price is the profit?
(a) 30% (b) 70%
(c) 100% (d) 250%

22. A main gains 10% by selling a certain article for a certain price. If he sells it at double the price then the profit made is

(a) 120% (b) 60%
(c) 100% (d) 80%

23. If the selling price is doubled, the profit triples. Find the profit per cent.

(a) $66\frac{2}{3}$ (b) 100
(c) $105\frac{1}{3}$ (d) 120

24. A cycle agent buys 30 bicycles, of which 8 are first grade and the rest are second grade, for ₹ 3150. Find at what price he must sell the first grade bicycles so that if he sells the second grade bicycles at three quarters of the price, he may make a profit of 40% on his out lay?

(a) ₹ 200 (b) ₹ 240
(c) ₹ 180 (d) ₹ 210

25. I purchase 120 exercise books at the rate of ₹ 3 each and sold $\frac{1}{3}$rd of them at the rate of ₹ 4 each, $\frac{1}{2}$ of them at the rate of ₹ 5 each and the rest at the cost price. My profit percentage was

(a) 44% (b) $44\frac{4}{9}\%$
(c) $44\frac{2}{3}\%$ (d) 45%

26. A man bought two old scooters for ₹ 18000. By selling one at a profit of 25% and the other at a loss of 20%, he neither gains nor loses. Find the cost price of each scooter.

(a) 12000, 6000 (b) 8000, 10000
(c) 11000, 7000 (d) 13000, 5000

27. Mohan bought 20 dining tables for ₹ 12000 and sold them at a profit equal to the selling price of 4 dining tables. The selling price of each dining table is

(a) ₹ 700 (b) ₹ 750
(c) ₹ 725 (d) ₹ 775

28. Veenu purchased two types of wheat, one costing ₹ 20 per kg and the other ₹ 25 per kg in quantities having ratio 5:4. He mixed the two and sold the mixture at ₹ 24 per kg. In the transaction he got

(a) 2% loss (b) 2% gain
(c) 8% gain (d) 8% loss

29. By selling a transistor for ₹ 572, a shopkeeper earns a profit equivalent to 30% of the cost price of the transistor. What is the cost price of the transistor?

(a) ₹ 340 (b) ₹ 400
(c) ₹ 440 (d) ₹ 350

30. A man bought goods worth ₹ 6000 and sold half of them at a gain of 10%. At what gain per cent must he sell the remainder to get a gain of 25% on the whole?

(a) 40% (b) 30%
(c) 25% (d) 20%

Answers

1. (d)	**2.** (c)	**3.** (d)	**4.** (a)	**5.** (a)	**6.** (b)	**7.** (b)	**8.** (a)	**9.** (b)	**10.** (a)
11. (c)	**12.** (a)	**13.** (c)	**14.** (a)	**15.** (c)	**16.** (d)	**17.** (a)	**18.** (a)	**19.** (d)	**20.** (c)
21. (b)	**22.** (a)	**23.** (b)	**24.** (c)	**25.** (b)	**26.** (b)	**27.** (b)	**28.** (c)	**29.** (c)	**30.** (a)

Hints and Solutions

1. (d) C.P. of 100 oranges = ₹ 350

S.P of 100 oranges = ₹ $\frac{48}{12}\times 100$ = ₹ 400

$\therefore$ Profit % = $\frac{50}{350}\times 100 = \frac{100}{7} = \mathbf{14\frac{2}{7}\%}$.

2. (c) Let S.P. = x

$\therefore$ Profit = 25% of $x = \frac{x}{4}$

$\Rightarrow$ C.P. = $x - \frac{x}{4} = \frac{3x}{4}$

Profit per cent on C.P. = $\left(\frac{x}{4}\div\frac{3x}{4}\right)\times 100$

$= \frac{x}{4}\times\frac{4}{3x}\times 100 = \mathbf{33\frac{1}{3}\%}$.

3. (d) Let the C.P. of 1 mango = ₹ 1. Then,

S.P. of 320 mangoes = C.P. of 400 mangoes = ₹ 400

C.P. of 320 mangoes = ₹ 320

$\therefore$ Profit % = $\frac{80}{320}\times 100 = \mathbf{25\%}$.

4. (a) Let the C.P. of the article be ₹ x. Then,

S.P. at 6% profit – S.P. at $12\frac{1}{2}\%$ loss = ₹ 51.80

$$\Rightarrow \frac{x\times(100+6)}{100} - \frac{x\times(100-12.5)}{100} = 51.80$$

$\Rightarrow \dfrac{106x}{100} - \dfrac{87.5x}{100} = 51.80$

$\Rightarrow \dfrac{18.5x}{100} = 51.80 \Rightarrow x = \dfrac{51.80 \times 100}{18.5} = \mathbf{280}.$

5. (a) C.P. = ₹ 950 – 10% of ₹ 950 = 90% of ₹ 950 = $\dfrac{90}{100} \times 950$ = ₹ 855

Repairs cost = ₹ 45

∴ Net C.P. = ₹ 855 + ₹ 45 = ₹ 900, Profit = 25%

∴ S.P. = ₹$\left(\dfrac{900 \times 125}{100}\right)$ = ₹ **1125.**

6. (b) Let the C.P. of sugar be ₹ x per kg.

Then, C.P. = ₹ 1000 x

Let the sugar sold at 8% gain by y kg.

Then, sugar sold at 18% gain = $(1000 - y)$kg

∴ $\left(\dfrac{108}{100} \times xy\right) + \left[\dfrac{118}{100} \times x(1000 - y)\right] = \dfrac{114}{100} \times 1000x$

$\Rightarrow \dfrac{108y}{100} + 1180 - \dfrac{118y}{100} = 1140$

$\Rightarrow \dfrac{10y}{100} = 40 \Rightarrow y = 400$

∴ Quantity sold at 18% gain = (1000 – 400)kg = **600 kg.**

7. (b) Let the C.P. of 1 article = ₹ 1. Then,

S.P. of x articles = C.P. of 20 articles = ₹ 20

C.P. of x articles = ₹ x

Profit = ₹$(20 - x)$. Profit % = 25

$\Rightarrow \dfrac{(20 - x)}{x} \times 100 = 25 \Rightarrow 2000 - 100x = 25x$

$\Rightarrow 125x = 2000 \Rightarrow x = \mathbf{16}.$

8. (a) Let the initial cost price of the article be ₹ 100, Profit = 20%

S.P. of the 1st dealer = ₹ $\dfrac{100 \times 120}{100}$ = ₹ 120

Let the second profit % = x. Then,

S.P. of the 2nd dealer = ₹ $\dfrac{120 \times (100 + x)}{100}$

Final S.P. of the article after passing through two hands at a profit = 38%

$= \dfrac{100 \times (100 + 38)}{100}$ = ₹ 138

Given, $\dfrac{120 \times (100 + x)}{100} = 138$

$\Rightarrow 12000 + 120x = 13800$

$\Rightarrow 120x = 1800 \Rightarrow x = \mathbf{15\%}.$

9. (b) Suppose he purchased x articles of each type.

Then, Total C.P. = $\dfrac{x}{5} + \dfrac{x}{4}$ = ₹ $\dfrac{9x}{20}$

Total S.P. = $\dfrac{2}{9} \times 2x$ = ₹ $\dfrac{4x}{9}$

Total loss = $\dfrac{9x}{20} - \dfrac{4x}{9}$

$\Rightarrow \dfrac{81x - 80x}{180} = 3 \Rightarrow x = 3 \times 180 = 540$

∴ Total number of articles = 2 × 540 = **1080.**

10. (a) Let the C.P. of the table be ₹ x.

Then, S.P. of the table at 20% profit

$= \dfrac{120}{100} \times$ ₹ x = ₹ $\dfrac{6x}{5}$

C.P. of the table at 10% loss = $\dfrac{90}{100} \times$ ₹ x = ₹ $\dfrac{9x}{10}$

S.P. of the table at 35% profit now

$= \dfrac{135}{100} \times$ ₹ $\dfrac{9x}{10}$

Given, $\dfrac{6x}{5} + 105 = \dfrac{135}{100} \times \dfrac{9x}{10}$

$\Rightarrow \dfrac{6x}{5} + 105 = \dfrac{243}{200}x \Rightarrow \dfrac{243}{200}x - \dfrac{6}{5}x = 105$

$\Rightarrow \left(\dfrac{243 - 240}{200}\right)x = 105$

$\Rightarrow x = \dfrac{105 \times 200}{3}$ = ₹ 7000.

11. (c) Let the percentage of profit or loss on the article be x.

C.P. at x% loss when S.P. is ₹ 50 = ₹ $\dfrac{50 \times 100}{(100 - x)}$

C.P. at x% profit when S.P. is ₹ 70 = ₹ $\dfrac{70 \times 100}{(100 + x)}$

∴ $\dfrac{50 \times 100}{(100 - x)} = \dfrac{70 \times 100}{(100 + x)}$

$\Rightarrow 5000\,(100 + x) = 7000\,(100 - x)$

$\Rightarrow 500 + 5x = 700 - 7x \Rightarrow 12x = 200$

$\Rightarrow x = \dfrac{200}{12} = \mathbf{16\tfrac{2}{3}\%}.$

12. (a) Let the teas be mixed in the ratio $x : y$.
Then, the total C.P. of the tea = ₹ $(60x + 65y)$
The total S.P. of the mixture = ₹ $(x + y) \times 68.20$

Given, $(60x + 65y) \times \frac{110}{100} = 68.2x + 68.2y$

$\Rightarrow 660x + 715y = 682x + 682y$

$\Rightarrow 33y = 22x \Rightarrow 2x = 3y \Rightarrow x : y = \mathbf{3 : 2}$.

13. (c) Let the C.P. of the article be ₹ x. Then,
S.P. of the article at a loss of 20%

$= \frac{80}{100} \times x = ₹ \frac{80x}{100}$

S.P. of the article at a profit of 5%

$= \frac{105}{100} \times x = ₹ \frac{105x}{100}$

According to the question,

$\frac{80x}{100} + 100 = \frac{105x}{100} \Rightarrow \frac{105x}{100} - \frac{80x}{100} = 100$

$\Rightarrow x = \frac{100 \times 100}{25} = ₹ \mathbf{400}$.

14. (a) Let the C.P. of a goat be ₹ x. Then,
C.P. of the other goat = ₹ $(1008 - x)$
∵ S.P. of the both the goats is the same,

$x \times \frac{(100-20)}{100} = (1008 - x) \times \frac{144}{100}$

$\Rightarrow 80x = 1008 \times 144 - 144x$

$\Rightarrow 224x = 1008 \times 144$

$\Rightarrow x = ₹ \frac{1008 \times 144}{224} = ₹ \mathbf{648}$.

15. (c) Let the value (C.P.) of the consignment be ₹ x.

Then, total S.P. = ₹ $\left(\frac{2}{3} \times x \times \frac{106}{100} + \frac{1}{3} \times x \times \frac{97}{100}\right)$

$= ₹ \left(\frac{53x}{75} + \frac{97x}{300}\right) = ₹ \left(\frac{212x + 97x}{300}\right)$

$= ₹ \frac{309x}{300}$

Given, $\frac{309x}{300} - x = 540$

$\Rightarrow \frac{309x - 300x}{300} = 540 \Rightarrow \frac{9x}{300} = 540$

$\Rightarrow x = \frac{540 \times 300}{9} = \mathbf{18000}$.

16. (d) Let the S.P. of the first horse = ₹ x. Then,
S.P. of the second horse = ₹ $(1485 - x)$

C.P. of the first horse = $\frac{x \times 100}{(100-20)} = ₹ \frac{5x}{4}$

C.P. of the second horse = $\frac{(1485 - x) \times 100}{(100 + 25)}$

$= ₹ \frac{4(1485 - x)}{5}$

Given, $\frac{5x}{4} = (1485 - x)$

$\Rightarrow 5x = 4 \times (1485 - x) \Rightarrow 9x = 4 \times 1485$

$\Rightarrow x = 4 \times \frac{1485}{9} = ₹ 660$

$\therefore$ Total C.P. = $\frac{5 \times 660}{4} + \frac{4(1485 - 660)}{5}$

= ₹ 825 + ₹ 660 = ₹ 1485

Since S.P. = C.P., there is no profit no loss.

17. (a) Let the cost of the watch = ₹ x. Then,
Cost of the clock = ₹ $(390 - x)$
Given, 10% of x + 15% of $(390 - x) = 51.50$

$\Rightarrow \frac{x}{10} + \frac{15 \times (390 - x)}{100} = 51.50$

$\Rightarrow 10x + 5850 - 15x = 5150$

$\Rightarrow 5x = 700 \Rightarrow x = 140$

$\therefore$ Watch of clock = ₹ (390 – 140) = ₹ 250

$\therefore$ Required difference = ₹ 250 – ₹ 140 = **₹ 110.**

18. (a) Let the C.P. be ₹ x. Then,
$900 - x = 2(x - 450)$

$\Rightarrow x = 600$

i.e., C.P. = ₹ 600, Profit = 25%

$\therefore$ S.P. of the article = ₹ $\left(\frac{600 \times 125}{100}\right)$ = **₹ 750.**

19. (d) C.P. of A = ₹ 2000, Profit = 6%

$\therefore$ S.P. of $A = \frac{106}{100} \times$ ₹ 2000 = ₹ 2120

$\Rightarrow$ C.P. of B = ₹ 2120, Loss = 5%

$\therefore$ S.P. of $B = \frac{95}{100} \times$ ₹ 2120 = **₹ 2014.**

20. (c) Let the C.P. for Sanjay be ₹ x.
Then, S.P. for Sanjay = x + 46% of x

$= x + \frac{46}{100}x = \frac{146x}{100}$

$\therefore$ C.P. for Salman $= \frac{146x}{100} + 40$

$\because$ Salman made neither profit nor loss.

$\frac{146x}{100} + 40 = 1500$

$\Rightarrow \frac{146x}{100} = 1460 \Rightarrow x = \frac{1460 \times 100}{146} = ₹\ \mathbf{1000}.$

21. (b) Let the original cost price be ₹ 100

Then, profit = ₹ 320 and S.P. = ₹ 420

New C.P. $= \frac{125}{100} \times ₹\ 100 = ₹\ 125$

$\therefore$ New profit = ₹ 420 – ₹ 125 = ₹ 295

Required % $= \frac{295}{420} \times 100 = \mathbf{70\%}.$

22. (a) Let the C.P. = ₹ 100, then S.P. $= \frac{110}{100} \times ₹\ 100$

$= ₹\ 110$

New S.P. = 2 × ₹ 110 = ₹ 220

$\therefore$ Profit = ₹ 220 – ₹ 100 = ₹ 120

Profit % $= \frac{120}{100} \times 100 = \mathbf{120\%}.$

23. (b) Let the C.P. = ₹ x and S.P. = ₹ y.

Then, profit = ₹ $(y - x)$

If S.P. = $2y$, then profit = $3(y - x)$

Given, $2y - x = 3(y - x)$

$\Rightarrow y = 2x \Rightarrow$ S.P. = 2 × C.P. $\Rightarrow$ Profit is 100%.

24. (c) Let the S.P. of one A grade cycle = ₹ n.

Then, the S.P. of one B grade cycle = ₹ $\frac{3n}{4}$

C.P. of 30 cycles = ₹ 3150, profit = 40%.

$\therefore$ S.P. of 30 cycles $= \frac{140}{100} \times ₹3150 = ₹3150 \times 1.4$

Given, $8n + 22 \times \frac{3n}{4} = 3150 \times 1.4$

$\Rightarrow 32n + 66n = 17640 \Rightarrow 98n = 17640 \Rightarrow n = 180$

$\therefore$ He should sell the first grade bicycle at ₹ 180 each.

25. (b) Total C.P. = 120 × ₹ 3 = ₹ 360

Total S.P. $= \frac{1}{3} \times 120 \times \text{Rs } 4 + \frac{1}{2} \times 120 \times \text{Rs } 5 + \left(1 - \frac{1}{3} - \frac{1}{2}\right) \times 120 \times \text{Rs } 3$

= ₹ 160 + ₹ 300 + ₹ 60 = ₹ 520

$\therefore$ Profit % $= \frac{(520 - 360)}{360} \times 100$

$= \frac{160}{360} \times 100\% = \frac{400}{9}\% = \mathbf{44\frac{4}{9}\%}.$

26. (b) Let the C.P. of the 1st scooter = ₹ x. Then, C.P. of the 2nd scooter = ₹ $(18000 - x)$

S.P. of 1st scooter = ₹ $x \times \frac{125}{100} = ₹\ \frac{5x}{4}$

S.P. of 2nd scooter = ₹ $(18000 - x) \times \frac{80}{100}$

$= ₹\ \frac{4}{5}(18000 - x)$

Since Mohan neither gains nor loses, his Total S.P. = Total C.P.

$\Rightarrow \frac{5x}{4} + \frac{4}{5}(18000 - x) = 18000$

$\Rightarrow \frac{5x}{4} - \frac{4x}{5} = 18000 - 14400 = 3600$

$\Rightarrow \frac{25x - 16x}{20} = 3600 \Rightarrow 9x = 3600 \times 20$

$\Rightarrow x = \frac{3600 \times 20}{9} = 8000$

$\therefore$ C.P. of 1st scooter = ₹ 8000

C.P. of 2nd scooter = ₹ 10,000.

27. (b) Profit = (S.P. of 20 tables) – (C.P. of 20 tables)

$\Rightarrow$ S.P. of 4 tables = S.P. of 20 tables – C.P. of 20 tables

$\Rightarrow$ S.P. of 16 tables = C.P. of 20 tables = ₹ 12000

$\Rightarrow$ S.P. of 1 dining table = ₹ $\frac{12000}{16}$ = ₹ **750.**

28. (c) Let the quantities of the two types of wheat be $5x$ kg and $4x$ kg.

Then, C.P. of the two types of wheat

= $5x$ × ₹ 20 + $4x$ × ₹ 25

= ₹ $100x$ + ₹ $100x$ = ₹ $200x$

S.P. of the mixture = $(5x + 4x)$ × ₹ 24

= $9x$ × ₹ 24 = ₹ $216x$

$\therefore$ Profit % $= \frac{(216x - 200x)}{200x} \times 100$

$= \frac{16x}{200x} \times 100 = \mathbf{8\%}.$

29. (c) Let the C.P. of the transistor be ₹ x.

Profit = 30% of ₹ x = ₹ $\frac{3x}{10}$

S.P. = ₹ 572

$\therefore\quad x+\frac{3x}{10}=572 \Rightarrow \frac{13x}{10}=572$

$\Rightarrow\quad x=₹\ \frac{572\times10}{13}=$ **₹ 440.**

30. (a) C.P. of 1st half of goods $=\frac{1}{2}\times ₹\ 6000=₹\ 3000$

S.P. of 1st half of the goods $=\frac{110}{100}\times ₹\ 3000$

$=₹\ 3300$

C.P. of 2nd half of goods = ₹ 3000,

Let profit % = x

Then, S.P. of 2nd half of goods

$=\left(\frac{100+x}{100}\right)\times ₹\ 3000$

Total C.P. = ₹ 6000, Total gain = 25%

$\therefore$ Total S.P. $=\frac{125}{100}\times ₹\ 6000=₹\ 7500$

Given, $3300+30(100+x)=7500$

$\Rightarrow 30(100+x)=4200$

$\Rightarrow 100+x=140$

$\Rightarrow$ $x=$ **40%.**

Section-B
DISCOUNT

KEY FACTS

1. **Marked Price (M.P.):** The printed price or the tagged price of an article is called the **marked price** or **list price**.
2. **Discount:** The deduction allowed on the marked price is called **discount.** *It is always calculated on the M.P. of an article.*
3. **Net Price:** The selling price at which the article is sold to the customer after deducting the discount from the M.P. is called the **net price.**
4. (i) S.P. = M.P. – discount

   (ii) Rate of discount = Discount% $=\frac{\text{Discount}}{\text{M.P.}}\times100$

   (iii) S.P. $=\frac{\text{M.P.}\times(100-\text{Discount\%})}{100}$

   (iv) M.P. $=\frac{\text{S.P.}\times100}{(100-\text{discount\%})}$

Solved Examples

Ex. 1. ***A shopkeeper offers 2.5% discount on cash purchases. What cash amount would Rohan pay for a cycle, the marked price of which is ₹ 650?***

Sol. Marked price = ₹ 650, Discount = 2.5%

$\therefore$ Cash amount paid by Rohan = ₹ 650 – 2.5% of ₹ 650

= ₹ 650 – 0.025 × ₹ 650 = ₹ 650 – ₹ 16.25 = **₹ 633.75.**

Ex. 2. ***A pair of articles was bought for ₹ 37.40 at a discount of 15%. What must be the marked price of each article?***

Sol. S.P. of each article $=₹\left(\frac{37.40}{2}\right)=₹\ 18.70$

Discount rate = 15%

$\therefore$ M.P. of each article $=₹\ \frac{18.70\times100}{(100-15)}=₹\ \frac{1870}{85}=$ **₹ 22.**

Ex. 3. ***Find the selling price of an article if a shopkeeper allows two successive discount of 5% each on the marked price of ₹ 80.***

Sol. Price after 1st discount $= ₹\ 80 - 5\%$ of $₹\ 80 = 95\%$ of $₹\ 80 = 0.95 \times ₹\ 80 = ₹\ 76$

Selling price after 2nd discount $= 95\%$ of $₹\ 76 = 0.95 \times ₹\ 76 =$ **₹ 72.20**.

Ex. 4. ***A dealer buys an article marked at ₹ 25000 with 20% and 5% off. He spends ₹ 1000 on its repairs and sells it for ₹ 25000. What is his gain or loss %?***

Sol. Marked price = ₹ 25000

Price of the article after 1st discount $= \frac{80}{100} \times ₹\ 25000 = ₹\ 20000$

Price of the article after 2nd discount $= \frac{95}{100} \times ₹\ 20000 = ₹\ 19000$

$\Rightarrow$ Net C.P. of the article $= ₹\ 19000 + ₹\ 1000 = ₹\ 20000$

S.P. = ₹ 25000

$\therefore$ Gain % $= \frac{(25000 - 20000)}{20000} \times 100 = \frac{5000}{20000} \times 100 = \mathbf{25\%}$.

Ex. 5. ***The marked price of a watch is ₹ 800. A customer gets two successive discounts on the marked price, the first being 10%. What is the second discount if the customer pays ₹ 612 for it?***

Sol. M.P. = ₹ 800, First discount = 10%

$\therefore$ Net price after the first discount $= 90\%$ of $₹\ 800 = 0.9 \times ₹\ 800 = ₹\ 720$

Final price after the second discount = ₹ 612

$\therefore$ Second discount $= ₹\ 720 - ₹\ 612 = ₹\ 108$

$\Rightarrow$ Second discount rate $= \frac{108}{720} \times 100\% = \mathbf{15\%}$.

Ex. 6. ***A shopkeeper allows a discount of 10% on his goods. For cash payments, he further allows a discount of 20%. Find a single discount equivalent of the above offer.***

Sol. Let the M.P. = ₹ 100,

$\therefore$ S.P. $= 80\%$ of 90% of $₹\ 100 = ₹\left(\frac{80}{100} \times \frac{90}{100} \times 100\right) = ₹\ 72$

$\therefore$ Single discount $= ₹\ 100 - ₹\ 72 = ₹\ 28$

Single discount rate $= \frac{28}{100} \times 100 = \mathbf{28\%}$.

Ex. 7. ***A shopkeeper sells his goods at 10% discount on the marked price. What price should be marked on an article that costs him ₹ 900 to gain 10%?***

Sol. Let the M.P. of the article be ₹ x, Discount = 10%

$\therefore$ S.P. of the article $= 90\%$ of $₹\ x = \frac{90}{100} \times ₹\ x = ₹\ \frac{9x}{10}$ ...(*i*)

Now given C.P. = ₹ 900 and Gain = 10%.

$\therefore$ S.P. $= ₹\ \frac{900 \times 110}{100} = ₹\ 990$ ...(*ii*)

From (*i*) and (*ii*)

$\frac{9x}{10} = 990 \Rightarrow x = ₹\ \frac{990 \times 10}{9} =$ **₹ 1100**.

Ex. 8. *The marked price of an article is ₹ 200. A discount of $12\frac{1}{2}$% is allowed on the marked price and a profit of 25% is made. What is the cost price of the article?*

Sol. M.P. = ₹ 200, Discount = 12.5%

∴ S.P. = ₹ 200 – 12.5% of ₹ 200 = ₹ 200 – ₹ 25 = ₹ 175

Profit = 25%

∴ C.P. = ₹ $\left(\frac{175\times100}{125}\right)$ = **₹ 140.**

Ex. 9. *A dealer marks an item 40% above the cost price and offers a discount of 25% on the marked price. What is his profit percentage?*

Sol. Let the C.P. of the item be ₹ 100

Then, M.P. = ₹ 140, Discount = 25%

∴ S.P. = ₹ 140 – 25% of ₹ 140

= 75% of ₹ 140 = $\frac{75}{100}$ × ₹ 140 = ₹ 105

∴ Profit = ₹ 105 – ₹ 100

Profit % = $\frac{5}{100}\times100=$ **5%.**

Question Bank–11(b)

1. At a clearance sale, all goods are on sale at 35% discount. If I buy a dress marked ₹ 800, how much would I need to pay?

(a) ₹ 280 (b) ₹ 350
(c) ₹ 520 (d) ₹ 600

2. After allowing a discount of 26% on the marked price of an article, it is sold at ₹ 444. Find its marked price.

(a) ₹ 500 (b) ₹ 550
(c) ₹ 640 (d) ₹ 600

3. Find the rate of discount when an article marked at ₹ 1500 is sold at ₹ 1140.

(a) 25% (b) 76%
(c) 24% (d) 20%

4. The marked price of an article is ₹ 500. It is sold at successive discounts of 20% and 10%. The selling price of the article (in ₹) is

(a) 350 (b) 375
(c) 360 (d) 400

5. For the purchase of a motorcar, a man has to pay ₹ 17000 when a single discount of 15% is allowed. How much will he have to pay for it if two successive discounts of 5% and 10% respectively are allowed?

(a) ₹ 17000 (b) ₹ 17010
(c) ₹ 17100 (d) ₹ 18000

6. By giving a discount of 10% on the marked price of ₹ 1100 of a cycle, a dealer gains 10%. The cost price of the cycle is

(a) ₹ 1100 (b) ₹ 900
(c) ₹ 1089 (d) ₹ 891

7. Ashok buys a car at 20% discount of its value and sells it for 20% more than its value. What will be his profit percentage?

(a) 40% (b) 50%
(c) $66\frac{2}{3}$% (d) 20%

8. A dealer buys a table listed at ₹ 1500 and gets successive discounts of 20% and 10%. He spends ₹ 20 on transportation and sells it a profit of 10%. Find the selling price of the table.

(a) ₹ 1800 (b) ₹ 1650
(c) ₹ 1188 (d) ₹ 1210

9. A dealer buys a car listed at ₹ 200000 at successive discounts of 5% and 10%. If he sells the car for ₹ 179550, then his profit is

(a) 10% (b) 9%
(c) 5% (d) 4%

10. A fan is listed at ₹ 1400 and the discount offered is 10%. What additional discount must be given to bring the net selling price to ₹ 1200?
(a) $16\frac{2}{3}\%$ (b) 5%
(c) $4\frac{16}{21}\%$ (d) 6%

11. A trader marks his goods at 40% above the cost price but allows a discount of 20% on the marked price. His profit percentage is
(a) 20% (b) 10%
(c) 8% (d) 12%

12. A shopkeeper buys an article for ₹ 180. After allowing a discount of 10% on his marked price, he wants to earn a profit of 20%. His marked price must be
(a) ₹ 210 (b) ₹ 240
(c) ₹ 270 (d) ₹ 300

13. A pen is listed for ₹ 12. A discount of 15% is given on it. A second discount is given bringing the price down to ₹ 8.16. The rate of second discount is
(a) 15% (b) 18%
(c) 20% (d) 25%

14. A single discount equivalent to successive discounts of 30%, 20% and 10% is
(a) 50% (b) 51%
(c) 49.4% (d) 49.6%

15. A shopkeeper offers his customers 10% discount and still makes a profit of 26%. What is the actual cost of an article for him marked ₹ 280?
(a) ₹ 175 (b) ₹ 200
(c) ₹ 215 (d) ₹ 225

16. An article is listed at ₹ 900 and two successive discounts of 8% and 8% are given on it. How much would the seller gain or lose, if he gives a single discount of 16%, instead of two discounts?
(a) Gain, ₹ 4.76 (b) Loss, ₹ 5.76
(c) Gain, ₹ 5.76 (d) Loss, ₹ 4.76

17. A wholesaler sells 20 pens at the marked price of 16 pens to a retailer. The retailer in turn sells them giving a discount of 2% on the marked price. The gain per cent to the retailer is
(a) 20% (b) 25%
(c) 22.5% (d) 24%

18. The marked price of an electric iron is ₹ 690. The shopkeeper allows a discount of 10% and gains 8%. If no discount is allowed, his gain per cent would be
(a) 20% (b) 24%
(c) 25% (d) 28%

19. What would be the printed price of a watch purchased at ₹ 380, so that after giving 5% discount, there is 25% profit?
(a) ₹ 400 (b) ₹ 450
(c) ₹ 500 (d) ₹ 600

20. Successive discounts of $12\frac{1}{2}\%$ and $7\frac{1}{2}\%$ are given on the marked price of a cupboard. If the customer pays ₹ 2590, then what is the marked price?
(a) ₹ 3108 (b) ₹ 3148
(c) ₹ 3200 (d) ₹ 3600

21. The difference between a discount of 35% and two successive discounts of 20% on a certain bill was ₹ 22. Find the amount of the bill.
(a) ₹ 1100 (b) ₹ 2000
(c) ₹ 2200 (d) ₹ 1000

22. A shopkeeper purchased 150 identical pieces of calculators at the rate of ₹ 250 each. He spent an amount of ₹ 2500 on transport and packing. He fixed the labelled price of each calculator at ₹ 320. However, he decided to give a discount of 5% on the labelled price. What is the percentage profit earned by him?
(a) 14% (b) 15%
(c) 16% (d) 20%

23. By selling an article at $\frac{2}{5}$th of the marked price, there is a loss of 25%. The ratio of the marked price and the cost price of the article is
(a) 2 : 5 (b) 5 : 2
(c) 8 : 15 (d) 15 : 8

24. Two shopkeepers sell a radio of similar brand and type at the same list price of ₹ 1000. The first allows two successive discounts of 20% and 10% and the second allows two successive discounts 15% and 15%. Find the difference in the discounts offered by the two shopkeepers.
(a) ₹ 3.50 (b) ₹ 2.50
(c) ₹ 1.50 (d) ₹ 1.75

25. A discount of 15% on one article is the same as a discount of 20% on another article. The cost of the two articles can be
(a) ₹ 40, ₹ 20 (b) ₹ 60, ₹ 40
(c) ₹ 80, ₹ 60 (d) ₹ 50, ₹ 30

Answers

1. (c)	**2.** (d)	**3.** (c)	**4.** (c)	**5.** (c)	**6.** (b)	**7.** (b)	**8.** (d)	**9.** (c)	**10.** (c)
11. (d)	**12.** (b)	**13.** (c)	**14.** (d)	**15.** (b)	**16.** (b)	**17.** (c)	**18.** (a)	**19.** (c)	**20.** (c)
21. (b)	**22.** (a)	**23.** (d)	**24.** (b)	**25.** (c)					

Hints and Solutions

1. (c) M.P. = ₹ 800, Discount = 35%

$\therefore$ S.P. = ₹ 800 – 35% of ₹ 800

$= 65\%$ of ₹ 800 $= \frac{65}{100} \times$ ₹ 800 = **₹ 520.**

2. (d) S.P. = ₹ 444, Discount = 26%

Let M.P. = ₹ x. Then,

$x - 26\%$ of $x = 444$

$\Rightarrow 74\%$ of $x = 444$

$\Rightarrow x = \frac{444}{74} \times 100 =$ **₹ 600.**

3. (c) Discount = ₹ 1500 – ₹ 1140 = ₹ 360

$\therefore$ Discount rate $= \left(\frac{360}{1500} \times 100\right)\% =$ **24%.**

4. (c) M.P. = ₹ 500, 1st discount = 20%

Net price after 1st discount = 80% of ₹ 500

$= \frac{80}{100} \times$ ₹ 500 = ₹ 400

2nd discount = 10%

$\therefore$ Final S.P. after 2nd discount $= \frac{90}{100} \times$ ₹ 400

= **₹ 360**

5. (c) S.P. = ₹ 17000, discount = 15%

$\therefore$ M.P. $= \frac{\text{S.P.}}{(100 - \text{discount}\%)} \times 100$

= ₹ $\frac{17000 \times 100}{85}$ = ₹ 20000.

Now 1st discount = 5% and 2nd discount = 10%

$\therefore$ S.P. = 95% of 90% of ₹ 20000

$= \frac{95}{100} \times \frac{90}{100} \times$ ₹ 20000 = **₹ 17100.**

6. (b) M.P. = ₹ 1100, discount = 10%

$\therefore$ S.P. = 90% of ₹ 1100 $= \frac{90}{100} \times$ ₹ 1100 = ₹ 990

Gain = 10%

$\therefore$ C.P. = ₹ $\frac{990 \times 100}{110}$ = **₹ 900.**

7. (b) Let the value of the car (M.P.) = ₹ 100

Then, C.P. = 80% of ₹ 100 = ₹ 80

S.P. = 120% of ₹ 100 = ₹ 120

$\therefore$ Profit % $= \frac{40}{80} \times 100 =$ **50%.**

8. (d) C.P. of the table

= 80% of 90% of ₹ 1500 + ₹ 20

$= \frac{80}{100} \times \frac{90}{100} \times$ ₹ 1500 + ₹ 20

= ₹ 1080 + ₹ 20 = ₹ 1100

Profit = 10%

$\therefore$ S.P. = ₹ $\frac{110 \times 1100}{100}$ = **₹ 1210.**

9. (c) C.P. of the car = 95% of 90% of ₹ 200000

= ₹ 171000

S.P. of the car = ₹ 179550

$\therefore$ Profit % $= \left(\frac{(179550 - 171000)}{171000} \times 100\right)\%$

$= \left(\frac{8550}{171000} \times 100\right)\% =$ **5%.**

10. (c) M.P. = ₹ 1400, discount rate = 10%

$\therefore$ Net price after 1st discount = 90% of ₹ 1400

$= \frac{90}{100} \times$ ₹ 1400 = ₹ 1260

Final selling price = ₹ 1200

$\therefore$ Additional discount = ₹ 1260 – ₹ 1200 = ₹ 60

$\therefore$ Additional discount rate $= \left(\frac{60}{1260} \times 100\right)\% = \mathbf{4\frac{16}{21}\%}$

11. (d) Let the C.P. of the goods = ₹ 100. Then,

Marked price of the goods = ₹ 140

Discount = 20%

∴ S.P. of the goods = 80% of ₹ 140 = ₹ 112

∴ Profit % = $\frac{(112-100)}{100}\times 100 = \mathbf{12\%}$.

12. (b) C.P. = ₹ 180, Profit = 20%

∴ S.P. = ₹ $\frac{180\times 120}{100}$ = ₹ 216

Discount = 10%

∴ M.P. = ₹ $\frac{216\times 100}{90}$ = **₹ 240.**

13. (c) Similar to Q. 10.

14. (d) Let the M.P. = ₹ 100

Then, S.P. after 1st discount of 30%
= 70% of ₹ 100 = ₹ 70

S.P. after 2nd discount of 20%
= 80% of ₹ 70 = ₹ 56

Net S.P. after 3rd discount of 10%
= 90% of ₹ 56 = ₹ 50.4

∴ Total discount = ₹ 100 – ₹ 50.4 = ₹ 49.6

Discount rate = $\frac{49.6}{100}\times 100 = \mathbf{49.6\%}$.

15. (b) M.P. = ₹ 280, Discount = 10%

∴ S.P. = 90% of ₹ 280 = ₹ 252

Profit = 26%

∴ C.P. = ₹ $\frac{252\times 100}{126}$ = **₹ 200.**

16. (b) M.P. = ₹ 900

S.P. of the article when two successive discounts of 8% each are given = 92% of ₹ 900
= ₹ 761.76

S.P. of the article when a single discount of 16% is given = 84% of ₹ 900 = ₹ 756

∴ In the second case, the seller would lose
= (₹ 761.76 – ₹ 756) = **₹ 5.76.**

17. (c) Let the M.P. of 1 pen = ₹ 1. Then,

C.P. of retailer, *i.e.,* C.P. of 20 pens
= M.P. of 16 pens
= ₹ 16

S.P. of 20 pens = 98% of ₹ 20 = ₹ 19.60

∴ Profit % = $\frac{(19.6-16)}{16}\times 100$

$= \frac{3.6}{16}\times 100 = \mathbf{22.5\%}$.

18. (a) M.P. of the iron = ₹ 690, Discount = 10%

∴ S.P. = 90% of ₹ 690 = ₹ 621

Gain = 8%

∴ C.P. = ₹ $\frac{621\times 100}{108}$ = ₹ 575

If no discount is allowed, S.P. = ₹ 690

∴ Gain % = $\frac{(690-575)}{575}\times 100 = \mathbf{20\%}$.

19. (c) C.P. = ₹ 380, Profit = 25%

∴ S.P. = ₹ $\left(\frac{380\times 125}{100}\right)$ = ₹ 475

Discount = 5%

∴ M.P. = ₹ $\left(\frac{475\times 100}{95}\right)$ = **₹ 500.**

20. (c) Let the M.P. of the cupboard be ₹ *x*. Then,
(100 – 12.5)% of (100 – 7.5)% of ₹ *x* = 2590

$\Rightarrow \frac{87.5}{100}\times\frac{92.5}{100}\times x = 2590$

$\Rightarrow x = \frac{2590\times 100\times 100}{87.5\times 92.5}$ = **₹ 3200.**

21. (b) Let the amount of the bill be ₹ *x*. Then,
65% of *x* – (80% of 80% of *x*) = 22

$\Rightarrow \frac{65}{100}\times x - \frac{80}{100}\times\frac{80}{100}\times x = 22$

$\Rightarrow \frac{6500x - 6400x}{100} = 22 \Rightarrow 100x = 220000$

$\Rightarrow x = \mathbf{2200}$.

22. (a) Cost price of each calculator = ₹ $\left(250 + \frac{2500}{150}\right)$

= ₹ $\left(250 + \frac{50}{3}\right)$ = ₹ $\frac{800}{3}$ = ₹ $266\frac{2}{3}$

Selling price of each calculator = ₹ $\left(\frac{95}{100}\times 320\right)$
= ₹ 304

∴ Profit % = $\frac{\left(304 - \frac{800}{3}\right)}{\frac{800}{3}}\times 100$

$= \frac{\frac{912-800}{3}}{\frac{800}{3}}\times 100 = \frac{112}{800}\times 100$

= **14%.**

23. (d) Let the C.P. of the article be ₹ 100. Then,

Loss = 25% $\Rightarrow$ S.P. = ₹ 75

Given, $\frac{2}{5}$th of Marked price = ₹ 75

$\Rightarrow$ M.P. = ₹ $75 \times \frac{5}{2}$ $\Rightarrow$ ₹ $\frac{375}{2}$

$\therefore$ M.P. : C.P. = $\frac{375}{2}:100$ = 375 : 200 = **15 : 8.**

24. (b) S.P. of the 1st shopkeeper

= 80% of 90% of ₹ 1000

$= \frac{80}{100} \times \frac{90}{100} \times$ ₹ 1000

= ₹ 720

S.P. of the 2nd shopkeeper

= 85% of 85% of ₹ 1000

$= \frac{85}{100} \times \frac{85}{100} \times$ ₹ 1000

= ₹ 722.50

$\therefore$ Difference in discount = ₹ 722.50 – ₹ 720

= ₹ 2.50

25. (c) Let the cost of the two articles be x and y. Then,

15% of x = 20% of y $\Rightarrow$ $\frac{x}{y} = \frac{20}{15} = \frac{4}{3}$.

$\therefore$ x and y should be in the ratio 4 : 3.

Self Assessment Sheet–11

1. A referigerator and a camera were sold for ₹ 12,000 each. The referigerator was sold at a loss of 20% of the cost and the camera at a gain of 20% of the cost. The entire transaction results in which one of the following ?

(a) No loss or gain (b) Loss of ₹ 1,000
(c) Gain of ₹ 1000 (d) Loss of ₹ 2,000

2. A man sells a book at a profit of 20%. If he had bought it at 20% less and sold it for ₹ 18 less, he would have gained 25%. The cost price of the book is :

(a) ₹ 80 (b) ₹ 70
(c) ₹ 60 (d) ₹ 90

3. If a shopkeeper sells $\frac{1}{3}$ of his goods at a profit of 14%, $\frac{3}{5}$ of the goods at a profit of 17.5% and remaining at a profit of 20%, then his profit on the whole is equal to :

(a) 15.5% (b) 16 %
(c) 16. 5% (d) 17%

4. A shopkeeper sells an item at a loss of $12\frac{1}{2}$%. In order the gain 6% profit, he must increase the selling price by ₹ 92.50. The cost price of the item is :

(a) ₹ 500 (b) ₹ 450
(c) ₹ 550 (d) ₹ 600

5. A shopkeeper sold a TV set for ₹ 17,940 with a discount of 8% and earned a profit of 19.6%. What would have been the percentage of profit earned if no discount was offered ?

(a) 24.8% (b) 25%
(c) 26.4% (d) Cannot be determined
(e) None of these

6. A trader lists his articles 20% above cost price and allows a discount of 10% on cash payment. His gain per cent is :

(a) 10% (b) 8%
(c) 6% (d) 5%

7. A machine is sold at a profit of 10%. Had it been sold for ₹ 40 less, there would have been a loss of 10% . What was the cost price ?

(a) ₹ 175 (b) ₹ 200
(c) ₹ 225 (d) ₹ 250

8. Vishal goes to a shop to buy a radio costing ₹ 2568. The rate of sales tax is 7%. He tells the shopkeeper to reduce the price of the radio to such an extent that he has to pay ₹ 2568, inclusive of sales tax. Find the reduction needed in the price of the radio.

(a) ₹ 179.76 (b) ₹ 170
(c) ₹ 168 (d) ₹ 169

9. A house costs C rupees. Later it was sold for a profit of 25%. What is the capital gains tax if it is 50% of the profit ?

(a) $C/24$ (b) $C/8$
(c) $C/4$ (d) $C/2$

10. If selling price of an article is $\frac{4}{3}$ of its cost price, the profit in the transaction is :

(a) $\frac{1}{3}$% (b) $20\frac{1}{2}$%
(c) $33\frac{1}{3}$% (d) $25\frac{1}{2}$%

Answers

1. (b) **2.** (d) **3.** (c) **4.** (a) **5.** (e) The answer is 23.07 % **6.** (b) **7.** (b) **8.** (c)
9. (b) **10.** (c)

Chapter

12 SIMPLE INTEREST

KEY FACTS

1. (i) The money borrowed from a lender is called **principal.**

(ii) The additional money paid by the borrower to the lender after a specified period of time is called **interest.**

(iii) The total money paid by the borrower to the lender is called **amount.**

Amount = Principal + Interest.

2. Formulae used.

(i) Simple Interest (S.I.) $= \dfrac{\text{Principal (P)} \times \text{Rate (R)} \times \text{Time (T)}}{100}$ (ii) $R = \dfrac{S.I. \times 100}{P \times T}$

(iii) $T = \dfrac{S.I. \times 100}{P \times R}$ (iv) $P = \dfrac{S.I. \times 100}{R \times T}$ (v) Amount $= P + S.I. = P + \dfrac{PRT}{100}$

Solved Examples

Ex. 1. ***Mr. Abhishek borrowed ₹ 600 and returned ₹ 856.50 at the end of 9 years and 6 months. What was the interest per annum he paid at simple interest?***

Sol. Principal = ₹ 600, Amount = ₹ 856.50,

Time = 9 yrs 6 months $= 9\frac{1}{2}$ yrs $= \frac{19}{2}$ yrs.

Simple interest = ₹ 856.50 – ₹ 600 = ₹ 256.50

$\therefore$ Required interest $= \dfrac{S.I. \times 100}{P \times T} = \dfrac{256.50 \times 100 \times 2}{600 \times 19} =$ **4.5% p.a.**

Ex. 2. ***A certain sum is invested on simple interest. If it trebles in 10 years, what is the rate of interest?***

Sol. Let the sum be ₹ x. Then, Amount = ₹ $3x$, $T = 10$ years

$\therefore$ S.I. = Amount – Sum = ₹ $3x$ – ₹ x = ₹ $2x$

$\therefore$ Rate of Interest $= \dfrac{S.I \times 100}{P \times T} = \dfrac{2x \times 100}{x \times 10} =$ **20% p.a.**

Ex. 3. ***Two equal sums of money were invested, one at 4% and the other at $4\frac{1}{2}$%. At the end of 7 years, the simple interest received from the latter exceeded that received from the former by ₹ 31.50. What was each sum?***

Sol. Let each of the equal sums be ₹ x. Then,

$$\frac{x \times 4.5 \times 7}{100} - \frac{x \times 4 \times 7}{100} = 31.50 \Rightarrow \frac{31.5x}{100} - \frac{28x}{100} = 31.5$$

$\Rightarrow \dfrac{3.5x}{100} = 31.5 \Rightarrow x = ₹\ \dfrac{31.5 \times 100}{3.5} = \textbf{₹ 900}.$

Ex. 4. ***A sum of ₹ 5000 was lent partly at 6% and partly at 9% simple interest. If the total interest received after 1 year was ₹ 390, what was the ratio in which the money was lent at 6% and 9%?***

Sol. Let the sum lent at 6% p.a. be ₹ x. Then,

Sum lent at 9% p.a. is ₹ $(5000 - x)$

Given, $\dfrac{x \times 6 \times 1}{100} + \dfrac{(5000 - x) \times 9 \times 1}{100} = 390 \Rightarrow \dfrac{6x}{100} + 450 - \dfrac{9x}{100} = 390$

$\Rightarrow \dfrac{3x}{100} = 60 \Rightarrow x = 2000$

$\therefore$ Required ratio = 2000 : 3000 = **2 : 3**.

Ex. 5. ***A sum of ₹ 10000 is lent partly at 8% and remaining at 10% per annum. If the yearly interest on the average is 9.2%, what are the two parts?***

Sol. $\dfrac{\text{Total interest} \times 100}{\text{Total sum}}$ = Average yearly interest $\Rightarrow$ Total interest = $\dfrac{9.2 \times 10000}{100}$ = ₹ 920

Let the sum lent at 8% p.a. be ₹ x. Then,

The sum lent at 10% p.a. is ₹ $(10000 - x)$

$\therefore \quad \dfrac{x \times 8 \times 1}{100} + \dfrac{(10000 - x) \times 10 \times 1}{100} = 920 \Rightarrow \dfrac{8x}{100} + 1000 - \dfrac{10x}{100} = 920$

$\Rightarrow \dfrac{2x}{100} = 80 \Rightarrow x = 4000$

$\therefore$ The two parts are ₹ 4000 and ₹ 6000.

Ex. 6. ***A certain sum of money amounts to ₹ 756 in 2 years and to ₹ 873 in $3\frac{1}{2}$ years at a certain rate of simple interest. What is the rate of interest per annum?***

Sol. Amount in 2 years = ₹ 756

Amount is $3\frac{1}{2}$ years = ₹ 873

$\therefore$ Interest for $1\frac{1}{2}$ years = ₹ 873 – ₹ 756 = ₹ 117

$\therefore$ Interest for 1 year = ₹ $\dfrac{117 \times 2}{3}$ = ₹ 78

$\therefore$ Interest for 2 years = 2 × ₹ 78 = ₹ 156

$\therefore$ Principal = Amount – Interest (for 2 years) = ₹ 756 – ₹ 156 = ₹ 600

$\therefore$ Rate of interest = $\dfrac{78 \times 100}{600 \times 1}$ = **13% p.a.**

Ex. 7. ***Pratap borrowed a sum of money from Arun at simple interest, at the rate of 12% per annum for the first three years, 16% per annum for the next five years and 20% per annum for a period beyond eight years. If at the end of 11 years, the total interest is ₹ 6080 more than the sum, what was the sum borrowed?***

Sol. Let the sum borrowed be ₹ x. Then,

$\dfrac{x \times 12 \times 3}{100} + \dfrac{x \times 16 \times 5}{100} + \dfrac{x \times 20 \times (11 - 8)}{100} = x + 6080$

$\Rightarrow \dfrac{36x}{100} + \dfrac{80x}{100} + \dfrac{60x}{100} = x + 6080 \Rightarrow \dfrac{176x}{100} = x + 6080$

$\Rightarrow 176x = 100x + 608000 \Rightarrow 76x = 608000 \Rightarrow x = \dfrac{608000}{76} = \textbf{₹ 8000}.$

Ex. 8. ***A person invests money in three different schemes for 6 years, 10 years and 12 years at 10%, 12% and 15% simple interest respectively. At the completion of each scheme, he gets the same interest. What is the ratio of his investments?***

Sol. Let the required ratio be $x : y : z$. Then,

S.I. on ₹ x for 6 years at 10% p.a. = S.I. on ₹ y for 10 years at 12% p.a.

$$\Rightarrow \quad \frac{x \times 6 \times 10}{100} = \frac{y \times 10 \times 12}{100} \quad \Rightarrow \quad \frac{x}{y} = \frac{2}{1}$$

S.I. on ₹ y for 10 years at 12% p.a. = S.I. on ₹ z for 12 years at 15% p.a.

$$\Rightarrow \quad \frac{y \times 10 \times 12}{100} = \frac{z \times 12 \times 15}{100} \quad \Rightarrow \quad \frac{y}{z} = \frac{3}{2}$$

$x : y = 2 : 1$ and $y : z = 3 : 2$

$\Rightarrow$ $x : y = 6 : 3$ and $y : z = 3 : 2$

$\Rightarrow$ $x : y : z =$ **6 : 3 : 2.**

Ex. 9. ***A sum of ₹ 725 is lent in the beginning of a year at a certain rate of interest. After 8 months, a sum of ₹ 362.50 more is lent but at the rate twice the former. At the end of the year, ₹ 33.50 is earned as interest from both the loans. What was the original rate of interest?***

Sol. ∴ Let the original rate be R%.

Then, new rate = $(2R)$% and time = 4 months = $\frac{4}{12} = \frac{1}{3}$ year

$$\therefore \quad \left(\frac{725 \times R \times 1}{100}\right) + \left(\frac{362.50 \times 2R \times 1}{100 \times 3}\right) = 33.50$$

$$\Rightarrow \quad (2175 + 725)R = 33.50 \times 100 \times 3 = 10050 \Rightarrow \quad R = \frac{10050}{2900} = 3.46\%$$

∴ Original rate = **3.46%**.

Ex. 10. ***A man invested ₹ 1000 on simple interest at a certain rate and ₹ 1500 at 2% higher rate. The total interest in three years is ₹ 390. What is the rate of interest for ₹ 1000?***

Sol. Let the interest rate at which ₹ 1000 is invested is r%. Then,

₹ 1500 is invested at $(r + 2)$%

$$\text{Given, } \frac{1000 \times r \times 3}{100} + \frac{1500 \times (r+2) \times 3}{100} = 390 \Rightarrow 30r + 45r + 90 = 390$$

$\Rightarrow$ $75r = 300$ $\Rightarrow$ $r =$ **4%**.

Question Bank–12

1. Mr. Sharma takes loan of ₹ 25000 and repays an amount of ₹ 31000 at the end of 2 years. What is the rate of simple interest at which he repays the loan?

(a) 8% p.a. (b) 6% p.a.

(c) 12% p.a. (d) 9% p.a.

2. A part of ₹ 1500 was lent at 10% p.a. and the rest at 7% p.a. simple interest. The total interest earned in three years was ₹ 396. The sum lent at 10% was

(a) ₹ 900 (b) ₹ 800

(c) ₹ 700 (d) ₹ 600

3. In what time will ₹ 72 become ₹ 81 at $6\frac{1}{4}$% p.a. simple interest?

(a) 2 years (b) 3 years

(c) 2 years 6 months (d) 4 years

4. A money lender finds that due to a fall in the annual rate of interest from 8% to $7\frac{3}{4}$%, his yearly income diminishes by ₹ 61.50. His capital is

(a) ₹ 22400 (b) ₹ 23800

(c) ₹ 24600 (d) ₹ 26000

5. An investment triples itself in 30 years. The rate of simple interest is
(a) 6% (b) $6\frac{2}{3}\%$
(c) $7\frac{1}{3}\%$ (d) 10%

6. ₹ 800 becomes ₹ 956 in 3 years at a certain rate of simple interest. If the rate of interest is increased by 4%, what will ₹ 800 amount to in 3 years?
(a) ₹ 1020.80 (b) ₹ 1025
(c) ₹ 1052 (d) ₹ 1050

7. In how many years, ₹ 150 will produce the same interest at 8% p.a. as ₹ 800 produce in 3 years at $4\frac{1}{2}\%$ p.a.?
(a) 6 (b) 8
(c) 9 (d) 12

8. What will be the ratio of simple interest earned by certain amount at the same rate of interest for 6 years and that for 9 years?
(a) 1 : 3 (b) 2 : 3
(c) 1 : 4 (d) 3 : 2

9. A sum of money at simple interest amounts to ₹ 815 in 3 years and to ₹ 854 in 4 years. The sum is
(a) ₹ 650 (b) ₹ 690
(c) ₹ 698 (d) ₹ 700

10. A person invested part of ₹ 45000 at 4% and the rest at 6%. If his annual income from both is equal, then what is the average rate of interest?
(a) 4.6% (b) 4.8%
(c) 5% (d) 5.2%

11. ₹ 6000 becomes ₹ 7200 in 4 years at a certain rate of interest. If the rate becomes 1.5 times of itself, the amount of the same principal in 5 years will be
(a) ₹ 8000 (b) ₹ 8250
(c) ₹ 9000 (d) ₹ 9250

12. A certain sum of money becomes three times of itself in 20 years at simple interest. In how many years does it become double of itself at the same rate?
(a) 8 years (b) 10 years
(c) 12 years (d) 14 years

13. The sum of money that will give ₹ 1 as simple interest per day at the rate of 5% per annum is
(a) ₹ 730 (b) ₹ 3650
(c) ₹ 7300 (d) ₹ 36500

14. Simple interest on ₹ 500 for 4 years at 6.25% per annum is equal to the simple interest on ₹ 400 at 5% per annum for a certain period of time. The period time is
(a) 4 years (b) 5 years
(c) $6\frac{1}{4}$ years (d) $8\frac{2}{3}$ years

15. *A* borrows ₹ 800 at the rate of 12% per annum simple interest and *B* borrows ₹ 910 at the rate of 10% per annum simple interest. In how many years will their amounts of debts be equal?
(a) 18 years (b) 20 years
(c) 22 years (d) 24 years

16. If the simple interest on a certain sum of money for 15 months at $7\frac{1}{2}\%$ p.a. exceeds the simple interest on the same sum for 8 months at $12\frac{1}{2}\%$ p.a. by ₹ 32.50, the sum is
(a) ₹ 312 (b) ₹ 312.50
(c) ₹ 3120 (d) ₹ 3120.50

17. A man invests $\frac{1}{3}$ of his capital at 7% p.a., $\frac{1}{4}$ at 8% p.a. and the remainder at 10% p.a. If his annual income is ₹ 561, the capital is
(a) ₹ 5400 (b) ₹ 6000
(c) ₹ 6600 (d) ₹ 7200

18. Simple interest on a certain amount is $\frac{9}{16}$ of the principal. If the number representing the rate of interest in per cent and time in years be equal, then the time for which the principal amount is lent out is
(a) $5\frac{1}{2}$ years (b) $6\frac{1}{2}$ years
(c) 7 years (d) $7\frac{1}{2}$ years

19. A person lends 40% of his sum of money at 15% per annum, 50% of rest at 10% per annum and the rest at 18% per annum rate of interest. What would be the annual rate of interest, if the interest is calculated on the whole sum?
(a) 13.4% (b) 14.33%
(c) 14.4% (d) 13.33%

20. The rate of simple interest in two banks *A* and *B* are in the ratio 5 : 4. A person wants to deposit his total savings in the two banks in such a way that he receives equal half yearly interest from both. He should deposit the savings in Bank *A* and *B* in the ratio.
(a) 5 : 2 (b) 2 : 5
(c) 4 : 5 (d) 5 : 4

21. A person lent a certain sum of money at 4% simple interest and in 5 years, the interest amounted to ₹ 520 less than the sum lent. The sum lent was

(a) ₹ 600 (b) ₹ 650
(c) ₹ 700 (d) ₹ 750

22. *A* lent ₹ 600 to *B* and some amount to *C* at the rate of $8\frac{1}{3}\%$ per annum simple interest. After 5 years he got the total interest of ₹ 400 from *B* and *C* together. The amount of money lent by *A* to *C* was

(a) ₹ 300 (b) ₹ 360
(c) ₹ 400 (d) ₹ 420

23. Arun borrowed a sum of money from Jayant at the rate of 8% simple interest for the first four years, 10% per annum for the next 6 years and 12% per annum beyond 10 years. If he pays a total of ₹ 12160 as interest only at the end of 15 years, how much money did he borrow?

(a) ₹ 12000 (b) ₹ 10000
(c) ₹ 8000 (d) ₹ 9000

24. Mohan lent some amount of money at 9% simple interest and an equal amount of money at 10% simple interest each for 2 years. If his total interest was ₹ 760, what amount was lent in each case?

(a) ₹ 1700 (b) ₹ 1800
(c) ₹ 1900 (d) ₹ 2000

25. What sum of money will amount to ₹ 520 in 5 years and to ₹ 568 in 7 years at simple interest?

(a) ₹ 400 (b) ₹ 120
(c) ₹ 510 (d) ₹ 220

Answers

1. (c)	**2.** (a)	**3.** (a)	**4.** (c)	**5.** (b)	**6.** (c)	**7.** (c)	**8.** (c)	**9.** (c)	**10.** (b)
11. (b)	**12.** (b)	**13.** (c)	**14.** (c)	**15.** (c)	**16.** (c)	**17.** (c)	**18.** (d)	**19.** (c)	**20.** (c)
21. (b)	**22.** (b)	**23.** (c)	**24.** (d)	**25.** (a)					

Hints and Solutions

1. (c) Principal = ₹ 25000, Amount = ₹ 31000, Time = 2 years

$\therefore$ S.I. = ₹ 31000 – ₹ 25000 = ₹ 6000

Rate $= \frac{6000 \times 100}{25000 \times 2} =$ **12% p.a.**

2. (a) Let the sum lent at 10% p.a. be ₹ x. Then, the sum lent at 7% p.a. is ₹ $(1500 - x)$.

Given, $\frac{x \times 10 \times 3}{100} + \frac{(1500 - x) \times 7 \times 3}{100} = 396$

$\Rightarrow \frac{30x}{100} + 315 - \frac{21x}{100} = 396$

$\Rightarrow \frac{9x}{100} = 81 \Rightarrow x = \frac{81 \times 100}{9} =$ **₹ 900.**

3. (a) $P =$ ₹ 72, Amount = ₹ 81, Rate p.a. $= \frac{25}{4}\%$,

S.I. = ₹ 81 – ₹ 72 = ₹ 9

$\therefore$ Time (in years) $= \frac{9 \times 100 \times 4}{72 \times 25} =$ **2 years.**

4. (c) Let his capital be ₹ x. Then,

$$\frac{x \times 8 \times 1}{100} - \frac{x \times 31 \times 1}{4 \times 100} = 61.50$$

$$\Rightarrow \frac{8x}{100} - \frac{31x}{400} = 61.50$$

$$\Rightarrow \frac{32x - 31x}{400} = 61.50$$

$\Rightarrow x = 61.5 \times 400 =$ **₹ 24600.**

5. (b) Let the principal = ₹ x, Amount = ₹ $3x$, Time = 30 years

$\therefore$ S.I. = ₹ $3x$ – ₹ x = ₹ $2x$

$\therefore$ Rate $= \left(\frac{2x \times 100}{x \times 30}\right)\%$ p.a. $= \frac{20}{3}\%$ p.a.

$= \mathbf{6\frac{2}{3}\%}$ **p.a.**

6. (c) Principal = ₹ 800, Amount = ₹ 956

$\Rightarrow$ S.I. = ₹ 956 – ₹ 800 = ₹ 156

Time = 3 years $\therefore$ Rate % p.a. $= \frac{156 \times 100}{800 \times 3} = 6.5.$

Now, Principal = ₹ 800, Rate % p.a. = 10.5, Time = 3 years

$\therefore$ Amount = ₹ 800 + ₹ $\left(\frac{800 \times 10.5 \times 3}{100}\right)$

= ₹ 800 + ₹ 252 = **₹ 1052.**

7. (c) $P = ₹\ 800, R = 4\frac{1}{2}\%$ p.a., $T = 3$ years

$\therefore$ S.I. $= ₹\ \frac{800 \times 9 \times 3}{2 \times 100} = ₹\ 108$

Now, S.I. $= ₹\ 108, P = ₹\ 150, R = 8\%$ p.a.

$\therefore$ Time $= \left(\frac{108 \times 100}{150 \times 8}\right)$ years $=$ **9 years.**

8. (c) Let the principal for both the case be ₹ P and the rate of interest be $R\%$ p.a. Then,

Required ratio $= \frac{P \times R \times 6}{100} : \frac{P \times R \times 4}{100} = 6:4 = \mathbf{3:2}.$

9. (c) Let the sum of money be ₹ x.

Amount for 3 years = ₹ 815

Amount for 4 years = ₹ 854

$\therefore$ S.I. for 1 year = ₹ 854 – ₹ 815 = ₹ 39

$\therefore$ S.I. for 3 years = 3 × ₹ 39 = ₹ 117

$\therefore$ Principal = ₹ 815 – ₹ 117 = **₹ 698.**

10. (b) Let the amount invested on 4% be ₹ x.

Then, $\frac{x \times 4 \times 1}{100} = \frac{(45000 - x) \times 6 \times 1}{100}$

$\Rightarrow 2x = 3 \times 45000 - 3x$

$\Rightarrow 5x = 3 \times 45000$

$\Rightarrow x = ₹\ 27000$

$\therefore$ Total interest $= \frac{2 \times 27000 \times 4}{100} = ₹\ 2160$

$\therefore$ Rate of interest $= \frac{2160 \times 100}{45000} = \mathbf{4.8\%}.$

11. (b) $P = ₹\ 6000$, Amount = ₹ 7200, Time = 4 years

S.I. $= A - P = ₹\ 7200 - ₹\ 6000 = ₹\ 1200$

$\therefore$ Rate $= \left(\frac{1200 \times 100}{6000 \times 4}\right)\%$ p.a. = 5% p.a.

New rate = 1.5 × 5 = 7.5% p.a.

$\therefore$ Reqd. amount $= ₹\ 6000 + ₹\ \frac{6000 \times 7.5 \times 5}{100}$

= ₹ 6000 + ₹ 2250

= **₹ 8250.**

12. (b) Let the principal = ₹ x. Amount = ₹ $3x$,

Time = 20 years

$\therefore$ S.I. = ₹ $3x$ – ₹ x = ₹ $2x$

Rate $= \left(\frac{2x \times 100}{x \times 20}\right)\%$ p.a. = 10% p.a.

Now, Principal = ₹ x, Amount = ₹ $2x$,

Rate = 10% p.a.

S.I. = ₹ $2x$ – ₹ x = ₹ x

$\therefore$ Time $= \frac{x \times 100}{x \times 10} =$ **10 years.**

13. (c) Annual interest = 365, Rate = 5% p.a.

$\therefore$ Sum $= \frac{365 \times 100}{5 \times 1} =$ **₹ 7300.**

14. (c) Let the required period of time be t years.

Then, $\frac{500 \times 4 \times 6.25}{100} = \frac{400 \times t \times 5}{100}$

$\Rightarrow t = \frac{500 \times 4 \times 6.25}{400 \times 5}$ years

$= 6.25$ years or $\mathbf{6\frac{1}{4}}$ **years.**

15. (c) Let the required time be t years. Then,

$800 + 800 \times \frac{12}{100} \times t = 910 + 910 \times \frac{10}{100} \times t$

$\Rightarrow (96t - 91t) = 110$

$\Rightarrow 5t = 110 \quad \Rightarrow \quad t =$ **22 years.**

16. (c) Let the required sum be ₹ x. Then,

$\frac{x \times 15 \times 15}{12 \times 2 \times 100} - \frac{x \times 8 \times 25}{12 \times 2 \times 100} = 32.50$

$\Rightarrow \frac{225x - 200x}{2400} = 32.50$

$\Rightarrow 25x = 32.50 \times 2400$

$\Rightarrow x = \frac{32.50 \times 2400}{25} =$ **₹ 3120.**

17. (c) Let the capital be ₹ x. Then,

$\frac{\frac{x}{3} \times 7 \times 1}{100} + \frac{\frac{x}{4} \times 8 \times 1}{100} + \frac{\left(1 - \frac{1}{3} - \frac{1}{4}\right) \times x \times 10 \times 1}{100} = 561$

$\Rightarrow \frac{7x}{300} + \frac{8x}{400} + \frac{50x}{1200} = 561$

$\Rightarrow \dfrac{28x + 24x + 50x}{1200} = 561$

$\Rightarrow \dfrac{102x}{1200} = 561$

$\Rightarrow x = \dfrac{561 \times 1200}{102} =$ ₹ **6600.**

18. (d) Let the principal be ₹ x. Then, S.I. = ₹ $\dfrac{9x}{16}$

Let time (in years) = Rate % p.a. = y. Then,

$\dfrac{9x}{16} = \dfrac{x \times y \times y}{100} \Rightarrow y^2 = \dfrac{900}{16}$

$\Rightarrow y = \dfrac{30}{4} = \dfrac{15}{2}$ years = $\mathbf{7\frac{1}{2}}$ **years.**

19. (c) Let the whole sum be ₹ 100. Then,

Sum at 15% p.a. = ₹ 40,

Remaining sum = ₹ 60

∴ Sum at 10% p.a. = 50% of ₹ 60 = ₹ 30 and sum at 18% p.a. = ₹ 30

∴ S.I. on ₹ 100 for 1 year

$= \left(40 \times \dfrac{15}{100} \times 1\right) + \left(30 \times \dfrac{10}{100} \times 1\right) + \left(30 \times \dfrac{18}{100} \times 1\right)$

= ₹ (6 + 3 + 5.4) = ₹ 14.4.

Hence, required rate = **14.4%**.

20. (c) Let the rates of simple interest in Bank A and Bank B be $5R$% and $4R$% p.a. respectively and the amounts invested in the banks respectively be ₹ x and ₹ y.

Then, $x \times \dfrac{1}{2} \times \dfrac{5R}{100} = y \times \dfrac{1}{2} \times \dfrac{4R}{100}$

$\Rightarrow \dfrac{x}{y} = \dfrac{4}{5} \Rightarrow x : y = \mathbf{4 : 5}$.

21. (b) Let the sum lent be ₹ x. Then,

$x - 520 = \dfrac{x \times 4 \times 5}{100}$

$\Rightarrow x - \dfrac{x}{5} = 520 \Rightarrow \dfrac{4x}{5} = 520$

$\Rightarrow x =$ ₹ **650.**

22. (b) Let the amount lent by A to C be ₹ x. Then,

$\dfrac{600 \times 25 \times 5}{3 \times 100} + \dfrac{x \times 25 \times 5}{3 \times 100} = 400$

$\Rightarrow 250 + \dfrac{5x}{12} = 400 \Rightarrow \dfrac{5x}{12} = 150$

$\Rightarrow x = \dfrac{150 \times 12}{5} =$ ₹ **360.**

23. (c) Let the money borrowed by Arun = ₹ x. Then,

$\dfrac{x \times 8 \times 4}{100} + \dfrac{x \times 10 \times 6}{100} + \dfrac{x \times 12 \times 5}{100} = 12160$

$\Rightarrow 152x = 1216000 \Rightarrow x =$ ₹ **8000.**

24. (d) Let the amount lent in each case be ₹ x. Then,

$\dfrac{x \times 9 \times 2}{100} + \dfrac{x \times 10 \times 2}{100} = 760$

$\Rightarrow 38x = 76000 \Rightarrow x = \mathbf{2000}$.

25. (a) Interest for two years = ₹ 568 – ₹ 520 = ₹ 48

∴ S.I. for 1 year = ₹ $\dfrac{48}{2}$ = ₹ 24

∴ S.I for 5 years = 5 × ₹ 24 = ₹ 120

∴ Principal = Amount – S.I.

= ₹ 520 – ₹ 120 = ₹ **400.**

Self Assessment Sheet–12

1. Out of a sum of ₹ 640, a part was lent at 6% simple interest and the other at 9% simple interest. If the interest on the first part after 3 years is equal to the interest on the second part after 6 years then what is the second part ?

(a) ₹ 120 (b) ₹ 140
(c) ₹ 180 (d) ₹ 160

2. What sum will amount to ₹ 6600 in 4 years at 8% p.a. simple interest ?

(a) ₹ 6000 (b) ₹ 5000
(c) ₹ 6200 (d) None of these

3. ₹ 1200 amounts to ₹ 1632 in 4 years at a certain rate of simple interest. If the rate of interest is increased by 1%, it would amount to how much ?

(a) ₹ 1635 (b) ₹ 1644
(c) ₹ 1670 (d) ₹ 1680

4. A certain sum lent out at simple interest doubles itself in 20 years. Number of years in which this sum trebles itself at the same rate of interest is :

(a) 30 (b) 40
(c) 25 (d) 20

5. A sum of money becomes $\frac{41}{40}$ of itself in $\frac{1}{4}$ year at a certain rate of simple interest. The rate of interest per annum is :
(a) 10%
(b) 1%
(c) 2.5%
(d) 5%

6. Simple interest on a certain amount is $\frac{16}{25}$ of the principal. If the number representing the rate of interest in per cent and time in years be equal, then time, for which the principal amount is lent out, is :
(a) $5\frac{1}{2}$ years
(b) $6\frac{1}{2}$ years
(c) 6 years
(d) 8 years

7. A man borrowed ₹ 1000 to build a house. He pays 5% per annum simple interest. He lets the house and receives a rent of ₹ 12.50 per month. In how many years he is expected to clear the debt ?
(a) 7
(b) 15
(c) 10
(d) 8

8. The difference between the interests received from two different banks on ₹ 5000 for 2 years is ₹ 25. The difference between their rates is :
(a) 1%
(b) 2.5%
(c) 0.5%
(d) 0.25%

9. x, y, z are three sums of money such that y is the simple interest on x and z is the simple interest on y for the same time and rate. Which of the following is correct ?
(a) $xyz = 1$
(b) $z^2 = xy$
(c) $x^2 = yz$
(d) $y^2 = zx$

10. ₹ 1500 is invested at a rate of 10% simple interest and interest is added to the principal after every 5 years. In how many years will it amount to ₹ 2500 ?
(a) $6\frac{1}{9}$ years
(b) $6\frac{1}{4}$ years
(c) 7 years
(d) None of these

Answers

1. (d)	2. (b)	3. (d)	4. (b)	5. (a)	6. (d)	7. (c)	8. (d)	9. (d)	10. (a)

Chapter 13

TIME AND WORK

KEY FACTS

1. If a man finishes total work in d days, then in 1 day he does $\frac{1}{d}$ of the total work.

For example: If a man finishes a work in 6 days, then his 1 day's work = $\frac{1}{6}$.

2. Conversely, if the work in 1 day that a man does is given, then the total number of days taken to finish the work = $\frac{1}{\text{one days' work}}$.

For example: If a man does $\frac{1}{4}$ of the total work in 1 day, then the total number of days required to complete the total work $= \frac{1}{\frac{1}{4}} = 1 \div \frac{1}{4} = 1 \times \frac{4}{1} = 4$ days.

Solved Examples

Ex. 1. ***A person can do a job as fast as his two sons working together. If one son does the job in 6 days and the other in 12 days, how many days does it take the father to do the job?***

Sol. Let the father complete the job in x days.

Given, Father's 1 days' job = One son's one days' job + Second son's one days' job

$$\Rightarrow \frac{1}{x} = \frac{1}{6} + \frac{1}{12} = \frac{2+1}{12} = \frac{3}{12} = \frac{1}{4}$$

$\therefore$ Father can do the job in **4 days**.

Ex. 2. ***A and B can do a piece of work in 8 days. B and C can do the same work in 12 days. If A, B and C can complete the same work in 6 days, in how many days can A and C complete the some work?***

Sol. $(A+B+C)$'s one days' work = $\frac{1}{6}$; $(A+B)$'s one days' work = $\frac{1}{8}$;

$(B+C)$'s one days' work = $\frac{1}{12}$

$\Rightarrow C$'s one days' work $= \frac{1}{6} - \frac{1}{8} = \frac{4-3}{24} = \frac{1}{24}$

Similarly, A's one days' work $= \frac{1}{6} - \frac{1}{12} = \frac{2-1}{12} = \frac{1}{12}$

$\therefore$ $(A+C)$'s = one days' work $= \frac{1}{24} + \frac{1}{12} = \frac{1+2}{24} = \frac{3}{24} = \frac{1}{8}$

$\therefore$ $(A + C)$ can finish the work in **8 days.**

Ex. 3. ***A and B can do a work in 8 days. B and C together in 6 days while C and A together in 10 days. If they all work together, in how many days will the work be completed.***

Sol. $(A+B)$'s 1 days' work $= \frac{1}{8}$; $(B+C)$'s 1 days' work $= \frac{1}{6}$; $(C+A)$'s 1 days' work $= \frac{1}{10}$

$\therefore$ $(A+B)$'s + $(B+C)$'s + $(C+A)$'s 1 days' work $= \frac{1}{8} + \frac{1}{6} + \frac{1}{10} = \frac{47}{120}$

$\Rightarrow$ $2(A+B+C)$'s $= \frac{47}{120}$

$\Rightarrow$ $(A+B+C)$'s 1 days' work $= \frac{47}{240}$

$\therefore$ If A, B and C work together, the work will be completed in $\frac{240}{47} = \mathbf{5\frac{5}{47}}$ **days.**

Ex. 4. ***Kamal can do a work in 15 days. Bimal is 50% more efficient than Kamal. What is the number of days, Bimal will take to do the some piece of work?***

Sol. If Kamal does 1 work in 15 days,

Bimal will do $1\frac{1}{2}$ work in 15 days, *i.e.*,

Bimal will do 1 work in $15 \times \frac{2}{3}$ days = **10 days.**

Ex. 5. ***A is three times more efficient worker than B and is therefore able to complete a work in 60 days less than B. What is the number of days that A and B together will take to complete the work?***

Sol. Let B complete the work in x days. Then,

A will complete the same work in $\frac{x}{3}$ days.

Given, $x - \frac{x}{3} = 60 \Rightarrow \frac{2x}{3} = 60 \Rightarrow x = 60 \times \frac{3}{2} = 90$ days.

$\therefore$ A will complete the work in 30 days.

$(A+B)$'s 1 days' work $= \frac{1}{90} + \frac{1}{30} = \frac{1+3}{90} = \frac{4}{90}$

$\Rightarrow$ $(A + B)$ will complete the whole work in $\frac{90}{4}$ = **22.5 days.**

Ex. 6. ***A and B can do a work in 45 days and 40 days respectively. They began the work together but A left after sometime and B completed the remaining work in 23 days. After how many days of start of the work did A leave?***

Sol. A's 1 days' work $= \frac{1}{45}$, B's 1 days' work $= \frac{1}{40}$

Suppose A left the work after x days of the start of the work. Then,

Work done by A in x days $= \dfrac{x}{45}$

Work done by B in $(x + 23)$ days $= \dfrac{(x+23)}{40}$

Given, $\dfrac{x}{45} + \dfrac{(x+23)}{40} = 1 \Rightarrow \dfrac{8x + 9x + 207}{360} = 1$

$\Rightarrow \quad 17x = 360 - 207 = 153 \Rightarrow x = \dfrac{153}{17} = \mathbf{9\ days}$

$\therefore$ A left after 9 days of start of work.

Ex. 7. ***A certain number of men can do a work in 60 days. If there were 8 more men, it could be completed in 10 days less. How many men were there in the beginning?***

Sol. Let the number of men in the beginning be x.

x men can do the work in 60 days.

$(x + 8)$ men can do the work in 50 days.

More men $\Rightarrow$ Less time $\Rightarrow$ Inverse variation.

$\therefore$ The proportion can be written as $x : (x + 8) :: 50 : 60$

$\Rightarrow \quad \dfrac{x}{x+8} = \dfrac{50}{60} \quad \Rightarrow \quad 60x = 50(x + 8)$

$\Rightarrow \quad 60x = 50x + 400 \quad \Rightarrow \quad 10x = 400 \Rightarrow x = \mathbf{40}.$

$\therefore$ There were 40 men in the beginning.

Ex. 8. ***A can finish a work in 24 days, B in 9 days and C in 12 days. B and C start the work but are forced to leave after 3 days. In how many days was the remaining work done by A?***

Sol. $(B + C)$'s 1 days' work $= \dfrac{1}{9} + \dfrac{1}{12} = \dfrac{4+3}{36} = \dfrac{7}{36}$

$\therefore$ $(B + C)$'s 3 days' work $= 3 \times \dfrac{7}{36} = \dfrac{7}{12}$

$\therefore$ Remaining part of the work $= 1 - \dfrac{7}{12} = \dfrac{5}{12}$

A will finish the whole work (1) in 24 days.

$\therefore$ A will do $\dfrac{5}{12}$ part of the whole work in $\dfrac{5}{12} \times 24 = \mathbf{10\ days}$.

Ex. 9. ***A and B working separately can do a piece of work in 10 and 15 days respectively. If they work on alternate days beginning with A, in how many days will the work be completed?***

Sol. Work done by A in 1 day $= \dfrac{1}{10}$

Work done by B in 1 day $= \dfrac{1}{15}$

Work done by A and B in 2 days $= \dfrac{1}{10} + \dfrac{1}{15} = \dfrac{3+2}{30} = \dfrac{5}{30} = \dfrac{1}{6}$

$\because$ $\dfrac{1}{6}$ work is done in 2 days

1 work will be done in $2 \times 6 = \mathbf{12\ days}$.

Ex. 10. ***A can do $\frac{1}{2}$ of a piece of work in 5 days, B can do $\frac{3}{5}$ of the same work in 9 days and C can do $\frac{2}{3}$ of that work in 8 days. In how many days can the three of them together do the work?***

Sol. A can do $\frac{1}{2}$ of the work in 5 days $\Rightarrow$ A can do the whole work in $5\times\frac{2}{1}$ = 10 days

B can do $\frac{3}{5}$ of the work in 9 days $\Rightarrow$ B can do the whole work in $9\times\frac{5}{3}$ = 15 days

C can do $\frac{2}{3}$ of that work in 8 days $\Rightarrow$ C can do the whole work in $8\times\frac{3}{2}$ days = 12 days

$\therefore$ $(A+B+C)$'s 1 days' work $=\frac{1}{10}+\frac{1}{15}+\frac{1}{12}=\frac{6+4+5}{60}=\frac{15}{60}=\frac{1}{4}$

Hence, $A + B + C$ can complete the whole work in **4 days**.

Ex. 11. ***If 8 men or 12 women can do a piece of work in 10 days, then what is the number of days required by 4 men and 4 women to finish the work?***

Sol. 8 men = 12 women $\Rightarrow$ 4 men = 6 women

$\therefore$ 4 men + 4 women = 6 women + 4 women = 10 women.

Given, 12 women can do a piece of work in 10 days

$\Rightarrow$ 1 women can do the same work in (12×10) days

$\Rightarrow$ 10 women can do the same work in $\frac{12\times10}{10}$ days = **12 days.**

Question Bank–13(a)

1. A and B together can complete a piece of work in 16 days while B and C together can complete the same work in 12 days and A and C together in 24 days. Find out the number of days that A will take to complete the work, working alone.

(a) 36 days (b) 48 days
(c) 96 days (d) 80 days

2. A and B together can do a piece of work in 8 days. B alone can do it in 12 days. B alone works at it for 4 days. In how many more days after that could A alone complete it.

(a) 15 days (b) 18 days
(c) 16 days (d) 20 days

3. A man and a boy can do a piece of work in 24 days. If the man works alone for last 4 days, it is complete in 5 days. How long would the boy take to do it alone?

(a) 120 days (b) 20 days
(c) 80 days (d) 36 days

4. A and B can do a piece of work in 30 days. While B and C can do the same work in 24 days and C and A in 20 days. They all work together for 10 days when B and C leave. How many days more will A take to finish the work?

(a) 18 days (b) 24 days
(c) 30 days (d) 36 days

5. Babu and Asha can do a job together in 7 days. Asha is $1\frac{3}{4}$ times as efficient as Babu. The same job can be done by Asha alone in

(a) $\frac{49}{4}$ days (b) $\frac{49}{3}$ days
(c) 11 days (d) $\frac{28}{3}$ days

6. I can do a piece of work in 8 days, which can be done by you in 10 days. How long would it take to do it if we work together?

(a) $4\frac{4}{9}$ days (b) $5\frac{3}{9}$ days
(c) $5\frac{1}{2}$ days (d) $4\frac{7}{9}$ days

7. X is twice as good as a workman as Y. X finished a piece of work in 3 hours less than Y. In how many hours could they have finished that piece of work together?
(a) 3 (b) 2
(c) 4 (d) 5

8. A can cultivate $\frac{2}{5}$ of a land in 6 days and B can cultivate $\frac{1}{3}$ of the same land in 10 days. Working together, A and B can cultivate $\frac{4}{5}$ of the land in
(a) 4 days (b) 5 days
(c) 8 days (d) 10 days

9. The time taken by 4 men to complete a job is double the time taken by 5 children to complete the same job. Each man is twice as fast as a woman. How long will 12 men, 10 children and 8 women take to complete a job, given that a child would finish the job in 20 days.
(a) 2 days (b) $2\frac{1}{8}$ days
(c) 4 days (d) 1 day

10. A, B and C can do a piece of work in 11 days, 20 days, and 55 days respectively working alone. How soon can the work be done if A is assisted by B and C on alternate days?
(a) 7 days (b) 8 days
(c) 9 days (d) 10 days

11. A can do a piece of work in 20 days and B in 40 days. A begins the work and there after they work alternately one on each day. On which day will the work be completed?
(a) 24th day (b) 25th day
(c) 26th day (d) 27th day

12. A can complete a work in 9 days, B in 10 days and C in 15 days. B and C together started the work but left the work after 2 days. The time taken to complete the remaining work by A will be
(a) 6 days (b) 8 days
(c) 9 days (d) 5 days

13. A can do a piece of work in 4 hrs., B and C together in 3 hours and A and C together in 2 hours. How long will B alone take to do it?
(a) 10 hours (b) 12 hours
(c) 8 hours (d) 24 hours

14. Ram can do a piece of work in 6 days and Shyam can finish the same work in 12 days. How much work will be finished if both work together for 2 days?
(a) One-fourth of the work
(b) One-third of the work
(c) Half of the work
(d) Whole of the work

15. A is twice as fast a workman as B and together they finish a piece of work in 14 days. In how many days can A alone finish the work?
(a) 18 days (b) 21 days
(c) 24 days (d) 27 days

16. A does half as much work as B in one-sixth of the time. If together they take 10 days to complete a work, how much time shall B alone take to do it?
(a) 70 days (b) 30 days
(c) 40 days (d) 50 days

17. A work could be completed in 100 days by some workers. However, due to absence of 10 workers, it was completed in 110 days. The original number of workers was
(a) 100 (b) 110
(c) 55 (d) 50

18. A man and a boy received ₹ 800 as wages for 5 days for the work they did together. The man's efficiency in the work was three times that of the boy. What are the daily wages of the boy?
(a) ₹ 76 (b) ₹ 56
(c) ₹ 44 (d) ₹ 40

19. A certain number of persons can complete a piece of work in 55 days. If there were 6 persons more, the work could be finished in 11 days less. How many persons were originally there?
(a) 17 (b) 24
(c) 30 (d) 22

20. A can finish a work in 18 days and B can do the same work in half the time taken by A. What part of the same work can they finish in a day, working together?
(a) $\frac{1}{6}$ (b) $\frac{2}{5}$
(c) $\frac{1}{9}$ (d) $\frac{2}{7}$

21. A, B and C can complete a work in 10, 12 and 15 days respectively. They started the work together, but A left the work before 5 days of its completion.

B also left the work 2 days after *A* left. In how many days was the work completed.

(a) 4 (b) 5

(c) 7 (d) 8

22. *A* and *B* together can complete a work in 12 days. *A* alone can complete it in 20 days. If *B* does the work only for half a day daily, then in how many days will *A* and *B* together complete the work?

(a) 10 days (b) 11 days

(c) 15 days (d) 20 days

23. *A*, *B* and *C* can do a piece of work in 20, 30 and 60 days respectively. In how many days can *A* do the work if he is assisted by *B* and *C* on every third day?

(a) 12 days (b) 15 days

(c) 16 days (d) 18 days

24. *P* can complete a work in 12 days working 8 hours a day. *Q* can complete the same work in 8 days working 10 hours a day. If both *P* and *Q* work together, working 8 hours a day, in how many days can they complete the work?

(a) $5\frac{5}{11}$ (b) $5\frac{6}{11}$

(c) $6\frac{5}{11}$ (d) $6\frac{6}{11}$

25. If '*b*' men can do a piece of work in '*c*' days, then the number of days taken by '*d*' men to do $\frac{1}{m}$th of the same piece of work will be

(a) $\frac{bc}{d+m}$ (b) $\frac{bc}{dm}$

(c) $\frac{b+c}{d+m}$ (d) $\frac{b+c}{dm}$

26. If 16 men or 20 women can do a piece of work in 25 days, in what time will 28 men and 15 women do it?

(a) $14\frac{2}{7}$ days (b) $33\frac{1}{3}$ days

(c) $18\frac{3}{4}$ days (d) 10 days

27. If one man or two women or three boys can do a piece of work in 22 days, then the same piece of work will be done by 1 man, 1 boy and 1 woman in

(a) 8 days (b) 12 days

(c) 6 days (d) 11 days

28. *A* does half as much work as *B* in three fourth of the time. If together they take 18 days to complete a work, how much time shall *B* take to do it?

(a) 40 days (b) 35 days

(c) 30 days (d) 45 days

29. A man completes $\frac{5}{8}$ of a job in 10 days. At this rate, how many more days will it take him to finish the job?

(a) 5 (b) 6

(c) 7 (d) $7\frac{1}{2}$

30. 10 men and 15 women finish a work in 6 days. One man alone finishes that work in 100 days. In how many days will a women finish the work?

(a) 125 days (b) 150 days

(c) 90 days (d) 225 days

Answers

1. (c)	**2.** (c)	**3.** (a)	**4.** (a)	**5.** (c)	**6.** (a)	**7.** (b)	**8.** (c)	**9.** (d)	**10.** (b)
11. (d)	**12.** (a)	**13.** (b)	**14.** (c)	**15.** (b)	**16.** (c)	**17.** (b)	**18.** (d)	**19.** (b)	**20.** (a)
21. (c)	**22.** (c)	**23.** (b)	**24.** (a)	**25.** (b)	**26.** (d)	**27.** (b)	**28.** (c)	**29.** (b)	**30.** (d)

Hints and Solutions

1. (c) 1 day's work of $(A+B)+(B+C)+(C+A)$

$$=\frac{1}{16}+\frac{1}{12}+\frac{1}{24}=\frac{9}{48}$$

$\Rightarrow$ 1 day's work of $(A+B+C)=\frac{1}{2}\times\frac{9}{48}=\frac{9}{96}$

$\therefore$ 1 day's work of *A* alone $=\frac{9}{96}-\frac{1}{12}=\frac{1}{96}$

$\therefore$ *A* can do the whole work alone in **96 days.**

2. (c) *A* alone can do $\frac{1}{8}-\frac{1}{12}=\frac{3-2}{24}=\frac{1}{24}$ of the work in 1 day, *i.e.*, *A* alone can do the same work in 24 days.

B's 1 days' work $=\frac{1}{12}$

$\Rightarrow$ B's 4 days' work $= \dfrac{4}{12} = \dfrac{1}{3}$

$\therefore$ Remaining $\dfrac{2}{3}$ of the work is done by A alone in $\dfrac{2}{3} \times 24$ = **16 days**

3. (a) (Man+boy)'s one days' work $= \dfrac{1}{24}$

(Man+boy)'s 20 days' work $= \dfrac{20}{24} = \dfrac{5}{6}$

Remaining $\dfrac{1}{6}$ of the work is done by the man alone in 5 days, *i.e.*, the man alone can do $\dfrac{1}{30}$ of the work in 1 day.

$\therefore$ Boy's one days' work $= \dfrac{1}{24} - \dfrac{1}{30} = \dfrac{1}{120}$.

$\therefore$ Boy alone would take **120 days** to complete the work.

4. (a) $(A+B)$'s one days' work $= \dfrac{1}{30}$

$(B+C)$'s one days' work $= \dfrac{1}{24}$ ...(1)

$(C+A)$'s one days' work $= \dfrac{1}{20}$

$\therefore$ $(A+B+C)$'s one days' work

$$= \frac{1}{2}\left[\frac{1}{30} + \frac{1}{24} + \frac{1}{20}\right] = \frac{1}{2} \times \left[\frac{20+25+30}{600}\right]$$

$$= \frac{75}{1200} = \frac{1}{16} \quad ...(2)$$

$(A+B+C)$'s 10 days' work $= \dfrac{10}{16} = \dfrac{5}{8}$

From (1) and (2), A's one days' work

$$= \frac{1}{16} - \frac{1}{24} = \frac{1}{48}$$

$\therefore$ Remaining $\dfrac{3}{8}$ of the work is done by A alone in $\dfrac{3}{8} \times 48$ = **18 days.**

5. (c) If Babu can do a job in x days, then Asha can do the same job in $\dfrac{4x}{7}$ days.

$\therefore$ (Asha and Babu)'s one days' job $= \dfrac{1}{x} + \dfrac{7}{4x}$

Also given, (Asha and Babu)'s one days' job $= \dfrac{1}{7}$

$\therefore$ $\dfrac{1}{x} + \dfrac{7}{4x} = \dfrac{1}{7}$ $\Rightarrow$ $\dfrac{4+7}{4x} = \dfrac{1}{7}$ $\Rightarrow$ $x = \dfrac{77}{4}$

$\therefore$ Asha alone can do the job in $\dfrac{4}{7} \times \dfrac{77}{4}$ days = **11 days.**

6. (a) Yours' and mine one days' work $= \dfrac{1}{8} + \dfrac{1}{10} = \dfrac{9}{40}$.

$\therefore$ I and you together can finish the work in $\dfrac{40}{9}$, *i.e.,* $\mathbf{4\frac{4}{9}}$ **days.**

7. (b) X finished a work in 3 hours less than Y. X is twice as good as a workman as Y. Therefore, Y finished the same work in 6 hours.

$\therefore$ $(X+Y)$'s one hours' work $= \dfrac{1}{3} + \dfrac{1}{6} = \dfrac{3}{6} = \dfrac{1}{2}$

$\therefore$ X and Y together can finish the whole work in **2 hours.**

8. (c) A can cultivate the full part of the land in $\dfrac{6 \times 5}{2}$ days = 15 days

B can cultivate the full part of the land in (10 × 3) = 30 days

$\therefore$ Part of the land cultivated by $(A + B)$ in 1 day

$$= \left(\frac{1}{15} + \frac{1}{30} = \frac{2+1}{30} = \frac{3}{30}\right) = \frac{1}{10}$$

$\dfrac{1}{10}$ part of the land is cultivated by $(A+B)$ in 1 day

$\therefore$ $\dfrac{4}{5}$ part of the land will be cultivated by $(A + B)$ in $10 \times \dfrac{4}{5}$ days = **8 days.**

9. (d) Time taken by 5 children to complete the job

$= \dfrac{20}{5}$ days = 4 days

$\therefore$ Time taken by 4 men to complete the job
= (4 × 2) days = 8 days

Time taken by 4 women to the complete the job
= (8 × 2) days = 16 days

Therefore, 5 children's 1 days' work $= \dfrac{1}{4}$

4 men's 1 days' work = $\frac{1}{8}$

4 women's 1 days' work = $\frac{1}{16}$

∴ (12 men's + 10 children's + 8 women's) 1 days' work = 3 × (4 men's 1 days' work) + 2 × (5 children's 1 days' work) + 2 × (4 women's 1 days' work)

$= 3\times\frac{1}{8}+2\times\frac{1}{4}+2\times\frac{1}{16}$

$= \frac{3}{8}+\frac{1}{2}+\frac{1}{8}=\frac{8}{8}=1$

∴ Required time = **1 day**.

10. (b) $(A+B)'$s 1 days' work $=\left(\frac{1}{11}+\frac{1}{20}\right)=\frac{20+11}{220}=\frac{31}{220}$

$(A+C)'$s 1 days' work $=\left(\frac{1}{11}+\frac{1}{55}\right)=\frac{6}{55}$

∴ Work done in 2 days $=\frac{31}{220}+\frac{6}{55}=\frac{55}{220}=\frac{1}{4}$

$\frac{1}{4}$th of the work is done in 2 days

∴ Whole work is done in **8 days**.

11. (d) Work done by $(A+B)$ in 2 days $=\left(\frac{1}{20}+\frac{1}{40}\right)=\frac{3}{40}$

$\frac{3}{40}$ of the work is done by (A + B) in 2 days

∴ Whole work is done by (A + B) in $2\times\frac{40}{3}$ days

$=\frac{80}{3}$ days $= 26\frac{2}{3}$ days

∴ Work will be completed on **27th day**.

12. (a) $(B+C)'$s 1 days' work $=\left(\frac{1}{10}+\frac{1}{15}\right)$

$=\frac{3+2}{30}=\frac{5}{30}=\frac{1}{6}$

$(B+C)'$s 2 days' work $= 2\times\frac{1}{6}=\frac{1}{3}$

Remaining work $= 1-\frac{1}{3}=\frac{2}{3}$

Whole (1) work can be completed by A in 9 days

∴ $\frac{2}{3}$ of the work can be completed by A in $\left(9\times\frac{2}{3}\right)$ days = **6 days**.

13. (b) A's 1 hours' work $=\frac{1}{4}$

$(B+C)'$s 1 hours' work $=\frac{1}{3}$

$(A+C)'$s 1 hours' work $=\frac{1}{2}$

∴ C's 1 hours' work $=\frac{1}{2}-\frac{1}{4}=\frac{1}{4}$

B's 1 hours' work $=\frac{1}{3}-\frac{1}{4}=\frac{1}{12}$

Hence, B will complete the whole work in **12 days**.

14. (c) Ram's 1 days' work $=\frac{1}{6}$

Shyam's 1 days' work $=\frac{1}{12}$

The part of work in 2 days' by both

$= 2\left(\frac{1}{6}+\frac{1}{12}\right)=2\left(\frac{2+1}{12}\right)=2\times\frac{3}{12}=\frac{\mathbf{1}}{\mathbf{2}}.$

15. (b) If B finishes the work in x days, then A finishes the work in $\frac{x}{2}$ days.

$\Rightarrow B'$s one days' work $=\frac{1}{x}$

and A's one days' work $=\frac{2}{x}$

Given $\frac{1}{x}+\frac{2}{x}=\frac{1}{14} \Rightarrow \frac{3}{x}=\frac{1}{14}$

$\Rightarrow x = 14\times 3 = 42$ days

∴ A alone can finish the work in $\frac{42}{2}$ = **21 days.**

16. (c) Let the time taken by B alone to complete the work be x days.

Then, time taken by A to complete $\frac{1}{2}$ of the work $=\frac{x}{6}$ days

∴ Work of A for 1 day $=\frac{1}{2}\times\frac{6}{x}=\frac{3}{x}$

and work of B for 1 day $=\frac{1}{x}$

Given, $\frac{3}{x}+\frac{1}{x}=\frac{1}{10}$

$\Rightarrow \frac{4}{x}=\frac{1}{10} \quad\Rightarrow\quad x =$ **40 days.**

17. (b) Let the number of workers be x.
x workers shall take 100 days, but
$(x - 10)$ workers shall take 110 days, *i.e.,*
Less workers $\Rightarrow$ more time.
Hence, it is an inverse variation.
The proportion can be written as
$x : (x - 10) : : 110 : 100$

$\Rightarrow \frac{x}{x-10} = \frac{110}{100} \quad \Rightarrow \quad 100x = 110x - 1100$

$\Rightarrow 10x = 1100 \quad \Rightarrow \quad \mathbf{x = 110}$.

18. (d) 5 days' wages for a man and a boy = ₹ 800

$\therefore$ 1 days' wage for a man and a boy = ₹ $\frac{800}{5}$ = ₹ 160

Let the wage of the boy per day be ₹ x.Then,
Wage of the man per day = ₹ $3x$
Given $3x + x = 160$

$\Rightarrow 4x = 160 \quad \Rightarrow \quad \mathbf{x = 40}$.

19. (b) Type Solved Example 7.

20. (a) A can finish a work in 18 days
$\Rightarrow$ B can finish the same work in 9 days.

A's 1 days' work $= \frac{1}{18}$

B's 1 days' work $= \frac{1}{9}$

$\therefore$ $(A+B)$'s 1 days' work $= \frac{1}{18} + \frac{1}{9} = \frac{1+2}{18} = \frac{3}{18} = \mathbf{\frac{1}{6}}$.

21. (c) Let the work be completed in x days.

A's 1 days' work $= \frac{1}{10}$

$\Rightarrow$ A's $(x - 5)$ days' work $= \frac{(x-5)}{10}$

B's 1 days' work $= \frac{1}{12}$

$\Rightarrow$ B's $(x - 3)$ days' work $= \frac{(x-3)}{12}$

C's 1 days' work $= \frac{1}{15}$

$\Rightarrow$ C's x days' work $= \frac{x}{15}$

$\Rightarrow \frac{x-5}{10} + \frac{x-3}{12} + \frac{x}{15} = 1$

$\Rightarrow \frac{6(x-5)+5(x-3)+4x}{60} = 1$

$\Rightarrow 6x - 30 + 5x - 15 + 4x = 60$

$\Rightarrow 15x = 60 + 45 = 105 \quad \Rightarrow \quad x = 7$

$\therefore$ The work was completed in **7 days**.

22. (c) B's 1 days' work $= \frac{1}{12} - \frac{1}{20} = \frac{5-3}{60} = \frac{2}{60} = \frac{1}{30}$

$\therefore$ B's half days' work $= \frac{1}{2} \times \frac{1}{30} = \frac{1}{60}$

Hence, $(A+B)$'s 1 days' work $= \left(\frac{1}{20} + \frac{1}{60}\right)$

$= \frac{3+1}{60} = \frac{4}{60} = \frac{1}{15}$

$\therefore$ A and B will together complete the work in **15 days**.

23. (b) A's 2 days' work $= 2 \times \frac{1}{20} = \frac{1}{10}$

$(A+B+C)$'s 1 days' work $= \frac{1}{20} + \frac{1}{30} + \frac{1}{60}$

$= \frac{3+2+1}{60} = \frac{6}{60} = \frac{1}{10}$

Since A is assisted by B and C on every third day.

Work done in 3 days $= \frac{1}{10} + \frac{1}{10} = \frac{1}{5}$

$\frac{1}{5}$ of the work is done in 3 days

$\therefore$ Whole work is done in $3 \times \frac{5}{1} = \mathbf{15}$ **days**.

24. (a) P can complete the work in (12 × 8) hours = 96 hours

$\therefore$ P's 1 hours' work $= \frac{1}{96}$

Q can complete the work in (8 × 10) hours = 80 hours

$\therefore$ Q's 1 hours' work $= \frac{1}{80}$

$\therefore$ $(P+Q)$'s 1 hours' work $= \frac{1}{96} + \frac{1}{80} = \frac{5+6}{480} = \frac{11}{480}$

$\therefore$ $(P+Q)$ can complete the whole work in $\frac{480}{11}$ hours.

$\therefore$ Working 8 hours a day, P and Q can complete the whole work in $\left(\frac{480}{11} \times \frac{1}{8}\right)$ days $= \frac{60}{11}$ days

$= \mathbf{5\frac{5}{11}}$ **days**.

25. (b) b men can do a piece of work in c days.

$\therefore$ 1 man can do the some piece of work in bc days

$\therefore$ d men can do the same piece of work in $\frac{bc}{d}$ days

$\therefore$ d men can do $\frac{1}{m}$th of the same piece of work in $\frac{bc}{dm}$ **days.**

26. (d) 16 men = 20 women $\Rightarrow$ 4 men = 5 women

$\therefore$ 28 men + 15 women = 35 women + 15 women = 50 women

20 women can complete a work in 25 days.

$\therefore$ 1 woman can complete the work in (25×20) days

$\therefore$ 50 women can complete the work in $\left(\frac{25 \times 20}{50}\right)$ days = **10 days.**

27. (b) 1 man's 1 days' work $= \frac{1}{22}$

1 woman's 1 days' work $= \frac{1}{2} \times \frac{1}{22} = \frac{1}{44}$

1 boy's 1 days' work $= \frac{1}{3} \times \frac{1}{22} = \frac{1}{66}$

$\therefore$ (1 man's + 1 woman's + 1 boy's) 1 days' work

$= \frac{1}{22} + \frac{1}{44} + \frac{1}{66} = \frac{6+3+2}{132} = \frac{11}{132} = \frac{1}{12}$

$\therefore$ 1 man, 1 boy and 1 woman can do the work in **12 days.**

28. (c) Let B complete the work in x days.

$\Rightarrow$ B's 1 days' work $= \frac{1}{x}$

$\therefore$ A complete $\frac{1}{2}$ of the work in $\frac{3x}{4}$ days

$\Rightarrow$ A's 1 days' work $= \frac{\frac{1}{2}}{\frac{3x}{4}} = \frac{2}{3x}$

Given, $\frac{1}{x} + \frac{2}{3x} = \frac{1}{18}$

$\Rightarrow \frac{3+2}{3x} = \frac{1}{18} \Rightarrow \frac{5}{3x} = \frac{1}{18}$

$\Rightarrow 3x = 90 \Rightarrow x = \mathbf{30}$

29. (b) $\frac{5}{8}$ of the job is done in 10 days

$\therefore$ Whole job is done is $\left(10 \times \frac{8}{5}\right)$ days

$\therefore$ $\frac{3}{8}$ of the job is done in $\left(10 \times \frac{8}{5} \times \frac{3}{8}\right)$ days = **6 days.**

30. (d) 1 man can finish the whole work in 100 days

$\therefore$ 1 man's 1 days' work $= \frac{1}{100}$

$\Rightarrow$ 10 men's 1 days' work $= 10 \times \frac{1}{100} = \frac{1}{10}$

Let the time taken by one woman to finish the work be x days.

Then, 1 woman's 1 days' work $= \frac{1}{x}$

$\Rightarrow$ 15 women's 1 days' work $= \frac{15}{x}$

Given, (10 men's + 15 women's) 1 days' work $= \frac{1}{6}$

$\Rightarrow \frac{1}{10} + \frac{15}{x} = \frac{1}{6} \Rightarrow \frac{15}{x} = \frac{1}{6} - \frac{1}{10} = \frac{5-3}{30} = \frac{2}{30} = \frac{1}{15}$

$\Rightarrow x = $ **225 days.**

PIPES AND CISTERNS

KEY FACTS

1. A cistern or a water tank is connected with two types of pipes. One which fills it up is called a **inlet** and the other one which empties it out is called an **outlet.**
2. If a pipe fills a water tank in 10 hours, then in 1 hour it fills $\frac{1}{10}$th part of it. In other words, we can say that the work done by the pipe in 1 hour $= \frac{1}{10}$.

3. Similarly, if an outlet empties a tank in 8 hours, then in 1 hour it empties $\frac{1}{8}$th part of the tank. We can say that the work done by the outlet in 1 hour is $\left(-\frac{1}{8}\right)$.

Note: The work done by the inlet is always positive whereas the work done by the outlet is always negative.

Solved Examples

Ex. 1. ***How long will two pipes together take to fill a cistern which they can separately fill in 20 and 25 minutes?***

Sol. Part of cistern filled by 1st pipe in 1 min = $\frac{1}{20}$

Part of cistern filled by 2nd pipe in 1 min = $\frac{1}{25}$

$\therefore$ Part of cistern filled by both pipes together in 1 min = $\frac{1}{20}+\frac{1}{25}=\frac{5+4}{100}=\frac{9}{100}$

$\therefore$ The cistern will be filled by both the pipes together in $\frac{100}{9}$ min = $\mathbf{11\frac{1}{9}}$ **min.**

Ex. 2. ***An electric pump can fill a tank in 3 hours. Because of a leak in the tank, it took $3\frac{1}{2}$ hours to fill the tank. In how much time can the leak drain all the water of the tank?***

Sol. Part of the tank filled by the pump in 1 hour = $\frac{1}{3}$

Net part of the tank filled by the pump and the leak in 1 hour = $\frac{1}{\frac{7}{2}}=\frac{2}{7}$

$\therefore$ Part of the tank emptied by the leak in 1 hour = $\frac{1}{3}-\frac{2}{7}=\frac{1}{21}$

$\therefore$ The tank will be emptied by the leak in **21 hours.**

Ex. 3. ***Two taps can fill a tank in 15 and 12 minutes respectively. A third tap can empty it in 20 minutes. If all the taps are opened at the same time, then in how much time will the tank be filled?***

Sol. Part of the tank filled when all the three taps are opened simultaneously= $\left(\frac{1}{15}+\frac{1}{12}-\frac{1}{20}\right)$

$=\frac{4+5-3}{60}=\frac{6}{60}=\frac{1}{10}$

$\therefore$ The tank will be filled in **10 minutes.**

Ex. 4. ***There are two taps to fill a tank and a third to empty it. When the third tap is closed, they can fill the tank in 10 minutes and 12 minutes respectively. If all the three taps are opened, the tank is filled in 15 minutes. If the first two taps are closed, in what time can the third tap empty the tank when it is full?***

Sol. Let the third tap take x minutes to empty the tank. Then,

$$\frac{1}{10}+\frac{1}{12}-\frac{1}{x}=\frac{1}{15} \Rightarrow \frac{1}{x}=\frac{1}{10}+\frac{1}{12}-\frac{1}{15}=\frac{6+5-4}{60}=\frac{7}{60}$$

$\therefore$ The third tap takes $\frac{60}{7}$ min = $\mathbf{8\frac{4}{7}}$ **min** to empty the tank.

Ex. 5. ***Two pipes A and B can fill a cistern in $37\frac{1}{2}$ minutes and 45 minutes respectively. Both the pipes are opened. The cistern will be filled in just half an hour, if pipe B is turned off after x minutes. What is the value of x?***

Sol. Let A work till the end, *i.e.*, 30 minutes and B work for x minutes.

Then, part of cistern filled by pipe A in 30 min = $\frac{30}{\frac{75}{2}}$

Part of cistern filled by pipe B in x min = $\frac{x}{45}$

$$\frac{30}{\frac{75}{2}}+\frac{x}{45}=1 \quad \Rightarrow \quad \frac{60}{75}+\frac{x}{45}=1$$

$$\Rightarrow \quad \frac{x}{45}=1-\frac{4}{5}=\frac{1}{5} \quad \Rightarrow \quad x=\frac{45}{5}=\mathbf{9\ min.}$$

Ex. 6. ***Two pipes A and B can fill a cistern in 20 minutes and 25 minutes respectively. Both are opened together but at the end of 5 minutes B is turned off. What is the total time taken to fill the cistern?***

Sol. Part of the cistern filled by $(A + B)$ in one minute = $\left(\frac{1}{20}+\frac{1}{25}\right)=\frac{5+4}{100}=\frac{9}{100}$

$\therefore$ Part of the cistern filled by $(A + B)$ in 5 minutes = $5\times\frac{9}{100}=\frac{9}{20}$

Remaining part = $1-\frac{9}{20}=\frac{11}{20}$

$\frac{1}{20}$ part of the cistern is filled by A in a minute

$\therefore$ $\frac{11}{20}$ part of the cistern is filled by A in $20\times\frac{11}{20}$ minutes = 11 minutes

$\therefore$ Total time taken to fill the cistern = (11 + 5) minutes = 16 min.

Question Bank–13(b)

1. 10 identical taps fill a tank in 24 minutes. To fill the tank in 1 hour, how many taps are required to be used?

(a) 2 (b) 4
(c) 6 (d) 8

2. Two pipes can fill a tank in 20 minutes and 30 minutes respectively. If both the pipes are opened simultaneously, then the tank will be filled in

(a) 10 minutes (b) 12 minutes
(c) 15 minutes (d) 25 minutes

3. A cistern can be filled up by one pipe in 12 hours and by another in 8 hours. Both the pipes are kept open for $2\frac{1}{2}$ hours. The part of the cistern filled up is

(a) $\frac{25}{48}$ (b) $\frac{5}{6}$
(c) $\frac{25}{36}$ (d) $\frac{12}{25}$

4. A tap can fill a tank in 25 minutes and another tap can empty it in 50 minutes. If both are opened together simultaneously, then the tank will be filled in

(a) 20 minutes (b) 30 minutes
(c) 40 minutes (d) 50 minutes

5. Tap A can fill a water tank in 12 min, tap B will fill the same tank in 10 min and tap C can empty the tank in 6 min. If all the three taps are opened together, in how many minutes will the tank be completely filled up or emptied?

(a) 15 min (b) 30 min
(c) 45 min (d) 60 min

6. One pipe can fill a tank three times as fast as another pipe. If together the two pipes can fill the tank in 36 minutes, then the slower pipe alone would be able to fill the tank in

(a) 81 minutes (b) 108 minutes
(c) 144 minutes (d) 192 minutes

7. A cistern has two pipes. One pipe can fill it with water in 8 hours and the other can empty it in 5 hours. In how many hours will the cistern be emptied if both the pipes are opened together when $\frac{3}{4}$th of the cistern is already full of water.

(a) $13\frac{1}{3}$ hours (b) 10 hours

(c) 6 hours (d) $3\frac{1}{3}$ hours

8. A cistern has two taps which fill it in 12 minutes and 15 minutes respectively. There is also a waste pipe in the cistern. When all the pipes are opened, the empty cistern is full in 20 minutes. How long will the waste pipe take to empty the full cistern?

(a) 12 minutes (b) 10 minutes

(c) 8 minutes (d) 16 minutes

9. Two pipes P and Q would fill a cistern in 24 and 32 minutes respectively. Both pipes are kept open. When should the first pipe be turned off so that the cistern may be just filled in 16 minutes?

(a) After 10 minutes (b) After 12 minutes

(c) After 14 minutes (d) After 11 minutes

10. A tank has 3 pipes. The first pipe can fill $\frac{1}{2}$ part of the tank in 1 hour and the second pipe can fill $\frac{1}{3}$ part in 1 hour. The third pipe is for making the tank empty. When all the three pipes are open, $\frac{7}{12}$ part of the tank is filled in 1 hour. How much time will the third pipe take to empty the completely filled tank?

(a) 3 hours (b) 4 hours

(c) 5 hours (d) 6 hours

11. A pump can fill a tank with water in 2 hours. Because of a leak in the tank, it takes $2\frac{1}{3}$ hours to fill the tank. The leak can empty the filled tank in

(a) $2\frac{1}{3}$ hours (b) 7 hours

(c) 8 hours (d) 14 hours

12. A pipe can fill a tank in x hours and another pipe can empty it in y ($y > x$) hours. If both the pipes are open, in how many hours will the tank be filled?

(a) $(x - y)$ hours (b) $(y - x)$ hours

(c) $\frac{xy}{(x-y)}$ hours (d) $\frac{xy}{(y-x)}$ hours

13. A tank can be filled by a tap in 20 minutes and by another tap in 60 minutes. Both the taps are kept open for 10 minutes and then the first tap is shut off. After this, the tank will be completely filled in

(a) 10 minutes (b) 12 minutes

(c) 15 minutes (d) 20 minutes

14. A water tank is $\frac{2}{5}$th full. Pipe A can fill the tank in 10 minutes and pipe B can empty it in 6 minutes. If both the pipes are open, how long will it take to empty or fill the tank completely?

(a) 6 minutes to empty (b) 6 minutes to fill

(c) 9 minutes to empty (d) 9 minutes to fill

15. Pipes A and B can fill a tank in 5 and 6 hours respectively. Pipe C can empty it in 12 hours. The tank is half full. All the three pipes are in operation simultaneously. After how much time the tank will be full?

(a) $3\frac{9}{17}$ hours (b) 11 hours

(c) $2\frac{8}{11}$ hours (d) $1\frac{13}{17}$ hours.

Answers

1. (b)	**2.** (b)	**3.** (a)	**4.** (d)	**5.** (d)	**6.** (c)	**7.** (b)	**8.** (b)	**9.** (b)	**10.** (b)
11. (d)	**12.** (c)	**13.** (d)	**14.** (a)	**15.** (d)					

Hints and Solutions

1. (b) In 24 minutes, the no. of taps required to fill a tank = 10

In 1 minute, the no. of taps required to fill the tank = (10 × 24)

In 60 minutes, the no. of taps required

$$= \frac{10 \times 24}{60} = \textbf{4 taps.}$$

2. (b) Part of tank filled by 1st pipe in 1 minute = $\frac{1}{20}$

Part of tank filled by 2nd pipe in 1 minute = $\frac{1}{30}$

∴ Part of tank filled by both the pipes in 1 minute

$$= \left(\frac{1}{20} + \frac{1}{30}\right) = \frac{5}{60} = \frac{1}{12}$$

Time taken to fill the tank = **12 minutes.**

3. (a) Part of the cistern filled in 1 hour

$$= \left(\frac{1}{12}+\frac{1}{8}\right) = \frac{2+3}{24} = \frac{5}{24}$$

$\therefore$ Part of the cistern filled in $2\frac{1}{2}$ hours, *i.e.,* $\frac{5}{2}$ hours

$$= \frac{5}{24} \times \frac{5}{2} = \mathbf{\frac{25}{48}}.$$

4. (d) Part of the tank filled when both the taps are opened simultaneously $= \left(\frac{1}{25}-\frac{1}{50}\right) = \frac{1}{50}$

$\therefore$ If both the taps are opened together, the tank will be filled in **50 minutes**.

5. (d) Part of the tank filled when the three taps are opened together

$$= \left(\frac{1}{12}+\frac{1}{10}-\frac{1}{6}\right) = \frac{5+6-10}{60} = \frac{11-10}{60} = \frac{1}{60}$$

$\therefore$ If all the three taps are opened together, then the tank will be filled in **60 minutes**.

6. (c) Suppose the faster pipe A can fill the tank in x minutes. Then,

Slower pipe B will fill it in $3x$ minutes.

In one minute both the pipes can fill $\frac{1}{x}+\frac{1}{3x} = \frac{4}{3x}$ of the tank.

Also, given that in one minute both the pipes can fill $\frac{1}{36}$ of the tank.

$\therefore \frac{4}{3x} = \frac{1}{36} \Rightarrow x = 48$

The slower pipe will take $3 \times 48 =$ **144 minutes**.

7. (b) Part of the tank filled by the first pipe in 1 hour $= \frac{1}{8}$

Part of the tank emptied by the second pipe in 1 hour $= \frac{1}{5}$

$\therefore$ Part of the tank emptied when both the pipes are opened together $= \frac{1}{5}-\frac{1}{8} = \frac{3}{40}$

$\therefore \frac{3}{40}$ part of the tank is emptied in 1 hour

$\therefore \frac{3}{4}$th of the tank will be emptied in $\left(\frac{40}{3}\times\frac{3}{4}\right)$ hours = **10 hours**.

8. (b) Let the waste pipe take x minutes to empty the full cistern. Then,

$$\left(\frac{1}{12}+\frac{1}{15}-\frac{1}{x}\right) = \frac{1}{20}$$

$$\Rightarrow \frac{1}{x} = \frac{1}{12}+\frac{1}{15}-\frac{1}{20} = \frac{5+4-3}{60} = \frac{6}{60} = \frac{1}{10}$$

$\therefore$ The waste pipe takes **10 minutes** to empty the full cistern.

9. (b) Q works for 16 minutes

Part of cistern filled by Q in 16 minutes $= \frac{16}{32} = \frac{1}{2}$

$\Rightarrow \frac{1}{2}$ part of the cistern should be filled by P

So the time taken by P to fill the half part $= \frac{24}{2}$

= **12 min.**

10. (b) Let the time taken by the third pipe to completely empty the tank $= x$ hours. Then,

$$\left(\frac{1}{2}+\frac{1}{3}-\frac{1}{x}\right) = \frac{7}{12}$$

$$\Rightarrow \frac{5}{6}-\frac{1}{x} = \frac{7}{12} \Rightarrow \frac{1}{x} = \frac{5}{6}-\frac{7}{12} = \frac{3}{12} = \frac{1}{4}$$

$\therefore$ The third pipe shall take **4 hours** to empty the completely filled tank.

11. (d) Part of the tank filled by the pump in 1 hour $= \frac{1}{2}$

Total part of the tank filled by the pump and leak in 1 hour $= \frac{1}{\frac{7}{3}} = \frac{3}{7}$

$\therefore$ Part of the tank emptied by the leak in 1 hour

$$= \frac{1}{2}-\frac{3}{7} = \frac{7-6}{14} = \frac{1}{14}$$

$\Rightarrow$ The leak will empty the tank in **14 hours**.

12. (c) Work done by the pipe filling the tank in 1 hour $= \frac{1}{x}$

Work done by the pipe emptying the tank in 1 hour $= \frac{1}{y}$

Net part of the tank filled in 1 hour

$$= \frac{1}{x}-\frac{1}{y} = \frac{y-x}{xy}$$

$\therefore$ The tank will be filled in $\mathbf{\left(\frac{xy}{y-x}\right)}$ **hours**.

13. (d) Part of the tank filled by both the taps in 10 minutes $= 10\times\left(\frac{1}{20}+\frac{1}{60}\right)=10\times\left(\frac{3+1}{60}\right)=\frac{2}{3}$

Remaining part $= 1-\frac{2}{3}=\frac{1}{3}$

$\frac{1}{60}$th part of the tank is filled by the second pipe in 1 min.

$\therefore$ $\frac{1}{3}$rd of the tank will be filled by the second pipe in $60\times\frac{1}{3}$ = **20 min.**

14. (a) Pipe A in 1 minute fills $\frac{1}{10}$ part and pipe B in 1 minute empties $\frac{1}{6}$ part.

$\therefore$ Work done by pipe $(A + B)$ in 1 min $=\frac{1}{10}-\frac{1}{6}$

$=-\frac{1}{15}$

$\Rightarrow$ $\frac{1}{15}$ parts gets emptied in 1 min

$\Rightarrow$ $\frac{2}{5}$ part is emptied in $15\times\frac{2}{5}$ min = **6 min.**

15. (d) Part of the tank filled in one hour

$=\frac{1}{5}+\frac{1}{6}-\frac{1}{12}$

$=\frac{12+10-5}{60}=\frac{17}{60}$ of the tank is filled

Therefore, $\frac{1}{2}$ of the tank is filled in $\frac{60}{17}\times\frac{1}{2}=\frac{30}{17}$ hours $= \mathbf{1\frac{13}{17}}$ **hours.**

Self Assessment Sheet–13

1. A alone can complete a piece of work in 6 days and B alone can complete the same piece of work in 12 days. In how many days can A and B together complete the same piece of work ?

(a) 5 days (b) 4 days
(c) 3 days (d) 2 days
(e) None of these

2. Kamal can do a work in 15 days. Bimal is 50% more efficient than Kamal. The number of days, Bimal will take to do the same piece of work, is :

(a) 10 (b) $10\frac{1}{2}$
(c) 12 (d) 14

3. A man and a boy complete a work together in 24 days. If for the last 6 days man alone does the work then it is completed in 26 days. How long the boy will take to complete the work alone ?

(a) 72 days (b) 20 days
(c) 26 days (d) 36 days

4. A and B together can do a piece of work in 6 days. A alone can do it in 10 days. What time will B require to do it alone ?

(a) 20 days (b) 15 days
(c) 25 days (d) 30 days

5. 'A' can do a piece of work in 25 days and B in 20 days. They work together for 5 days and then 'A' goes away. In how many days will 'B' finish the remaining work ?

(a) 17 days (b) 11 days
(c) 10 days (d) None of these

6. Two man undertake to do a piece of work for ₹ 1400. First man alone can do this work in 7 days while the second man alone can do this work in 8 days. If they working together complete this work in 3 days with the help of a boy, how should money be divided ?

(a) ₹ 600, ₹ 500, ₹ 300
(b) ₹ 600, ₹ 550, ₹ 250
(c) ₹ 600, ₹ 525, ₹ 275
(d) ₹ 500, ₹ 525, ₹ 375

7. A and B can do a piece of work in 8 days, B and C can do the same work in 12 days. If A, B and C can complete the same work in 6 days, in how many days can A and C complete the same work ?

(a) 8 (b) 10
(c) 12 (d) 16

8. Two taps can fill a tub in 5 minutes and 7 minutes respectively. A pipe can empty it in 3 minutes. If all the three are kept open simultaneously, when will the tub be full ?

(a) 60 min (b) 85 min
(c) 90 min (d) 105 min

9. A man completes $\frac{4}{9}$ of a job in 12 days. At this rate, how many more days will it take him to finish the job?

(a) 15 (b) 10
(c) 9 (d) $7\frac{1}{2}$

10. Three workers, working all days can do a work in 10 days, but one of them having other employment can work only half time. In how many days the work can be finished.

(a) 15 days (b) 16 days
(c) 12 days (d) 12.5 days

11. A tap fills a tank in 12 hours and the other empties it in 24 hours. If both are opened simultaneously, then the tank will be filled in :

(a) 42 hours (b) 20 hours
(c) 24 hours (d) 22 hours

12. If 15 pumps of equal capacity can fill a tank in 7 days, then how many extra pumps will be required to fill the tank in 5 days ?

(a) 6 (b) 7
(c) 14 (d) 21

13. A tap can fill a tank in 6 hours. After half the tank i filled, three more, similar taps are opened. What i the total time taken to fill the tank completely ?

(a) 3 h 15 min (b) 3 h 45 min
(c) 4 h (d) 4 h 15 min

14. If three taps are opened together, a tank is filled i 12 hours. One of the taps can fill it in 10 hours an another in 15 hours. How does the third tap work

(a) empties in 12 hours (b) empties in 14 hours
(c) fills in 12 hours (d) fills in 14 hours

15. *A* can do a work in 18 days, *B* in 9 days and *C* in days. *A* and *B* start working together and after 2 day *C* joins them. What is the total number of days take to finish the work ?

(a) 4.33 (b) 4.5
(c) 4.66 (d) None

Answers

1. (b)	**2.** (a)	**3.** (d)	**4.** (b)	**5.** (b)	**6.** (c)	**7.** (a)	**8.** (d)	**9.** (a)	**10.** (c)
11. (c)	**12.** (a)	**13.** (b)	**14.** (a)	**15.** (d) Correct answer is 4 days					

Chapter 14

DISTANCE, TIME AND SPEED

KEY FACTS

1. **Speed** of a moving object is the distance moved by it in unit time.
2. **Uniform speed:** If an object covers equal distances in equal intervals of times, then its speed is to be constant or uniform.
3. **Relation between Distance, Speed and Time**

   (i) Distance = Time × Speed (ii) Speed = $\frac{\text{Distance}}{\text{Time}}$ (iii) Time = $\frac{\text{Distance}}{\text{Speed}}$

4. (i) To convert a speed in km/h to a speed in m/s, multiply by $\frac{5}{18}$.

   (ii) To convert a speed in m/s to a speed in km/h, multiply by $\frac{18}{5}$.

Solved Examples

Ex. 1. ***A person crosses a 600 m long street in 5 minutes. What is his speed in km per hour?***

Sol. Speed = $\frac{\text{Distance}}{\text{Time}} = \frac{600}{5\times60}$ m/sec = $\left(2\times\frac{18}{5}\right)$ km/hr = **7.2 km/hr**.

Ex. 2. ***A truck covers a distance of 550 metres in 1 minute whereas a bus covers a distance of 33 km in 45 minutes. What is the ratio of their speeds?***

Sol. Required ratio = Speed of truck : Speed of bus

$$= \frac{550}{60}\text{ m/sec} : \frac{33\times1000}{45\times60}\text{ m/sec} = \frac{55}{6}\text{ m/sec} : \frac{110}{9}\text{ m/sec} = \frac{55}{6}\times\frac{9}{110} = \frac{3}{4} = \mathbf{3:4}.$$

Ex. 3. ***A man walks at a speed of 4 km/hr and runs at a speed of 8 km/hr. How much time will the man require to cover a distance of 24 km if he completes half of his journey walking and half running?***

Sol. Required time = $\frac{12}{4}$ hrs. + $\frac{12}{8}$ hrs. = 3 hrs. + 1.5 hrs. = **4.5 hrs.**

Ex. 4. ***In a 800 m race, A defeated B by 15 seconds. If A's speed was 8 km /hour, what was B's speed?***

Sol. Speed of A = 8 km/hr = $8\times\frac{5}{18}$ m/sec = $\frac{20}{9}$ m/sec

Distance covered = 800 m

$\therefore$ Time taken by A to complete the race $= \dfrac{800}{\frac{20}{9}} = \dfrac{800 \times 9}{20} = 360$ sec.

Time taken by B to complete the race = (360 + 15) sec = 375 sec

$\therefore$ Speed of $B = \dfrac{800}{375}$ m/sec $= \dfrac{800}{375} \times \dfrac{18}{5}$ km/hr $= \dfrac{192}{25}$ km/hr $= 7\dfrac{17}{25}$ **km/hr**.

Ex. 5. ***A scooterist completes a certain journey in 10 hours. He covers half the distance at 30 km/hr and the rest at 70 km/hr. What is the total distance of the journey?***

Sol. Let the total distance of the journey be x km. Then,

$$\frac{\frac{x}{2}}{30} + \frac{\frac{x}{2}}{70} = 10 \Rightarrow \frac{x}{30} + \frac{x}{70} = 20 \Rightarrow \frac{7x + 3x}{210} = 20$$

$\Rightarrow$ $10x = 20 \times 210 \Rightarrow x =$ **420 km.**

Ex. 6. ***A person pedals from his house to his office at a speed of x_1 km/hour and returns by the same route at a speed of x_2 km/hour. What is his average speed?***

Sol. Let the distance from office to home = y km.

Then, the time taken for the round trip $= \dfrac{y}{x_1} + \dfrac{y}{x_2} = \dfrac{y(x_1 + x_2)}{x_1 x_2}$

Total distance travelled in that round trip = $2y$ km

$\therefore$ Average speed $= 2y \div \left(\dfrac{y(x_1 + x_2)}{x_1 x_2}\right) = \dfrac{2y x_1 x_2}{y(x_1 + x_2)} = \mathbf{\dfrac{2x_1 x_2}{x_1 + x_2}}$.

Ex. 7. ***A man can reach a certain place in 30 hours. If he reduces his speed by $\frac{1}{15}$th, he goes 10 km less in that time. Find his speed in km/hr?***

Sol. Let the man's speed be x km/hr.

Distance covered in 30 hours = $30x$ km

New speed $= \left(x - \dfrac{x}{15}\right)$ km/hr $= \dfrac{14x}{15}$ km/hr

Distance covered in 30 hours $= \dfrac{14x}{15} \times 30 = 28x$ km Given, $30x - 28x = 10$ km $\Rightarrow$ $x =$ **5 km/hr.**

Ex. 8. ***A tractor is moving with a speed of 20 km/hour, x km ahead of a truck moving with a speed of 35 km/hour. If it takes 20 minutes for a truck to overtake the tractor, then what is x equal to?***

Sol. In 20 minutes, the truck covers a distance $= 35 \times \dfrac{20}{60}$ km $= \dfrac{35}{3}$ km

In 20 minutes, the tractor covers a distance $= 20 \times \dfrac{20}{60}$ km $= \dfrac{20}{3}$ km

$\therefore \dfrac{35}{3} = \dfrac{20}{3} + x \Rightarrow x = \dfrac{15}{3} =$ **5 km.**

Ex. 9. ***A train running between two stations A and B arrives at its destination 10 minutes late when its speed is 5 km/hr and 50 minutes late when its speed is 30 km/hr. How far is station A from B?***

Sol. Let the distance between A and B be x km. Then, $\dfrac{x}{50} - \dfrac{10}{60} = \dfrac{x}{30} - \dfrac{50}{60}$

$\Rightarrow \quad \frac{x}{30}-\frac{x}{50}=\frac{5}{6}-\frac{1}{6} \quad \Rightarrow \quad \frac{5x-3x}{150}=\frac{4}{6}$

$\Rightarrow \quad \frac{2x}{150}=\frac{2}{3} \quad \Rightarrow \quad x=\frac{2\times150}{6}$ = **50 km.**

Ex. 10. ***A person drives his car for 3 hours at a speed of 40 km/hr and for 4.5 hours at a speed of 60 km/hr. At the end of it, he finds that he has covered $\frac{3}{5}$ of the total distance. What is the uniform speed with which he should further drive to cover the remaining distance in 4 hours?***

Sol. Let the total distance of the journey be D km. Then,

Distance covered by the person = (3 × 40 + 4.5 × 60) km = 120 km + 270 km = 390 km.

Given, $\frac{3}{5}\times D=390 \Rightarrow D=390\times\frac{5}{3}$ = 650 km

Remaining distance = 650 km – 390 km = 260 km

∴ Required uniform speed = $\frac{260}{4}$ = **65 km/hr.**

Ex. 11. ***Walking at three-fourth of his usual speed, a man covers a certain distance in 2 hours more than the time he takes to cover the distance at his usual speed. What is the time taken by him cover the same distance with his usual speed?***

Sol. Let the usual speed be x km/hour and time taken by usual speed be y hours. Then,

Distance covered = $(x \times y)$ km

New speed = $\frac{3}{4}x$ km/hour,

New time taken = $(y + 2)$ hours

∴ Distance covered = $\frac{3}{4}x(y+2)$ km

Since distance covered is the same. $xy=\frac{3}{4}x(y+2)$

$\Rightarrow \quad 4xy = 3xy + 6x \quad \Rightarrow \quad xy = 6x \quad \Rightarrow \quad y = 6$ **hours.**

Ex. 12. ***The speed of a bus is 54 km/hr excluding stoppage and 45 km/hr including stoppages. For how many minutes does the bus stop per hour?***

Sol. L.C.M. of 54 and 45 = 270

∴ Time taken by the bus to cover 270 km excluding stoppages = $\frac{270}{54}$ = 5 hours

Time taken by the bus to cover 270 km including stoppages = $\frac{270}{45}$ = 6 hours

∴ Time for which the bus stops per hour = $\frac{6-5}{6}$ hours = $\frac{1}{6}\times60$ min = **10 min.**

Ex. 13. ***A student rides on a bicycle at 8 km/hour and reaches his school 2.5 minutes late. The next day he increases his speed to 10 km/ hour and reaches the school 5 minutes early. How far is the school from his house?***

Sol. Let the distance of the school from the house = x km

Then $\frac{x}{8}-\frac{2.5}{60}=\frac{x}{10}+\frac{5}{60} \Rightarrow \frac{x}{8}-\frac{x}{10}=\frac{1}{12}+\frac{1}{24}=\frac{3}{24}=\frac{1}{8} \Rightarrow \frac{x}{40}=\frac{1}{8} \Rightarrow x = $ **5 km.**

Question Bank–14(a)

1. A car moves with a speed of 80 km/hour. What is the speed of the car in metres per second?

(a) $22\frac{2}{9}$ m/sec (b) 8 m/sec
(c) $20\frac{1}{9}$ m/sec (d) $8\frac{2}{3}$ m/sec

2. A man covers half of his journey at 6 km/hr and the remaining half at 3 km/hr. His average speed is

(a) 3 km/hr (b) 4 km/hr
(c) 4.5 km/hr (d) 9 km/hr

3. The speeds of A and B are in the ratio 3 : 4. A takes 20 minutes more than B to reach a destination. In what time does A reach the destination?

(a) $1\frac{1}{3}$ hours (b) 2 hours
(c) $1\frac{2}{3}$ hours (d) $2\frac{2}{3}$ hours

4. Two trains approach each other at 30 km/hr and 27 km/hr from two places 342 km apart. After how many hours do they meet?

(a) 5 hours (b) 6 hours
(c) 7 hours (d) 12 hours

5. A runs 100 metres in 11 seconds and B runs 100 metres in 12 seconds. The start which must be given to B to make the race to the completed in 11 seconds a draw will be

(a) 8 m (b) $8\frac{1}{4}$ m
(c) $8\frac{1}{3}$ m (d) $8\frac{1}{2}$ m

6. A man can walk uphill at the rate of $2\frac{1}{2}$ km/hr and downhill at the rate of $3\frac{1}{4}$ km/hr. If the total time required to walk a certain distance up the hill and return to the starting point was 4 hours 36 minutes, then what was the distance he walked up the hill?

(a) $6\frac{1}{2}$ km (b) $5\frac{1}{2}$ km
(c) $4\frac{1}{2}$ km (d) 4 km

7. A person wishes to reach his destination 90 km away in three hours but for the first half of the journey his speed was 20 km/hr. His average speed for the rest of the journey should be

(a) 40 km/hour (b) 0.75 km/min
(c) 1 km/min (d) 65 km/hour

8. A car travels a distance of 45 km at a speed of 15 km/hr. It covers the next 50 km of its journey at the speed of 25 km/hr and the last 25 km of its journey at the speed of 15 km/hr. What is the average speed of the car?

(a) 40 km/hr (b) 24 km/hr
(c) 15 km/hr (d) 18 km/hr

9. A car travels a distance of 170 km in 2 hours, partly at a speed of 100 km/hour and partly at 50 km/hour. The distance travelled at the speed of 50 km/hour is

(a) 50 km (b) 40 km
(c) 30 km (d) 60 km

10. A car travelling with $\frac{5}{7}$ of its usual speed covers 42 km in 1 hour 40 min 48 sec. What is the usual speed of the car?

(a) $17\frac{6}{7}$ km/hr (b) 25 km/hr
(c) 30 km/hr (d) 35 km/hr

11. A car can finish a certain journey in 10 hours at the speed of 48 km/hour. In order to cover the same distance in 8 hours, the speed of the car must be increased by

(a) 6 km/hour (b) 7.5 km/hour
(c) 12 km/hour (d) 15 km/hr

12. Excluding stoppages, the speed of a train is 45 km/hr and including stoppages it is 36 km/hr. For how many minutes does the train stop per hour?

(a) 10 (b) 12
(c) 15 (d) 18

13. A locomotive driver travelling at 72 km/hr finds a signal 210 metres ahead of him indicating that he should stop. He instantly applies brakes to stop the train. The train retards uniformly and stops 10 metres before the signal post. What time did he take to stop the train?

(a) 5 seconds (b) 10 seconds
(c) 15 seconds (d) 20 seconds

14. A runs twice as fast as B and B runs thrice as fast as C. The distance covered by C in 72 minutes will be covered by A in

(a) 18 min (b) 24 min
(c) 16 min (d) 12 min

15. A car runs at a speed of 40 km/hr when not serviced and runs at 60 km/hr when serviced. After servicing

the car covers a certain distance in 5 hours. How much time will the car take to cover the same distance when not serviced?

(a) 8 hours (b) 7.5 hours
(c) 6 hours (d) 7 hours

16. A train starts from Agra to Mathura at a speed of 60 km/hr and reaches there in 45 min. If on return its speed is reduced by 10%, how long will it take to reach Agra from Mathura?

(a) 1 hr (b) 50 min
(c) 1 hr 20 min (d) 49 min

17. Samir drove at the speed of 45 km/hr from home to a resort. Returning over the same route he got stuck in traffic and took an hour longer. Also he could drive only at the speed of 40 km/hr. How many kilometres did he drive each way?

(a) 250 km (b) 360 km
(c) 310 km (d) 275 km

18. A certain distance is covered at a certain speed. If half the distance is covered in double the time, the ratio of the two speeds is

(a) 4 : 1 (b) 1 : 4
(c) 2 : 1 (d) 1 : 2

19. Bombay Express left Delhi for Bombay at 14 : 30 hours travelling at a speed of 60 km/hr and Rajdhani Express left Delhi for Bombay on the same day at 16 : 30 hours travelling at a speed of 80 km/hr. How far from Delhi will the two trains meet?

(a) 120 km (b) 360 km
(c) 480 km (d) 500 km

20. *A* and *B* start simultaneously from a certain point in North and South directions on motorcycles. The speed of *A* is 80 km/hr and that of *B* is 65 km/hr. What is the distance between *A* and *B* after 12 minutes?

(a) 14.5 km (b) 29 km
(c) 36.2 km (d) 39 km

21. The distance between Bandel and Asansol is 100 km. *A* leaves Bandel for Asansol and walks at the rate of 3 km/hr. 3 hrs later, *B* starts from Asansol for Bandel and walks at 3.5 km/hr. Find the distance from Asansol where they would meet?

(a) 51 km (b) 49 km
(c) 53 km (d) 52 km

22. A star is 8.1×10^{13} km away from the earth. Suppose light travels at the speed of 3.0×10^5 km per second, how long will it take light from the star to reach the earth?

(a) 7.5×10^3 hrs. (b) 7.5×10^4 hrs
(c) 2.7×10^{10} sec (d) 2.7×10^{11} sec.

23. *A* takes 2 hours more than *B* to walk *d* km. If *A* doubles his speed then he can make it in 1 hour less than *B*. How much time does *B* require for walking *d* km?

(a) $\frac{d}{2}$ hours (b) 3 hours
(c) 4 hours (d) $\frac{2d}{3}$ hours

24. A boy is running at a speed of *p* km/hr to cover a distance of 1 km. But, due to slippery ground, his speed is reduced by *q* km/hr ($p > q$). If he takes *r* hours to cover the distance, then,

(a) $\frac{1}{r} = (p-q)$ (b) $r = (p-q)$
(c) $\frac{1}{r} = (p+q)$ (d) $r = (p+q)$

25. *A* walks at a uniform rate of 4 km an hour and 4 hours after his start, *B* cycles after him at an uniform rate of 10 km an hour. How far from the starting point will *B* catch *A*?

(a) 16.7 km (b) 18.6 km
(c) 21.5 km (d) 26.7 km

26. An express train travelled at an average speed of 100 km/hr, stopping for 3 minutes after every 75 km. How long did it take to reach its destination 600 km from the starting point?

(a) 6 hrs 21 min (b) 6 hrs 24 min
(c) 6 hrs 27 min (d) 6 hrs 30 min

27. Two trains start from stations *A* and *B* and travel towards each other at 50 km/hr and 60 km/hr respectively. At the time of their meeting the second train has travelled 120 km more than the first. The distance between *A* and *B* is

(a) 990 km (b) 1200 km
(c) 1320 km (d) 1440 km

28. In a kilometre race, *A* beats *B* by 30 seconds and *B* beats *C* by 15 seconds. If *A* beats *C* by 180 metres, the time taken by *A* to run 1 kilometre is

(a) 250 seconds (b) 205 seconds
(c) 200 seconds (d) 210 seconds

29. A man takes 6 hours 30 min in going by a cycle and coming back by scooter. He would have lost 2 hours

10 min by going on cycle both ways. How long would it take him to go by scooter both ways?

(a) 2 hrs (b) $4\frac{1}{3}$ hrs

(c) $3\frac{1}{3}$ hrs (d) $5\frac{1}{3}$ hrs

30. A student rides on a bicycle at 8 km/hour and reaches his school 2.5 minutes late. The next day he increases the speed to 10 km/hr and reaches school 5 minutes early. How far is the school from the house?

(a) 6 km (b) 4 km

(c) 5 km (d) 4.5 km

Answers

1. (a)	**2.** (b)	**3.** (a)	**4.** (b)	**5.** (c)	**6.** (a)	**7.** (c)	**8.** (d)	**9.** (c)	**10.** (d)
11. (c)	**12.** (b)	**13.** (b)	**14.** (d)	**15.** (b)	**16.** (b)	**17.** (b)	**18.** (a)	**19.** (c)	**20.** (b)
21. (b)	**22.** (b)	**23.** (c)	**24.** (a)	**25.** (d)	**26.** (a)	**27.** (c)	**28.** (b)	**29.** (b)	**30.** (c)

Hints and Solutions

1. (a) Speed $= 80\text{ km/hr} = \left(80\times\frac{5}{18}\right)\text{m/sec} = \mathbf{22\frac{2}{9}\ m/sec.}$

2. (b) Let the total distance be $2x$ km. Then,

Time taken $= \frac{x}{6}+\frac{x}{3}=\frac{3x}{6}=\frac{x}{2}$ hours

$\therefore$ Average speed $= \frac{\text{Total distance}}{\text{Time taken}} = \frac{2x}{\frac{x}{2}}$ km/hr

$= \mathbf{4\ km/hr}.$

3. (a) Let the time taken by A be x hrs. Then,

Time taken by $B = \left(x-\frac{20}{60}\right)$ hrs $= \left(x-\frac{1}{3}\right)$ hrs

Ratio of speeds = Inverse ratio of times

$\Rightarrow 3:4 = \left(x-\frac{1}{3}\right):x \quad \Rightarrow \quad \frac{x-\frac{1}{3}}{x}=\frac{3}{4}$

$\Rightarrow \frac{3x-1}{3x}=\frac{3}{4} \quad \Rightarrow \quad 12x-4=9x$

$\Rightarrow 3x=4 \quad \Rightarrow \quad x=\frac{4}{3}=\mathbf{1\frac{1}{3}\ hrs.}$

4. (b) Suppose the two trains meet after x hours. Then,

$30\times x+27\times x=342$

$\Rightarrow 57x=342 \quad \Rightarrow \quad x=\frac{342}{57}=\mathbf{6\ hours}.$

5. (c) Speed of $A = \frac{100}{11}$ m/s, Speed of $B = \frac{100}{12}$ m/s

Required distance (given to B)

= (Distance covered by A in 11 seconds)

– (Distance covered by B in 11 seconds)

$= \left(\frac{100}{11}\times 11\right)\text{m} - \left(\frac{100}{12}\times 11\right)\text{m}$

$= 100\text{ m} - 91\frac{2}{3}\text{ m}$

$= \mathbf{8\frac{1}{3}\ m.}$

6. (a) Let the total distance walked uphill and downhill be $2x$ km.

Then, $\frac{x}{2\frac{1}{2}}+\frac{x}{3\frac{1}{4}}=\left(4+\frac{36}{60}\right)$ hrs

$\Rightarrow \frac{2x}{5}+\frac{4x}{13}=4+\frac{3}{5}=\frac{23}{5}$

$\Rightarrow \frac{26x+20x}{65}=\frac{23}{5}$

$\Rightarrow \frac{46x}{65}=\frac{23}{5}$

$\Rightarrow x=\frac{23}{5}\times\frac{65}{46}=\mathbf{6.5\ km.}$

7. (c) Time taken to travel the first half of the journey

$= \frac{45}{20}$ hrs $= \frac{9}{4}$ hrs

Total time = 3 hrs

$\therefore$ Time left for the second half of the journey

$= \left(3-\frac{9}{4}\right)$ hrs $= \frac{3}{4}$ hrs

Distance to be covered = 45 km

$\therefore$ Required speed $= \frac{45}{\frac{3}{4}}$ km/hr

$= 60\text{ km/hr} = \mathbf{1\ km/min.}$

8. (d) Total time taken $= \frac{45}{15}$ hrs $+ \frac{50}{25}$ hrs $+ \frac{25}{15}$ hrs

$$= 3 \text{ hrs} + 2 \text{ hrs} + \frac{5}{3} \text{ hrs} = \frac{20}{3} \text{ hrs}$$

$\therefore$ Average speed of the car

$$= \frac{45+50+25}{\frac{20}{3}} \text{ km/hr}$$

$$= \frac{120 \times 3}{20} \text{ km/hr} = \textbf{18 km/hr}.$$

9. (c) Let the distance travelled at 100 km/hr be x km. Then, distance travelled at 50 km/hr is $(170 - x)$ km.

Given, $\frac{x}{100} + \frac{(170-x)}{50} = 2$

$\Rightarrow \frac{x+2(170-x)}{100} = 2 \Rightarrow x + 340 - 2x = 200$

$\Rightarrow x = 140$ km.

$\therefore$ Distance travelled at 50 km/hr = (170 – 140) km

= **30 km.**

10. (d) Let the usual speed of the car be x km/hr.

Time = 1 hour 40 min 48 sec $= 1 + \frac{40}{60} + \frac{48}{3600}$ hrs

$$= 1 + \frac{2}{3} + \frac{1}{75} \text{ hrs}$$

$$= \frac{75+50+1}{75} \text{ hrs}$$

$$= \frac{126}{75} \text{ hrs}$$

Given, $\frac{5}{7} x \times \frac{126}{75} = 42 \Rightarrow x = \frac{42 \times 7 \times 75}{5 \times 126}$

= **35 km/hr.**

11. (c) Distance of the entire journey = (48 × 10) km

= 480 km

If the time taken = 8 hours, then

increased speed $= \frac{480}{8}$ km/hr = 60 km/hr

$\therefore$ Increase in speed = 60 km/hr – 48 km/hr

= **12 km/hr.**

12. (b) Due to stoppages the train travels (45 – 36) = 9 km less in 1 hour and the time taken to travel 9 km is the time taken at the stoppages.

$\therefore$ Time taken to cover 9 km at 45 km/hr $= \frac{9}{45} \times 60$

= **12 min.**

13. (b) To stop the train the driver has to cover 200 m at 72 km/hr.

72 km/hr $= \left(72 \times \frac{5}{18}\right)$ m/sec = 20 m/sec.

$\therefore$ Time taken to cover 200 m $= \frac{200}{20}$ seconds

= **10 seconds.**

14. (d) Let the speed of A be x km/min.

Speed of $B = \frac{x}{2}$ km/min and

speed of $C = \frac{x}{6}$ km/min

Distance covered by C in 72 minutes $= \frac{x}{6} \times 72$

$= 12x$ km

$\therefore$ $12x$ distance will be covered by A in

$\frac{12x}{x}$ = **12 min.**

15. (b) Distance covered by the car in 5 hours after servicing = (60 × 5) km = 300 km

$\therefore$ Time required to cover the same distance when the car is not serviced $= \frac{300}{40}$ hrs = **7.5 hours.**

16. (b) Let the distance from Agra to Mathura be x km.

$\therefore$ Distance = Speed × Time $= 60 \times \frac{45}{60} = 45$ km.

When it returns its speed reduces by 10%, *i.e.*, its speed = 90% of 60

$= \left(\frac{9}{10} \times 60\right)$ km/hr = 54 km/hr

$\therefore$ Time taken $= \frac{\text{Distance covered}}{\text{Speed}}$

$= \left(\frac{45}{54} \times 60\right)$ min = **50 min.**

17. (b) Let the distance from home to resort be x km. Then,

Time taken to go from home to resort $= \frac{x}{45}$

Time taken to come back from resort to home $= \frac{x}{40}$

Given, $\frac{x}{40} = \frac{x}{45} + 1 \Rightarrow \frac{x}{40} - \frac{x}{45} = 1$

$\Rightarrow \frac{9x - 8x}{360} = 1 \Rightarrow x = \mathbf{360.}$

18. (a) Let x km be covered in y hours.

Then, speed $= \frac{x}{y}$ km/hr

In the second case, $\frac{x}{2}$ km is covered in $2y$ hours.

$\therefore$ New speed $= \left(\frac{x}{2} \times \frac{1}{2y}\right)$ km/hr $= \frac{x}{4y}$ km/hr

$\therefore$ Ratio of speeds $= \frac{x}{y} : \frac{x}{4y} = 1 : \frac{1}{4} = \mathbf{4:1}$.

19. (c) Let the trains meet at a distance of x km from Delhi. Then, $\frac{x}{60} - \frac{x}{80} = 2 \Rightarrow 4x - 3x = 480$

$\Rightarrow x = \mathbf{480\ km}$.

20. (b) Distance covered by A in 12 min

$= \left(80 \times \frac{12}{60}\right)$ km = 16 km

Distance covered by B in 12 min

$= \left(65 \times \frac{12}{60}\right)$ km = 13 km

$\therefore$ Distance between A and B after 12 min

= 16 km + 13 km = **29 km.**

21. (b) In 3 hours, distance covered by A = 9 km.

Suppose B and A meet at a point C in time t. Then,

$3t + 3.5t = 91 \quad \Rightarrow \quad t = \frac{91}{6.5} = 14$ hrs.

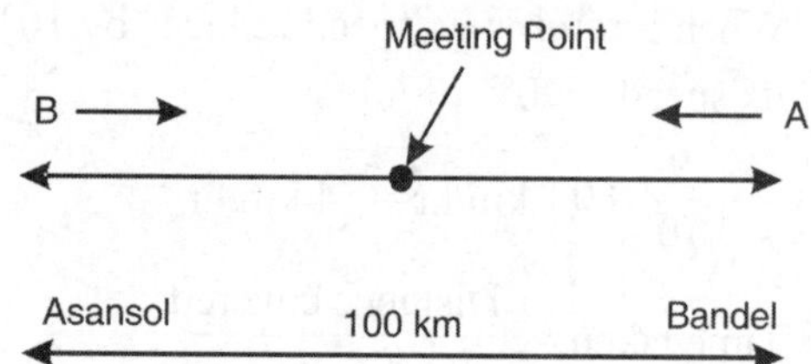

$\therefore$ Distance from Asansol to the meeting point

$= (3.5 \times 14)$km = **49 km.**

22. (b) Time taken $= \frac{\text{Distance}}{\text{Speed}} = \frac{8.1 \times 10^{13}}{3.0 \times 10^{5}}$ seconds

$= \left(2.7 \times 10^{8} \times \frac{1}{60} \times \frac{1}{60}\right)$ hrs

$= \frac{2.7 \times 10^{6}}{3600}$ hrs $= \left(\frac{2.7 \times 100 \times 10^{4}}{36}\right)$ hrs

$= \mathbf{(7.5 \times 10^{4})}$ **hrs.**

23. (c) Suppose B takes t hours to walk d km.

Then, A takes $(t + 2)$ hours to walk d km.

With double speed, A will take $\frac{1}{2}(t+2)$ hours.

$\therefore \quad t - \frac{1}{2}(t+2) = 1 \Rightarrow 2t - (t+2) = 2 \Rightarrow t = 4$

$\therefore$ B takes **4 hours** to walk d km.

24. (a) Running speed $= p$ km/hr, Reduced speed $= q$ km/hr

$\therefore$ Actual speed of the boy $= (p - q)$ km/hr

Time taken $= r$ hours, Distance covered = 1 km.

$\therefore \quad 1 = (p - q)r \quad \Rightarrow \quad \frac{1}{r} = (p - q)$

25. (d) Suppose B catches A after t hours. Then,

Distance travelled by A in $(t + 4)$ hours

= Distance travelled by B in t hours.

$\Rightarrow (t + 4) \times 4 = 10 \times t$

$\Rightarrow 6t = 16 \quad \Rightarrow \quad t = \frac{8}{3}$ hrs.

$\therefore$ Distance travelled by B in $\frac{8}{3}$ hours

$= \left(10 \times \frac{8}{3}\right)$ km $= \frac{80}{3}$ km = **26.7 km.**

26. (a) Time taken to cover 600 km $= \left(\frac{600}{100}\right)$ hrs = 6 hrs

Number of stoppages $= \frac{600}{75} - 1 = 7$

Total time taken in stoppages $= (3 \times 7)$ min

= 21 min.

$\therefore$ Total time taken = **6 hrs 21 min.**

27. (c) Let the two trains meet after x hours. Then,

$60x - 50x = 120 \Rightarrow 10x = 120 \Rightarrow x = 12$ hrs

Distance AB = (Dist. covered by slower train) + (Dist. covered by faster train)

$= [(50 \times 12) + (60 \times 12)]$ km

= 600 km + 720 km = **1320 km.**

28. (b) A beats B by 30 seconds, B beats C by 15 seconds

$\therefore$ A beats C by (30 + 15) seconds = 45 seconds

$\Rightarrow$ Time taken by C to travel 180 m = 45 seconds

$\Rightarrow$ Time taken by C to cover the distance of 1 km (1000 m) $= \frac{45}{180} \times 1000 = 250$ seconds

$\therefore$ Required time taken by A to cover the distance of 1 km = (250 – 45) sec = **205 sec.**

29. (b) Let time taken in going only by scooter be x hrs.

Then, time taken for one way by cycle

$$= \left(6\frac{1}{2} - x\right) \text{ hrs} = \left(\frac{13}{2} - x\right) \text{ hrs}$$

Total time taken for going on cycle both ways

$$= 6 \text{ hrs } 30 \text{ min} + 2 \text{ hrs } 10 \text{ min}$$

$$= 8 \text{ hrs } 40 \text{ min} = 8\frac{40}{60} = 8\frac{2}{3} \text{ hrs}$$

$$\therefore 2\left(\frac{13}{2} - x\right) = 8\frac{2}{3} \Rightarrow 13 - 2x = 8\frac{2}{3}$$

$$\Rightarrow 2x = 13 - 8\frac{2}{3} = 4\frac{1}{3} \text{ hrs}$$

$\therefore$ It will take $4\frac{1}{3}$ **hrs** to go on scooter both ways.

30. (c) Let the distance between school and his house be x km. Then,

$$\frac{x}{8} = \left(t + \frac{5}{2}\right) \times \frac{1}{60} \text{ hrs and } \frac{x}{10} = (t - 5) \times \frac{1}{60} \text{ hrs}$$

$$\Rightarrow \left(\frac{x}{8} - \frac{x}{10}\right) = \left(\frac{5}{2} + 5\right) \times \frac{1}{60}$$

$$\Rightarrow \frac{x}{40} = \frac{15}{2} \times \frac{1}{60} = \frac{1}{8} \Rightarrow x = \mathbf{5\ km}.$$

PROBLEMS ON TRAINS

KEY FACTS

1. The time taken by a train to pass a stationary object (as a man or a pole) is equal to the time taken by the train to cover a distance equal to the length of the train.

   $$\text{Time} = \frac{\text{Length of train}}{\text{Speed of train}} = \frac{l}{v} \text{ seconds}$$

   Where l and v are the length and speed of train in m and m/s respectively.

2. The time taken by a train of length l metres to pass a stationary object of length l_1 metres is equal to the time taken by the train to cover a distance of $(l + l_1)$ metres with a speed of v m/s (say).

   $$\text{Then, Time} = \frac{(l + l_1)}{v} \text{ seconds.}$$

Solved Examples

Ex. 1. ***A train 132 m long passes a telegraph pole in 6 seconds. Find the speed of the train in km/hr.***

Sol. $\text{Speed of train} = \frac{\text{Length of train}}{\text{Time taken to pass the pole}} = \frac{132}{6} \text{ m/sec} = 22 \text{ m/sec} = \left(22 \times \frac{18}{5}\right) \text{ km/hr} = \mathbf{79.2\ km/hr}.$

Ex. 2. ***A train covers a distance of 12 km in 10 minutes. If it takes 6 seconds to pass a telegraph post, then what is the length of the train?***

Sol. $\text{Speed of the train} = \frac{12 \times 1000}{10 \times 60} \text{ m/sec} = 20 \text{ m/sec.}$

Length of the train = Speed of the train × Time taken by the train to pass a telegraph post = (20 × 6) m = **120 m.**

Ex. 3. ***A goods train runs at a speed of 72 km/hr and crosses a 250 m long platform in 26 seconds. What is the length of the goods train?***

Sol. $\text{Speed of the train} = 72 \text{ km/hr} = 72 \times \frac{5}{18} \text{ m/sec} = 20 \text{ m/sec}$

Let the length of the goods train be x m. Then

$$\frac{250+x}{20}=26 \Rightarrow 250+x=520 \Rightarrow x=\mathbf{270\ m.}$$

Ex. 4. ***A train moves past a telegraph post and a bridge 264 m long in 8 seconds and 20 seconds respectively. What is the speed of the train in km/hr?***

Sol. Let the speed of the train be x m/s

Distance covered in 8 seconds = $8x$ m

$\therefore$ Length of the train = $8x$ m

$\Rightarrow\quad 8x + 264 = \text{Distance covered in 20 seconds} = 20 \times x \Rightarrow\quad 12x = 264$

$\Rightarrow\quad x = 22 \text{ m/s} = 22\times\frac{18}{5} \text{ km/hr} = \mathbf{79.2\ km/hr}.$

Ex. 5. ***The length of a train and that of a platform are equal. If with the speed of 54 km/hr, the train crosses the platform in $1\frac{1}{2}$ minutes, then what is the length of the platform in metres?***

Sol. Let the length of the train and platform be x m each.

Speed of the train = 54 km/hr = $\left(54\times\frac{5}{18}\right)$ m/s = 15 m/s

Then, $\frac{x+x}{15} = (1.5 \times 60)$ seconds $\Rightarrow\ 2x = 1.5 \times 60 \times 15 = 1350$ m

$\Rightarrow\quad x = \mathbf{675\ m.}$

Question Bank–14(b)

1. A 180 - metre long train crosses a man standing on the platform in 6 seconds. What is the speed of the train?

(a) 90 km/hr (b) 108 km/hr
(c) 120 km/hr (d) 88 km/hr

2. A train running at a speed of 84 km/hour crosses an electric pole in 9 seconds. What is the length of the train in metres?

(a) 126 (b) 630
(c) 210 (d) 70

3. A train 110 m long takes three seconds to pass a standing man. How long is the platform if the train passes through it in 15 seconds moving with the same speed?

(a) 440 m (b) 400 m
(c) 550 m (d) 450 m

4. A train speeding at 120 km/hr crosses an electric pole in 9 seconds and a platform in 24 seconds. What is the length of the platform?

(a) 500 m (b) 800 m
(c) 300 m (d) 1100 m

5. A train crosses a pole in 15 seconds while it crosses a 100 m long platform in 25 seconds. The length of the train is

(a) 125 m (b) 135 m
(c) 150 m (d) 175 m

6. A train passes a platform 90 m long in 30 seconds and a man standing on the platform in 15 seconds. The speed of the train is

(a) 12.4 km/hr (b) 14.6 km/hr
(c) 18.4 km/hr (d) 21.6 km/hr

7. A train takes 18 seconds to pass through a platform 162 m long and 15 seconds to pass through another platform 120 m long. The length of the train (in metres) is

(a) 70 (b) 80
(c) 90 (d) 105

8. A train 150 m long, takes 30 seconds to cross a bridge 500 m long. How much time will the train take to cross a platform 370 m long?

(a) 36 sec (b) 30 sec
(c) 24 sec (d) 18 sec

9. A 120 metre long train is running at a speed of 90 km/hour. It will cross a railway platform 230 m long in

(a) $4\frac{4}{5}$ seconds (b) $9\frac{1}{5}$ seconds
(c) 7 seconds (d) 14 seconds

10. A passenger train running at a speed of 80 km/hr leaves the railway station 6 hours after a good train leaves and overtakes it in 4 hours. What is the speed of the goods train?

(a) 48 km/hr (b) 60 km/hr
(c) 32 km/hr (d) 80 km/hr

11. A train X starts from a place at a speed of 50 km/hr. After one hour, another train Y starts from the same place at a speed of 70 km/hr. After how much time will Y cross X?

(a) 3 hrs (b) $2\frac{3}{4}$ hrs
(c) $3\frac{1}{2}$ hrs (d) $2\frac{1}{4}$ hrs

12. A man in a train notices that he can count 21 telephone posts in one minute. If they are known to be 50 metres apart, then at what speed is the train travelling?

(a) 57 km/hr (b) 60 km/hr
(c) 63 km/hr (d) 55 km/hr

Answers

1. (b)	**2.** (c)	**3.** (c)	**4.** (a)	**5.** (c)	**6.** (d)	**7.** (c)	**8.** (c)	**9.** (d)	**10.** (c)
11. (c)	**12.** (b)								

Hints and Solutions

1. (b) Speed $= \frac{\text{Distance}}{\text{Time}} = \frac{\text{Length of train}}{\text{Time taken}}$

$$= \frac{180}{6} \text{ m/sec} = 30\times\frac{18}{5} \text{ km/hr} = \mathbf{108 \text{ km/hr}}.$$

2. (c) Length of the train

= Distance covered by the train in 9 seconds

$$= 84\times\frac{5}{18}\times 9 \text{ m} = \mathbf{210 \text{ m}}.$$

3. (c) Speed of the train $= \frac{110}{3}$ m/s

$\therefore$ Length of the platform

= Speed of train × Time taken to pass the platform

$$= \left(\frac{110}{3}\times 15\right) \text{m} = \mathbf{550 \text{ m}}.$$

4. (a) Speed of train $= 120 \text{ km/hr} = 120\times\frac{5}{18} \text{ m/s} = \frac{100}{3}$ m/s

Time taken to cross the electric pole = 9 sec

$\therefore$ Length of the train $= \left(\frac{100}{3}\times 9\right)$ m = 300 m

Let the length of the platform be x m.

Then a distance of $(x + 300)$ m is to be covered in 24 seconds at a speed of $\frac{100}{3}$ m/s.

$$\therefore (x + 300) = \left(\frac{100}{3}\times 24\right) \text{m}$$

$\Rightarrow x + 300 = 800$

$\Rightarrow x = \mathbf{500 \text{ m}}.$

5. (c) Let the length of the train be x metres. Then,

$$\frac{x}{15} = \frac{x+100}{25} \Rightarrow 25x = 15x + 1500$$

$\Rightarrow 10x = 1500 \Rightarrow x = \mathbf{150 \text{ m}}.$

6. (d) Let the speed of the train be x m/sec and length of the train be y m. Then,

$\frac{\text{Length of the train}}{\text{Speed of the train}}$ = Time taken by the train to pass the man

$$\Rightarrow \frac{y}{x} = 15 \Rightarrow y = 15x \quad \ldots(1)$$

and $\frac{\text{Length of train + Length of platform}}{\text{Speed of train}}$

= Time taken by the train to cross the platform

$$\Rightarrow \frac{y+90}{x} = 30 \Rightarrow y + 90 = 30x \quad \ldots(2)$$

Put the value of y from (1) in (2)

$15x + 90 = 30x \Rightarrow 15x = 90 \Rightarrow x = 6$ m/sec

$\therefore$ Speed of train = $\left(6 \times \frac{18}{5}\right)$ km/hr = **21.6 km/hr.**

7. (c) Let the length of the train be x metres. Then,

Speed of train

$= \frac{\text{Length of train + Length of platform}}{\text{Time taken}}$

$\therefore \frac{x+162}{18} = \frac{x+120}{15}$

$\Rightarrow 15x + 2430 = 18x + 2160$

$\Rightarrow 3x = 270 \Rightarrow$ **$x = 90$ m.**

8. (c) Length of the train = 150 m

Length of the bridge = 500 m

$\therefore$ Total length = (500 + 150)m = 650 m

$\therefore$ Speed of the train = $\frac{650}{30}$ m/sec = $21\frac{2}{3}$ m/sec

Now total length of the train and new bridge = (370 + 150)m = 520 m

$\therefore$ Time taken = $\frac{520 \times 3}{65}$ sec = **24 sec.**

9. (d) Speed = 90 km/hour = $90 \times \frac{5}{18}$ = 25 m/sec

Distance covered

= Length of train + Length of platform

= (120 + 230) metres = 350 metres

Time taken = $\frac{350}{25}$ sec = **14 seconds.**

10. (c) Distance covered by the passenger train in 4 hours = (80 × 4) km = 320 km

This is the distance covered by the goods train in 10 hours.

$\therefore$ Speed of goods train = $\frac{320}{10}$ km/hr = **32 km/hr.**

11. (c) Suppose they cross after x hours.

Distance covered by Y in $(x - 1)$ hours

= Distance covered by X in x hours

$\Rightarrow 70(x - 1) = 50x \Rightarrow$ **$x = 3\frac{1}{2}$ hrs.**

12. (b) Distance between the 21 telephone posts

= 20 × 50 m = 1000 m

Time taken to cross these posts = 1 minute

= 60 seconds

$\therefore$ Speed of the train = $\frac{1000}{60}$ m/sec

$= \frac{1000}{60} \times \frac{18}{5}$ km/hr

= **60 km/hr.**

Self Assessment Sheet–14

1. The distance of the sun from the earth is one hundred forty million four hundred thousand kilometers and light travels from the former to the latter in 7 minutes and fifty eight seconds. The velocity of light per second is :

(a) 3×10^5 km/sec (b) 0.3×10^5 km/sec

(c) 30×10^5 km/sec (d) None of these

2. A man covered a distance of 2000 km in 18 hours partly by bus at 72 km/hr and partly by train at 160 km/hr. The distance covered by bus is :

(a) 860 km (b) 640 km

(c) 1280 km (d) 720 km

3. A man takes 4 hours 20 minutes in walking to a certain place and riding boat. If he walks on both sides he loses 1 hour. The time he would take by riding both sides is :

(a) 3 h 20 min (b) 2 h 20 min

(c) 4 h 20 min (d) 1 h 20 min

4. A car covers 4 successive 3 km stretches at speeds of 10 km/hr, 20 km/hr, 30 km/hr and 60 km/hr respectively. The average speed of the car for the entire journey is

(a) 15 km/hr (b) 35 km/hr

(c) 20 km/hr (d) 25 km/hr

5. Moving at $\frac{5}{6}$ of its usual speed, a train is 10 minutes late. Its usual time to cover the journey is :

(a) 40 min (b) 50 min

(c) 35 min (d) 55 min

6. If a train runs at 40 km/hr, it reaches its destination late by 11 minutes, but if it runs at 50 km/hr, it is late by 5 minutes only. The correct time for the train to complete the journey is :

(a) 13 min (b) 15 min

(c) 19 min (d) 21 min

7. The distance between two cities A and B is 330 km. A train starts from A at 8 a.m. and travels towards B

at 60 km/hr. Another train starts from *B* at 9 a.m. and travels towards *A* at 75 km/hr. At what time do they meet ?

(a) 10 a.m. (b) 10.30 a.m.

(c) 11 a.m. (d) 11.30 a.m.

8. A train passes a station platform in 36 seconds and a man standing on the platform in 20 seconds. If the speed of the train is 54 km/hr, what is the length of the platform ?

(a) 120 m (b) 240 m

(c) 300 m (d) None of these

9. A motor starts with the speed of 70 km/hr with its speed increasing every two hours by 10 km/hr. In how many hours will it cover 345 km ?

(a) $2\frac{1}{4}$ h

(b) $4\frac{1}{2}$ h

(c) 4 h 5 min

(d) Cannot be determined

(e) None of these

10. A person sets to cover a distance of 12 km in 45 minutes. If he covers $\frac{3}{4}$ of the distance in $\frac{2}{3}$rd time, what should be his speed to cover the remaining distance in the remaining time ?

(a) 16 km / hr (b) 8 km / hr

(c) 12 km / hr (d) 14 km / hr

Answers

1. (a) [**Hint.** Distance of sun from the earth = $(143 \times 10^6 + 4 \times 10^3)$ km $= 1434 \times 10^5$ km

Time taken for the light to reach the earth = (7 × 60 + 58) sec = 478 sec]

2. (d) **3.** (a) **4.** (c) **5.** (b) **6.** (c) **7.** (c) **8.** (b) **9.** (b) **10.** (c)

Chapter 15

RATIO AND PROPORTION

KEY FACTS

1. A ratio is formed when two quantities with same units are compared by division, *i.e.*, $a : b$ or $\frac{a}{b}$.
2. It has no unit.
3. If a quantity increases (or decreases) in the ratio $a : b$, then new quantity is $\frac{b}{a}$ of the original quantity.
4. An equality of two ratios is called a proportion.
5. The first and fourth terms are called **extremes** and second and third terms are called **means**.
6. Four quantities a, b, c, d are said to be in proportion, if $ad = bc$ *i.e.*, **product of extremes** = **product of means**.
7. In the proportion $a : b :: c : d$, d is called the **fourth proportional**.
8. Three quantities are said to be in continued proportion, if $a : b = b : c$ or $\frac{a}{b} = \frac{b}{c}$, *i.e.*, $b^2 = ac$.

   Here b is called the **mean proportional** between a and c. c is called the **third proportional**.

Solved Examples

Ex. 1. ***If*** $a : b = \frac{2}{9} : \frac{1}{3}$, $b : c = \frac{2}{7} : \frac{5}{14}$ ***and*** $d : c = \frac{7}{10} : \frac{3}{5}$, ***then find*** $a : b : c : d$.

Sol. $\frac{a}{b} = \frac{2}{9} \div \frac{1}{3} = \frac{2}{9} \times \frac{3}{1} = \frac{2}{3}$, $\frac{b}{c} = \frac{2}{7} \div \frac{5}{14} = \frac{2}{7} \times \frac{14}{5} = \frac{4}{5}$

$\frac{d}{c} = \frac{7}{10} \div \frac{3}{5} = \frac{7}{10} \times \frac{5}{3} = \frac{7}{6} \Rightarrow \frac{c}{d} = \frac{6}{7} \Rightarrow a = \frac{2b}{3}$, $c = \frac{5b}{4}$, $d = \frac{7c}{6} = \frac{7}{6} \times \frac{5b}{4} = \frac{35b}{24}$

$\therefore a : b : c : d = \frac{2b}{3} : b : \frac{5b}{4} : \frac{35b}{24}$ (LCM = 24)

$= \frac{2b}{3} \times 24 : b \times 24 : \frac{5b}{4} \times 24 : \frac{35b}{24} \times 24$ = **16 : 24 : 30 : 35.**

Ex. 2. ***If*** $A : B : C = 2 : 3 : 4$, ***then what is*** $\frac{A}{B} : \frac{B}{C} : \frac{C}{A}$ ***equal to?***

Sol. Let $A = 2k$, $B = 3k$, $C = 4k$.

Then, $\frac{A}{B} = \frac{2k}{3k} = \frac{2}{3}$, $\frac{B}{C} = \frac{3k}{4k} = \frac{3}{4}$, $\frac{C}{A} = \frac{4k}{2k} = \frac{2}{1}$

$\therefore \frac{A}{B} : \frac{B}{C} : \frac{C}{A} = \frac{2}{3} : \frac{3}{4} : \frac{2}{1}$ (LCM = 12)

$= \frac{2}{3} \times 12 : \frac{3}{4} \times 12 : \frac{2}{1} \times 12$ = **8 : 9 : 24.**

Ex. 3. ***If A : B = 1 : 2, B : C = 3 : 4 and C : D = 5 : 6, find D : C : B : A***

Sol. $\frac{A}{B}=\frac{1}{2} \Rightarrow A=\frac{1}{2}B$, $\frac{B}{C}=\frac{3}{4} \Rightarrow C=\frac{4B}{3}$

$\frac{C}{D}=\frac{5}{6} \Rightarrow D=\frac{6C}{5}=\frac{6}{5}\times\frac{4B}{3}=\frac{8B}{5}$

$\therefore D:C:B:A=\frac{8B}{5}:\frac{4B}{3}:B:\frac{B}{2}$

$=\frac{8B}{5}\times 30:\frac{4B}{3}\times 30:B\times 30:\frac{B}{2}\times 30=$ **48 : 40 : 30 : 15.**

Ex. 4. ***Divide 170 into three parts such that the first part is 10 more than the second and its ratio with the third part is 2 : 5.***

Sol. Let the first part = x. Then,

Second part = $x - 10$

$\frac{\text{First part}}{\text{Third part}}=\frac{2}{5} \Rightarrow \text{Third part}=\frac{5x}{2}$

Given, $x + x - 10 + \frac{5x}{2}=170 \Rightarrow 2x + 2x - 20 + 5x = 340$

$\Rightarrow 9x = 360 \Rightarrow x = 40$

$\therefore$ The three parts are 40, 40 – 10, $5\times\frac{40}{2}$, *i.e.*, **40, 30, 100.**

Ex. 5. ***The ratio of a father's age to his son's age is 4 : 1. The product of their ages is 196. What is the ratio of their ages after 5 years?***

Sol. Let the father's age and son's age be $4x$ years and x years respectively.

Given, $4x \times x = 196 \Rightarrow x^2 = \frac{196}{4}=49 \Rightarrow x = 7$

$\therefore$ Required ratio $=\frac{4x+5}{x+5}=\frac{4\times 7+5}{7+5}=\frac{33}{12}=\frac{11}{4}=$ **11 : 4.**

Ex. 6. ***If the ratio of boys to girls in a class is B and the ratio of girls to boys is G, then 3(B + G) is***

(a) equal to 3 *(b) less than 3* *(c) more than 3* *(d) less than* $\frac{1}{3}$

Sol. Let the number of boys be x and the number of girls = y

Then, $B=\frac{x}{y}$ and $G=\frac{y}{x}$

$\therefore 3(B + G) = 3\left(\frac{x}{y}+\frac{y}{x}\right)=\frac{3(x^2+y^2)}{xy}>3$

Ex. 7. ***In a mixture of 45 litres, the ratio of milk and water is 4 : 1. How much water must be added to make the mixture ratio 3 : 2?***

Sol. Quantity of milk in 45 litres of mixture $=\frac{4}{5}\times 45$ litres = 36 litres

$\therefore$ Quantity of water in the mixture = 9 litres

Let x litres of water be added to the mixture.

Then, $\frac{36}{9+x}=\frac{3}{2} \Rightarrow 72 = 27 + 3x \Rightarrow 3x = 45 \Rightarrow x =$ **15 litres.**

Ex. 8. ***In a class, the number of girls is 20% more than that of boys. The strength of the class is 66. If 4 more girls are admitted to the class, what will be the ratio of the number of boys to that of the girls?***

Sol. Let the number of boys be x.

Then, number of girls = 120% of x $=\frac{120}{100}\times x=\frac{6x}{5}$

Given, $x+\frac{6x}{5}=66 \Rightarrow \frac{11x}{5}=66 \Rightarrow x=30$

$\therefore$ Number of boys = 30, number of girls $=\frac{6\times30}{5}=36$

If 4 more girls are admitted then new number of girls = 40

$\therefore$ Required ratio = 30 : 40 = **3 : 4**.

Ex. 9. ***When ₹ 4572 is divided among A, B and C such that three times of A's share is equal to 4 times of B's share is equal 6 times C's share. What is A's share ?***

Sol. Given, $3A = 4B = 6C \Rightarrow \frac{A}{B}=\frac{4}{3}$ and $\frac{B}{C}=\frac{6}{4}=\frac{3}{2}$

$\therefore A : B : C = 4 : 3 : 2 \Rightarrow A = 4k, B = 3k, C = 2k$

$\therefore 4k + 3k + 2k = 4572 \Rightarrow 9k = 4572 \Rightarrow k = 508$

$\therefore$ A's share = 4 × 508 = **₹ 2032.**

Ex. 10. ***A bag contains 50 paise, 1 rupee and 2 rupee coins in the ratio 2 : 3 : 4. If the total amount is ₹ 240, what is the total number of coins?***

Sol. Let the number of 50 p coins be $2x$, 1 rupee coins be $3x$ and 2 rupee coins be $4x$.

Then, $2x\times\frac{1}{2} + 3x \times 1 + 4x \times 2 = 240$

$\Rightarrow x + 3x + 8x = 240 \Rightarrow 12x = 240 \Rightarrow x = 20$

$\therefore$ Total number of coins = $2x + 3x + 4x = 9x = 9 \times 20 =$ **180.**

Question Bank–15

1. If $A : B = 3 : 4$, $B : C = 5 : 7$ and $C : D = 8 : 9$, then the ratio $A : D$ is
   (a) 3 : 7 (b) 7 : 3
   (c) 21 : 10 (d) 10 : 21

2. If $\frac{1}{x}:\frac{1}{y}:\frac{1}{z}=2:3:5$, then $x : y : z$ = ?
   (a) 2 : 3 : 5 (b) 15 : 10 : 6
   (c) 5 : 3 : 2 (d) 6 : 10 : 15

3. If $\frac{3a+5b}{3a-5b}=5$, then $a : b$ is equal to
   (a) 2 : 1 (b) 5 : 3
   (c) 3 : 2 (d) 5 : 2

4. If $A : B = \frac{1}{2}:\frac{3}{8}$, $B : C = \frac{1}{3}:\frac{5}{9}$ and $C : D = \frac{5}{6}:\frac{3}{4}$, then the ratio $A : B : C : D$ is
   (a) 6 : 4 : 8 : 10 (b) 6 : 8 : 9 : 10
   (c) 8 : 6 : 10 : 9 (d) 4 : 6 : 8 : 10

5. If $a : b = 2 : 3$ and $b : c = 4 : 5$, find $a^2 : b^2 : bc$
   (a) 4 : 9 : 45 (b) 16 : 36 : 45
   (c) 16 : 36 : 20 (d) 4 : 36 : 20

6. If $a : b = 2 : 5$, then $(3a + 4b) : (5a + 6b)$ is equal to
   (a) 13/20 (b) 26/33
   (c) 16/27 (d) 18/35

7. If $\frac{2a+b}{a+4b}=3$, then find the value of $\frac{a+b}{a+2b}$.
   (a) $\frac{5}{9}$ (b) $\frac{2}{7}$
   (c) $\frac{10}{9}$ (d) $\frac{10}{7}$

8. The number of boys is more than the number of girls by 12% of the total strength of the class. The ratio of the number of boys to that of the girls is
   (a) 11 : 14 (b) 14 : 11
   (c) 25 : 28 (d) 28 : 25

9. The ratio of the number of ladies to that of gents at a party was 3 : 2. When 20 more gents joined the party, the ratio was reversed. The number of ladies present at the party was
   (a) 36 (b) 32
   (c) 24 (d) 16

10. The ratio of the ages of two boys is 5 : 6. After 2 years, the ratio of their ages will be 7 : 8. The ratio of their ages after 10 years will be
    (a) 15 : 16 (b) 17 : 18
    (c) 11 : 12 (d) 22 : 24

11. The sum of three numbers is 116. The ratio of the second to the third is 9 : 16 and the first to third is 1 : 4. The second number is
    (a) 30 (b) 32
    (c) 34 (d) 36

12. A man encashes a cheque of ₹ 600 from a bank. The bank pays him money in 10 rupee notes and 5 rupee notes only, totalling 72. The ratio of the number of 10 rupee notes to that of 5 rupee notes is
(a) 1 : 2 (b) 2 : 1
(c) 2 : 3 (d) 3 : 2

13. If 378 coins consist of one rupee, 50 paise and 25 paise coins whose values are in the ratio 13 : 11 : 7, then the number of 50 paise coins will be
(a) 132 (b) 128
(c) 136 (d) 133

14. An amount of money is to be divided among P, Q and R in the ratio 4 : 7 : 9. If the difference between the shares of Q and R is ₹ 500, what will be the difference between the shares of P and Q?
(a) ₹ 500 (b) ₹ 1000
(c) ₹ 750 (d) ₹ 850

15. The ratio of A to B is 4 : 5 and that of B to C is 2 : 3. If A equals 800, then C equals
(a) 1000 (b) 1200
(c) 1500 (d) 2000

16. In an examination, the number of those who passed and the number of those who failed were in the ratio 25 : 4. If five more had appeared and the number of failures were 2 less than earlier, the ratio of those passed to failures would be 22 : 3. The number who appeared at the examination is
(a) 145 (b) 150
(c) 155 (d) 180

17. The incomes of A and B are in the ratio 5 : 3. The expenses of A, B and C are in the ratio 8 : 5 : 2. If C spends ₹ 2000 and B saves ₹ 700, then A saves
(a) ₹ 1500 (b) ₹ 1000
(c) ₹ 500 (d) ₹ 250

18. The speeds of three cars are in the ratio 2 : 3 : 4. What is the ratio between the times taken by these cars to travel the same distance?
(a) 4 : 3 : 2 (b) 2 : 3 : 4
(c) 4 : 9 : 16 (d) 6 : 4 : 3

19. The sum of the squares of three numbers which are in the ratio 2 : 3 : 4 is 725. What are these numbers?
(a) 10, 15, 20 (b) 14, 21, 28
(c) 20, 15, 30 (d) 20, 30, 40

20. ₹ 2430 are distributed among three persons so that their shares be diminished by ₹ 5, ₹ 10 and ₹ 15 respectively, the remainder shall be in the ratio 3 : 4 : 5. The share of C is
(a) ₹ 1015 (b) ₹ 605
(c) ₹ 810 (d) ₹ 1415

21. Three person's A, B and C whose salaries together are ₹ 14400 and expenditures are 80%, 85% and 75% of their salaries respectively. If their savings are in the ratio 8 : 9 : 20, then find their respective salaries.
(a) ₹ 3000, ₹ 5000, ₹ 6400
(b) ₹ 3400, ₹ 4800, ₹ 6200
(c) ₹ 3200, ₹ 4800, ₹ 6400
(d) ₹ 3000, ₹ 5200, ₹ 6200

22. A pot contains 81 litres of pure milk. $\frac{1}{3}$ of the milk is replaced by the same amount of water. Again $\frac{1}{3}$ of the mixture is replaced by that amount of water. The ratio of milk and water in the new mixture is
(a) 1 : 2 (b) 1 : 1
(c) 2 : 1 (d) 4 : 5

23. A mixture contains milk and water in the ratio 5 : 1. On adding 5 litres of water, the ratio of milk to water becomes 5 : 2. The quantity of milk in the original mixture is
(a) 16 litres (b) 22.75 litres
(c) 25 litres (d) 32.5 litres

24. If 2 kg of metal of which $\frac{1}{3}$ is zinc and rest copper be mixed with 3 kg of metal of which $\frac{1}{4}$ is zinc and rest copper, what is the ratio of zinc to copper in the mixture?
(a) 13 : 35 (b) 12 : 30
(c) 15 : 40 (d) 17 : 43

25. The number of employees in companies A, B and C are in the ratio 3 : 2 : 4 respectively. If the number of employees in the three companies is increased by 20%, 30% and 15% respectively, what will be the new ratio of employees working in companies A, B and C respectively?
(a) 18 : 13 : 24 (b) 13 : 18 : 23
(c) 17 : 13 : 23 (d) 18 : 13 : 23

26. The ratio of a 2-digit number to the sum of its digits is 7 : 1. If the digit in the ten's place is 1 more than the digit in the one's place, then the number is
(a) 65 (b) 43
(c) 32 (d) 21

27. The ratio of the number of boys to that of girls in a group becomes 2 : 1 when 15 girls leave. But afterwards when 45 boys also leave, the ratio becomes 1 : 5. Originally the number of girls in the group was
(a) 20 (b) 30
(c) 40 (d) 50

28. A grocer buys kinds of rice X and Y; one (X) at the rate of ₹ a per kg and the other (Y) at the rate of ₹ b per kg. He mixes them and obtains a mixture of ₹ c per kg. What is the ratio of the variety X to that of variety Y in the mixture?

(a) $(b+c):(c+a)$ (b) $(b-c):(c+a)$
(c) $(b-c):(c-a)$ (d) $(b+c):(c-a)$

29. The ratio of first and second class train fares between two stations is 3 : 1 and that of the number of passengers travelling between these stations by first and second class is 1 : 50. If on a particular day ₹ 1325 be collected from the passengers travelling between these stations, then the amount collected from the second class passengers is

(a) ₹ 1250 (b) ₹ 1000
(c) ₹ 850 (d) ₹ 750

30. Two liquids A and B are in the ratio 5 : 1 in container 1 and 1 : 3 in container 2. In what ratio should the contents of the two containers be mixed so as to obtain a mixture of A and B in the ratio 1 : 1?

(a) 2 : 3 (b) 4 : 3
(c) 3 : 2 (d) 3 : 4

Answers

1. (d)	**2.** (b)	**3.** (d)	**4.** (c)	**5.** (b)	**6.** (a)	**7.** (c)	**8.** (b)	**9.** (c)	**10.** (a)
11. (d)	**12.** (b)	**13.** (a)	**14.** (c)	**15.** (c)	**16.** (a)	**17.** (a)	**18.** (d)	**19.** (a)	**20.** (a)
21. (c)	**22.** (d)	**23.** (c)	**24.** (d)	**25.** (d)	**26.** (d)	**27.** (c)	**28.** (c)	**29.** (a)	**30.** (d)

Hints and Solutions

1. (d) $\frac{A}{D} = \frac{A}{B}\times\frac{B}{C}\times\frac{C}{D}=\frac{3}{4}\times\frac{5}{7}\times\frac{8}{9}=\frac{10}{21}$

$\Rightarrow A:D=\mathbf{10:21}$.

2. (b) Let $\frac{1}{x}=2k$, $\frac{1}{y}=3k$, $\frac{1}{z}=5k$

$\Rightarrow x=\frac{1}{2k}, y=\frac{1}{3k}, z=\frac{1}{5k}$

Then, $x:y:z=\frac{1}{2k}:\frac{1}{3k}:\frac{1}{5k}$

$=\frac{1}{2k}\times 30k:\frac{1}{3k}\times 30k:\frac{1}{5k}\times 30k$

$= 15 : 10 : 6.$ ($\because$ LCM $= 30k$)

3. (d) $\frac{3a+5b}{3a-5b}=5 \Rightarrow 3a+5b=5(3a-5b)$

$\Rightarrow 3a+5b=15a-25b \Rightarrow 12a=30b$

$\Rightarrow \frac{a}{b}=\frac{30}{12}=\frac{5}{2} \Rightarrow a:b=\mathbf{5:2}$.

4. (c) $A:B=\frac{1}{2}:\frac{3}{8}=\frac{1}{2}\div\frac{3}{8}=\frac{1}{2}\times\frac{8}{3}=\frac{4}{3}=4:3=8:6$

$B:C=\frac{1}{3}:\frac{5}{9}=\frac{1}{3}\div\frac{5}{9}=\frac{1}{3}\times\frac{9}{5}=\frac{3}{5}=3:5=6:10$

$C:D=\frac{5}{6}:\frac{3}{4}=\frac{5}{6}\div\frac{3}{4}=\frac{5}{6}\times\frac{4}{3}=\frac{10}{9}=10:9$

$\therefore A:B:C:D=\mathbf{8:6:10:9}$.

5. (b) $a:b=2:3 \Rightarrow \frac{a}{b}=\frac{2}{3} \Rightarrow a=\frac{2b}{3}$

$\Rightarrow a^2=\frac{4b^2}{9}$

$b:c=4:5 \Rightarrow \frac{b}{c}=\frac{4}{5} \Rightarrow b=\frac{4c}{5}$

$\Rightarrow b^2=\frac{4bc}{5} \Rightarrow bc=\frac{5b^2}{4}$

$\therefore a^2:b^2:bc=\frac{4b^2}{9}:b^2:\frac{5b^2}{4}=\frac{4}{9}:1:\frac{5}{4}$

$=\frac{4}{9}\times 36:1\times 36:\frac{5}{4}\times 36$ ($\therefore$ LCM $= 36$)

$=\mathbf{16:36:45}$.

6. (a) $a:b=2:5 \Rightarrow \frac{a}{b}=\frac{2}{5}$

$(3a+4b):(5a+6b)=\frac{3a+4b}{5a+6b}=\frac{\frac{3a+4b}{b}}{\frac{5a+6b}{b}}$

(on dividing both numerator and denominator by b)

$\frac{\frac{3a}{b}+4}{\frac{5a}{b}+6}=\frac{3\times\frac{2}{5}+4}{5\times\frac{2}{5}+6}=\frac{\frac{6}{5}+4}{2+6}=\frac{26}{5}\times\frac{1}{8}=\mathbf{\frac{13}{20}}$.

7. (c) $\frac{2a+b}{a+4b}=3 \Rightarrow 2a+b=3a+12b$

$\Rightarrow a=-11b \Rightarrow \frac{a}{b}=-11$

Now, $\frac{a+b}{a+2b}=\frac{\frac{a}{b}+1}{\frac{a}{b}+2}=\frac{-11+1}{-11+2}=\frac{-10}{-9}=\mathbf{\frac{10}{9}}$.

8. (b) Let the number of boys be x and the number of girls be y.

Then, $x = y + 12\% (x + y)$

$\Rightarrow x = y + 0.12x + 0.12y \Rightarrow x - 0.12x = 1.12y$

$\Rightarrow 0.88x = 1.12y \Rightarrow \frac{x}{y}=\frac{112}{88}=\frac{14}{11} \Rightarrow x:y = \mathbf{14:11}$.

9. (c) Let the number of ladies and gents in the party be $3x$ and $2x$ respectively. Then,

$\frac{3x}{2x+20}=\frac{2}{3} \Rightarrow 9x = 4x + 40 \Rightarrow 5x = 40$

$\Rightarrow x = 8$

$\therefore$ Number of ladies in the party $= 3x = 3 \times 8 = \mathbf{24}$.

10. (a) Let the ages of the two boys be $5x$ and $6x$.

Given, $\frac{5x+2}{6x+2}=\frac{7}{8} \Rightarrow 40x + 16 = 42x + 14$

$\Rightarrow 2x = 2 \Rightarrow x = 1$

$\therefore$ Required ratio $= (5x + 10) : (6x + 10)$

$= (5 \times 1 + 10) : (6 \times 1 + 10) = \mathbf{15 : 16}$.

11. (d) Let the 1st number be x. Then, third number $= 4x$

Given, $\frac{\text{2nd number}}{\text{3rd number}}=\frac{9}{16} \Rightarrow \frac{\text{2nd number}}{4x}=\frac{9}{16}$

$\Rightarrow$ 2nd number $=\frac{9}{16}\times 4x=\frac{9}{4}x$

Given, $x+\frac{9x}{4}+4x=116$

$\Rightarrow 4x + 9x + 16x = 116 \times 4$

$\Rightarrow 29x = 116 \times 4 \Rightarrow x=\frac{116\times 4}{29}=16$

$\therefore$ 2nd number $=\frac{9}{16}\times 4\times 16=\mathbf{36}$.

12. (b) Let the number of 10 rupee notes $= x$. Then,

Number of 5 rupee notes $= 72 - x$

Given, $x \times 10 + 5 (72 - x) = 600$

$\Rightarrow 10x + 360 - 5x = 600 \Rightarrow 5x = 240 \Rightarrow x = 48$

$\therefore$ Required ratio $=\frac{48}{(72-48)}=\frac{48}{24}=\frac{2}{1}=\mathbf{2:1}$.

13. (a) Let, Number of one-rupee coins $= x$,

Number of 50 paise coins $= y$,

Number of 25 paise coins $= z$.

Given, $x + y + z = 378$ ...(1)

Also, $1 \times x : \frac{50}{100}\times y : \frac{25}{100}\times z = 13:11:7$

$\Rightarrow x : y/2 : z/4 = 13 : 11 : 7$

$\Rightarrow \frac{x}{13}=\frac{y/2}{11}=\frac{z/4}{7}=k$ (say)

$\Rightarrow x = 13k, y = 22k, z = 28k$.

Then, $13k + 22k + 28k = 378$ (From (1))

$\Rightarrow 63k = 378 \Rightarrow k = 6$

$\therefore$ Number of 50 paise coins $= 22 \times 6 = \mathbf{132}$.

14. (c) Let the shares of P, Q and R be ₹ $4x$, ₹ $7x$ and ₹ $9x$ respectively. Then, $9x - 7x = 500$

$\Rightarrow 2x = 500 \Rightarrow x = 250$

$\therefore$ Required difference $= 7x - 4x = 3x = 3 \times 250$

$=$ **₹ 750**.

15. (c) $\frac{A}{B}=\frac{4}{5}$ and $\frac{B}{C}=\frac{2}{3}$

To make both the values of B equal, take LCM of both values of B, *i.e.*, 5 and 2 which is equal to 10.

$\therefore \frac{A}{B}=\frac{4}{5}=\frac{8}{10}$ and $\frac{B}{C}=\frac{2}{3}=\frac{10}{15}$

$\Rightarrow A : B : C = 8 : 10 : 15$

$\Rightarrow A = 8x, B = 10x, C = 15x$

Given, $A = 800 \Rightarrow 8x = 800$

$\Rightarrow x = 100 \Rightarrow C = \mathbf{1500}$.

16. (a) Let the number of those who passed $= 25x$ and number of failures $= 4x$

$\therefore$ Total no. of candidates who appeared $= 29x$

New number of candidates who appeared

$= 29x + 5$

New number of failures $= 4x - 2$

$\therefore$ New number of passed candidates

$= (29x + 5) - (4x - 2) = 25x + 7$

Given, $\frac{25x+7}{4x-2}=\frac{22}{3} \Rightarrow 3(25x+7)=22(4x-2)$

$\Rightarrow 75x + 21 = 88x - 44 \Rightarrow 13x = 65 \Rightarrow x = 5$

$\therefore$ Total no. of the students who appeared at the examination $= 29 \times 5 = \mathbf{145}$.

17. (a) Let the expenses of A, B and C be ₹ $8x$, ₹ $5x$ and ₹ $2x$ respectively. Given, $2x = 2000$

$\Rightarrow x =$ ₹ 1000

$\Rightarrow$ B's expenses = 5 × ₹ 1000 = ₹ 5000,

A's expenses = ₹ 8000

Given, B's savings = ₹ 700

$\Rightarrow$ B's income = ₹ 5000 + ₹ 700 = ₹ 5700.

Given, A's income : B's income = 5 : 3

$$\Rightarrow \frac{A\text{'s income}}{5700} = \frac{5}{3}$$

$$\Rightarrow A\text{'s income} = \frac{5}{3} \times ₹\,5700 = ₹\,9500$$

$\therefore$ A's savings = ₹ 9500 – ₹ 8000 = **₹ 1500.**

18. (d) Speed is always inversely proportional to time,

i.e., $s \propto \frac{1}{t}$.

$$\therefore \text{ Ratio of times taken} = \frac{1}{2} : \frac{1}{3} : \frac{1}{4}$$

$$= \frac{1}{2} \times 12 : \frac{1}{3} \times 12 : \frac{1}{4} \times 12$$

$= \mathbf{6 : 4 : 3}$.

19. (a) Let the three numbers be $2x$, $3x$ and $4x$.

Given, $(2x)^2 + (3x)^2 + (4x)^2 = 725$

$\Rightarrow 4x^2 + 9x^2 + 16x^2 = 725 \Rightarrow 29x^2 = 725$

$\Rightarrow x^2 = 25 \Rightarrow x = 5$

$\therefore$ The numbers are **10, 15** and **20.**

20. (a) Let ₹ 2430 be divided among three persons with their shares as x, y and z respectively so that

$$x + y + z = 2430$$

Given, $x - 5 : y - 10 : z - 15 = 3 : 4 : 5$

$\Rightarrow x - 5 = 3k, y - 10 = 4k$ and $z - 15 = 5k$

$\Rightarrow x = 3k + 5, y = 4k + 10$ and $z = 5k + 15$

$\therefore 3k + 5 + 4k + 10 + 5k + 15 = 2430$

$\Rightarrow 12k + 30 = 2430 \Rightarrow 12k = 2400 \Rightarrow k = 200$

$\therefore$ C's share $= z = 5k + 15 = 5 \times 200 + 15 =$ **₹ 1015.**

21. (c) Total salaries of A, B and C = ₹ 14400

Let the salaries of A, B and C be ₹ x, ₹ y and ₹ z respectively.

$\therefore x + y + z = 14400$ ...(1)

Their spendings are 80%, 85% and 75% respectively.

$\therefore$ Their savings are 20%, 15% and 25% respectively and also their savings are given in the ratio 8 : 9 : 20.

$$\frac{\frac{20x}{100}}{\frac{15y}{100}} = \frac{8}{9} \text{ and } \frac{\frac{15y}{100}}{\frac{25z}{100}} = \frac{9}{20}$$

$$\Rightarrow \frac{20x}{15y} = \frac{8}{9} \Rightarrow x = \frac{8 \times 15}{9 \times 20} y = \frac{2}{3} y \quad ...(2)$$

$$\text{and } \frac{15y}{25z} = \frac{9}{20} \Rightarrow z = \frac{15 \times 20}{9 \times 25} y = \frac{4}{3} y \quad ...(3)$$

$\therefore$ From eq. (1), (2) and (3), we have

$$\frac{2}{3} y + y + \frac{4}{3} y = 14400$$

$$\Rightarrow \frac{9y}{3} = 14400 \Rightarrow y = ₹\,\frac{14400}{3} = \mathbf{₹\,4800}$$

$$\therefore x = \frac{2}{3} \times ₹\,4800 = \mathbf{₹\,3200}$$

$$\text{and } z = \frac{4}{3} \times ₹\,4800 = \mathbf{₹\,6400}.$$

22. (d) **Initially,** Milk = 81 litres and Water = 0 litre

After 1st operation,

$$\text{Milk} = \left(81 - \frac{1}{3} \times 81\right) \text{ litres} = (81 - 27) \text{ litres}$$

= 54 litres

$$\text{Water} = \left(0 + \frac{1}{3} \times 81\right) \text{ litres} = 27 \text{ litres}$$

After 2nd operation,

$$\text{Milk} = \left(54 - \frac{1}{3} \times 54\right) \text{ litres} = (54 - 18) \text{ litres}$$

= 36 litres

$$\text{Water} = \left(27 - \frac{1}{3} \times 27\right) \text{ litres} + \left(\frac{1}{3} \times 54 + \frac{1}{3} \times 27\right) \text{ litres}$$

= (27 – 9) litres + (18 + 9) litres = 45 litres

$\therefore$ Required ratio of milk and water in the new mixture = 36 : 45 = **4 : 5.**

23. (c) In the original mixture, let milk = $5x$ litres and water = x litres

$$\text{Then, } \frac{5x}{x + 5} = \frac{5}{2} \Rightarrow 10x = 5x + 25$$

$\Rightarrow 5x = 25 \Rightarrow x = 5$

$\therefore$ Quantity of milk in original mixture = 25 litres.

24. (d) **1st metal:**

Quantity of zinc = $\frac{1}{3} \times 2$ kg $= \frac{2}{3}$ kg

$\therefore$ Quantity of copper = $\left(2 - \frac{2}{3}\right)$ kg $= \frac{4}{3}$ kg

2nd metal:

Quantity of zinc = $\frac{1}{4}$ of 3 kg $= \frac{3}{4}$ kg

$\therefore$ Quantity of copper = $\left(3 - \frac{3}{4}\right)$ kg $= \frac{9}{4}$ kg

In the mixture, the required ratio

$= \left(\frac{2}{3} + \frac{3}{4}\right) : \left(\frac{4}{3} + \frac{9}{4}\right) = \frac{17}{12} : \frac{43}{12} = \mathbf{17 : 43}.$

25. (d) Let the number of employees be $3x$, $2x$ and $4x$ respectively.

Increased number of employees in

Company $A = 3x + 20\%$ of $3x$

$= \frac{120}{100} \times 3x = \frac{36x}{10} = \frac{18x}{5}$

Company $B = 2x + 30\%$ of $2x$

$= \frac{130}{100} \times 2x = \frac{26x}{10} = \frac{13x}{5}$

Company $C = 4x + 15\%$ of $4x$

$= \frac{115}{100} \times 4x = \frac{46x}{10} = \frac{23x}{5}$

$\therefore$ Required ratio $= \frac{18x}{5} : \frac{13x}{5} : \frac{23x}{5} = \mathbf{18 : 13 : 23}.$

26. (d) Let the digit at the unit's place in a 2-digit number be x.

Then, the digit at the ten's place $= x + 1$

$\therefore$ The number $= 10(x + 1) + x = 11x + 10$

Sum of digits $= x + 1 + x = 2x + 1$

Given, $\frac{11x + 10}{2x + 1} = \frac{7}{1}$

$\Rightarrow 11x + 10 = 14x + 7 \Rightarrow 3x = 3 \Rightarrow x = 1$

$\therefore$ The number $= 11 \times 1 + 10 = \mathbf{21}.$

27. (c) Let the number of boys and girls be x and y respectively.

Then, $\frac{x}{y - 15} = \frac{2}{1}$ and $\frac{x - 45}{y - 15} = \frac{1}{5}$

$\Rightarrow \frac{x}{2} = (y - 15)$ ...(1) and $5(x - 45) = y - 15$ ...(2)

From eq. (1) and (2)

$5(x - 45) = \frac{x}{2} \Rightarrow 10x - 450 = x$

$\Rightarrow 9x = 450 \Rightarrow x = 50$

$\therefore$ From eq. (1), $y - 15 = 25 \Rightarrow y = \mathbf{40}.$

28. (c) Cost of variety X rice = ₹ Xa

Cost of variety Y rice = ₹ Yb

Cost of mixture = ₹ $(X + Y)c$

Now, $Xa + Yb = (X + Y)c$

$\Rightarrow X(a - c) = Y(c - b) \Rightarrow \frac{X}{Y} = \frac{c - b}{a - c} = \frac{b - c}{c - a}$

$\therefore X : Y = (b - c) : (c - a).$

29. (a) Let the 1st class fare is ₹ $3x$ and the 2nd class fare is ₹ x.

Let the number of passengers travelling in 1st class $= y$

Then, the number of passengers travelling in 2nd class $= 50y$

$\therefore 3xy + 50xy = 1325$

$\Rightarrow 53xy = 1325 \Rightarrow xy =$ ₹ 25

$\therefore$ Amount collected from 2nd class passengers = 50 × ₹ 25 = **₹ 1250.**

30. (d) Let the required ratio be $x : y$

Liquid A in x litres of mixture in container 1 $= \frac{5x}{6}$ litres

Liquid B in x litres of mixture in container 1 $= \frac{x}{6}$ litres

Liquid A in y litres of mixture in container 2 $= \frac{y}{4}$ litres

Liquid B in y litres of mixture in container 2 $= \frac{3y}{4}$ litres

$\therefore$ Liquid A : Liquid $B = \left(\frac{5x}{6} + \frac{y}{4}\right) : \left(\frac{x}{6} + \frac{3y}{4}\right)$

Given, $\frac{\left(\frac{5x}{6} + \frac{y}{4}\right)}{\left(\frac{x}{6} + \frac{3y}{4}\right)} = \frac{1}{1} \Rightarrow \frac{5x}{6} + \frac{y}{4} = \frac{x}{6} + \frac{3y}{4}$

$\Rightarrow \frac{5x}{6} - \frac{x}{6} = \frac{3y}{4} - \frac{y}{4} \Rightarrow \frac{4x}{6} = \frac{2y}{4}$

$\Rightarrow \frac{2x}{3} = \frac{y}{2} \Rightarrow \frac{x}{y} = \mathbf{\frac{3}{4}}.$

Self Assessment Sheet–15

1. If $a : b = 3 : 4$ and $b : c = 8 : 9$, then $a : c = ?$
(a) 1 : 2 (b) 3 : 2
(c) 1 : 3 (d) 2 : 3

2. If $\frac{1}{x}:\frac{1}{y}:\frac{1}{z}=2:3:5$, then $x:y:z=?$
(a) 2 : 3 : 5 (b) 15 : 10 : 6
(c) 5 : 3 : 2 (d) 6 : 10 : 15

3. Two numbers are in the ratio $1\frac{1}{2} : 2\frac{2}{3}$. When each one of these is increased by 15, their ratio becomes $1\frac{2}{3}:2\frac{1}{2}$. The larger of the numbers is :
(a) 27 (b) 36
(c) 48 (d) 64

4. The total value of a collection of coins of denominations ₹ 1.00, 50 - paise, 25 - paise, 10 paise and 5 - paise is ₹ 380. If the number of coins of each denomination is the same, then the number of one - rupee coins is :
(a) 160 (b) 180
(c) 200 (d) 220

5. A flagstaff 17.5 m high casts a shadow of 40.25 m. The height of the building which casts a shadow 28.75 m long under similar conditions, will be
(a) 10 m (b) 12.5 m
(c) 17.5 m (d) 21. 25 m

6. If $x : y = 3 : 4$, then $(2x + 3y) : (3y - 2x) = ?$
(a) 2 : 1 (b) 3 : 2
(c) 3 : 1 (d) 21 : 1

7. By mistake, instead of dividing ₹ 117 among A, B and C in the ratio $\frac{1}{2} : \frac{1}{3} : \frac{1}{4}$, it was divided in the ratio of 2 : 3 : 4. Who gains the most and by how much?
(a) A, ₹ 28 (b) B, ₹ 3
(c) C, ₹ 20 (d) C, ₹ 25

8. The students in three classes are in the ratio 2 : 3: 5. If 40 students are increased in each class, the ratio changes to 4 : 5: 7. Originally the total number of students was
(a) 100 (b) 180
(c) 200 (d) 400

9. If $x : y = 7 : 3$, then the value of $\frac{xy + y^2}{x^2 - y^2}$ is :
(a) $\frac{3}{4}$ (b) $\frac{4}{3}$
(c) $\frac{3}{7}$ (d) $\frac{7}{3}$

10. ₹ 750 is divided among A, B and C in such a manner that $A : B = 5 : 2$ and $B : C = 7 : 13$. What is A's share ?
(a) ₹ 350 (b) ₹ 260
(c) ₹ 140 (d) ₹ 250

Answers

1. (d)	2. (b)	3. (c)	4. (c)	5. (b)	6. (c)	7. (d)	8. (c)	9. (a)	10. (a)

Chapter 16 AVERAGE

KEY FACTS

1. Average of n observations $= \dfrac{\text{Sum of } n \text{ observations}}{n}$

2. Sum of n observation $= n \times$ Average of n observations

Solved Examples

Ex. 1. ***What is the average of the first nine prime numbers?***

Sol. The first nine prime numbers are 2, 3, 5, 7, 11, 13, 17, 19, 23

$$\text{Required average} = \frac{2+3+5+7+11+13+17+19+23}{9} = \frac{100}{9} = \mathbf{11\frac{1}{9}}.$$

Ex. 2. ***The average of 7 consecutive numbers is 20. What is the largest number?***

Sol. Let the seven consecutive numbers be

$x, x+1, x+2, x+3, x+4, x+5, x+6$

$$\text{Given, } \frac{x+x+1+x+2+x+3+x+4+x+5+x+6}{7} = 20$$

$$\Rightarrow 7x + 21 = 140 \Rightarrow 7x = 140 - 21 = 119 \Rightarrow x = \frac{119}{7} = 17$$

$\therefore$ Largest number $= 17 + 6 = \mathbf{23}$.

Ex. 3. ***If the average of 5 observations $x, x+2, x+4, x+6$ and $x+8$ is 11, then what is the average of the last three observations?***

Sol. Given, $\dfrac{x+x+2+x+4+x+6+x+8}{5} = 11 \Rightarrow 5x + 20 = 55 \Rightarrow x = \dfrac{55-20}{5} = \dfrac{35}{5} = 7$

$$\text{Average of last three observations} = \frac{x+4+x+6+x+8}{3}$$

$$= \frac{3x+18}{3} = x + 6 = 7 + 6 = \mathbf{13}.$$

Ex. 4. ***If the average marks of three batches of 55, 60 and 45 students respectively is 50, 55 and 60, then what are the average marks of all the students?***

Sol. Total marks of first batch $= 55 \times 50 = 2750$

Total marks of second batch $= 60 \times 55 = 3300$

Total marks of third batch = $45 \times 60 = 2700$

Total number of students in three batche $= 55 + 60 + 45 = 160$

$\therefore$ Average marks of all the students $= \dfrac{2750 + 3300 + 2700}{160} = \dfrac{8750}{160} = \mathbf{54.6875.}$

Ex. 5. ***The average of 100 numbers is 44. The average of these 100 numbers and four other numbers is 50. What is the average of the four new numbers?***

Sol. Sum of 100 numbers = $100 \times 44 = 4400$

Sum of 104 numbers = $104 \times 50 = 5200$

Sum of 4 numbers = $5200 - 4400 = 800$

$\therefore$ Average of four new numbers = $\dfrac{800}{4} = \mathbf{200}$.

Ex. 6. ***The average age of 15 students of a class is 15 years. Out of these, the average age of 5 students is 14 years and that of the other nine students is 16 years. What is the age of the 15th student?***

Sol. Total age of 15 students = (15×15) years = 225 years

Total age of 5 students = (5×14) years = 70 years

Total age of other 9 students = (9×16) years = 144 years

$\therefore$ Age of the 15th student = $225 - (70 + 144) = 225 - 214 =$ **11 years.**

Ex. 7. ***The average weight of a class of 24 students is 35 kg. If the weight of the teacher be included, then the average rises by 400 g. What is the weight of the teacher?***

Sol. Total weight of all the students in the class = (24×35) kg = 840 kg

New average weight, when teacher is included = 35.4 kg

$\therefore$ Total weight of all the students and teacher = (25×35.4) kg = 885 kg

$\therefore$ Weight of the teacher = 885 kg – 840 kg = **45 kg.**

Ex. 8. ***There are 30 students in a class. The average age of the first 10 students is 12.5 years. The average age of the remaining 20 students is 13.1 years. What is the average age (in years) of the students of the whole class?***

Sol. Total age of first 10 students = (10×12.5) years = 125 years

Total age of remaining 20 students = (20×13.1) years = 262 years

$\therefore$ Average age of the whole class = $\dfrac{387}{30}$ years = **12.9 years.**

Ex. 9. ***A grocer has a sale of ₹ 6435, ₹ 6927, ₹ 6855, ₹ 7230 and ₹ 6562 for 5 consecutive months. How much sale must be have in the sixth month so that he gets an average sale of ₹ 6500?***

Sol. Let the sale of 6th month be ₹ x.

Then, total sales of 6 months = $6435 + 6927 + 6855 + 7230 + 6562 + x = 34009 + x$

Given, average sale required for 6 months = ₹ 6500

$\therefore$ Total required sale for 6 months = ₹ 6500×6 = ₹ 39000

$\therefore$ $34009 + x = 39000$

$\Rightarrow x = 39000 - 34009 =$ **₹ 4991.**

Ex. 10. ***Of the three numbers, the first is twice the second and the second is thrice the third. If the average of the three numbers is 10, find the largest number?***

Sol. Let the third number be x.

Then, second number = $3x$ and first number = $6x$

Given, $\frac{x+3x+6x}{3}=10 \Rightarrow 10x = 30 \Rightarrow x = 3$

$\therefore$ Largest number = $6x = 6 \times 3 =$ **18.**

Ex. 11. ***The average of a collection of 20 measurements was calculated to be 56 cm. But later it was found that a mistake occured in one of the measurements which was recorded as 64 cm but should have been 61 cm? What is the correct average?***

Sol. Incorrect total of 20 measurement = $20 \times 56 = 1120$

Correct total = $1120 - 64 + 61 = 1117$

$\therefore$ Correct average = $\frac{1117}{20}$ = **55.85 cm.**

Ex. 12. ***Three years ago the average age of A and B was 18 years with C joining them now, the average becomes 22 years. How old is C now?***

Sol. Let the present ages of A, B and C be x years, y years and z years respectively.

Given, $\frac{(x-3)+(y-3)}{2}=18 \Rightarrow x+y-6=36 \Rightarrow x+y=42$ and $\frac{x+y+z}{3}=22$

$\Rightarrow x+y+z=66 \quad \Rightarrow 42+z=66 \quad \Rightarrow z=$ **24 years.**

Question Bank–16

1. The average of the first five odd prime numbers is
(a) 7 (b) 7.8
(c) 8 (d) 8.7

2. Find the average of all the numbers between 6 and 34 which are divisible by 5
(a) 18 (b) 20
(c) 24 (d) 30

3. A, B, C, D, E are consecutive even numbers. The average of these five numbers is 82. What is the product of C and E?
(a) 6888 (b) 6720
(c) 7052 (d) 7224

4. If $37a + 37b = 5661$, what is the average of a and b ?
(a) 74.5 (b) 151
(c) 76.5 (d) 153

5. A student was asked to find the average of the following 12 numbers: 3, 11, 7, 9, 15, 13, 8, 19, 17, 21, 14 and x. He found the average to be 12. What should come in place of x?
(a) 3 (b) 7
(c) 17 (d) 31

6. The average of 2, 7, 6 and x is 5 and the average of 18, 1, 6, x and y is 10. What is the value of y?
(a) 10 (b) 30
(c) 20 (d) 5

7. If x_1, x_2, x_3, x_4, x_5 are five consecutive odd numbers, then their average is
(a) x_2 (b) x_3
(c) x_4 (d) x_5

8. The average of eight numbers is 38.4 and the average of seven of them is 39.2. What is the eight number?
(a) 0.8 (b) 32.8
(c) 34.8 (d) 33.8

9. The average age of 24 boys and a teacher is 15. When the teacher's age is excluded the average decreases by 1. What is the age of the teacher?
(a) 39 years (b) 38 years
(c) 40 years (d) 41 years

10. If the average weight of 6 students is 50 kg, that of 2 students is 51 kg and that of rest of 2 students is 55 kg, then the average weight of all the students is
(a) 61 kg (b) 51.5 kg
(c) 52 kg (d) 51.2 kg

11. The sum of five numbers is 555. The average of first two numbers is 75 and the third number is 115. What is the average of the last two numbers?
(a) 145 (b) 290
(c) 265 (d) 150

12. The average of 20 numbers is 12. The average of the first 12 numbers is 11 and that of the next 7 numbers is 10. The last number is
(a) 40 (b) 38
(c) 48 (d) 50

13. The average of 6 observations is 45.5. If one new observation is added to the previous observations, then the new average becomes 47. The new observation is
(a) 58 (b) 56
(c) 50 (d) 46

14. The average age of 3 friends is 23. Even if the age of the 4th friend is added, the average remains 23. What is the age of the 4th friend?
(a) 21 years (b) 23 years
(c) 32 years (d) 34 years

15. In a joint family, the average age of grandparents is 67 years, the average age of parents is 35 years and that of three grand children is 6 years. What is the average age of the family?
(a) $28\frac{4}{7}$ years (b) $31\frac{5}{7}$ years
(c) $32\frac{1}{2}$ years (d) $29\frac{4}{7}$ years

16. For a week, the average daily rainfall was 0.25 cm. There was no rainfall on Saturday. Rainfall on Sunday, Monday, Tuesday, Wednesday and Thursday was 0.4 cm, 0.03 cm, 0.45 cm, 0.27 cm and 0.5 cm respectively. What was the rainfall on Friday?
(a) 0.2 cm (b) 0.1 cm
(c) 0.05 cm (d) 0.15 cm

17. The average of n numbers is x. When 36 is subtracted from two of the numbers, the new average becomes $(x - 8)$. The value of n is
(a) 6 (b) 8
(c) 9 (d) 72

18. Of the four numbers whose average is 60, the first is $\frac{1}{4}$th of the sum of the last three. The first number is
(a) 15 (b) 45
(c) 48 (d) 60.25

19. Four years ago the average age of A, B and C was 25 years. Five years ago the average age of B and C was 20 years. A's present age is
(a) 60 years (b) 37 years
(c) 62 years (d) 15 years

20. Out of three numbers, the first is twice the second and is half of the third. If the average of three numbers is 56, then the difference of the first and third is
(a) 12 (b) 20
(c) 24 (d) 48

21. The average marks in Mathematics for 5 students was found to be 50. Later it was discovered that in case of one student, the marks 48 were misread as 84. The correct average is
(a) 40.2 (b) 40.8
(c) 42.8 (d) 48.2

22. The average of 75 numbers is calculated as 35. If each number is increased by 5, then the new average is
(a) 30 (b) 40
(c) 70 (d) 80

23. The average of 6 numbers is 30. If the average of the first four is 25 and that of the last three is 35, the fourth number is
(a) 25 (b) 30
(c) 35 (d) 40

24. The average of 11 numbers is 32. If the average of the first six numbers is 34 and that of the last six is 33, then the sixth number is
(a) 34 (b) 25
(c) 50 (d) 36

25. The average marks scored by 22 candidates in an examination are 45. The average marks of the first ten are 55 and those of last eleven are 40. The number of marks obtained by the 11th candidate. The number of marks obtained by the 11th candidate is
(a) 0 (b) 45
(c) 50 (d) 47.5

26. Out of 9 persons, 8 persons spent ₹ 30 each for their meals. The ninth one spent ₹ 20 more than the average expenditure of all nine. The total money spend by all of them was
(a) ₹ 260 (b) ₹ 290
(c) ₹ 292.50 (d) ₹ 400.50

27. Average age of 6 sons of a family is 8 years. Average age of sons together with their parents is 22 years. If the father is older than the mother by 8 years, the age of the mother (in years) is
(a) 44 (b) 52
(c) 60 (d) 68

28. The average age of a class is 15.8 years. The average age of the boys in the class is 16.4 years while that of the girls is 15.4 years. What is the ratio of boys to girls in the class?

(a) 1 : 2 (b) 2 : 3
(c) 3 : 4 (d) 3 : 5

29. The average expenditure of a man for the first five months of a year is ₹ 5000 and for the next seven months it is ₹ 5400. He saves ₹ 2300 during the year. His average monthly income is

(a) ₹ 5425 (b) ₹ 5500
(c) ₹ 5446 (d) ₹ 5600

30. The average monthly salary of the workers in a work shop is ₹ 8500. If the average monthly salary of 7 technicians is ₹ 10,000 and average monthly salary of the rest is ₹ 7800, the total number of workers in the work shop is

(a) 18 (b) 20
(c) 22 (d) 24

Answers

1. (b)	**2.** (b)	**3.** (c)	**4.** (c)	**5.** (b)	**6.** (c)	**7.** (b)	**8.** (b)	**9.** (a)	**10.** (d)
11. (a)	**12.** (b)	**13.** (b)	**14.** (b)	**15.** (b)	**16.** (b)	**17.** (c)	**18.** (c)	**19.** (b)	**20.** (d)
21. (c)	**22.** (b)	**23.** (a)	**24.** (c)	**25.** (a)	**26.** (c)	**27.** (c)	**28.** (b)	**29.** (a)	**30.** (c)

Hints and Solutions

1. (b) Reqd. average $= \dfrac{3+5+7+11+13}{5} = \dfrac{39}{5} = \mathbf{7.8}.$

2. (b) Numbers between 6 and 34 divisible by 5 are 10, 15, 20, 25, 30.

Required average $= \dfrac{10+15+20+25+30}{5}$

$= \dfrac{100}{5} = \mathbf{20}.$

3. (c) Let $A = x$, $B = x + 2$, $C = x + 4$, $D = x + 6$, $E = x + 8$.

Given, $\dfrac{A+B+C+D+E}{5} = 82$

$\Rightarrow x + x + 2 + x + 4 + x + 6 + x + 8 = 410$

$\Rightarrow 5x + 20 = 410 \Rightarrow 5x = 390 \Rightarrow x = 78$

$\therefore C = 78 + 4 = 82$ and $E = 78 + 8 = 86$

$\Rightarrow C \times E = 82 \times 86 = \mathbf{7052}.$

4. (c) $37a + 37b = 5661 \Rightarrow 37(a + b) = 5661$

$\Rightarrow a + b = \dfrac{5661}{37} = 153$

$\therefore$ Average $= \dfrac{a+b}{2} = \dfrac{153}{2} = \mathbf{76.5}.$

5. (b) $\dfrac{3+11+7+9+15+13+8+19+17+21+14+x}{12} = 12$

$\Rightarrow 137 + x = 144 \Rightarrow x = \mathbf{7}.$

6. (c) Given, $\dfrac{2+7+6+x}{4} = 5 \Rightarrow 15 + x = 20$

$\Rightarrow x = 5$

Also, $\dfrac{18+1+6+x+y}{5} = 10$

$\Rightarrow 25 + x + y = 50 \Rightarrow x + y = 25$

$\Rightarrow y = 25 - 5 = \mathbf{20}.$

7. (b) The five consecutive odd numbers are

$x_1, x_1 + 2, x_1 + 4, x_1 + 6, x_1 + 8$

$\therefore$ Average $= \dfrac{x_1 + x_1 + 2 + x_1 + 4 + x_1 + 6 + x_1 + 8}{5}$

$= \dfrac{5x_1 + 20}{5} = x_1 + 4 = \mathbf{x_3}.$

8. (b) Sum of eight numbers $= 38.4 \times 8 = 307.2$

$\because$ Total sum = Average × Number of items

Sum of seven numbers $= 39.2 \times 7 = 274.4$

$\therefore$ Eighth number $= 307.2 - 274.4 = \mathbf{32.8}.$

9. (a) Total age of 24 boys and one teacher $= 25 \times 15 = 375$ years

Average age of 24 boys $= 15 - 1 = 14$ years

$\therefore$ Total age of 24 boys $= 24 \times 14 = 336$ years

$\Rightarrow$ Teacher's age $= 375 - 336 =$ **39 years.**

10. (d) Total weight of 6 students $= 6 \times 50$ kg $= 300$ kg

Total weight of 2 students $= 2 \times 51$ kg $= 102$ kg

Total weight of 2 students $= 2 \times 55$ kg $= 110$ kg.

$\therefore$ Average weight of all the students

$= \dfrac{\text{Total weight}}{\text{Number of students}} = \dfrac{300+102+110}{10}$ kg

$= \dfrac{512}{10}$ kg $=$ **51.2 kg.**

11. (a) Let the five numbers be A, B, C, D, E.

Given, $A + B + C + D + E = 555,$

$\dfrac{A+B}{2} = 75 \Rightarrow A + B = 150$ and $C = 115$

$\therefore$ $150 + 115 + D + E = 555$

$\Rightarrow D + E = 555 - 265 = 290$

$\therefore$ Required Average $= \frac{D+E}{2} = \frac{290}{2} = \mathbf{145}$.

12. (b) Sum of 20 numbers $= 20 \times 12 = 240$

Sum of first 12 numbers $= 11 \times 12 = 132$

Sum of next 7 numbers $= 10 \times 7 = 70$

$\therefore$ Last number $= 240 - (132 + 70)$
$= 240 - 202 = \mathbf{38}$.

13. (b) Sum of 6 observations $= 6 \times 45.5 = 273$

Sum of (6 + 1) observations $= 7 \times 47 = 329$

$\therefore$ New observation $= 329 - 273 = \mathbf{56}$.

14. (b) Sum of the ages of 3 friends $= 3 \times 23 = 69$ years

Sum of the ages of 4 friends $= 4 \times 23 = 92$ years

$\therefore$ Age of 4th friend = 92 years – 69 years
= **23 years**.

15. (b) Total age of grandparents $= 2 \times 67$ years
= 134 years

Total age of parents $= 2 \times 35$ years = 70 years

Total age of three grand children $= 3 \times 6 = 18$ years

Total age of the family $= 134 + 70 + 18 = 222$ years

Number of member in the family = 7

$\therefore$ Required average $= \frac{222}{7}$ years $= \mathbf{31\frac{5}{7}}$ **years.**

16. (b) Total rainfall during the week $= 7 \times 0.25$ cm
= 1.75 cm

$\therefore$ Rainfall on Friday
$= 1.75 - (0.4 + 0.03 + 0.45 + 0.27 + 0.5 + 0)$
= 1.75 cm – 1.65 cm = **0.1 cm**.

17. (c) Sum of n numbers $= n \times x = nx$

Given, $\frac{nx - 36 - 36}{n} = x - 8 \Rightarrow nx - 72 = nx - 8n$

$\Rightarrow 8n = 72 \Rightarrow \mathbf{n = 9}$.

18. (c) Let the numbers be a, b, c, d. Then,

$\frac{a+b+c+d}{4} = 60 \Rightarrow a+b+c+d = 240$ ...(1)

and $a = \frac{1}{4}(b+c+d) \Rightarrow b+c+d = 4a$ ...(2)

$\therefore$ From eq (1) and (2)

$a + 4a = 240 \Rightarrow 5a = 240 \Rightarrow a = \mathbf{48}$.

19. (b) Let the present ages of A, B, C be x, y and z years respectively.

Given, $\frac{(x-4)+(y-4)+(z-4)}{3} = 25$

$\Rightarrow x + y + z - 12 = 75 \Rightarrow x + y + z = 87$ ...(1)

and $\frac{(y-5)+(z-5)}{2} = 20 \Rightarrow y + z - 10 = 40$

$\Rightarrow y + z = 50$ ...(2)

$\therefore$ From eq (1) and (2),

A's age $= x = (x+y+z) - (y+z) = (87 - 50)$ years
= **37 years.**

20. (d) Let the first number be x. Then,

Second number $= \frac{x}{2}$

Third number $= 2x$

Given, $\frac{x + \frac{x}{2} + 2x}{3} = 56$

$\Rightarrow \frac{7x}{6} = 56 \Rightarrow x = \frac{56 \times 6}{7} = 48$

$\therefore$ Difference bet. first and third $= 2x - x = x = \mathbf{48}$.

21. (c) Total marks of 5 students in Mathematics $= 5 \times 50$
$= 250$

Correct total $= 250 - 84 + 48 = 214$

$\therefore$ Correct average $= \frac{214}{5} = \mathbf{42.8}$.

22. (b) Sum of 75 numbers $= 75 \times 35 = 2625$

Total increase $= 75 \times 5 = 375$

$\therefore$ New sum $= 2625 + 375 = 3000$

$\Rightarrow$ New average $= \frac{3000}{75} = 40$.

23. (a) Sum of 6 numbers $= 6 \times 30 = 180$

Sum of first 4 numbers $= 4 \times 25 = 100$

Sum of last 3 numbers $= 3 \times 35 = 105$

$\therefore$ Fourth number $= (100 + 105) - 180$
$= 205 - 180$
$= \mathbf{25}$.

24. (c) Sixth number $= 34 \times 6 + 33 \times 6 - 32 \times 11 = \mathbf{50}$.

25. (a) Total marks scored by 22 candidates $= 22 \times 45$
$= 990$

Total marks scored by first 10 candidates $= 10 \times 55$
$= 550$

Total marks scored by last 11 candidates $= 11 \times 40$
$= 440$

$\therefore$ Marks scored by 11th candidate
$= 990 - (550 + 440) = 0$.

26. (c) Total amount spent by 8 persons $= 8 \times$ ₹ 30
= ₹ 240

Let the ninth person spend ₹ x. Then,

Average expenditure of all nine $= \frac{240 + x}{9}$

Money spent by ninth person

$= \frac{240+x}{9} + 20 = \frac{420+x}{9}$

Given, $\frac{420+x}{9} + 240 = 240 + x$

$\Rightarrow 420 + x = 9x \Rightarrow 8x = 420 \Rightarrow x = 52.5$

$\therefore$ Money spent by all of them = ₹ (240 + 52.5) = **₹ 292.50.**

27. (c) Total age of 6 sons = 6 × 8 = 48 years

Total age of sons together with their parents = 8 × 22 = 176 years

$\therefore$ Total age of parents = 176 – 48 = 128 years

Let the mother's age be x years. Then,

Father's age = $(x + 8)$ years

Given, $x + x + 8 = 128 \Rightarrow 2x = 120$

$\Rightarrow$ x = **60 years.**

28. (b) Let the number of boys and girls in the class be x and y respectively.

Total age of the class = $15.8 \times (x + y)$ = $(15.8x + 15.8y)$ years

Total age of the boys = $16.4x$ years

Total age of the girls = $15.4y$ years

$\therefore$ $16.4x + 15.4y = 15.8x + 15.8y$

$\Rightarrow 16.4x - 15.8x = 15.8y - 15.4y$

$\Rightarrow 0.6x = 0.4y \Rightarrow \frac{x}{y} = \frac{4}{6} = \mathbf{\frac{2}{3}}.$

29. (a) Expenditure for the whole year

= ₹ (5 × 5000 + 7 × 5400)

= ₹ 25000 + ₹ 37800 = ₹ 62800

Income for the whole year = ₹ 62800 + ₹ 2300 = ₹ 65100

$\therefore$ Average monthly income = ₹ $\frac{65100}{12}$ = **₹ 5425.**

30. (c) Let the total numbers of workers in workshop be x.

$\therefore$ $x \times 8500 = 7 \times 10000 + (x - 7) \times 7800$

$\Rightarrow 8500x = 70000 + 7800x - 54600$

$\Rightarrow 700x = 15400 \Rightarrow x = \frac{15400}{700} = \mathbf{22.}$

Self Assessment Sheet–16

1. The average height of the students in a class of 10 is 105 cm. If 20 more students with an average height of 120 cm join the class, what will the new average height be ?

(a) 105 cm (b) 110 cm
(c) 112 cm (d) 115 cm

2. The average consumption of petrol for a car for 7 months is 110 litres and for next 5 months it is 86 litres. The average monthly consumption is :

(a) 96 L (b) 98 L
(c) 100 L (d) 102 L

3. The average weight of 8 men is increased by 1.5 kg when one of the men who weights 65 kg is replaced by a new man. The weight of the new man is :

(a) 70 kg (b) 74 kg
(c) 76 kg (d) 77 kg

4. The average of x_1, x_2, x_3, x_4 is 16. Half the sum of x_2, x_3, x_4 is 23. What is the value of x_1?

(a) 17 (b) 18
(c) 19 (d) 20

5. If the total monthly income of 16 persons is ₹ 80800 and the income of one of them is 120% of the average income, then his income is :

(a) ₹ 5050 (b) ₹ 6060
(c) ₹ 6160 (d) ₹ 6600

6. The ratio of the arithmetic mean of two numbers to one of the numbers is 3 : 5. What is the ratio of the smaller number to the larger one ?

(a) 1 : 5 (b) 1 : 4
(c) 1 : 3 (d) 1 : 2
(e) None of the above

7. It rained as much as Wednesday as on all the other days of the week combined. If the average rainfall for the whole week was 3 cm, how much did it rain on Wednesday ?

(a) 2.625 cm (b) 3 cm
(c) 10.5 cm (d) 15 cm

8. The average of 50 numbers is 38. If two numbers 45 and 55 are discarded, the average of the remaining set of numbers is :

(a) 38.5 (b) 37.5
(c) 37.0 (d) 36.5

9. The average marks scored by Ravi in English, Science, Mathematics and History is less than 15 from that scored by him in English, History, Geography and Mathematics. What is the difference of marks in Science and Geography secured by him?

(a) 40 (b) 50
(c) 60 (d) Date inadequate
(e) None of these

10. Average age of seven persons in a group is 30 years. The average age of five person of this group is 31 years. What is the average age of the other two persons in the group ?

(a) 55 years
(b) 26 years
(c) 15 years
(d) Cannot be determined
(e) None of these

Answers

1. (d) **2.** (c) **3.** (d) **4.** (b) **5.** (b) **6.** (a) **7.** (c) **8.** (b) **9.** (c)
10. (d) The correct answer is 27.5 years

Unit Test–3

1. Sunil gets 10% more marks than Akbar. What percentage of marks does Akbar get less than Sunil?

(a) 9% (b) 10 %
(c) $9\frac{1}{11}\%$ (d) $11\frac{1}{9}\%$

2. A discount series of 10%, 20% and 40% is equal to a single discount of

(a) 50% (b) 56.80%
(c) 60% (d) 70.28%

3. A leak in the bottom of a tank can empty the full tank in 6 hours. An inlet pipe fills water at the rate of 4 litres per minute. When the tank is full, the inlet is opened and due to the leak, the tank is empty in 8 hours the capacity of the tank is :

(a) 5260 L (b) 5760 L
(c) 5846 L (d) 6970 L

4. A shopkeeper earns a profit of 15% after selling a book at 20% discount on the printed price. The ratio of the cost price and the printed price of the book is

(a) 20 : 23 (b) 23 : 20
(c) 16 : 23 (d) 23 : 16

5. A person invested part of ₹ 45000 at 4% and the rest at 6%. If his annual income from both are equal, then what is the average rate of interest ?

(a) 4.6% (b) 4.8%
(c) 5.0 % (d) 5.2%

6. What would be the printed price of a watch purchased at ₹ 380, so that after giving 5% discount, there is 25 % profit ?

(a) ₹ 400 (b) ₹ 450
(c) ₹ 500 (d) ₹ 600

7. There are some coins and rings of either gold or silver in a box. 60% of the objects are coins, 40% of the rings are of gold and 30% of the coins are of silver. What is the percentage of gold articles ?

(a) 16 (b) 27
(c) 58 (d) 70

8. A person gave 20% of his income to his elder son, 30% of the remaining to the younger son and 10% of the balance he donated to a trust. He is left with ₹ 10,080. His income was

(a) ₹ 50,000 (b) ₹ 40,000
(c) ₹ 30,000 (d) ₹ 20,000

9. The average weight of 4 persons five years ago was 45 years. By including a fifth person, the present average age of all the five is 49 years. The present age of the fifth person is :

(a) 64 years (b) 48 years
(c) 45 years (d) 40 years

10. Two numbers are in the ratio 3 : 5. If each number is increased by 10, the ratio becomes 5 : 7. The numbers are :

(a) 15, 25 (b) 30, 45
(c) 48, 60 (e) None of these

11. A person A sells a table costing ₹ 2000 to a person B and earns a profit of 6%. The person B sells it to another person C at a loss of 5%. At what price did B sell the table ?

(a) ₹ 2054 (b) ₹ 2050
(c) ₹ 2024 (d) ₹ 2014

12. Starting from his house one day, a student walks at a speed of $2\frac{1}{2}$ km/hour and reaches his school 6 minutes late. Next day he increases his speed by 1 km/hr and reaches the school 6 minutes early. How far is the school from his house ?

(a) 1 km (b) $1\frac{1}{2}$ km

(c) $1\frac{3}{4}$ km (d) 2 km

13. A train passes a telegraph post in 8 seconds and a 264 m long bridge in 20 seconds. What is the length of the train ?

(a) 180 m (b) 176 m

(c) 164 m (d) 158 m

14. Two taps can separately fill a cistern in 10 minutes and 15 minutes, respectively and when the waste pipe is open , they can together fill it in 18 minutes. The waste pipe can empty the full cistern in

(a) 7 min (b) 9 min

(c) 13 min (d) 23 min

15. If 72 men can build a wall 280 m long in 21 days, how many men will take 18 days to build a similar type of wall of length 100 m ?

(a) 30 (b) 10

(c) 18 (d) 28

16. '*A*' completes a work in 12 days. '*B*' completes the same work in 15 days. *A* started working along and after 3 days *B* joined him. How many days will they now take together to complete the remaining work?

(a) 6 (b) 8

(c) 5 (d) 4

(e) None of these

17. Three candidates in an election received 1136, 7636 and 11628 votes respectively. What percentage of the total votes did the winning candidate get ?

(a) 11.6% (b) 60%

(c) 76.4% (d) None of these

18. Due to increase of 15% in the price of milk, a family reduces its consumption of milk by 15%. What was the effect on expenditure of that family on account of milk ?

(a) 2.50% decrease (b) 2.25% decrease

(c) 3% decrease (d) 3.5% decrease

19. *X* can do a piece of work in 8 days and *Y* in 12 days. When *Z* joined them, they completed the work in 3 days. If the total money received for the work was ₹ 400, what would be the share of *X*, based upon his proportionate output ?

(a) ₹ 200 (b) ₹ 160

(c) ₹ 150 (d) ₹ 170

20. A jar is filled with a mixture of two liquids *A* and *B* in the proportion $A : B :: 5 : 3$, when 16 litres of the mixture is drawn off and the jar is filled with liquid *B*, then the proportion becomes $A : B :: 3 : 5$. How many litres of mixture does the jar hold ?

(a) 35 L (b) 50 L

(c) 45 L (d) 40 L

Answers

1. (c)	**2.** (b)	**3.** (b)	**4.** (c)	**5.** (b)	**6.** (c)	**7.** (c)	**8.** (d)	**9.** (c)	**10.** (a)
11. (d)	**12.** (c)	**13.** (b)	**14.** (b)	**15.** (a)	**16.** (c)	**17.** (d)	**18.** (b)	**19.** (c)	**20.** (d)

UNIT–4

GEOMETRY

- *Lines and Angles*
- *Triangles*
- *Polygons and Quadrilaterals*
- *Circles*

Chapter 17

LINES AND ANGLES

KEY FACTS

BASIC TERMS

1. A **point** indicates an exact location in space and has no dimensions.

Point P

2. A **line** is a straight path extending indefinitely in both the directions. It has no definite length.

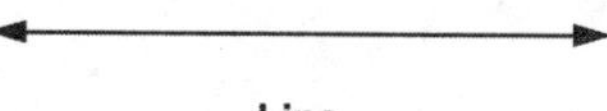

Line

3. **Line Segment** is a portion of a line having a definite length.

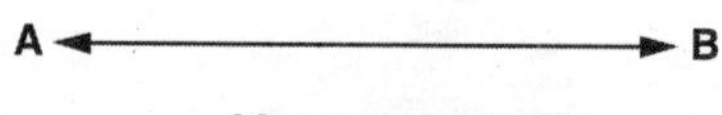

Line segment AB

4. A **ray** is a part of a line that has one fixed end point and extends indefinitely in the other direction.

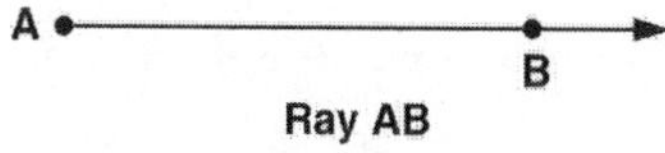

Ray AB

5. A **plane** is a flat surface which extends indefinitely in all directions. A plane has length and breadth but no thickness.

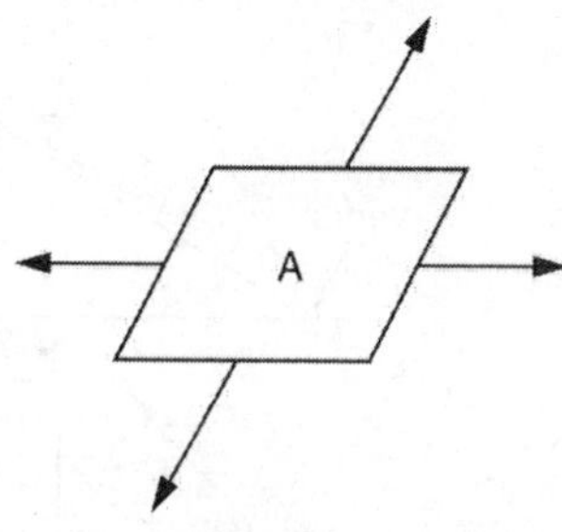

Plane A

6. Three or more points which lie on the same line are called **collinear points.**

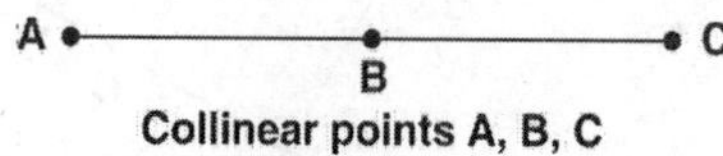

Collinear points A, B, C

7. Three or more lines in a plane passing through the same point are called **concurrent lines.**

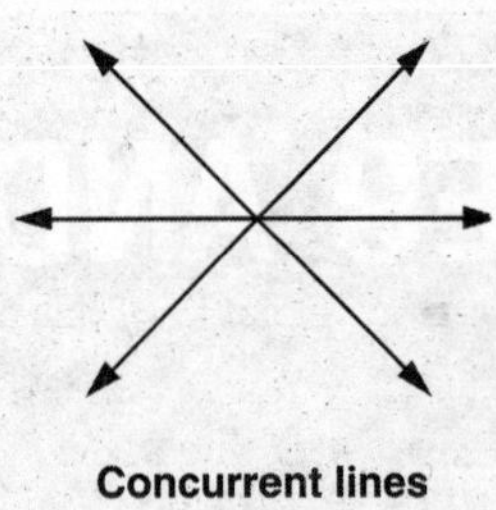

Concurrent lines

8. Two lines in a plane either intersect each other at exactly one point or are parallel to each other.

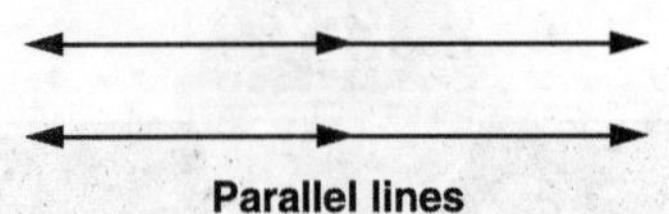

Parallel lines

Angle:

9. (a) An **angle** is an inclination between two rays with the same initial point. The initial point is the **vertex** and the two rays are the **arms** of the angle.
An angle is also thought of as the figure formed by rotating a single ray about its vertex.

(b) The unit of angle measure is degree denoted by °.
1 rotation = 360°, 1° = 60' (60 minutes), 1' = 60" (60 seconds).

Angle AOB vertex O
Arms OA and OB

10. Types of angles

(a) Acute angle — Between 0° and 90°. (b) Right angle — Equal to 90°.

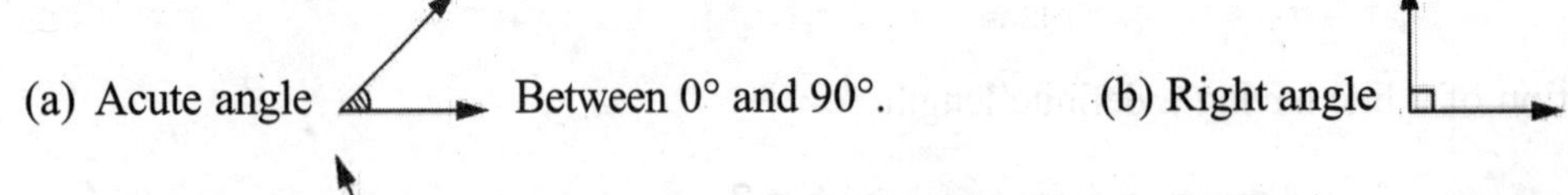

(c) Obtuse angle — Between 90° and 180°. (d) Straight angle — Equal to 180°.

(e) Reflex angle — Between 180° and 360°. (f) Complete angle — Equal to 360°.

11. Basic properties of angles

(a) **Adjacent angles** $\angle AOB$ and $\angle BOC$

(b) **Linear pair** $\angle AOC$ and $\angle BOC$

(c) **Vertically opposite angles** are equal

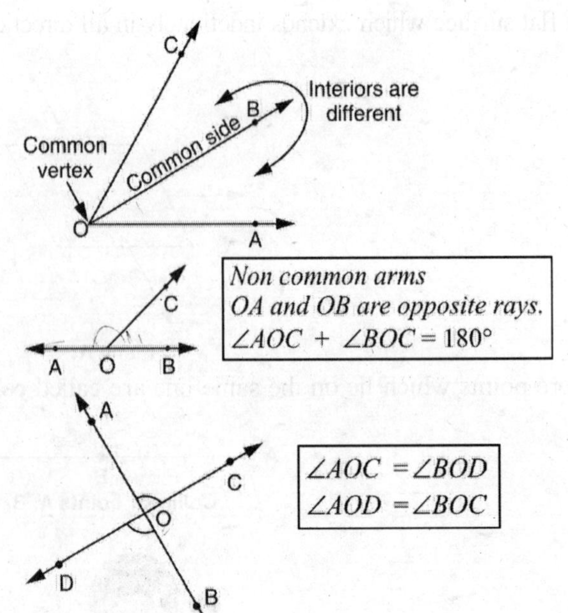

(d) **Complementary angles**

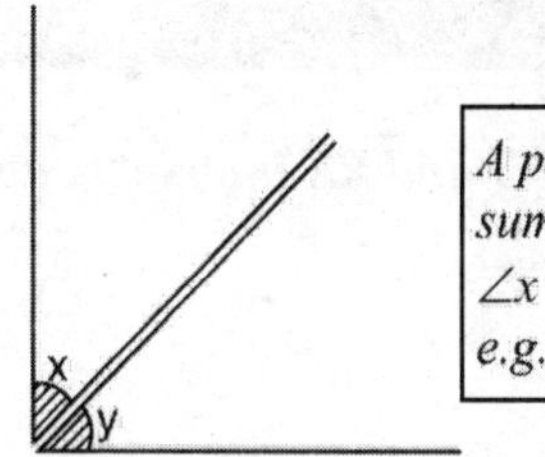

A pair of angles whose sum is 90°.
$\angle x + \angle y = 90°$
e.g., 35° *and* 55°

(e) **Supplementary angles**

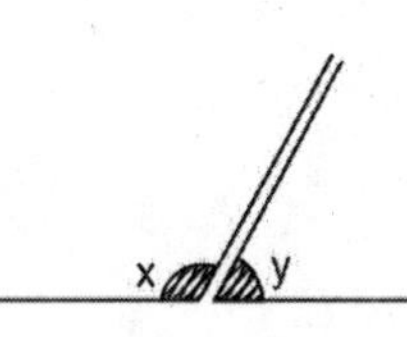

A pair of angles whose sum is 180°
$\angle x + \angle y = 180°$
e.g., 105° and 75°

(f) Sum of angles round a point is 360°.

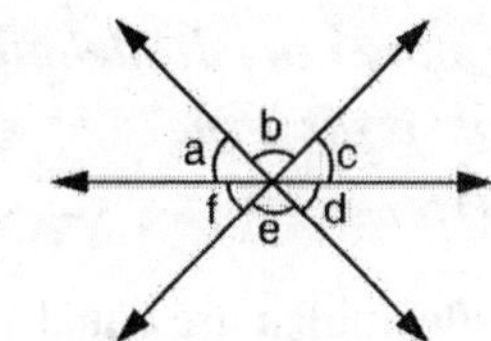

$\angle a+\angle b+\angle c+\angle d+\angle e+\angle f=360°$

(g) The sum of the adjacent angles on one side of a straight line is 180°

$\angle x+\angle y+\angle z+\angle a=360°$

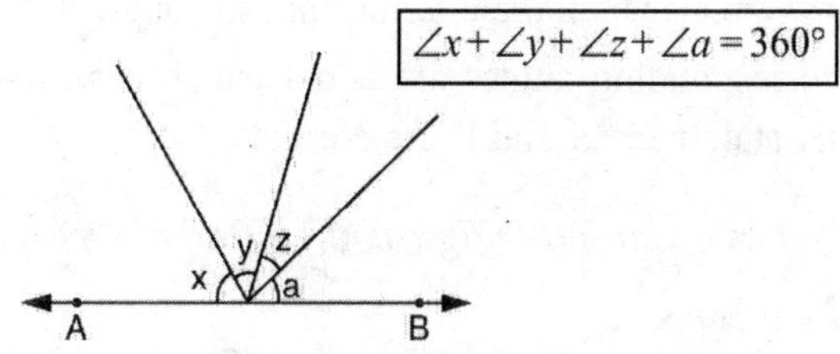

5. Two lines which intersect at right angle are called **perpendicular lines.**

$a \perp b$

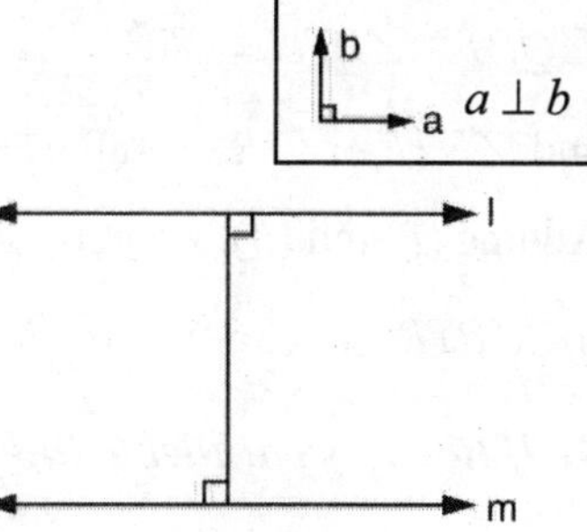

6. Lines in a plane which do not inersect are **parallel lines.**

The distance between two parallel lines is the same everywhere and is equal to the perpendicular distance between them.

7. A **transversal** is a line that cuts across (intersects) two or more lines in distinct points.

8. If a **transversal** cut two parallel lines then,

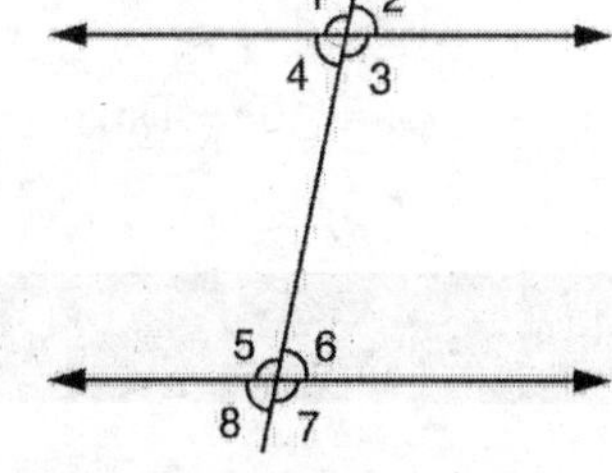

(i) ***alternate angles are equal,*** *i.e.,*

$\angle 3 = \angle 5, \angle 4 = \angle 6$

(ii) ***corresponding angles are equal,*** *i.e.,*

$\angle 1 = \angle 5, \angle 2 = \angle 6, \angle 4 = \angle 8, \angle 3 = \angle 7$

(iii) ***alternate exterior angles are equal,*** *i.e.,* $\angle 1 = \angle 7$ and $\angle 2 = \angle 8$

(iv) interior angles on the same side of the transversal (co-interior angles) are supplementary, *i.e.,* $\angle 3 + \angle 6 = 180°$ and $\angle 4 + \angle 5 = 180°$

9. Converse of property. If a transversal intersects two lines, then **they are parallel if any one of the following is true:**

(i) *pairs of corresponding angles are equal* (ii) *pairs of alternate angles are equal*

(iii) *co-interior angles are supplementary*

Solved Examples

Ex.1. ***In the given figure, PQ and RS intersect each other at O. If $\angle SOT = 75°$, find the values of a, b and c.***

Sol. $a = 4b$ and $2c = b + 75°$ *(vert. opp. ∠s)*

Also $a + 2c = 180°$ *(PQ is a straight line)*

$\Rightarrow 4b + b + 75° = 180°$

$\Rightarrow 5b = 105° \Rightarrow b = \mathbf{21°}$

$\therefore a = 4 \times 21° = \mathbf{84°}$ and

$$c = \frac{(21° + 75°)}{2} = \frac{96°}{2} = \mathbf{48°}.$$

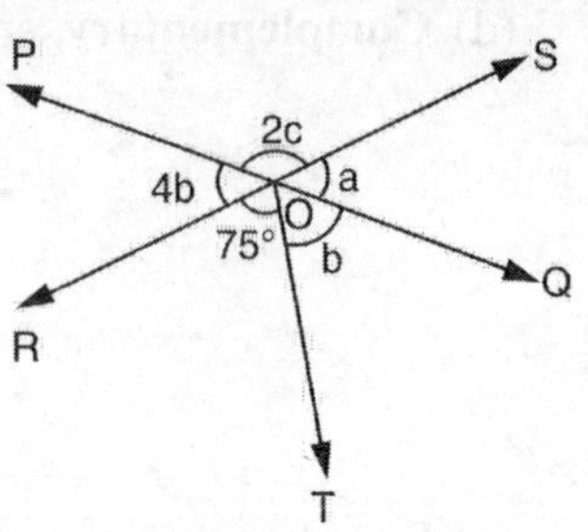

Ex. 2. ***In the given figure, lines m and n are not parallel and will intersect at some point to the right of the page. Which of the following statements is /are true?***

I. c = e ***II. b > d*** ***III. a = c***

Sol. I. $c = e$ because vertically opposite angles are equal.

II. $b > d$ because the intersecting lines form a triangle with b as an exterior angle. An exterior angle is always greater than the interior opposite angle.

III. $a = c$ shall imply that corresponding angles are equal which in turn would mean than $m \parallel n$. So, only statements I and II are correct.

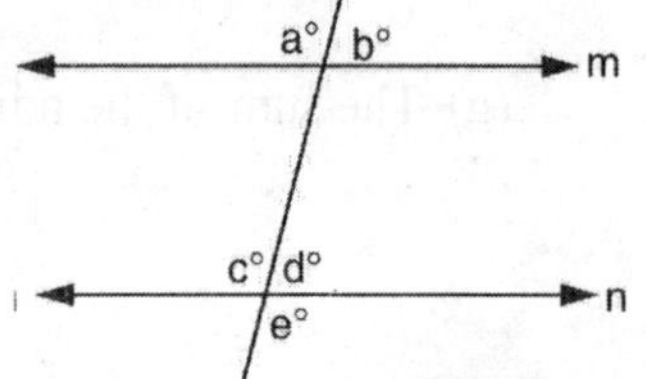

Ex. 3. ***If ray PQ and RS are parallel as given in the figure, then find a + b + c.***

Sol. Draw a ray TU parallel to PQ and RS. Now,

$\angle QPT + \angle PTU = 180°$ *(PQ || TU, co-int. ∠s)* ...(*i*)

and $\angle UTR + \angle TRS = 180°$ *(TU || RS, co-int. ∠s)* ...(*ii*)

Adding (*i*) and (*ii*), we get $\angle QPT + \angle PTU + \angle UTR + \angle TRS = 180° + 180° = 360°$

$a + \angle PTR + c = 360° \Rightarrow a + b + c = \mathbf{360°}$.

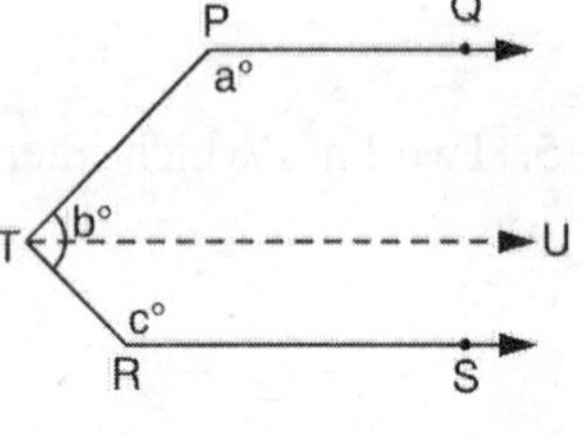

Ex. 4. ***If line l_1 is parallel to line l_2 in the given figure, what is the value of y?***

Sol. $l_1 \parallel l_2 \Rightarrow 115° + 2x + 3x = 180°$ *(co-int. ∠s)*

$\Rightarrow 5x = 180° - 115° = 65°$

$\Rightarrow x = 13°$

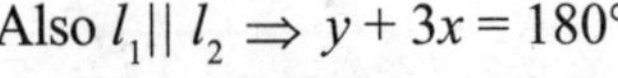

Also $l_1 \parallel l_2 \Rightarrow y + 3x = 180°$ *(co-int ∠s)*

$\Rightarrow y + 39° = 180° \Rightarrow y = 180° - 39° = \mathbf{141°}$.

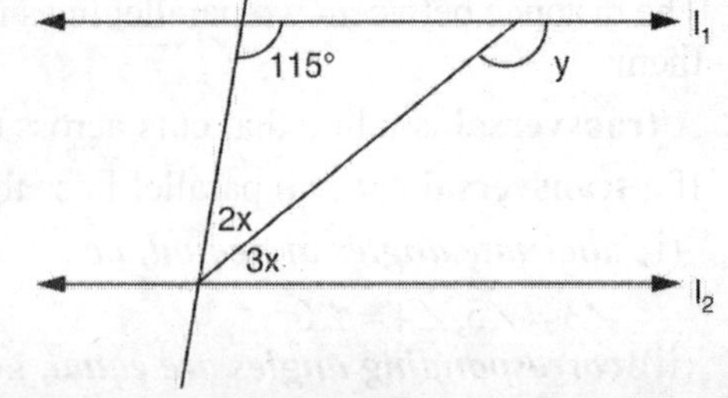

Question Bank–17

1. How many degrees are there in an angle which equals one-fifth of its supplement?
   (a) 15° (b) 30°
   (c) 75° (d) 150°

2. Two complementary angles are such that two times the measure of one is equal to three times the measure of the other. The measure of the larger angle is :
   (a) 72° (b) 108°
   (c) 36° (d) 54°

3. In the given figure, if $\angle BOC = 7x + 20°$ and $\angle COA = 3x$, then the value of x for which AOB becomes a straight line is :

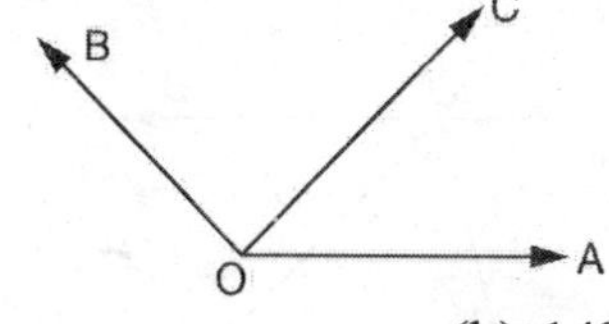

(a) 16° (b) 14°
(c) 20° (d) 21°

4. LM is a straight line and O is a point on LM. Line ON is drawn not coinciding with OL or OM. If $\angle MON$ is three times $\angle LON$, then $\angle MON$ is equal to:
(a) 45° (b) 60°
(c) 105° (d) 135°

5. Given that $AB \| DE$, $\angle ABC = 115°$, $\angle EDC = 140°$, then the value of x is:

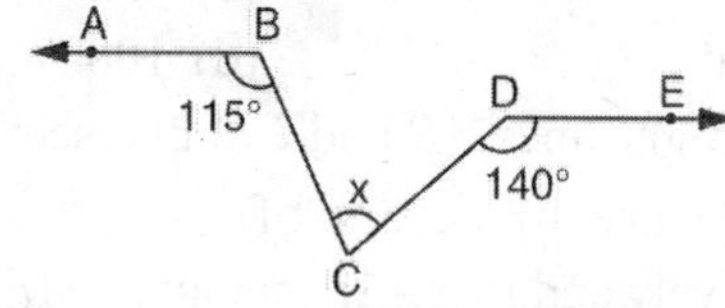

(a) 45° (b) 55°
(c) 65° (d) 75°

6. In the given figure $AC \| BD$, and $AE \| BF$. The measure of $\angle x$ is

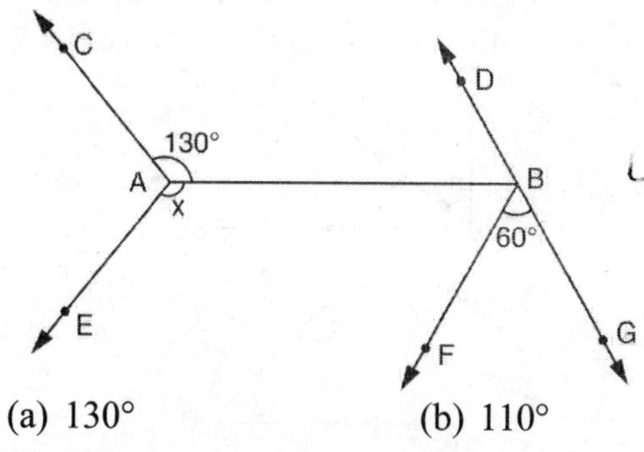

(a) 130° (b) 110°
(c) 70° (d) 50°

7. In the given figure $PQ \| RS$, $\angle RSF = 40°$, $\angle PQF = 35°$ and $\angle QFP = x°$. What is the value of x?

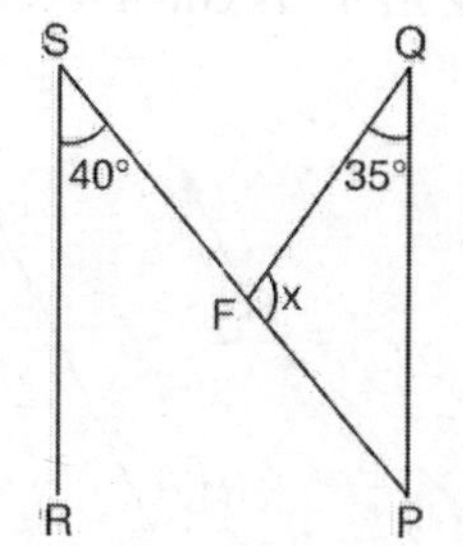

(a) 75° (b) 105°
(c) 135° (d) 140°

8. In the given figure, $AB \| CD$. Find x.

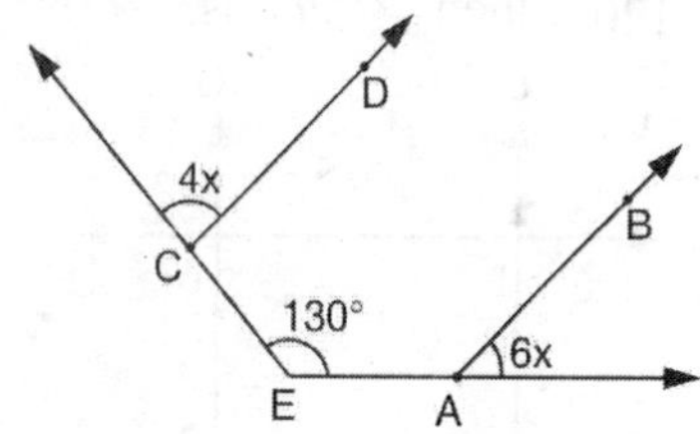

(a) 26° (b) 13°
(c) 23° (d) 46°

9. A and B are two points and C is any point collinear with A and B. If $AB = 10$, $BC = 5$, then AC is equal to:
(a) either 15 or 5 (b) necessarily 5
(c) necessarily 16 (d) none of these

10. Three lines intersect at a point generating six angles. If one of these angles is 90°, then the number of other distinct angles is:
(a) 1 or 2 (b) 1 or 3
(c) 2 or 3 (d) 2 or 4

11. Assume $p \| q$ in the figure shown. Then 'x' equals. **(*NSTSE 2009*)**

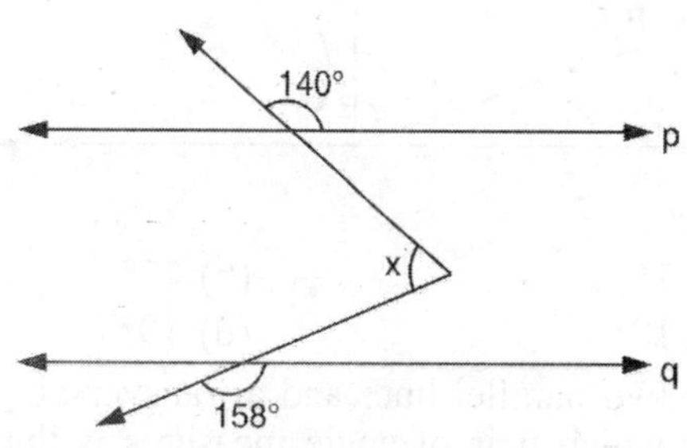

(a) 18° (b) 22°
(c) 62°
(d) cannot be determined

12. If X, Y and Z are three points such that $XY = 2YZ$ and $XZ = 3YZ$, then the three points are:
(a) not collinear
(b) collinear and X lies between Y and Z
(c) collinear and Y lies between X and Z
(d) collinear and Z lies between X and Y

13. If the arms of one angle are respectively parallel to the arms of another angle, then the two angles are:
(a) neither equal nor supplementary
(b) not equal but supplementary
(c) equal but not supplementary
(d) either equal or supplementary

14. In the given figure $AB \| CD$ and $EF \| GH$. If $\angle AIF = 120°$, then $\angle CJG$ is

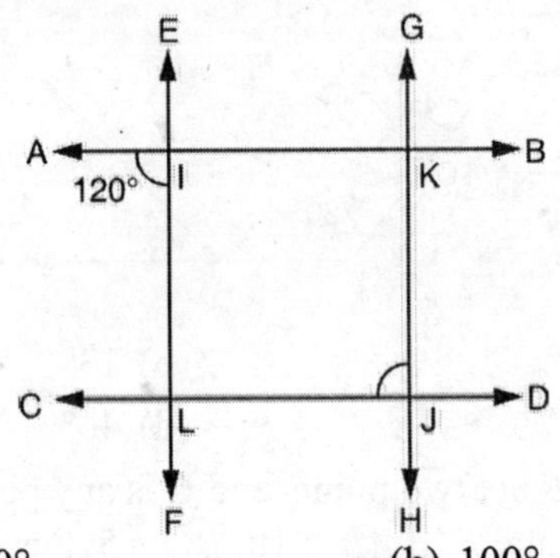

(a) 80° (b) 100°

(c) $\frac{1}{3}$ of a right angle (d) $\frac{2}{3}$ of a right angle

15. In the given figure, $\angle COE$ and $\angle BOD$ are right angles. If the measure of $\angle BOC$ is four times the measure of $\angle COD$, what is the measure of $\angle AOB$?

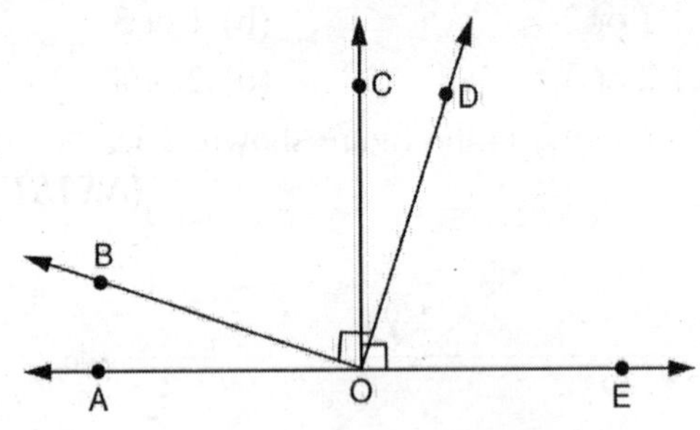

(a) 16° (b) 17°
(c) 18° (d) 19°

16. For two parallel lines and a transversal, $\angle 1 = 74°$. For which pair of angle measures is the sum the least?

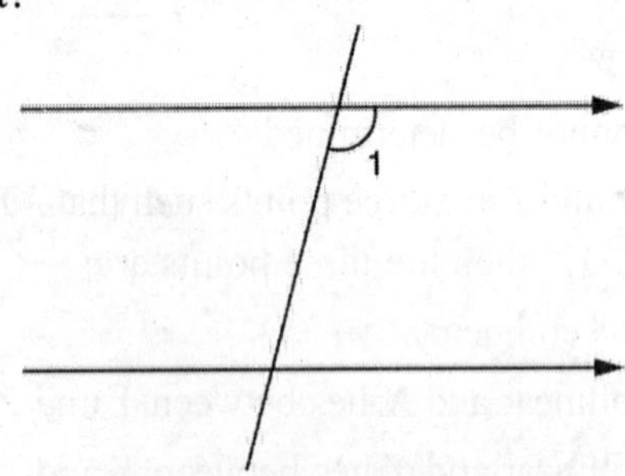

(a) $\angle 1$ and a corresponding angle
(b) $\angle 1$ and the corresponding co-interior angle
(c) $\angle 1$ and its supplement
(d) $\angle 1$ and its complement

17. Lines m and n are cut by a transversal so that $\angle 1$ and $\angle 5$ are corresponding angles. If $\angle 1 = 26x - 7°$ and $\angle 5 = 20x + 17°$. What value of x makes the lines m and n parallel?
(a) 5 (b) 4
(c) $4\frac{1}{2}$ (d) $3\frac{1}{4}$

18. In the given figure, ray AD is the bisector of $\angle CAB$ and the measure of $\angle ACD$ is y. What must be the measure of $\angle ADC$ in order for line AB to be parallel to line CD.

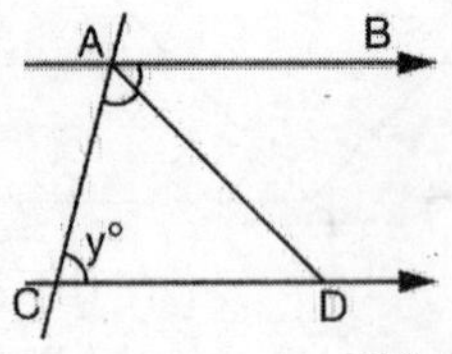

(a) $90° - y$ (b) $90° - y/2$
(c) $180° - y/2$ (d) $90° + y$

19. AB is a straight line and O is a point lying on AB. A line OC is drawn from O such that $\angle COA = 36°$. OD is a line within $\angle COA$ such that $\angle DOA = \frac{1}{3}\angle COA$. If OE is a line within the $\angle BOC$, $\angle BOE = \frac{1}{4}\angle BOC$, then $\angle DOE$ must be
(a) 60° (b) 132°
(c) 144° (d) 108°

20. The straight lines AB and CD intersect at E. If EF and EG are bisectors of $\angle DEA$ and $\angle AEC$ respectively and if $\angle AEF = x$ and $\angle AEG = y$, then
(a) $x + y > 90°$ (b) $x + y < 90°$
(c) $x + y = 90°$ (d) $x + y = 180°$

21. Given, $AB \| ED$, $AG \| CB$ and $AF \perp AB$. $\angle FAG = 38°$, $\angle CDE = 45°$. Find the value of x.

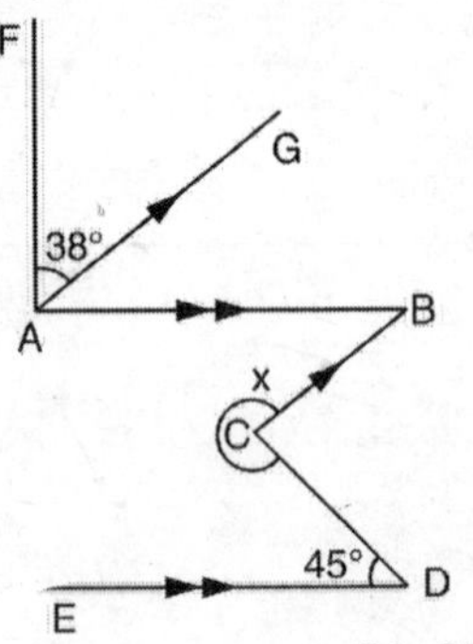

(a) 263° (b) 277°
(c) 289° (d) 308°

22. In the given figure, $DE \| BC$, $\angle ABC = 118°$, $\angle DAB = 42°$, then $\angle ADE$ is equal to:

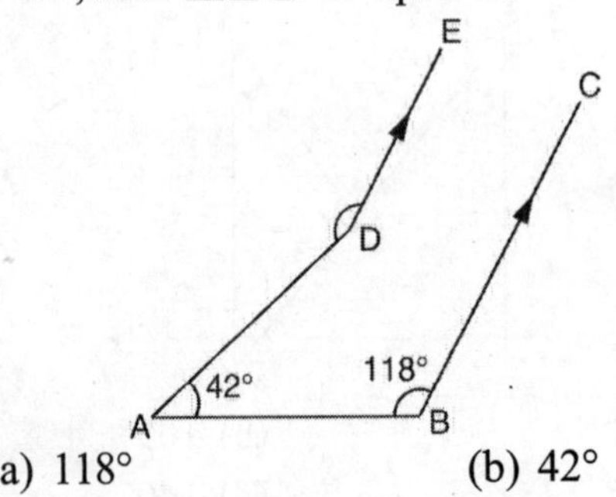

(a) 118° (b) 42°
(c) 138° (d) 160°

Answers

1. (b)	**2.** (d)	**3.** (a)	**4.** (d)	**5.** (d)	**6.** (b)	**7.** (b)	**8.** (b)	**9.** (a)	**10.** (a)
11. (c)	**12.** (c)	**13.** (d)	**14.** (d)	**15.** (c)	**16.** (d)	**17.** (b)	**18.** (b)	**19.** (b)	**20.** (c)
21. (a)	**22.** (d)								

Hints and Solutions

1. (b) Let the angle in degrees be x. Then, its supplement = $(180° - x)$

Given, $x = \frac{1}{5}(180° - x)$

$\Rightarrow 5x = 180° - x \Rightarrow 6x = 180° \Rightarrow x = \mathbf{30°}$.

2. (d) Let the complementary angles be x and $(90° - x)$

Then, $2x = 3(90° - x)$

$\Rightarrow 2x = 270° - 3x \Rightarrow 5x = 270°$

$\Rightarrow x = \frac{270°}{5} = 54°$

$\therefore$ The two angles are **54°, 36°**.

3. (a) $\angle AOB = \angle BOC + \angle COA$

$= 7x + 20° + 3x = 10x + 20°$

AOB becomes a straight line if $\angle AOB = 180°$, *i.e.*,

$10x + 20° = 180° \Rightarrow 10x = 160° \Rightarrow x = 16°$.

4. (d) LM is a straight line, $\angle MON = 3\ \angle LON$

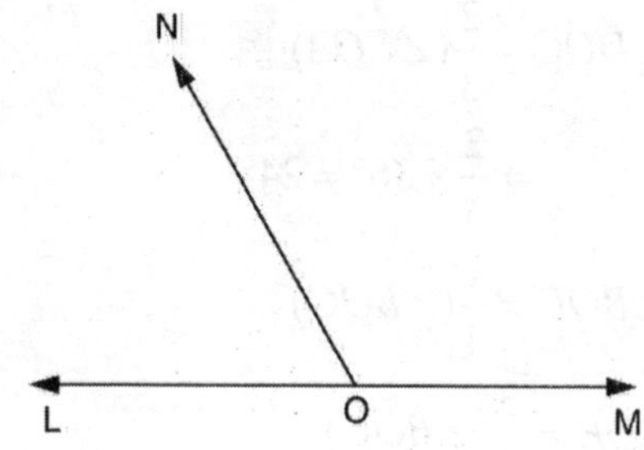

$\Rightarrow \angle LOM = 180°$

$\Rightarrow \angle LON + \angle NOM = 180°$

$\Rightarrow \angle LON + 3\angle LON = 180°$

$\Rightarrow 4\angle LON = 180° \Rightarrow \angle LON = \frac{180°}{4} = 45°$

$\therefore \angle MON = 3 \times 45° = \mathbf{135°}$.

5. (d) Through C draw a line OP parallel to AB and DE.

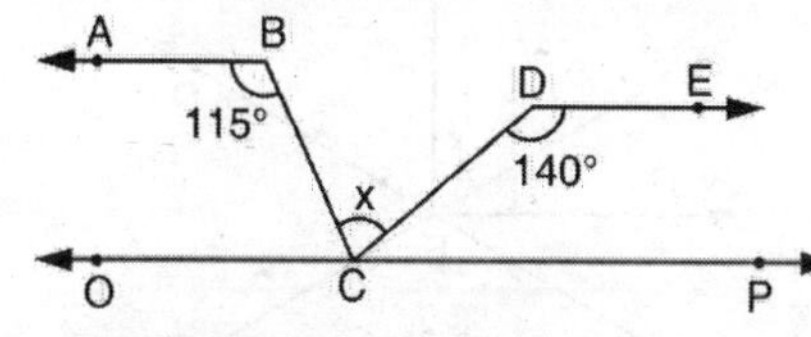

$\angle BCO = 180° - \angle ABC$

$= 180° - 115° = 65°$

($AB \parallel OC$, *co. int* $\angle s$ *are supp.*)

$\angle DCP = 180° - \angle EDC$

$= 180° - 140° = 40°$

($DE \parallel CP$, *co-int.* $\angle s$ *are supp.*)

OP being a straight line, $\angle OCP = 180°$

$\therefore \angle BCO + x + \angle DCP = 180°$

$\Rightarrow x = 180° - (\angle BCO + \angle DCP)$

$= 180° - (65° + 40°)$

$= 180° - 105° = \mathbf{75°}$.

6. (b) $\angle CAB + \angle DBA = 180°$

$\angle DBA = 180° - \angle CAB = 180° - 130° = 50°$

($AC \parallel BD$, *co-int.* $\angle s$ *are supp.*)

DBG being a straight line, $\angle DBG = 180°$

$\Rightarrow \angle DBA + \angle ABF + \angle FBG = 180°$

$\Rightarrow 50° + \angle ABF + 60° = 180°$

$\Rightarrow \angle ABF = 180° - 110° = 70°$

Now, $AE \parallel BF \Rightarrow \angle EAB + \angle ABF = 180°$ (*co-int.* $\angle s$)

$\Rightarrow x + 70° = 180° \Rightarrow x = \mathbf{110°}$.

7. (b) $\angle QPF = \angle RSP = 40°$ ($RS \parallel PQ$, *alt.* $\angle s$)

In $\Delta FQP, x + 35° + 40° = 180°$

$\Rightarrow x = 180° - 75° = \mathbf{105°}$.

8. (b) Through point E draw a line EF parallel to CD and AB.

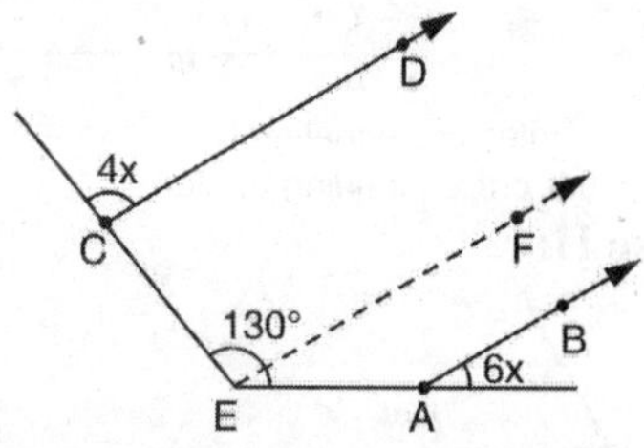

Now, $\angle CEF = 4x$ ($CD \parallel EF$, *corr.* $\angle s$)

$\angle AEF = 6x$ ($EF \parallel AB$, *corr.* $\angle s$)

Given, $\angle CEA = 130° \Rightarrow \angle CEF + \angle AEF = 130°$

$\Rightarrow 4x + 6x = 130° \Rightarrow 10x = 130° \Rightarrow x = \mathbf{13°}$.

9. (a) Since C is collinear with A and B, C lies either

C ← 5 →
A ← 10 → B

(i) to the left of point B or

(ii) to the right of point B.

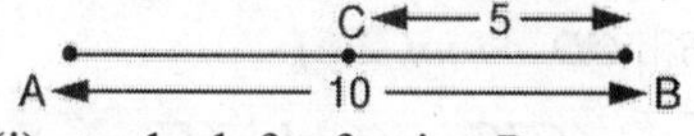

$\therefore$ In case (i) $AC = AB - BC = 10 - 5 = \mathbf{5}$.

In case (ii) $AC = AB + BC = 10 + 5 = \mathbf{15}$.

10. (a) Case I: If $\angle 1 = \angle 2$, then

$\angle 1 = \angle 2 = \angle 3 = \angle 4$

($\angle 1 = \angle 4, \angle 2 = \angle 3$, *vert. opp.* $\angle s$)

$\therefore$ Number of distinct angles is one.

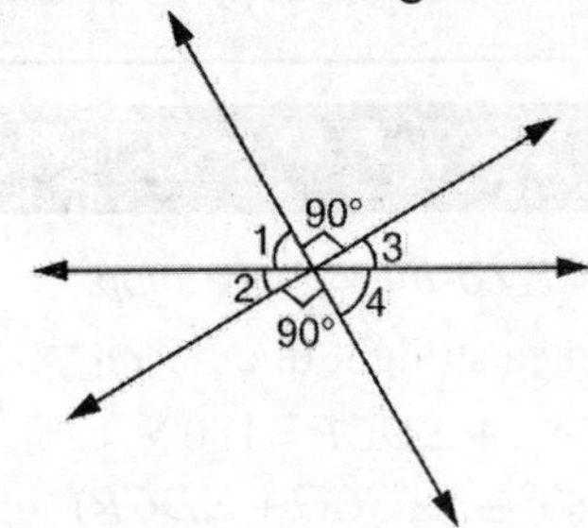

Case II: If $\angle 1 \neq \angle 2$, then

$\angle 1 = \angle 4$ and $\angle 2 = \angle 3$ (*vert. opp.* $\angle s$)

$\therefore$ Number of distinct angles is two.

11. (c) Similar to Solved Example 3.

12. (c) Given, $XY = 2YZ$ and $XZ = 3YZ$

X Y Z

$\therefore$ $XZ = 2YZ + YZ = XY + YZ$

$\Rightarrow$ *X*, *Y*, *Z* are collinear and *Y* lies between *X* and *Z*.

13. (d) Two conditions, as shown in the figures satisfy the given condition.

Case I:

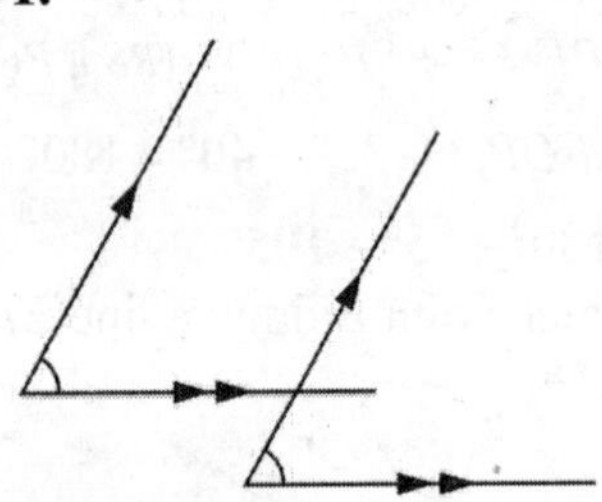

Angles are equal as they are corresponding angles.

Case II:

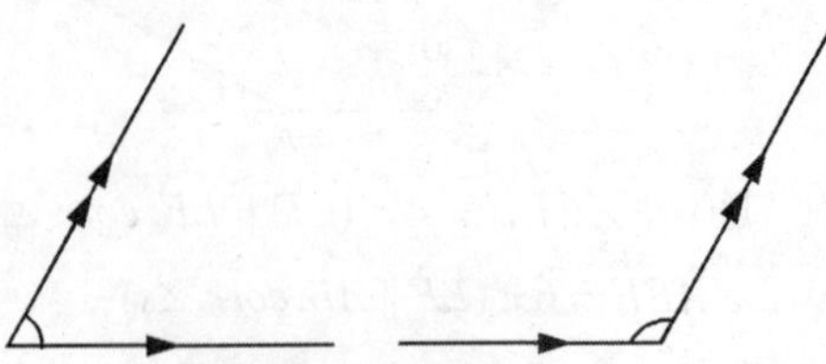

Angles are supplementary as they are co-interior angles.

14. (d) $\angle CLI = 180° - \angle AIL = 180° - 120° = 60°$

($AB \parallel CD$, *co-int.* $\angle s$)

Now, $\angle CJG = \angle CLI = 60° = \frac{2}{3} \times \mathbf{90°}$.

($EF \parallel GH$, *corr.* $\angle s$)

15. (c) Let $\angle COD = x$, then $\angle BOC = 4x$

Since, $\angle BOD = 90°$, therefore $x + 4x = 90°$

$\Rightarrow 5x = 90° \Rightarrow x = 18°$

$\angle BOC = 72°$ and

$\angle AOB = 90° - 72° = \mathbf{18°}$. ($\angle AOE = \angle COE = 90°$)

16. (d) $\angle 1 + \text{corr.} \angle = 74° + 74° = 148°$

$\angle 1 + \text{corr. co-int. } \angle = 180°$

$\angle 1 + \text{its supplement} = 180°$

$\angle 1 + \text{its complement} = 90°$

Hence option (d) is correct.

17. (b) For the lines *m* and *n* to be parallel, corresponding angles should be equal, *i.e.*,

$\angle 1 = \angle 5$.

$\Rightarrow 26x - 7° = 20x + 17°$

$\Rightarrow 6x = 24° \Rightarrow x = \mathbf{4°}$.

18. (b) For *AB* to be parallel to *CD*,

$\angle BAC + y = 180° \Rightarrow \angle BAC = 180° - y$

Since *AD* bisects $\angle BAC, \angle BAD = 90° - y/2$

Also, $AB \parallel CD$

$\Rightarrow \angle ADC = \angle BAD = 90° - y/2$. (*alt.* $\angle s$)

19. (b) Given: $\angle DOA = \frac{1}{3}(\angle COA)$

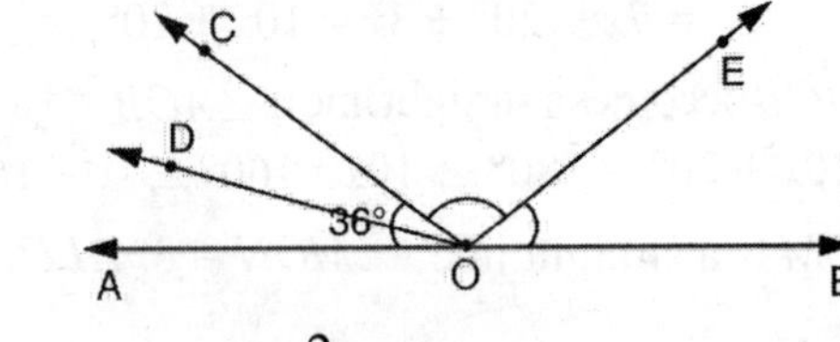

$\Rightarrow \angle DOC = \frac{2}{3}(\angle COA)$

$= \frac{2}{3} \times 36° = 24°$

$\angle BOE = \frac{1}{4}(\angle BOC)$

$\Rightarrow COE = \frac{3}{4}(\angle BOC)$

$= \frac{3}{4} \times (180° - 36°) = \frac{3}{4} \times 144° = 108°$

$\therefore \angle DOE = \angle DOC + \angle COE = 24° + 108° = \mathbf{132°}$.

20. (c) $\angle AED + \angle AEC = 180°$ (*Linear pair*)

$\Rightarrow \frac{1}{2}\angle AED + \frac{1}{2}\angle AEC = 90°$

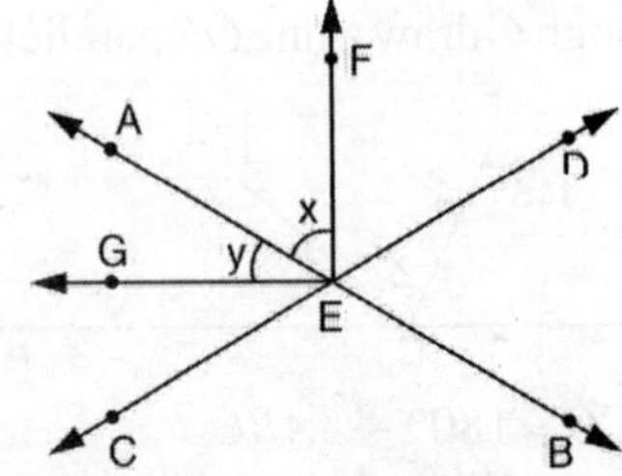

$\Rightarrow \angle AEF + \angle AEG = 90°$

$\Rightarrow x + y = \mathbf{90°}$.

21. (a) $\angle GAB = 90° - \angle FAG = 90° - 38°$
$= 52°$ $(AF \perp AB)$

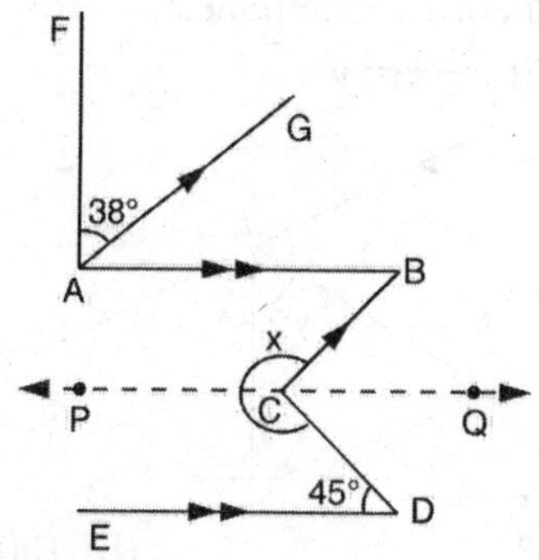

Now, $\angle ABC = \angle GAB = 52°$ $(AG \parallel CB, alt. \angle s)$
Through C, draw $PCQ \parallel AB$ and ED
$\angle BCQ = \angle ABC = 52°$ $(AB \parallel PQ, alt. \angle s)$
$\angle DCQ = \angle CDE = 45°$ $(PQ \parallel ED, alt. \angle s)$

$\therefore$ $x = 360° - \angle BCD$
$= 360° - (\angle BCQ + \angle DCQ)$
$= 360° - (52° + 45°) = 360° - 97° = \mathbf{263°}$.

22. (d) Produce ED to F. Now $EF \parallel CB$
$\Rightarrow \angle DFA = \angle CBF = 118°$ $(corr. \angle s)$

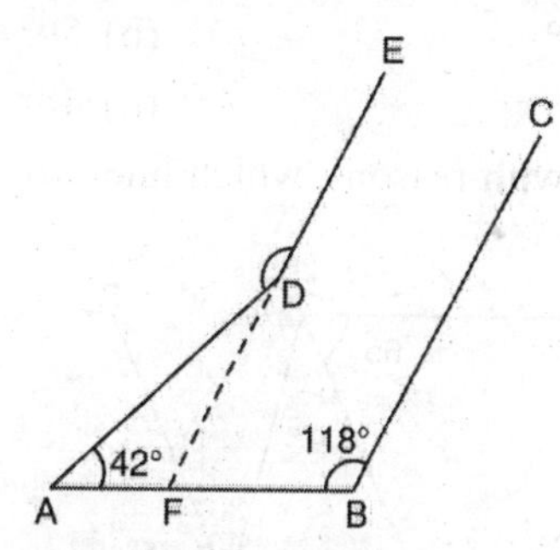

In Δ DAF, $\angle ADF = 180° - (118° + 42°)$
$= 180° - 160° = 20°$
$\angle ADE = 180° - \angle ADF = 180° - 20°$
$\therefore$ $= \mathbf{160°}$ (Linear pair)

Self Assessment Sheet–17

1. The bisector of an angle is produced backwards. It bisects which angle at the same vertex ?
(a) acute (b) obtuse
(c) reflex (d) complete

2. If $x + z + y = 240°$, find the value of each of the four angles in the order x, y, z, u.

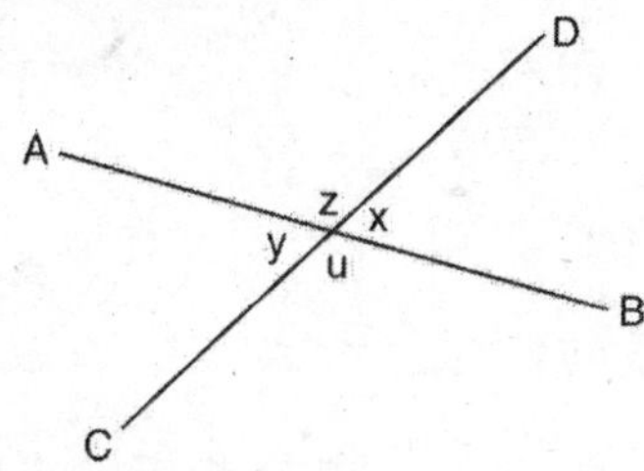

(a) 50°, 50°,130°,130°
(b) 60°, 60°,120°,120°
(c) 60°,70°,110°,120°
(d) 50°,60°,130°,120°

3. Find the complement of an angle whose measure is $3x - 8°$.
(a) $3x - 98°$ (b) $82° - 3x$
(c) $98° - 3x$ (d) $3x - 82°$

4. $AB \parallel CD$ and $\angle ACM$ is a straight line $\angle BAC$ is equal to

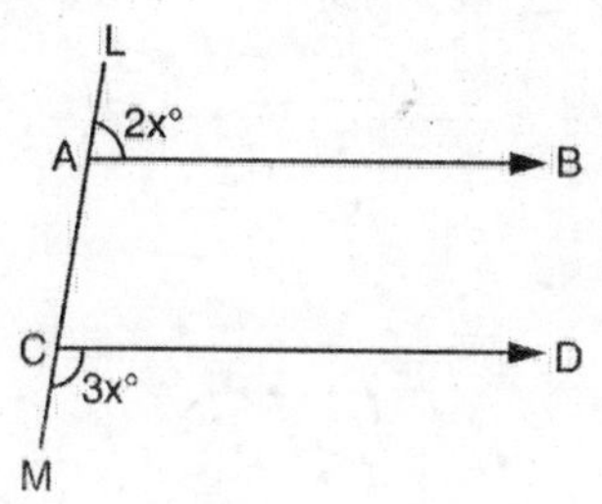

(a) 106° (b) 108°
(c) 110° (d) 109°

5. $\angle ABC$ is equal to

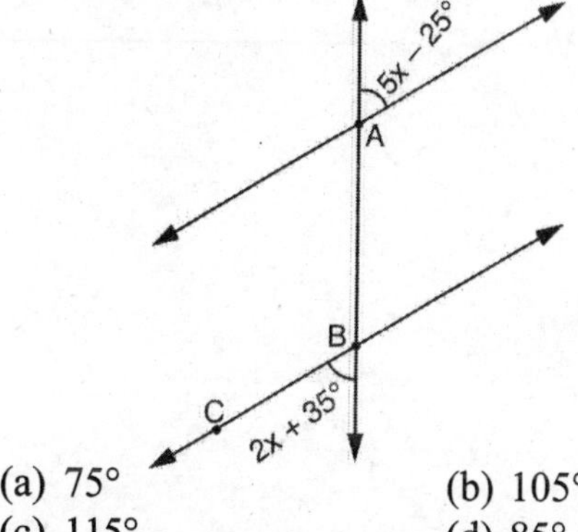

(a) 75° (b) 105°
(c) 115° (d) 85°

6. Two lines in a plane are cut by a transversal. Which condition does **NOT** imply that the two lines are parallel ?
(a) A pair of alternate interior angles are congruent.
(b) A pair of co-interior angles are supplementary.
(c) A pair of corresponding angles are congruent.
(d) A pair of alternate exterior angles are complementary.

7. $QP \parallel TS$ and $\angle QRS = 36°$. Calculate $\angle PQR$.

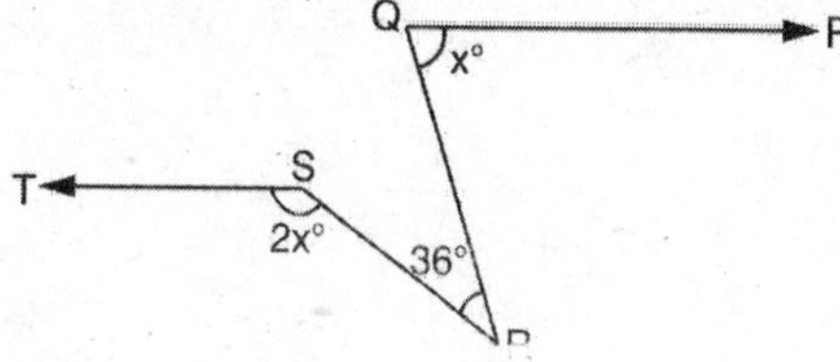

(a) 75° (b) 70° (c) 72° (d) 71°

8. Draw four lines OA, OB, OC, OD in order, $\angle AOB = 48°$, $\angle COD = 34°$. OP bisects $\angle AOB$, OQ bisects $\angle COD$. If OQ is perpendicular to OP, calculate $\angle BOC$.

(a) 49° (b) 50°
(c) 51° (d) 48°

9. State, with reasons, which lines are parallel.

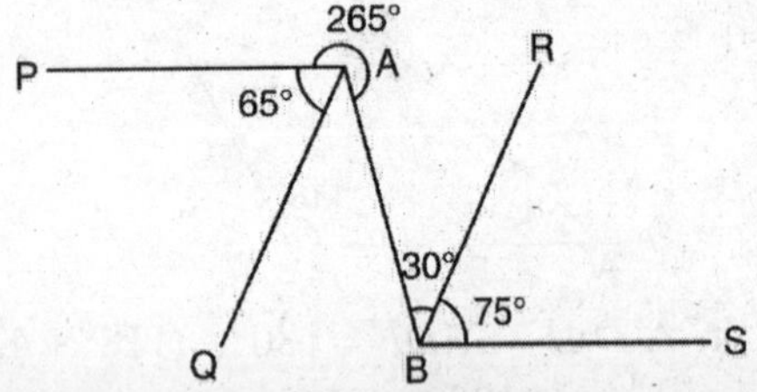

(a) AP, BS
(b) AQ, RB
(c) AP, RB
(d) Cannot be determined

10. Find y if $x + z = y$

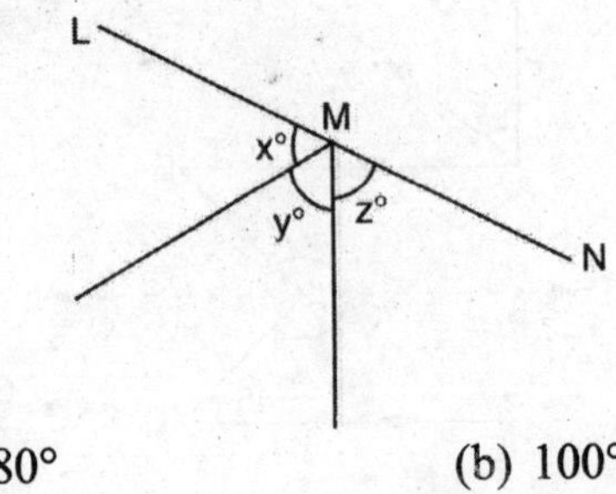

(a) 80° (b) 100°
(c) 90° (d) 110°

Answers

1. (c)	2. (b)	3. (c)	4. (b)	5. (b)	6. (d)	7. (c)	8. (a)	9. (b)	10. (c)

Chapter

18 TRIANGLES

KEY FACTS

1. The plane closed figure bounded by three line segments is called a **triangle.**
2. Sum of the angles of a triangle is equal to 180°, *i.e.,*

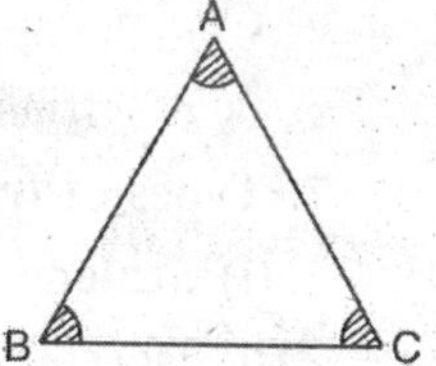

$$\angle A + \angle B + \angle C = 180^\circ.$$

3. Sum of any two sides of a triangle is always greater than the third side, *i.e.,*

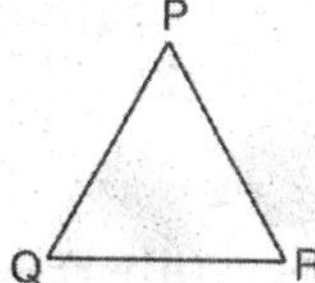

$$PQ + QR > PR$$
$$PQ + PR > QR$$
$$PR + QR > PQ$$

4. Exterior angle of a triangle is equal to the sum of its interior opposite angles, *i.e.,*

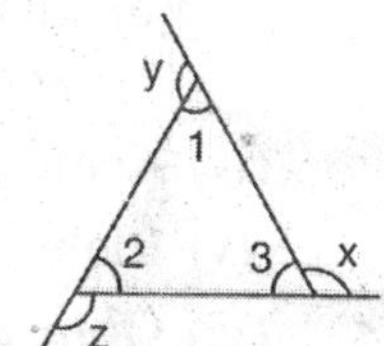

$$\angle x = \angle 1 + \angle 2,$$
$$\angle y = \angle 2 + \angle 3,$$
$$\angle z = \angle 1 + \angle 3$$

5. Side opposite to the greatest angle will be greatest in length and vice versa.
6. **Important terms of a triangle**
   (i) Median and centroid: A line joining the mid point of a side of a triangle to the opposite vertex is called a median. *D, E, F* are the mid points of sides *BC, AC* and *AB* respectively. Hence, *AD, BE* and *CF* are the medians. A median divides a triangle into two parts of equal area.
   The point where the three medians of a triangle meet is called the **centroid** of a triangle.

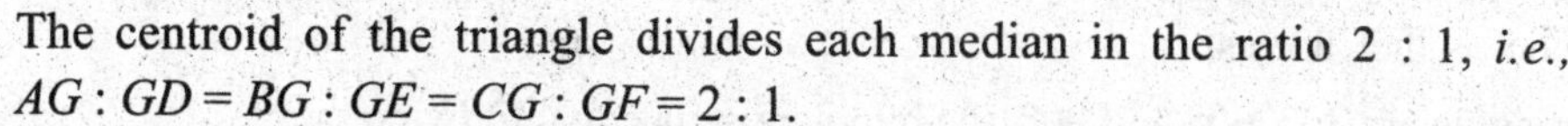

   The centroid of the triangle divides each median in the ratio 2 : 1, *i.e.,* $AG : GD = BG : GE = CG : GF = 2 : 1$.
   (ii) Perpendicular bisector and circumcentre: Perpendicular bisector to any side is the line that is perpendicular to that side and passes through its mid point. Perpendicular bisectors need not pass through the opposite vertex. The point of intersection of the three perpendicular bisectors of the triangle is called **circumcentre.**

The circumcentre of a triangle is equidistant from its three vertices. If we draw a circle with the circumcentre as the centre and the distance of any vertex from the circumcentre as radius, the circle passes through all the three vertices and the circle is called **circum circle.**

The circumcentre can be inside or outside the circle.

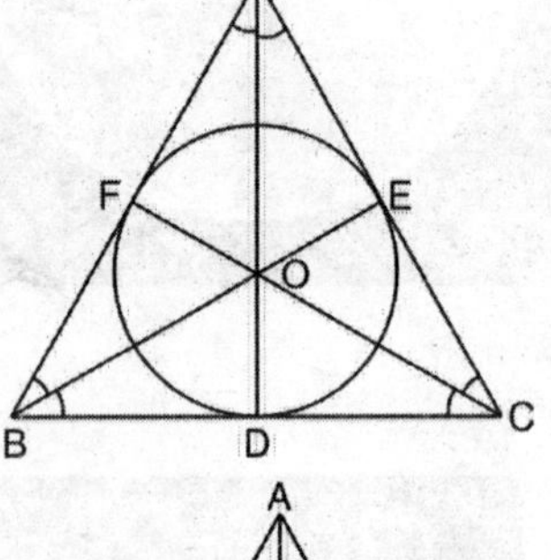

(iii) Angle bisector and incentre: An angle bisector is a line that divides the angle into two parts. The point of intersection of three angle bisectors of a triangle is called the **incentre**. It always lies inside the triangle. It is always equidistant from the sides of the triangle.

The circle drawn with incentre as centre and touching all the three sides of the triangle is called **incircle.**

(iv) Altitude and orthocentre: The perpendicular drawn from the vertex of a triangle to the opposite side is called an **altitude**. The point of intersection of three altitudes of a triangle is called **orthocentre**, which can lie inside or outside the triangle.

7. Types of triangles and their properties

(i) Scalene triangle: All three sides are unequal.

(ii) Isosceles triangle:

(a) Two sides are equal.

(b) The median drawn from a vertex to the opposite side is also the perpendicular bisector of that side.

(c) In an isosceles ΔABC, all the four points; the centroid, the orthocentre, the circumcentre and the incentre lie on the altitude drawn from vertex A to base BC.

(iii) Equilateral triangle:

(a) All the three sides and angles are equal.

(b) The median, angle bisector, altitude and perpendicular bisector of sides are all represented by the same straight lines.

(iv) Acute angled triangle: A triangle with each angle less than 90°.

(v) Right angled triangle: A triangle with one angle equal to 90°.

(vi) Obtuse angled triangle: A triangle with one angle greater than 90°.

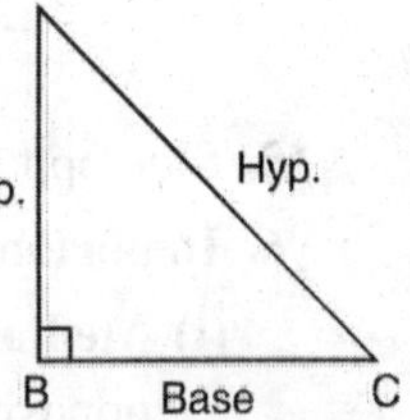

8. Pythagoras' theorem: In a right angled triangle, the square of the hypotenuse is equal to the sum of the squares of the other two sides.

$\text{Hypotenuse}^2 = \text{Base}^2 + \text{Perpendicular}^2$

or $\quad AC^2 = AB^2 + BC^2$.

9. In a right triangle the hypotenuse is the longest side. Mid point of the hypotenuse of a right triangle is equidistant from the 3 vertices. If in a triangle, the square of the largest side is equal to the sum of the squares of the remaining two sides, then the angle opposite to the largest side is a right angle.

10. Of all the line segments that can be drawn to a given line from a point outside it, the perpendicular line segment is the shortest.

11. If the angles of a right triangle are 30°, 60° and 90°, the hypotenuse is equal to twice the side opposite to the 30° angle, *i.e.*, $AC = 2BC$.

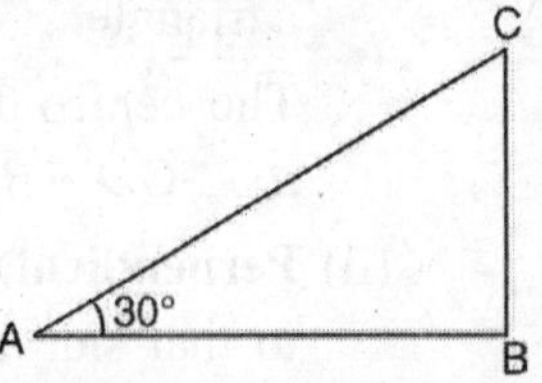

Congruence of Triangles.

12. Two triangles are congruent if pairs of corresponding sides and corresponding angles are equal. Thus,

$\Delta ABC \cong \Delta DEF$ if $\angle A = \angle D$, $\angle B = \angle E$,

$\angle C = \angle F$, $AB = DE$, $AC = DF$ and $BC = EF$.

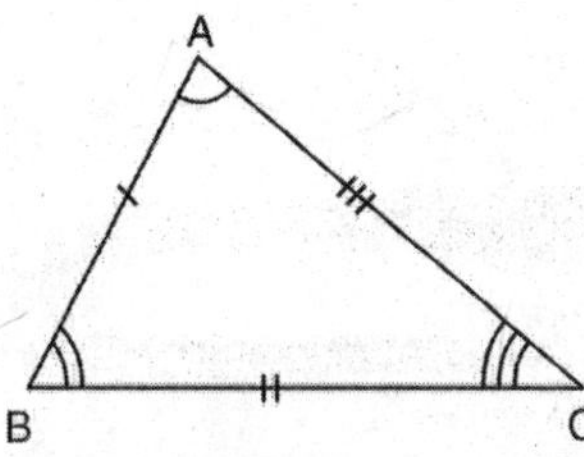

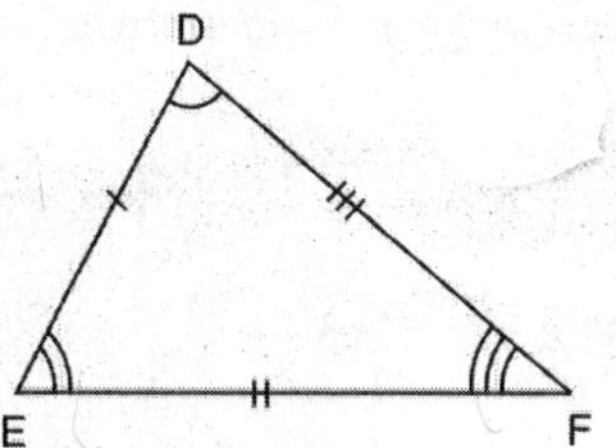

13. Conditions of congruence of triangles: Two triangles are congruent if

(i)

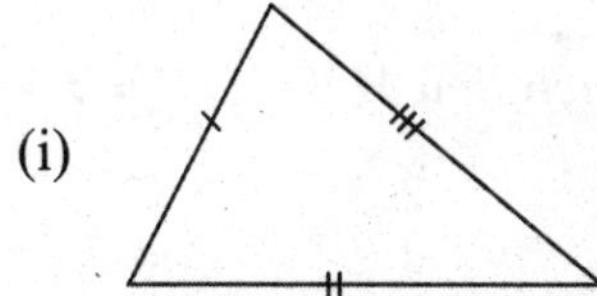

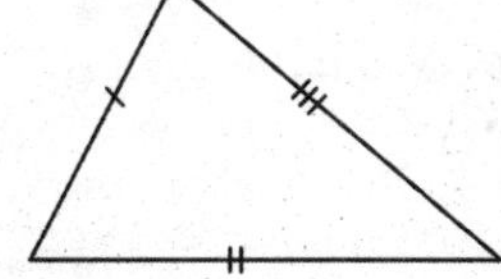

Three sides of one are respectively equal to the three sides of the other. **(SSS condition)**

(ii)

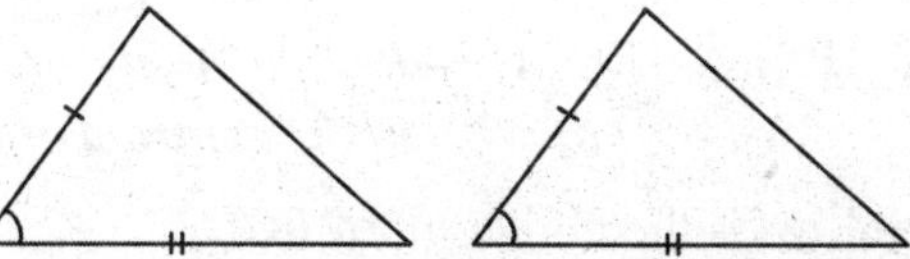

Two sides and included angle of one is respectively equal to the two sides and the included angle of the other. **(SAS condition)**

(iii)

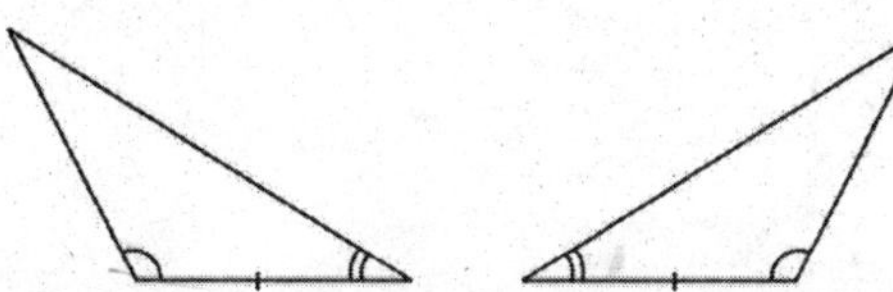

Two angles and the included side of one are respectively equal to the two angles and the included side of the other. **(ASA condition)**

Cor. The **AAS condition** follows from the ASA rule.

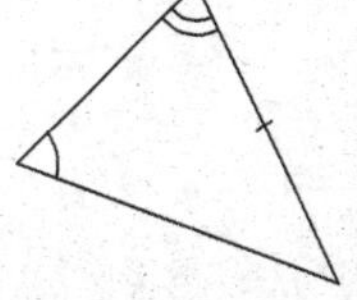

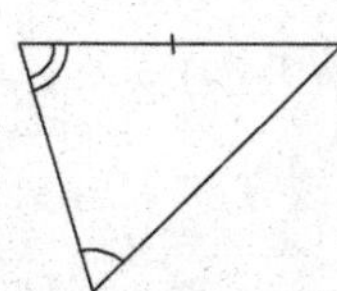

Two angles and a side of one are respectively equal to the two angles and corresponding side of the other. **(AAS condition)**

(iv)

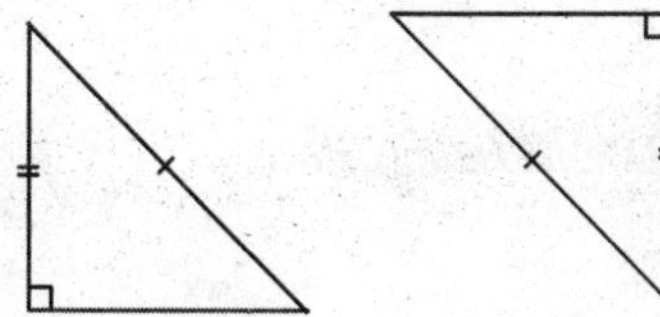

Two right triangles are congruent to each other if the hypotenuse and one side of a triangle are respectively equal to the hypotenuse and a corresponding side of the other. **(RHS condition)**

14. In an isosceles triangle, the angles opposite to the equal sides are equal. Thus, in ΔABC, $AB = AC$ $\Rightarrow \angle C = \angle B$. Conversely, in an isosceles triangles the sides opposite to the equal angles are equal, $\angle B = \angle C \Rightarrow AC = AB$

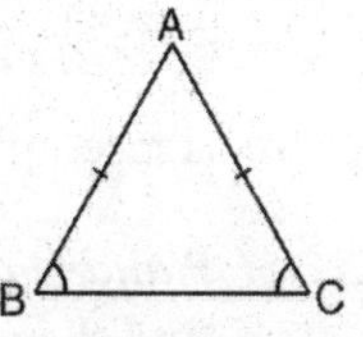

Solved Examples

Ex. 1. ***If AB is parallel to CD as given in the figure, then find the values of $\angle x, \angle y$ and $\angle z$.***

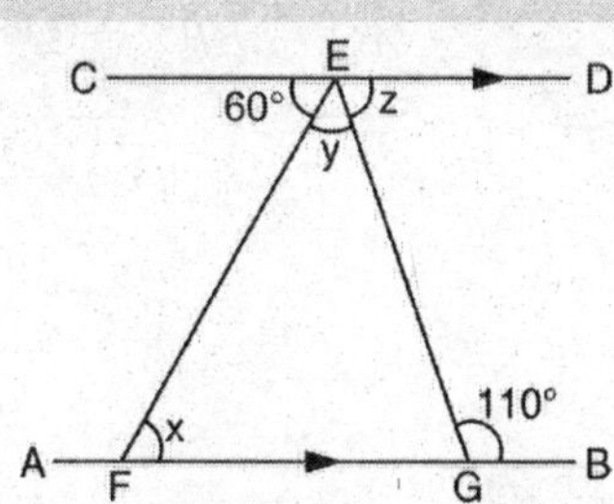

Sol. $\angle x = 60^\circ$ $(AB \parallel CD, alt.\angle s)$

$\angle EGF = 180^\circ - \angle EGB = 180^\circ - 110^\circ = 70^\circ$ *(Linear pair)*

$\angle y = 180^\circ - (\angle x + \angle EGF) = 180^\circ - (60^\circ + 70^\circ)$

$= 180^\circ - 130^\circ = 50^\circ$ *(Angle sum property of a Δ)*

$\angle z = \angle EGF = 70^\circ$ (alt. $\angle s$)

Ex. 2. ***In the following figure, find $\angle ADC$.***

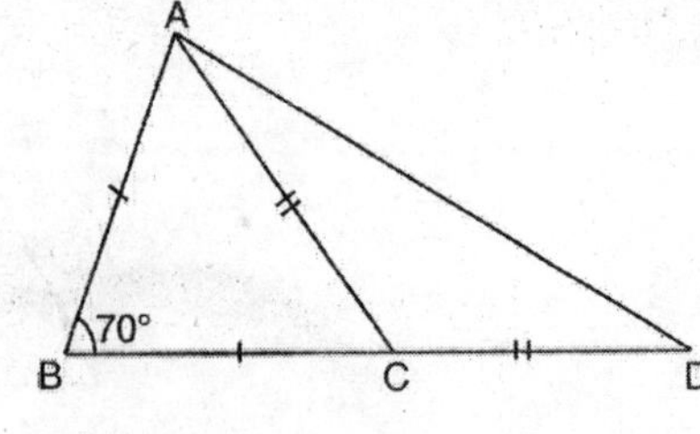

Sol. Given, $BA = BC$

$\Rightarrow \angle BCA = \angle BAC$ (*Angles opp. equal sides are equal*)

In ΔBAC,

$$\angle BCA = \angle BAC = \frac{1}{2}(180° - 70°) = \frac{1}{2} \times 110° = 55°$$

$$\angle ACD = 180° - \angle BCA = 180° - 55° = 125° \quad \textit{(Linear pair)}$$

Again, $CA = CD \Rightarrow \angle ADC = \angle CAD$ (*Isos.* Δ *Prop.*)

$$\therefore \angle ADC = \angle CAD = \frac{1}{2}(180° - 125°) = \frac{1}{2} \times 55° = \mathbf{27.5°}.$$

Ex. 3. ***In ΔABC, M is the mid point of BC. Length of AM is 9, N is a point on AM such that MN = 1. What is the distance of N from the centroid of the triangle?***

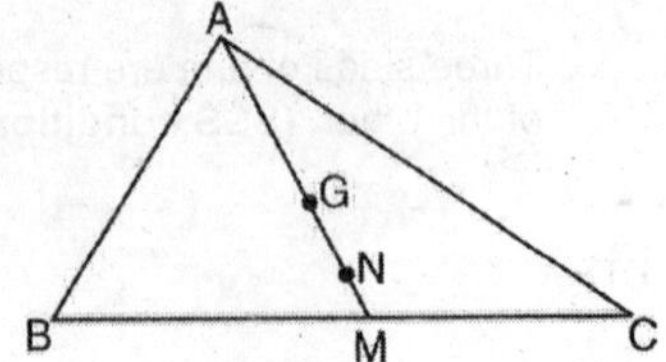

Sol. M is the mid point of BC.

$\Rightarrow AM$ is the median of ΔABC,

G is the centroid of ΔABC.

$\Rightarrow G$ divides AM in the ratio $2 : 1 \quad \Rightarrow AG : GM = 2 : 1$

Given, $AM = 9$, Let $AG = x$, Then $GM = (9 - x)$

$$x : (9 - x) = 2 : 1 \Rightarrow \frac{x}{9 - x} = \frac{2}{1} \Rightarrow x = 2(9 - x)$$

$$\Rightarrow x = 18 - 2x \Rightarrow 3x = 18 \Rightarrow x = 6 \qquad \therefore GM = 9 - 6 = 3$$

Now, $GN = GM - NM = 3 - 1 = \mathbf{2}$.

Ex. 4. ***In a ΔABC, if the bisector of the angle BAC meets BC in D, then which one of he following is correct?***

(a) $AB \le BD$ ***(b)*** $\angle BDA > \frac{1}{2} \angle BAC$

(c) ΔABD ***can never be isosceles*** ***(d) If*** $BD = AD$, ***then*** ΔABD ***will be equilateral.***

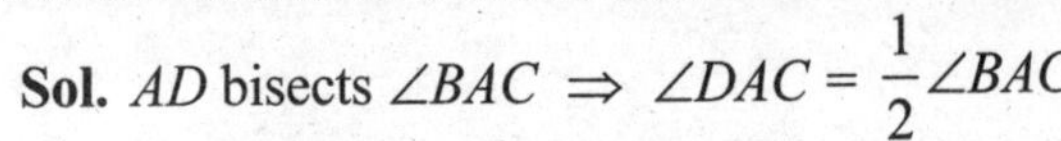

Sol. AD bisects $\angle BAC \Rightarrow \angle DAC = \frac{1}{2}\angle BAC$

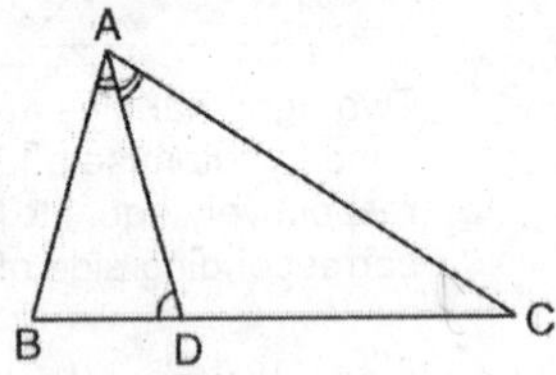

An exterior angle of a triangle is always greater than its interior opposite angle,

$$\therefore \angle BDA > \angle DAC \Rightarrow \angle BDA > \frac{1}{2}\angle BAC$$

Hence option (b) is correct.

Ex. 5. ***P and Q are the mid points of the sides CA and CB respectively of a triangle ABC, right angled at C. Then, find the value of 4 ($AQ^2 + BP^2$).***

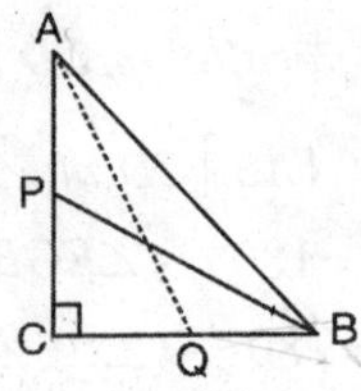

Sol. In rt. $\angle d\ \Delta ACQ,\ AQ^2 = QC^2 + AC^2$

In rt. $\angle d\ \Delta BCP,\ BP^2 = BC^2 + PC^2$

$$\therefore AQ^2 + BP^2 = (QC^2 + AC^2) + (BC^2 + PC^2)$$

$$= (AC^2 + BC^2) + (QC^2 + PC^2)$$

$$= AB^2 + PQ^2 = AB^2 + \left(\frac{1}{2}AB\right)^2$$

[($\because PQ = \frac{1}{2}AB$) Line joining the mid points of two sides of a Δ is always half the third side]

$$= AB^2 + \frac{1}{4}AB^2 = \frac{5}{4}AB^2$$

$$\Rightarrow 4(AQ^2 + BP^2) = 5AB^2.$$

Question Bank–18

1. In the given figure, $\angle A = 80°$, $\angle B = 60°$, $\angle C = 2x°$ and $\angle BDC = y°$. BD and CD bisect angles B and C respectively.

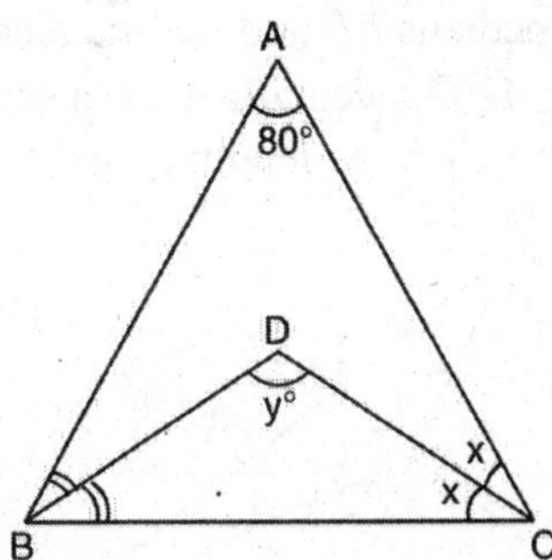

The values of x and y respectively are :

(a) 15° and 70° (b) 10° and 160°

(c) 20° and 130° (d) 20° and 125°

2. If CE is parallel to DB in the given figure, then the value of x will be

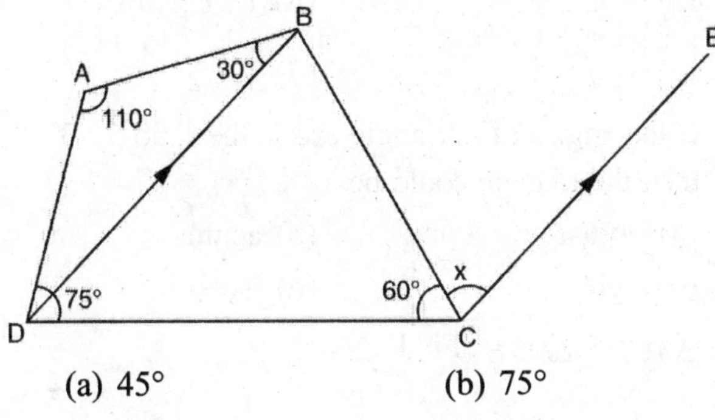

(a) 45° (b) 75°

(c) 30° (d) 85°

3. AB is parallel to CD. EF intersects them at M and N. The bisectors of $\angle M$ and $\angle N$ meet at Q. If $\angle AME = 80°$, then $\angle MQN$ is equal to

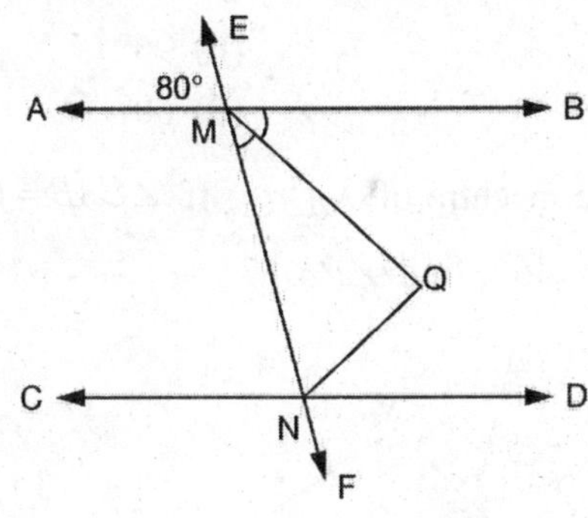

(a) 60° (b) 70°

(c) 80° (d) 90°

4. The diagram shows 2 isosceles triangles. What is the sum of a and b.

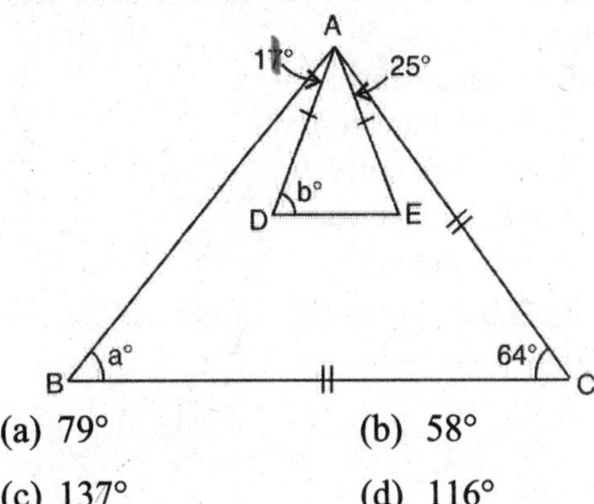

(a) 79° (b) 58°

(c) 137° (d) 116°

5. The given figure shows two overlapping triangles. The value of $a - b$ is

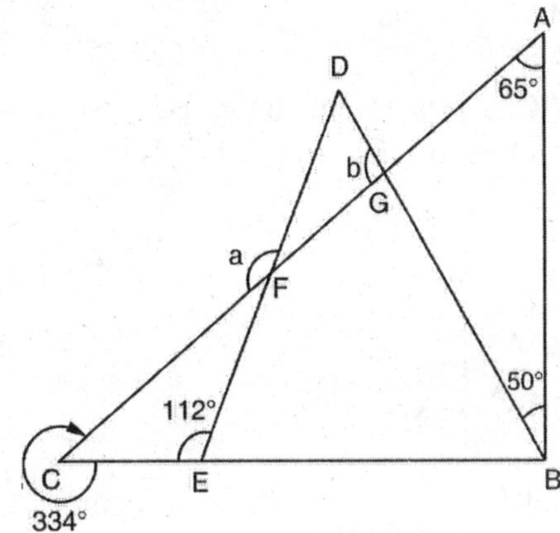

(a) 65° (b) 138°

(c) 73° (d) 48°

6. D, E, F are the mid points of BC, CA and AB of ΔABC. If AD and BE intersect in G, then $AG + BG + CG$ is equal to

(a) $AD = BE = CF$ (b) $\frac{2}{3}(AD + BE + CF)$

(c) $\frac{3}{2}(AD + BE + CF)$ (d) $\frac{1}{3}(AD + BE + CF)$

7. Consider the following statements relating to the congruency of two right triangles.

(1) Equality of two sides of one triangle with any two sides of the second makes the triangle congruent

(2) Equality of the hypotenuse and a side of one triangle with the hypotenuse and a side of the second respectively makes the triangle congruent.

(3) Equality of the hypotenuse and an acute angle of one triangle with the hypotenuse and an angle of the second respectively makes the triangle congruent.

Of these statements :

(a) 1, 2 and 3 are correct
(b) 1 and 2 are correct
(c) 1 and 3 are correct
(d) 2 and 3 are correct

8. $AB \perp BC$, $BD \perp AC$ and CE bisects $\angle C$. $\angle A = 30°$. Then what is $\angle CED$?

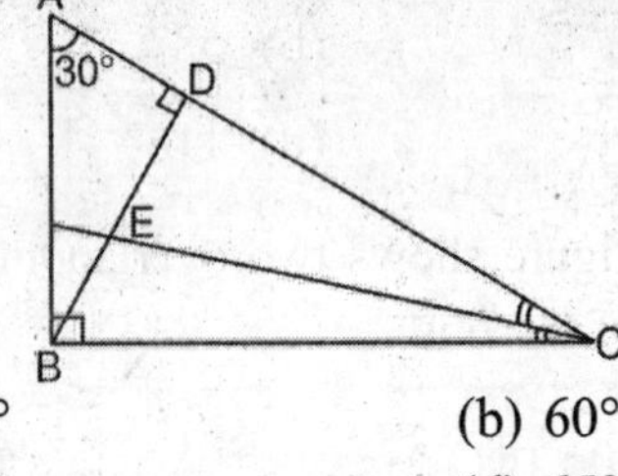

(a) 30° (b) 60°
(c) 45° (d) 65°

9. In ΔABC, $\angle B$ is a right angle, $AC = 6$ cm, D is the mid point of AC. The length of BD is

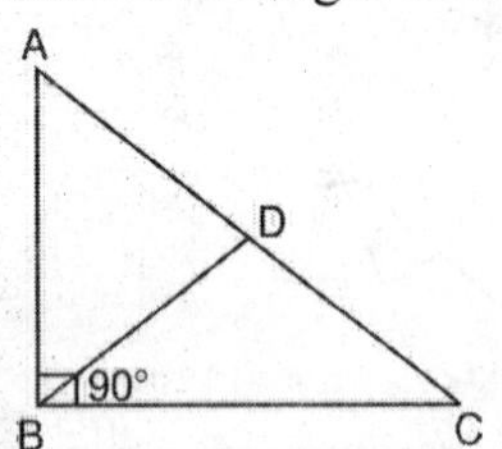

(a) 4 cm (b) $\sqrt{6}$ cm
(c) 3 cm (d) 3.5 cm

10. If the sides of a right triangle are x, $x + 1$ and $x - 1$, then the hypotenuse is

(a) 5 (b) 4
(c) 1 (d) 0

11. The degree measure of each of the three angles of a triangle is an integer. Which of the following could NOT be the ratio of their measures ?

(a) 2 : 3 : 4 (b) 3 : 4 : 5
(c) 5 : 6 : 7 (d) 6 : 7 : 8

12. In a $\Delta\, ABC$ right angled at B, which statement is true?

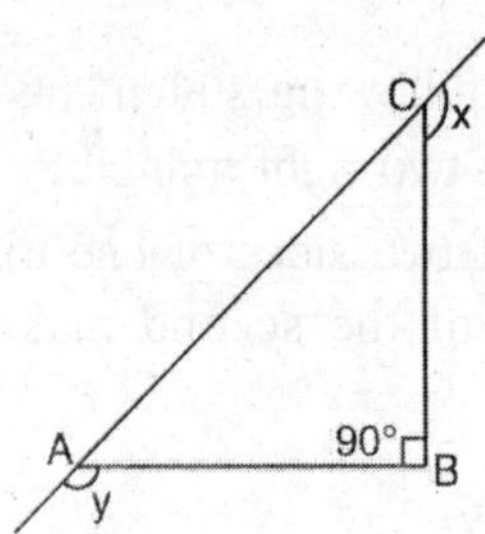

(a) $x + y = 180°$ (b) $x + y = 270°$
(c) $x + y = 300°$ (d) $x + y = 90°$

13. If PL, QM and RN are the altitudes of ΔPQR whose orthocentre is O, then P is the orthocentre of :

(a) ΔPQO (b) ΔPQL
(c) ΔQLO (d) ΔQRO

14. In ΔABC medians BE and CF intersect at G. If the straight line AGD meets BC in D in such a way that $GD = 1.5$ cm, then the length of AD is :

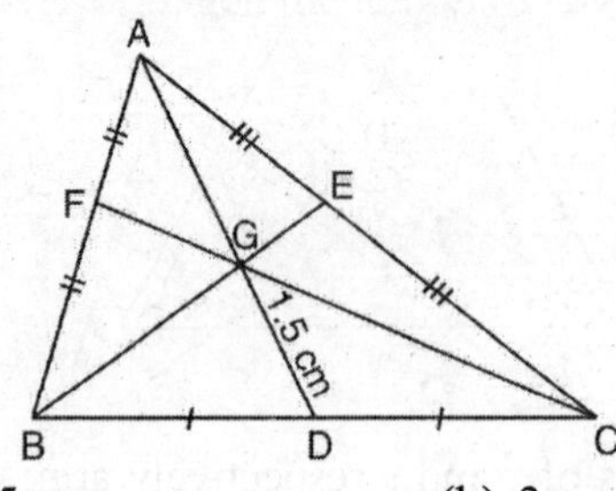

(a) 2.5 cm (b) 3 cm
(c) 4 cm (d) 4.5 cm

15. If the angles of a triangle are 30°, 60°, 90°, then what is the ratio of corresponding sides ?

(a) 1 : 2 : 3 (b) $1 : 1 : \sqrt{2}$
(c) $1 : \sqrt{3} : 3$ (d) $1 : \sqrt{2} : 2$

16. If the angles of a triangle are in the ratio 5 : 3 : 2, then the triangle could be

(a) obtuse (b) acute
(c) right (d) isosceles

17. $\Delta ACB \cong \Delta DFE$. Find $\angle F$

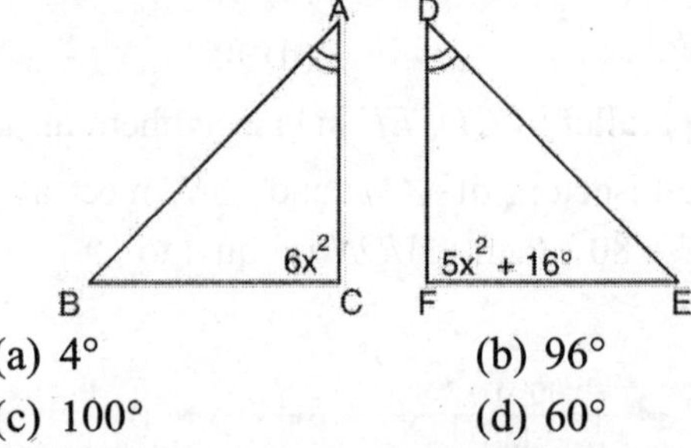

(a) 4° (b) 96°
(c) 100° (d) 60°

18. P is the incentre of ΔABC . If $\angle CAB = 66°$, and $\angle CPY = 46°$, find $\angle PBA$?

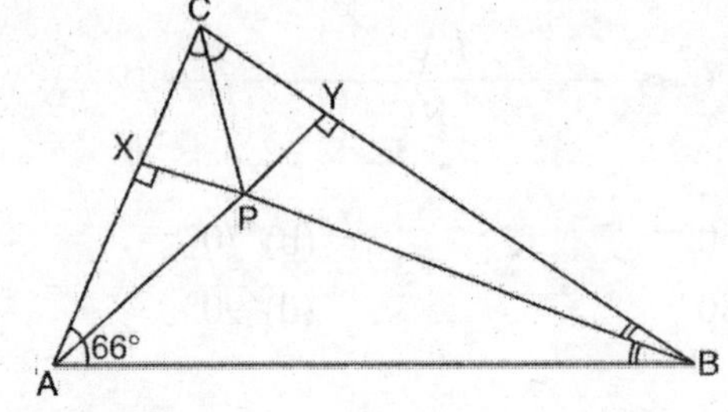

(a) 16° (b) 23°
(c) 13° (d) 18°

19. A man goes to a garden and runs in the following manner: From the starting point, he goes west 25 m, then due north 60 m, then due last 80 m and finally due south 12 m. The distance between the finishing point and the starting point is :

(a) 177 m (b) 103 m
(c) 83 m (d) 73 m

20. In the given figure, ABC is a triangle in which BC is produced to D. If $\angle A : \angle B : \angle C :: 3 : 2 : 1$ and $AC \perp CE$, then $\angle ECD$ is :

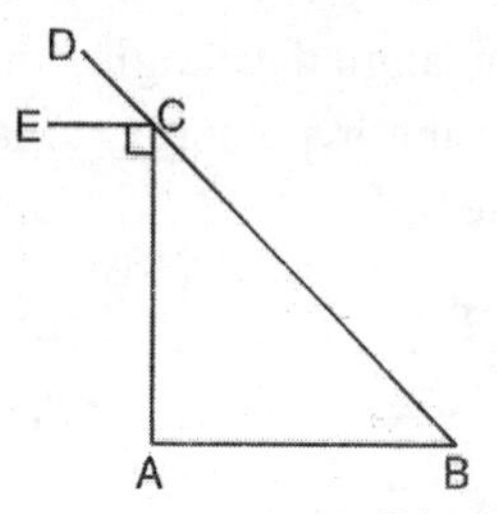

(a) 30° (b) 45°
(c) 60° (d) 72°

21. If $ABCD$ is a square; X is the mid point of AB and Y is the mid point of BC, then which of the following is NOT correct?

(a) The triangles ADX and BAY are congruent
(b) $\angle DXA = \angle AYB$
(c) $\angle ADX = \angle BAY$
(d) DX is not perpendicular to AY

22. In a ΔPQR, the sides PQ and PR are produced to S and T respectively. Bisectors of $\angle SQR$ and $\angle QRT$ meet at the point O. If $\angle P = 66^\circ$, then what is the value of $\angle QOR$?

(a) 47° (b) 50°
(c) 57° (d) 67°

23. In a ΔABC, $\angle C = 110°$. Which one of the following statements is correct ?

(a) $AB^2 > AC^2 + BC^2$
(b) $AB^2 < AC^2 + BC^2$
(c) $AC^2 > AB^2 + BC^2$
(d) $BC^2 > AB^2 + AC^2$

24. In which one of the following triangles does the orthocentre lie in the exterior of the triangle.

(a) ΔABC, wherein, $\angle A = \angle B = \angle C = 60°$
(b) ΔPQR, wherein, $\angle P = 40°$, $\angle Q = 30°$, $\angle R = 110°$
(c) ΔXYZ, wherein, $\angle X = 80°$, $\angle Y = 60°$, $\angle Z = 40°$
(d) ΔDEF, wherein, $\angle D = 52°$, $\angle E = 90°$, $\angle F = 38°$

Answers

1. (c)	**2.** (d)	**3.** (d)	**4.** (c)	**5.** (c)	**6.** (d)	**7.** (d)	**8.** (b)	**9.** (c)	**10.** (a)
11. (d)	**12.** (b)	**13.** (d)	**14.** (d)	**15.** (a)	**16.** (c)	**17.** (a)	**18.** (c)	**19.** (d)	**20.** (c)
21. (d)	**22.** (c)	**23.** (a)	**24.** (b)						

Hints and Solutions

1. (c) In $\Delta ABC, \angle A = 80^\circ, \angle B = 60^\circ$ and $\angle C = 2x$

$\Rightarrow 80^\circ + 60^\circ + 2x = 180^\circ$

$\Rightarrow 2x = 180^\circ - 140^\circ = 40^\circ \Rightarrow x = 20^\circ$

Since, BD bisects $\angle B, \angle DBC = 30^\circ$

$\therefore$ In ΔDBC, $y + 30^\circ + 20^\circ = 180^\circ$

$\Rightarrow y = 180^\circ - 50^\circ = \mathbf{130^\circ}$.

2. (d) In ΔADB,

$\angle ADB + \angle DAB + \angle ABD = 180^\circ$

$\Rightarrow \angle ADB + 110^\circ + 30^\circ = 180^\circ$

$\Rightarrow \angle ADB = 180^\circ - 140^\circ = 40^\circ$

$\therefore \angle BDC = 75^\circ - \angle ADB = 75^\circ - 40^\circ = 35^\circ$

In ΔBDC,

$\angle DBC = 180^\circ - (\angle BDC + \angle BCD)$

$= 180^\circ - (35^\circ + 60^\circ) = 180^\circ - 95^\circ = 85^\circ$

$x = \angle DBC = \mathbf{85^\circ}$. *(DB||CE, alt. ∠s are equal)*

3. (d) $\angle BMN = \angle AME = 80^\circ$ *(vert. opp. ∠s are equal)*

$\angle BMN + \angle MND = 180^\circ$

(AB || CD, co-int. ∠s are supp)

$\Rightarrow \angle MND = 180^\circ - \angle BMN$

$= 180^\circ - 80^\circ = 100^\circ$

MQ bisects $\angle BMN \Rightarrow \angle NMQ = 40^\circ$

NQ bisects $\angle MND \Rightarrow \angle QMN = 50^\circ$

$\therefore$ In ΔMQN,

$\angle MQN + \angle NMQ + \angle QNM = 180^\circ$

$\Rightarrow \angle MQN + 40^\circ + 50^\circ = 180^\circ$

$\Rightarrow \angle MQN = 180^\circ - 90^\circ = \mathbf{90^\circ}$.

4. (c) In ΔABC, $CA = CB \Rightarrow \angle CBA = \angle CAB$

(Angles opp. equal sides are equal)

$\therefore \angle CBA = \angle CAB = \frac{1}{2}(180^\circ - \angle BAC)$

$= \frac{1}{2}(180^\circ - 64^\circ) = \frac{1}{2} \times 116^\circ = 58^\circ$

$\Rightarrow a = 58^\circ$

Now $\angle CAB = \angle CAE + \angle EAD + \angle DAB$

$\Rightarrow 58^\circ = 25^\circ + \angle EAD + 11^\circ$

$\Rightarrow \angle EAD = 58^\circ - 36^\circ = 22^\circ$

$\therefore$ In $\Delta ADE, AD = AE$

$\Rightarrow \angle AED = \angle ADE = \frac{1}{2}(180^\circ - 22^\circ)$

$= \frac{1}{2} \times 158^\circ = 79^\circ$

$\Rightarrow b = 79^\circ$

Hence, $a + b = 58^\circ + 79^\circ =$ **137°**.

5. (c) In $\Delta CEF, \angle FCE = 360^\circ - 334^\circ = 26^\circ$

$\therefore \angle CFE = 180^\circ - (\angle FCE + \angle FEC)$

$= 180^\circ - (26^\circ + 122^\circ)$

$= 180^\circ - 138^\circ = 42^\circ$

Now $a + \angle CFE = 180^\circ$ *(Linear pair)*

$\Rightarrow a = 180^\circ - \angle CFE = 180^\circ - 42^\circ = 138^\circ$

In $\Delta ABG, \angle AGB = 180^\circ - (\angle GBA + \angle BAG)$

$= 180^\circ - (50^\circ + 65^\circ)$

$= 180^\circ - 115^\circ = 65^\circ$

$\therefore b = \angle AGB = 65^\circ$ *(vert. opp. ∠s)*

Hence, $a - b = 138^\circ - 65^\circ =$ **73°**.

6. (d) Since, D, E and F and the mid points of BC, CA and AB respectively; AD, BE and CF are the medians of ΔABC. G is the centroid of ΔABC.

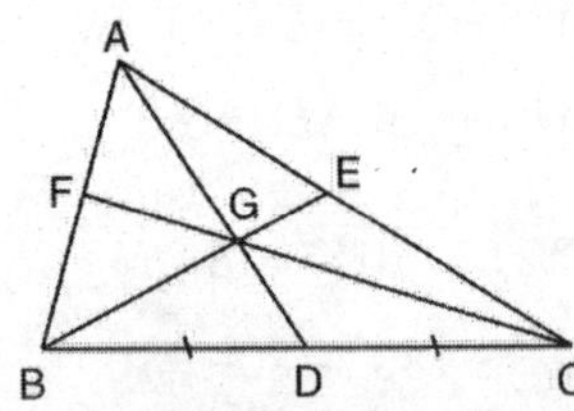

$\therefore$ G divides AD, BE and CF in the ratio 2 : 1, *i.e.*, $AG : GD = 2:1, BG : GE = 2:1, CG : GF = 2:1$

$\Rightarrow AG = \frac{1}{3}AD,\ BG = \frac{1}{3}BE,\ CG = \frac{1}{3}CF$

$\Rightarrow AG + BG + CG = \frac{1}{3}(AD + BE + CF)$.

7. (d) Statement 2 $\Rightarrow$ *RHS* congruency

Statement 3 $\Rightarrow$ *AAS* congruency

8. (b) In ΔABC,

$\angle ABC + \angle BAC + \angle ACB = 180^\circ$

$\Rightarrow 90^\circ + 30^\circ + \angle ACB = 180^\circ$

$\Rightarrow \angle ACB = 180^\circ - 120^\circ = 60^\circ$

Now CE bisects $\angle C \Rightarrow \angle ECD = 30^\circ$

Also, $BD \perp AC \Rightarrow \angle EDC = 90^\circ$

$\therefore$ In ΔDEC,

$\angle ECD + \angle EDC + \angle CED = 180^\circ$

$\Rightarrow 30^\circ + 90^\circ + \angle CED = 180^\circ$

$\Rightarrow \angle CED = 180^\circ - 120^\circ =$ **60°**.

9. (c) In a right angled triangle, the length of the median to the hypotenuse is half the length of the hypotenuse.

Hence, $BD = \frac{1}{2}AC = \frac{1}{2} \times 6\text{cm} = \mathbf{3cm}$.

10. (a)

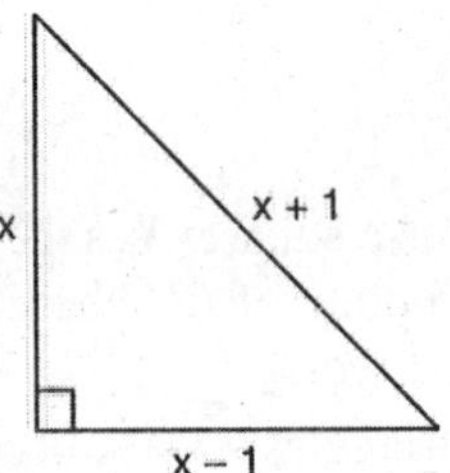

In a right Δ, hypotenuse is the largest side.

$\therefore$ Hypotenuse $= x + 1$

$\therefore (x+1)^2 = x^2 + (x-1)^2$ *(Pyth. Th.)*

$\Rightarrow x^2 + 2x + 1 = x^2 + x^2 - 2x + 1$

$\Rightarrow x^2 = 4x \Rightarrow x = 4$

$\therefore$ Hypotenuse = 4 + 1 = **5**.

11. (d) The sum of the terms of the given ratio should completely divide 180°, *i.e.*, angle sum of a triangle.

In option (d) sum of terms of ratio = 6 + 7 + 8 = 21, which does not completely divide 180°.

12. (b) In ΔABC, $\angle BAC + \angle ACB = 180^\circ - \angle ABC$

$= 180^\circ - 90^\circ = 90^\circ$

$x = 180^\circ - \angle ACB,\ y = 180^\circ - \angle BAC$

$\therefore x + y = 180^\circ - \angle ACB + 180^\circ - \angle BAC$

$= 360^\circ - (\angle ACB + \angle BAC)$

$= 360^\circ - 90^\circ =$ **270°**.

13. (d) Orthocentre is the point of intersection of altitudes. OL is an altitude for ΔQOR

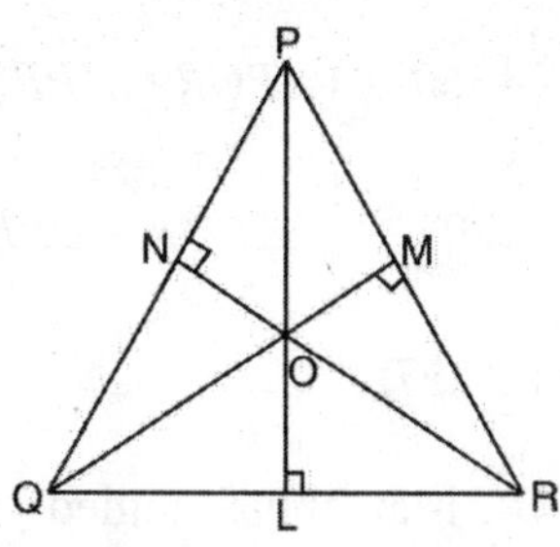

$\therefore$ P is the orthocentre for $\Delta\, QOR$.

14. (d) G is the centroid of $\Delta\, ABC$

$\Rightarrow AG : GD = 2 : 1$

$\Rightarrow AG : 1.5 = 2 : 1$

$\Rightarrow \frac{AG}{1.5} = \frac{2}{1} \Rightarrow AG = 2 \times 1.5 = 3$

$\therefore\ AD = AG + GD = 3 + 1.5 = \mathbf{4.5}$.

15. (a) Since the side opposite to the greatest angle is the greatest, the ratio of sides = ratio of corresponding angles = 30 : 60 : 90 = 1: 2 : 3.

16. (c) Let the angles of the triangle be $5x$, $3x$ and $2x$.

Then, $5x + 3x + 2x = 180^\circ \Rightarrow 10x = 180^\circ$

$\Rightarrow x = 18^\circ$

$\therefore$ The angles are $5 \times 18^\circ, 3 \times 18^\circ, 2 \times 18^\circ$, *i.e.*, 90°, 54° and 36°.

Hence, the triangle is a right triangle.

17. (a) Since $\Delta ACB \cong \Delta DFE$, $\angle C = \angle F$

$\Rightarrow 6x^2 = 5x^2 + 16 \Rightarrow x^2 = 16 \Rightarrow x = \mathbf{4}$.

18. (c) In ΔCPY, $\angle CYP = 90^\circ$, $\angle CPY = 46^\circ$

$\therefore\ \angle PCY = 180^\circ - (\angle CYP + \angle CPY)$

$= 180^\circ - (90^\circ + 46^\circ) = 180^\circ - 136^\circ = 44^\circ$

Since P is the incentre of $\Delta\, ABC$, CP bisects $\angle ACB \Rightarrow \angle ACB = 2 \times \angle PCY = 2 \times 44^\circ = 88^\circ$

In ΔABC,

$\angle CBA = 180^\circ - (\angle CAB + \angle ACB)$

$= 180^\circ - (66^\circ + 88^\circ) = 180^\circ - 154^\circ = 26^\circ$

BP bisects $\angle CBA \Rightarrow \angle PBA = \frac{1}{2} \times 26^\circ = \mathbf{13^\circ}$.

19. (d) Let A be the starting point and B the finishing point

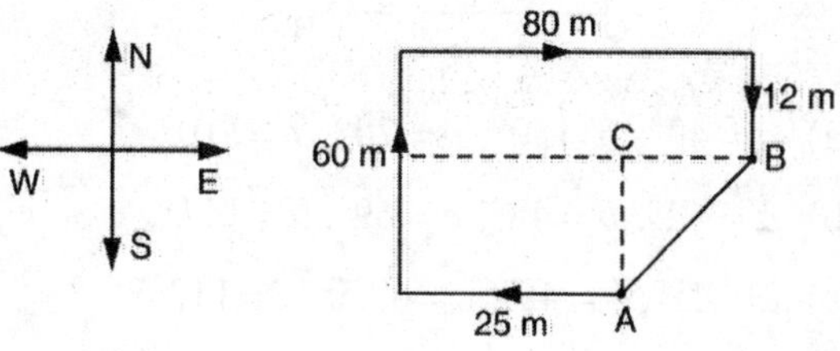

As can be seen from the figure, we have to find AB.

$BC = 80\text{ m} - 25\text{ m} = 55\text{m}$,

$AC = 60\text{ m} - 12\text{ m} = 48\text{ m}$

By Pythagoras' Theorem,

$AB^2 = AC^2 + BC^2 = 48^2 + 55^2$

$= 2304 + 3025 = 5329$

$\Rightarrow AB = \sqrt{5329} = \mathbf{73\,m}$.

20. (c) Let $\angle A = 3x$, $\angle B = 2x$, $\angle C = x$

$\Rightarrow \angle A + \angle B + \angle C = 180^\circ$

$\Rightarrow 3x + 2x + x = 180^\circ$

$\Rightarrow 6x = 180^\circ \Rightarrow x = 30^\circ \Rightarrow \angle C = 30^\circ$

$EC \perp CA \Rightarrow \angle ECA = 90^\circ$

$\therefore\ \angle ECD = 180^\circ - (\angle ECA + \angle ACB)$

$= 180^\circ - (90^\circ + 30^\circ) = 180^\circ - 120^\circ = \mathbf{60^\circ}$.

21. (d) In $\Delta\, ADX$ and $\Delta\, BAY$,

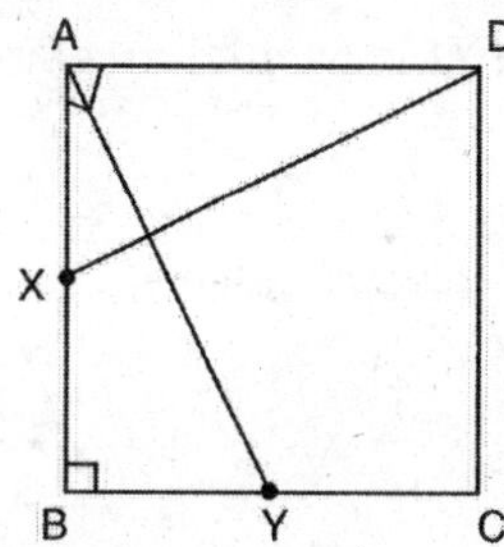

$AD = AB$ (*Sides of a square*)

$\angle DAX = \angle ABY = 90^\circ$

$AX = BY$ ($\because AB = BC \Rightarrow \frac{1}{2}AB = \frac{1}{2}BC$)

$\therefore\ \Delta ADX \cong \Delta BAY$

$\Rightarrow \angle DXA = \angle AYB$ and $\angle ADX = \angle BAY$ (cpct)

22. (c) In ΔPQR,

$\angle PQR + \angle PRQ + \angle QPR = 180^\circ$

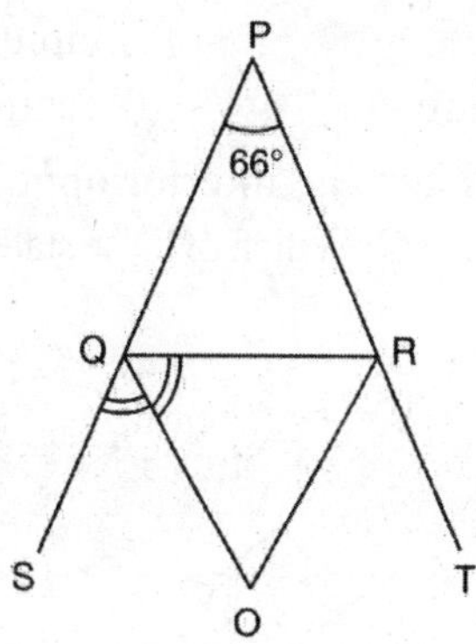

$\Rightarrow \angle PQR + \angle PRQ = 180^\circ - \angle QPR$

$= 180^\circ - 66^\circ$

$= 114^\circ$

Now, $\angle SQR = 180^\circ - \angle PQR$ ...(i)

and $\angle QRT = 180^\circ - \angle PRQ$ ....(ii)

Adding eq. (i) and (ii), we get

$\Rightarrow \angle SQR + \angle QRT = 360^\circ - (\angle PQR + \angle PRQ)$

In ΔQOR,

$\angle RQO + \angle QRO + \angle QOR = 180^\circ$

$\Rightarrow \angle QOR = 180^\circ - (\angle RQO + \angle QRO)$

$= 180^\circ - \left(\frac{1}{2}\angle SQR + \frac{1}{2}\angle QRT\right)$

($\because$ *QO bisects* $\angle SQR$, *RO bisects* $\angle QRT$)

$= 180^\circ - \frac{1}{2}(\angle SQR + \angle QRT)$

$= 180^\circ - \frac{1}{2}[360^\circ - (\angle PQR + \angle PRQ)]$

$= \cancel{180^\circ} - \cancel{180^\circ} + \frac{1}{2}(\angle PQR + \angle PRQ)$

$= \frac{1}{2} \times 114^\circ = \mathbf{57^\circ}$.

23. (a) Since ΔABC is an obtuse angled Δ and $\angle C$ i the greatest angle,

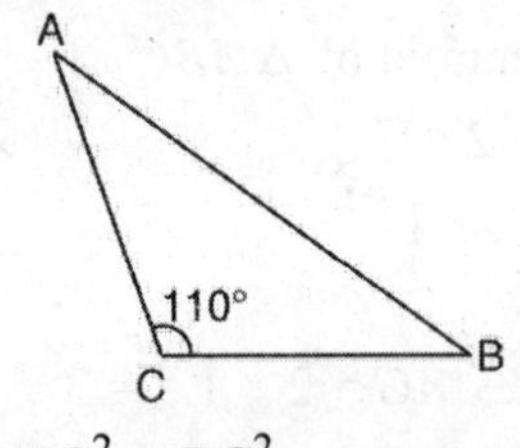

$AB^2 > AC^2 + BC^2$

24. (b) ΔPQR is an obtuse angled triangle, so the orthocentre lies in the exterior of the triangle.

Self Assessment Sheet–18

1. In the figure *CD* is parallel to *AB*. The angle *y* is equal to :

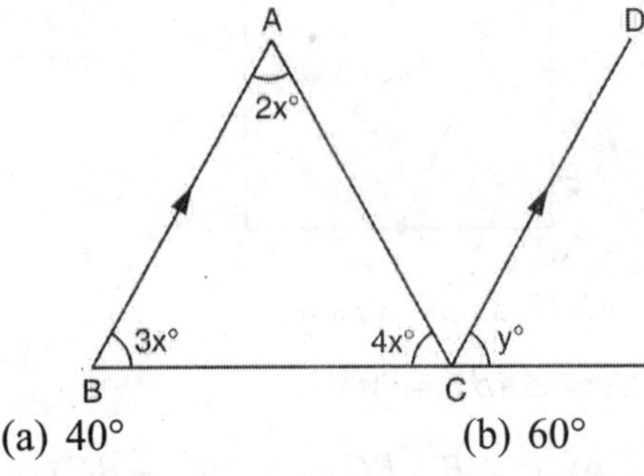

(a) 40° (b) 60°
(c) 80° (d) 100°

2. The side opposite to an obtuse angle of a triangle is :
(a) smallest (b) greatest
(c) half of the perimeter (d) none of these

3. The point of intersection of the right bisectors of a triangle is called
(a) in-centre (b) circumcentre
(c) orthocentre (d) centroid

4. $\angle A$ and $\angle B$ are the interior opposite angles of $\angle BCD$ in ΔABC. Which of these statements is true ?

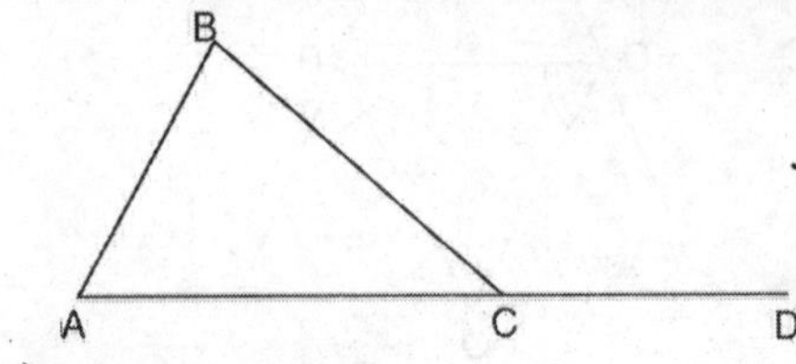

(a) $\angle BCA = \angle BCD - \angle A$
(b) $\angle A - 180^\circ = \angle B$
(c) $\angle A = 90^\circ - \angle B$
(d) $\angle B = \angle BCD - \angle A$

5. Ankita wants to prove $\Delta ABC \cong \Delta DEF$ using *SAS*. She knows $AB = DE$ and $AC = DF$. What additional piece of information does she need ?
(a) $\angle A = \angle D$ (b) $\angle C = \angle F$
(c) $\angle B = \angle E$ (d) $\angle A = \angle B$

6. In a ΔABC, the sum of exterior angles at *B* and *C* is equal to :
(a) $180^\circ - \angle BAC$ (b) $180^\circ + \angle BAC$
(c) $180^\circ - 2\angle BAC$ (d) $180^\circ + 2\angle BAC$

7. *PQ* = *QR* = *PS*. Calculate the size of the labelled angles.

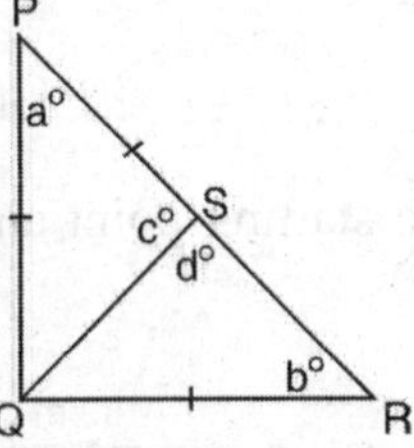

(a) $a = 40^\circ, b = 50^\circ, c = 70^\circ, d = 110^\circ$
(b) $a = 42^\circ, b = 48^\circ, c = 69^\circ, d = 111^\circ$
(c) $a = 45^\circ, b = 45^\circ, c = 67.5^\circ, d = 112.5^\circ$
(d) $a = 50^\circ, b = 40^\circ, c = 65^\circ, d = 115^\circ$

8. Which of the following is not a correct statement.

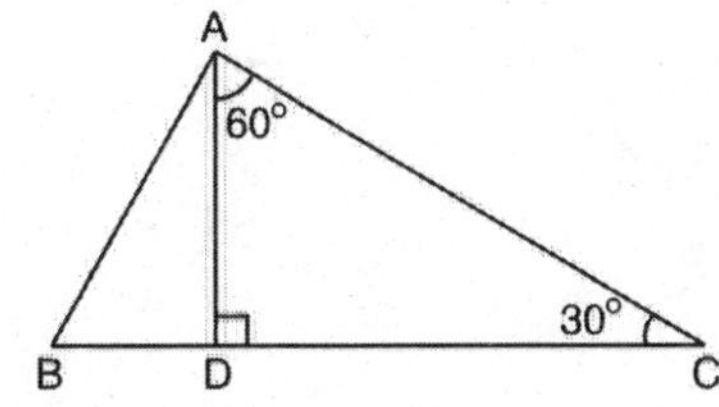

(a) $AC = 2\,AD$

(b) $\angle ACD = \angle CAB + \angle ABC$

(c) $BD^2 = AB^2 - AD^2$

(d) $AC > AB + BC$

9. If the straight line which bisects the vertical angle of a triangle is perpendicular to the base, the triangle is :

(a) equilateral (b) isosceles

(c) scalene (d) right- angled

10. Match correctly

(1) Each angle of an equilateral Δ	(a) greater than 2 rt. $\angle s$
(2) Measure of the vertical angle of a Δ if each base angle is double the vertical angle	(b) incentre
(3) Sum of any two exterior angles of a Δ	(c) 36°
(4) Point of intersection of the angle bisectors of a Δ	(d) 60°

Answers

1. (b) **2.** (b) **3.** (b) **4.** (d) **5.** (a) **6.** (b) **7.** (c) **8.** (d) **9.** (b)

10. (1) → (d) (2) → (c) (3) → (a) (4) → (b)

POLYGONS AND QUADRILATERALS

KEY FACTS

POLYGONS

1. A simple closed figure formed by three or more line segments is called a **polygon.**
2. If the measure of each interior angle of a polygon is less than 180°, then it is called a **convex** polygon.
3. If the measure of at least one interior angle of a polygon is greater than 180°, then it is a **concave** or **rentrant** polygon.

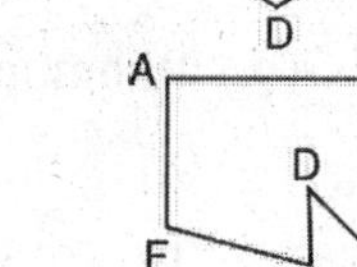

4. A polygon with all sides and all angles equal is called a **regular** polygon.
5. **Properties of polygons:** For a polygon of n sides,
   (i) Sum of interior angles $= (2n-4)\times 90°$
   (ii) Sum of exterior angles = 360° (always)
   (iii) Each interior angle $= \dfrac{(2n-4)}{n}\times 90°$(regular polygon).
   (iv) Each exterior angle $\dfrac{360°}{n}$ (in regular polygon)
   (v) Interior angle + exterior angle =180°(always)

QUADRILATERALS

1. **Quadrilaterals** are figures enclosed by four line segments.
2. The sum of the four interior angles of a quadrilateral is 360°.
3. **Types of quadrilaterals and their properties:**
   (i) **Parallelogram:**

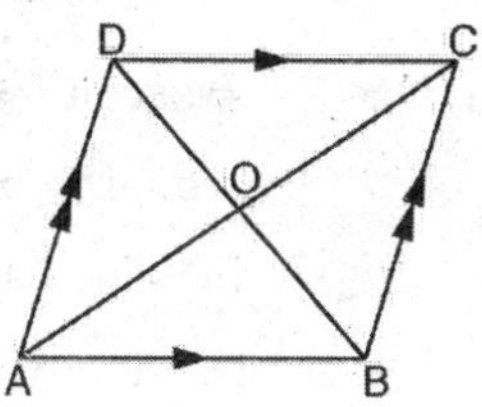

   (a) Opposite sides are equal and parallel. $AB \parallel CD$, $AB = CD$ and $BC \parallel AD$, $BC = AD$.
   (b) Diagonals bisect each other, $AO = OC$, $BO = OD$
   (c) Each diagonal divides it into two congruent triangles,
   $\Delta ABC \cong \Delta CDA$ and $\Delta DAB \cong \Delta BCD$
   (d) Sum of any two adjacent angles =180°
   $\angle A + \angle B = 180°, \angle B + \angle C = 180°,$
   $\angle C + \angle D = 180°, \angle D + \angle A = 180°$

(ii) **Rectangle:** A parallelogram whose all angles are 90°.

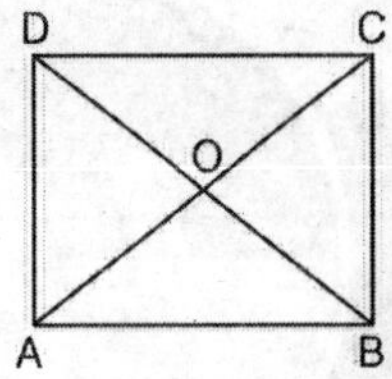

(a) Opposite sides are parallel and equal. $AB \parallel DC, AB = DC, AD \parallel BC, AD = BC$

(b) All the angles are equal to 90°,

$$\angle A = \angle B = \angle C = \angle D = 90°$$

(c) Diagonals are equal, $AC = BD$

(d) Diagonals bisect each other, $AO = OC = BO = OD$

(iii) **Square:** A rectangle whose adjacent sides are equal.

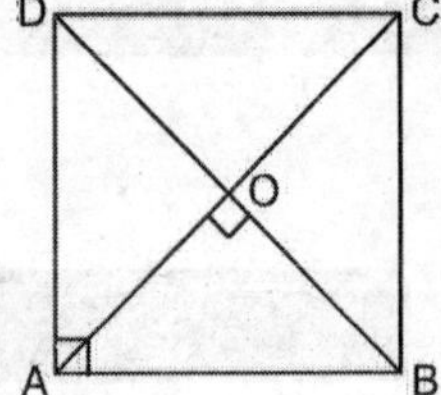

(a) All sides are equal

(b) Opposite sides are parallel

(c) All the angles are equal to 90°

(d) The diagonals are equal and bisect each other

(e) The diagonals intersect at right angles, *i.e.*,

$\angle AOB = \angle BOC = \angle COD = \angle DOA = 90°$

(f) The diagonals bisect the opposite angles, *i.e.*,

$\angle DAC = \angle CAB = 45°, \angle DCA = \angle BCA = 45°$

$\angle ABD = \angle DBC = 45°, \angle ADB = \angle BDC = 45°$

(iv) **Rhombus :** A parallelogram with all sides equal.

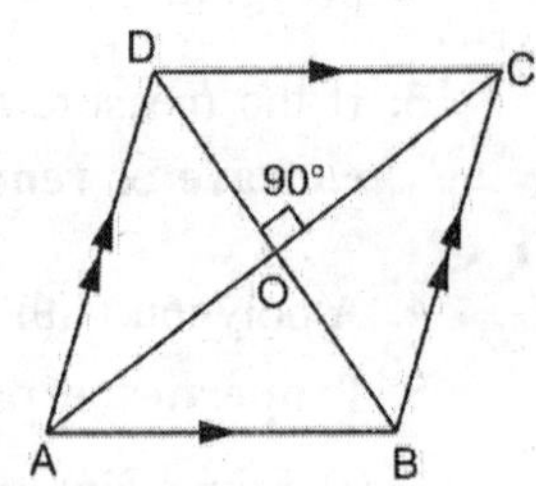

(a) Opposite sides are parallel.

(b) All the sides are equal.

(c) The opposite angles are equal.

(d) The diagonals are not equal. $AC \neq BD$.

(e) The diagonals intersect at right angles:

$\angle AOB = \angle BOC = \angle COD = \angle AOD = 90°$

(f) The diagonals bisect the opposite angles:

$\angle BAC = \angle DAC,\ \angle ACB = \angle ACD,\ \angle DBA = \angle DBC,\ \angle ADB = \angle BDC.$

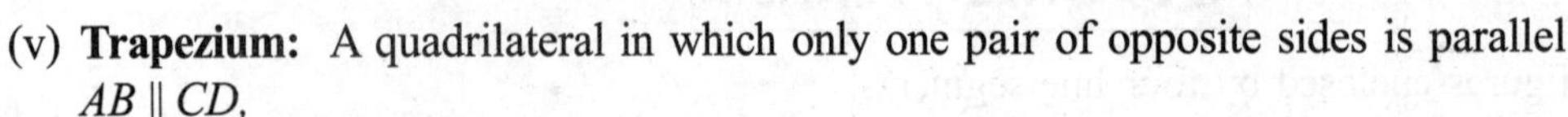

(v) **Trapezium:** A quadrilateral in which only one pair of opposite sides is parallel, $AB \parallel CD$,

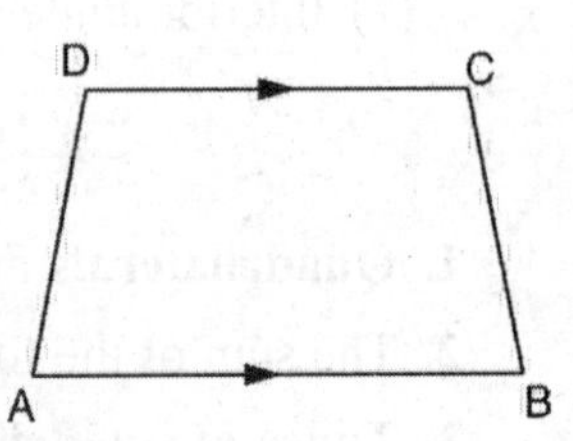

$\angle A + \angle D = 180°$

$\angle B + \angle C = 180°$

Isosceles trapezium: A trapezium whose non-parallel sides are equal.

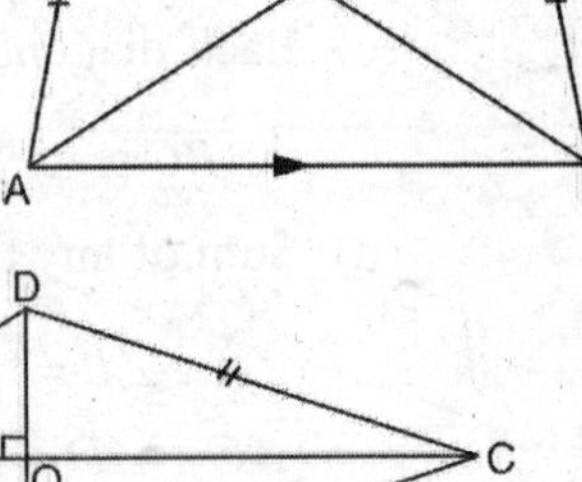

$AB \parallel DC, AD = BC$

(a) The base angles of an isosceles trapezium are equal, *i.e.*, $\angle DAB = \angle ABC$.

(b) The diagonals of an isosceles trapezium are equal, *i.e.*, Diagonal AC = Diagonal BD.

(vi) **Kite:** A kite is a quadrilateral in which two pairs of adjacent sides are equal. Opposite sides are not equal. $AD = AB, DC = CB$

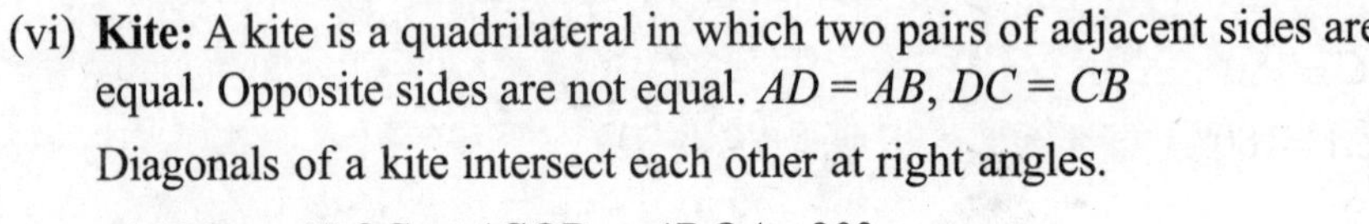

Diagonals of a kite intersect each other at right angles.

$\angle AOB = \angle BOC = \angle COD = \angle DOA = 90°$.

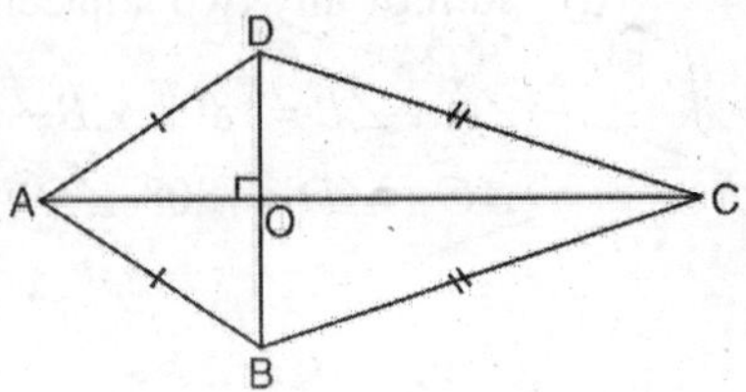

4. The given chart shows how special quadrilaterals are related

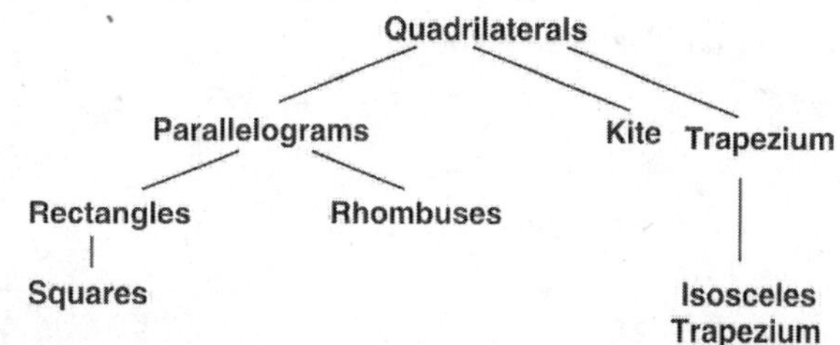

Solved Examples

Ex. 1. ***The ratio of an interior angle to an exterior angle of a regular polygon is 5 : 2. What is the number of sides of the polygon?***

Sol. Interior angle of a polygon of n sides $= \dfrac{(n-2)\times 180°}{n}$

Exterior angle of a polygon of n sides $= \dfrac{360°}{n}$

Given , $\dfrac{(n-2)\times 180°}{n} : \dfrac{360°}{n} = 5:2$

$\Rightarrow \dfrac{(n-2)\times 180°}{n} \times \dfrac{n}{360°} = \dfrac{5}{2} \;\Rightarrow\; \dfrac{(n-2)}{2} = \dfrac{5}{2} \;\Rightarrow\; 2(n-2) = 10$

$\Rightarrow\; 2n-4=10 \;\Rightarrow\; 2n=14 \;\Rightarrow\; n=7.$

Ex. 2. ***Which of the following statements is not correct?***

(a) If the sum of angles of a n sided polygon is n right angles, then n = 4.

(b) No regular polygon can have an interior angle equal to 110°.

(c) A regular polygon with 180 sides have each interior angle equal to 178°.

(d) No regular polygon can have interior angle equal to 140°.

Sol. (a) Sum of angles of an n sided polygon = $(n-2) \times 180°$

Given, $(n-2) \times 180° = n \times 90°$

$\Rightarrow 2(n-2) = n \Rightarrow 2n-4 = n \Rightarrow n = 4$

(b) Each interior angle of a regular polygon of n sides

$= \dfrac{(n-2)\times 180°}{n}$

$\Rightarrow \dfrac{(n-2)\times 180°}{n} = 110°$

$\Rightarrow 180°n - 360° = 110°n$

$\Rightarrow 70°n = 360° \Rightarrow n = \dfrac{360°}{70°} = 5\dfrac{1}{7},$

which is not a whole number. Hence the given angle cannot be the interior angle of a regular polygon.

(c) Interior angle of a regular polygon with 180 sides = $\dfrac{(180-2)\times 180°}{180} = 178\times 1° = 178°.$

(d) Check as in option (b)

Hence option (b) is not correct.

Ex. 3. ***The sides of a quadrilateral are extended to make the angles as shown in the figure. Find the value of x.***

Sol. Sum of the exterior angles of a quadrilateral = 360°

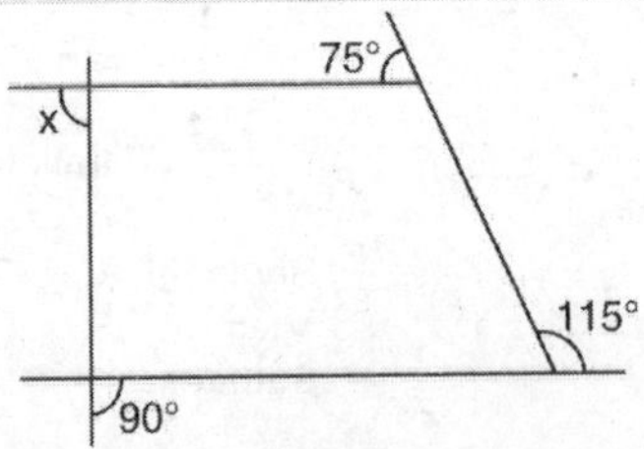

$\Rightarrow x + 75° + 115° + 90° = 360°$

$\Rightarrow x = 360° - 280° = \mathbf{80°}$.

Ex. 4. ***In the given figure a regular octagon and a regular hexagon are placed edge to edge as shown. What is the size of angle marked x?***

Sol. In the given figure,

$\angle CDE$ = Interior angle of a regular octagon $= \dfrac{(8-2)\times 180°}{8} = 135°$

$\angle CDS$ = Interior angle of a regular hexagon $= \dfrac{(6-2)\times 180°}{6} = 120°$

It can be seen from the figure,

$\angle CDE + \angle CDS + x = 360°$ (*Sum of* $\angle$s *round point D*)

$\Rightarrow 135° + 120° + x = 360°$

$\Rightarrow x = 360° - 255° = \mathbf{105°}$.

Ex. 5. ***In the given hexagon ABCDEF, AB||FE. Find the value of x.***

Sol. $AB \| FE \Rightarrow \angle BAF + \angle AFE = 180°$ (*co-int* $\angle$s)

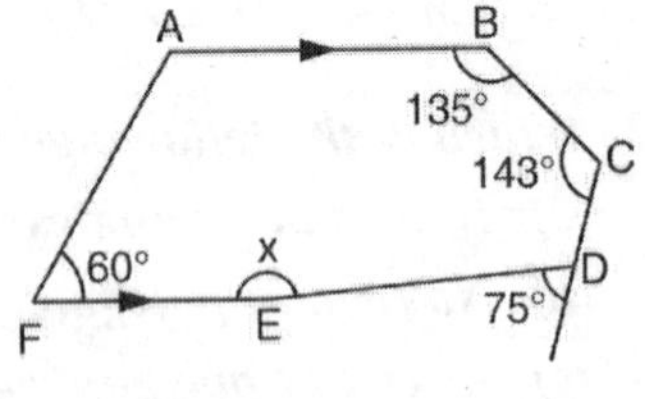

$\Rightarrow \angle BAF = 180° - 60° = 120°$

$\angle CDE = 180° - 75° = 105°$ (*Linear pair*)

Sum of the interior angles of a hexagon = $(6-2)\times 180° = 720°$

$\therefore x = \angle FED = 720° - (120° + 60° + 135° + 143° + 105°)$

$= 720° - 563° = \mathbf{157°}$.

Ex. 6. ***ABCD is a square with centre O. If X is on the side CD such that DX = DO, find the ratio $\angle DOX : \angle XOC$***

Sol. In square *ABCD, DO* (diagonal *DB*) bisects $\angle D$.

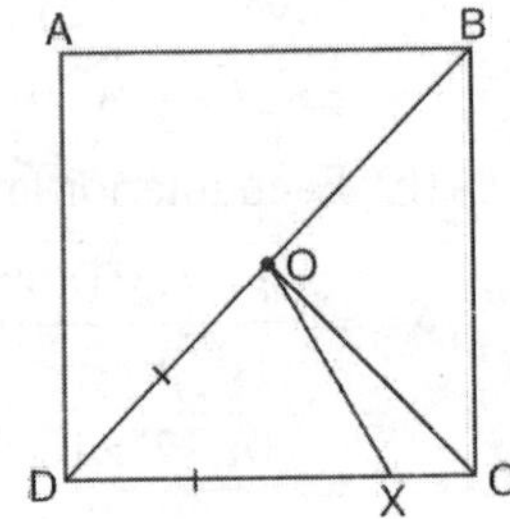

$\Rightarrow \angle ODC = 45°$

$DO = DX \Rightarrow \angle DXO = \angle DOX$ (*Angles opp. equal sides are equal*)

$\therefore$ In ΔDOX, $\angle DXO = \angle DOX = \frac{1}{2}(180° - 45°) = \frac{1}{2}\times 135° = 67.5°$

Since the diagonals of a square bisect each other at right angles,

$\angle DOC = 90°$

$\therefore \angle XOC = \angle DOC - \angle DOX = 90° - 67.5° = 22.5°$

$\therefore$ Required ratio $= \dfrac{\angle DOX}{\angle XOC} = \dfrac{67.5°}{22.5°} = \dfrac{3}{1} = \mathbf{3:1}$.

Question Bank–19

1. The sum of the interior angles of a polygon is 1620°. The number of sides of the polygon are:
   (a) 9 (b) 11
   (c) 15 (d) 12

2. The sum of the interior angles of a 12-sided regular polygon is equal to:
   (a) 180° (b) 360°
   (c) 1800° (d) 2160°

3. How many sides does a regular polygon have with its interior angle as eight times its exterior angle.
(a) 16 (b) 24
(c) 18 (d) 20

4. If s denotes the sum of all the interior angles of a polygon of n sides, then the number of right angles in s is :
(a) $(n-1)$ (b) $2(n-1)$
(c) $2(n-2)$ (d) $2n$

5. Find the sum of the degree measures of the internal angles in the polygon shown below:
(a) 600° (b) 720°
(c) 900° (d) 1080°

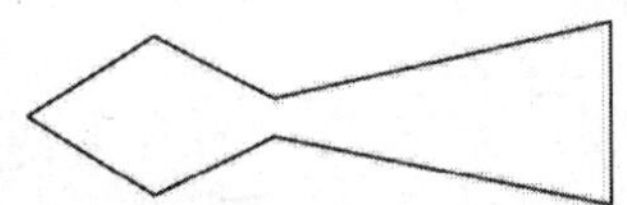

6. If the exterior angle of a regular polygon is equal to the acute angle of a right-angled isosceles triangle, then the polygon is a regular
(a) pentagon (b) hexagon
(c) octagon (d) heptagon

7. The following figure shows a polygon with all its exterior angles. The value of x is :

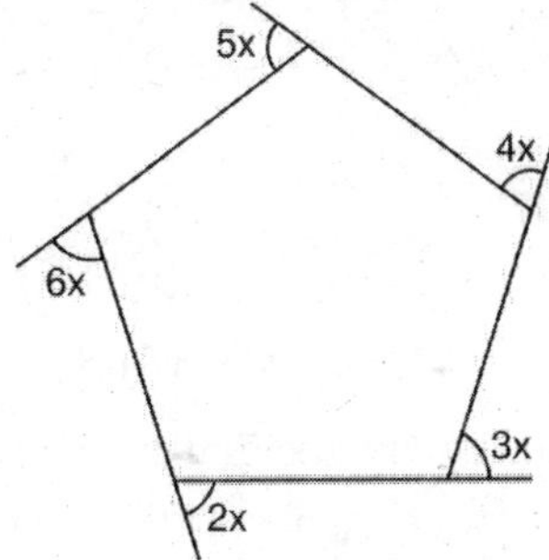

(a) 10° (b) 18°
(c) 20° (d) 36°

8. In the figure given below, PTU is a straight line. What is the value of x ?

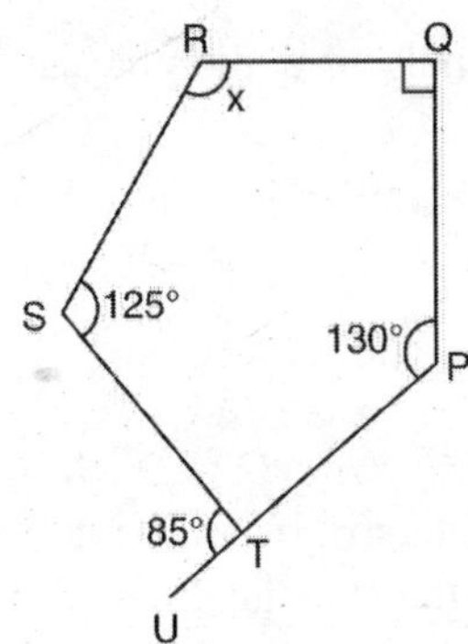

(a) 100° (b) 110°
(c) 120° (d) 130°

9. A regular hexagon, a square and an equilateral triangle are placed edge to edge as shown. The value of S is :

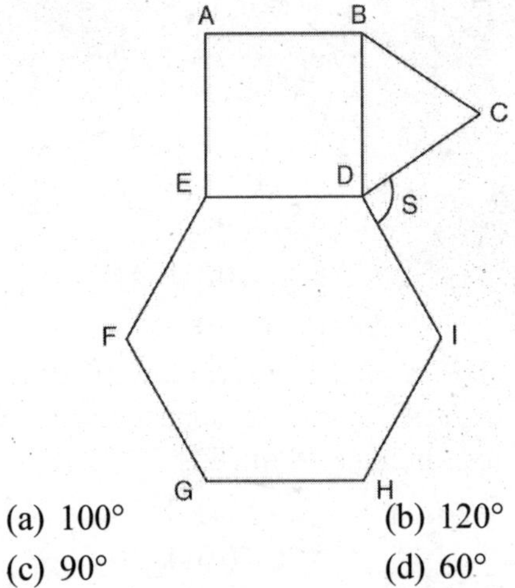

(a) 100° (b) 120°
(c) 90° (d) 60°

10. This is a regular hexagon. The angles x, y and z are respectively:

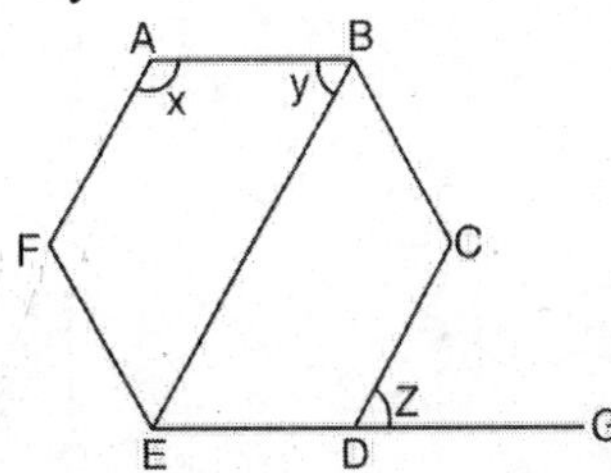

(a) 60°, 120°, 60° (b) 60°, 60°, 120°
(c) 120°, 60°, 60° (d) 60°, 60°, 100°

11. $ABCDE$ is a regular pentagon. CBF and EAF are straight lines. What kind of triangle is triangle ABF?

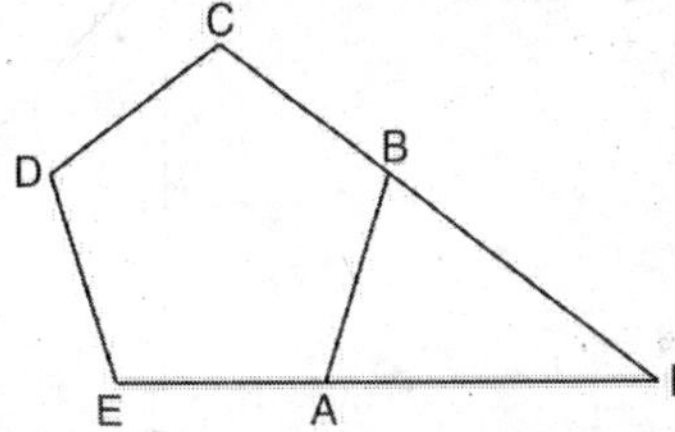

(a) Right (b) Equilateral
(c) Isosceles (d) None of these

12. The aperture of a camera is formed by 10 blades. The blades overlap to form a regular decagon. What is the measure of $\angle BAX$?

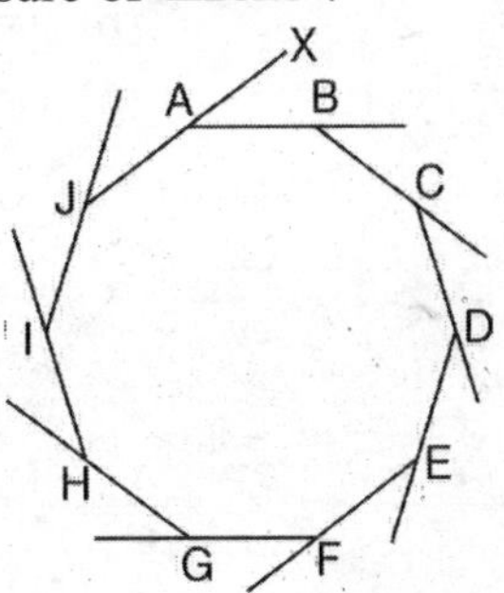

(a) 45° (b) 36°
(c) 144° (d) 44°

13. In the given figure, what is the measure of $\angle FED$?

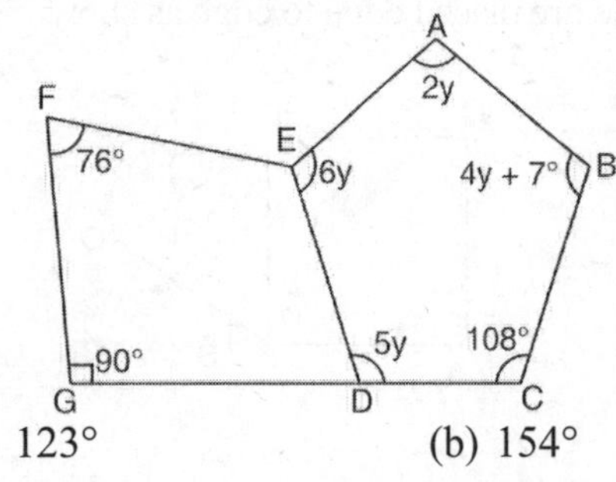

(a) 123° (b) 154°
(c) 139° (d) 85°

14. If $ABCD$ is a parallelogram with diagonals intersecting at O, then the number of distinct pairs of congruent triangles formed is

(a) 1 (b) 2
(c) 3 (d) 4

15. In the given figure, if $PQRS$ is a rectangle, then what type of a triangle is PTQ?

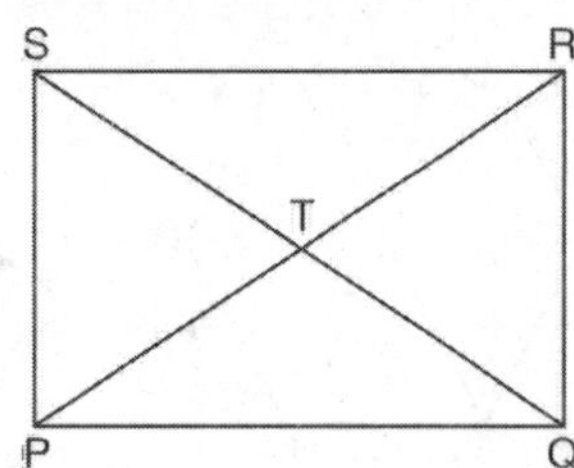

(a) scalene (b) equilateral
(c) right (d) isosceles

16. Under what conditions must $PQRS$ be a parallelogram?

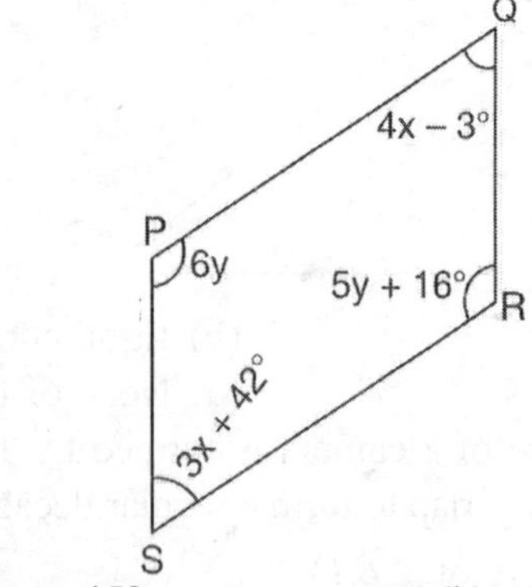

(a) $x = 45°$ (b) $y = 16°$
(c) $x = 45°, y = 16°$ (d) $x = 16°, y = 45°$

17. In an isosceles trapezium $ABCD$, $\angle C$ is equal to:

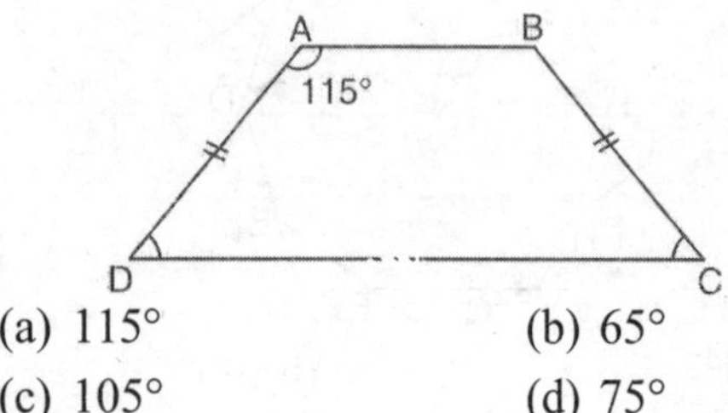

(a) 115° (b) 65°
(c) 105° (d) 75°

18. In the given diagram, $ABCD$ is a rhombus, AEC and BED are straight lines.
$p + q + r + s + t = ?$

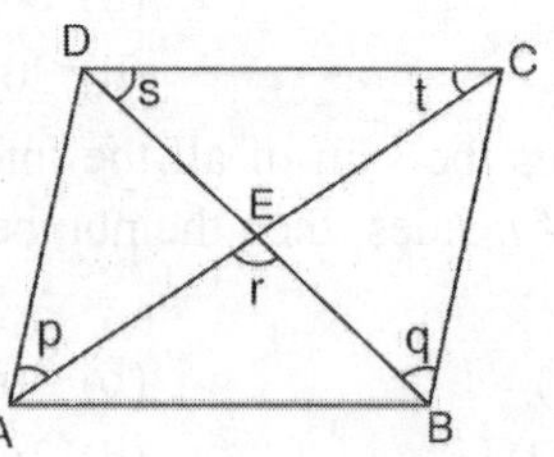

(a) 200° (b) 270°
(c) 360° (d) 540°

19. If two parallel lines are cut by two distinct transversals, then the quadrilateral formed by these four lines will always be a:

(a) parallelogram (b) rhombus
(c) square (d) trapezium

20. In ΔDEF shown below, points A, B, C are taken on DE, DF and EF respectively such that $EC = AC$ and $CF = BC, \angle D = 40°$. What is $\angle ACB$ in degrees?

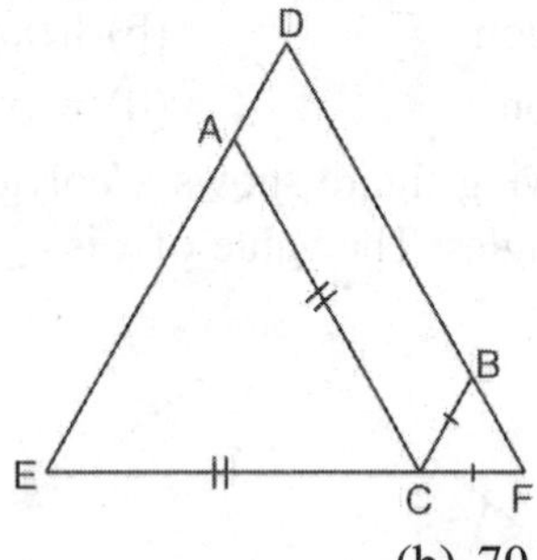

(a) 140 (b) 70
(c) 100 (d) 80

21. If the bisectors of the angles A and B of a quadrilateral $ABCD$ meet at O, then $\angle AOB$ is equal to:

(a) $\angle C + \angle D$ (b) $\frac{1}{2}(\angle C + \angle D)$
(c) $\frac{1}{2}\angle C + \frac{1}{3}\angle D$ (d) $\frac{1}{3}\angle C + \frac{1}{2}\angle D$

22. If the diagonals of a rhombus are equal, then the rhombus is a:

(a) parallelogram but not a square
(b) parallelogram but not a rectangle
(c) rectangle but not a square
(d) square

23. If two diameters of a circle intersect each other at right angles, then the quadrilateral formed by joining their end points is a :

(a) Rhombus (b) Rectangle
(c) Square (d) Parallelogram

24. What is the value of x in the given figure?

(a) $b-a-c$ (b) $b-a+c$

(c) $b+a-c$ (d) $(a+b+c)$

25. $ABCD$ is a parallelogram. If P be a point on CD such that $AP = AD$, then the measure of $\angle PAB + \angle BCD$ is:

(a) 180° (b) 225°

(c) 240° (d) 135°

26. If $ABCD$ is a parallelogram and E, F are the centroids of $\Delta s\ ABD$ and BCD respectively, then EF equals.

(a) AE (b) BE

(c) CE (d) DE

27. Given that $ABCD$ is a parallelogram whose diagonals intersect at point O. $\angle ABC = 110°$, $\angle ACB = 35°$ and $\angle ADB = 55°$. The term that best describes $ABCD$ is :

(a) Rectangle (b) Rhombus

(c) Square (d) Kite

28. If the angles A, B, C, D of a quadrilateral, taken in order are in the ratio 7 : 13 : 12 : 8, then $ABCD$ is :

(a) rhombus (b) parallelogram

(c) trapezium (d) kite

29. In the given pattern, eight isosceles trapeziums surround a regular octagon. The measure of $\angle B$ in the trapezium $ABCD$ is :

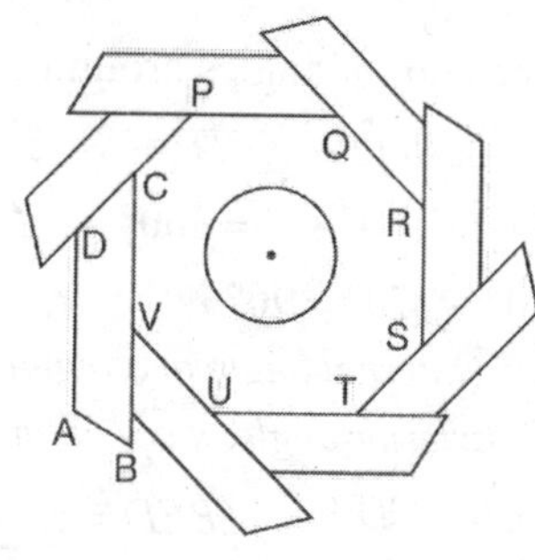

(a) 55° (b) 40°

(c) 45° (d) 60°

Answers

1. (b)	**2.** (c)	**3.** (c)	**4.** (c)	**5.** (c)	**6.** (c)	**7.** (b)	**8.** (a)	**9.** (c)	**10.** (c)
11. (c)	**12.** (b)	**13.** (c)	**14.** (d)	**15.** (d)	**16.** (c)	**17.** (b)	**18.** (b)	**19.** (d)	**20.** (c)
21. (b)	**22.** (d)	**23.** (c)	**24.** (a)	**25.** (a)	**26.** (a)	**27.** (b)	**28.** (c)	**29.** (c)	

Hints and Solutions

1. (b) Let the number of sides of the polygon be n, Then,
$(n-2) \times 180° = 1620°$

$$\Rightarrow (n-2) = \frac{1620°}{180°} = 9 \Rightarrow n = 9 + 2 = \mathbf{11}.$$

2. (c) Reqd. sum of 12-sided regular polygon
$= (12-2) \times 180° = \mathbf{1800°}$.

3. (c) For a n-sided regular polygon

Each interior angle $= \dfrac{(n-2)\times 180°}{n}$

Each exterior angle $= \dfrac{360°}{n}$

Given, $\dfrac{(n-2)\times 180°}{n} = 8 \times \dfrac{360°}{n}$

$\Rightarrow n-2 = 16 \Rightarrow n = \mathbf{18}$.

4. (c) s = Sum of the interior angles of a polygon of n sides
$= (n-2)\times 180° = 2(n-2) \times 90°$
$= 2(n-2)$ right angles.

5. (c) The figure shown is a heptagon (polygon with 7 sides)

$\therefore$ Sum of the degree measures of the internal angles of a heptagon
$= (7-2)\times 180° = 5 \times 180° = \mathbf{900°}$.

6. (c) The acute angle of a right angled isosceles $\Delta = 45°$

$\therefore$ Each exterior angle of the regular polygon = 45°
Hence, number of sides of the regular polygon
$= \dfrac{360°}{45°} = 8$

Therefore, the given polygon is a regular octagon.

7. (b) Sum of the exterior angles of a polygon =360°

$\Rightarrow 5x + 4x + 3x + 2x + 6x = 360°$

$\Rightarrow 20x = 360° \Rightarrow x = \mathbf{18°}$.

8. (a) $PQRST$ is a pentagon. Therefore, sum of the internal angles of $PQRST = (5-2)\ 180° = 540°$

$\angle PTS = 180° - 85° = 95°$ (*Linear pair*)

$\therefore\ x + 125° + 95° + 130° + 90° = 540°$

$\Rightarrow\ x + 440° = 540°$

$\Rightarrow\ x = 540° - 440° = \mathbf{100°}$.

9. (c) Each angle of a square = 90° $\Rightarrow\ \angle EDB = 90°$

Each angle of an equilateral $\Delta = 60°$

$\Rightarrow\ \angle BDC = 60°$

Each internal angle of a regular hexagon

$$= \frac{(6-2)\times 180°}{6} = 120°$$

$\Rightarrow\ \angle EDI = 120°$

Since the sum of angles around a point = 360°

$\angle EDB + \angle BDC + \angle EDI + S = 360°$

$\Rightarrow\ 90° + 60° + 120° + S = 360°$

$\Rightarrow\ S = 360° - 270° = \mathbf{90°}$.

10. (c) $x = 120°$ (*Internal angle of a regular hexagon*)

$z = 60°$ (*Exterior angle of a regular hexagon*)

Diag. $EB \parallel CD \Rightarrow \angle BED = z = 60°$ (*corr.* $\angle s$)

$\therefore\ y = \angle BED = \mathbf{60°}$. ($AB \parallel ED, alt. \angle s$)

11. (c) Each internal angle of a regular pentagon

$$= \frac{(5-2)\times 180°}{5} = 108°$$

$\therefore\ \angle CBA = \angle BAE = 108°$

$\Rightarrow\ \angle ABF = 180° - \angle CBA = 180° - 108° = 72°$

$\Rightarrow\ \angle BAF = 180° - \angle BAE = 180° - 108° = 72°$

$\therefore$ In $\Delta\ ABF$, $\angle ABF = \angle BAF$

$\Rightarrow\ AF = BF$ (*sides opp. equal angles are equal*)

$\Rightarrow\ \Delta\ ABF$ is an isosceles Δ.

12. (b) $\angle BAX$ is an exterior angle of the decagon

$$\therefore\ \angle BAX = \frac{360°}{10} = \mathbf{36°}.$$

13. (c) *ABCDE* being a regular pentagon,

$\angle A + \angle B + \angle C + \angle CDE + \angle DEA = 540°$

$\Rightarrow\ 2y + 4y + 7° + 108° + 5y + 6y = 540°$

$\Rightarrow\ 17y + 115° = 540° \Rightarrow 17y = 540° - 115° = 425°$

$\Rightarrow\ y = 25°$

$\therefore\ \angle CDE = 5 \times 25° = 125°$

$\Rightarrow\ \angle EDG = 180° - 125° = 55°$

In quad. *FEDG*,

$\angle FED = 360° - (\angle EFG + \angle FGD + \angle EDG)$

$= 360° - (76° + 90° + 55°)$

$= 360° - 221° = \mathbf{139°}$.

14. (d) $\Delta\ AOB \cong \Delta COD$,

$\Delta\ BOC \cong \Delta\ AOD$,

$\Delta\ ABC \cong \Delta\ CDA$,

$\Delta\ CAB \cong \Delta\ BCD$

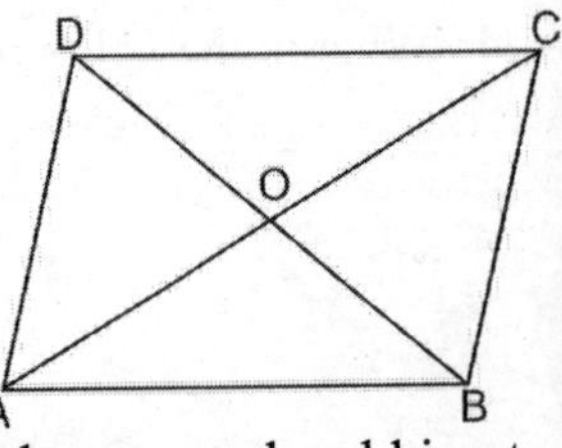

15. (d) Diagonals of a rectangle are equal and bisect each other.

$$\therefore\ PR = SQ \Rightarrow \frac{1}{2}PR = \frac{1}{2}SQ \Rightarrow PT = TQ$$

$\therefore\ \Delta\ PTQ$ is an isosceles Δ.

16. (c) Opposite angles of a parallelogram are equal

$\Rightarrow\ \angle S = \angle Q$ and $\angle P = \angle R$

$\Rightarrow\ 3x + 42° = 4x - 3°$ and $6y = 5y + 16°$

$\Rightarrow\ x = 45°$ and $y = 16°$.

17. (b) $\angle D = 180° - 115° = 65°$

(*AB* || *DC, co-int.* $\angle$*s are supplementary*)

$\angle C = \angle D = \mathbf{65°}$

(*Angles on the parallel sides are equal in an isoscele trapezium*)

18. (b) In the rhombus *ABCD*,

$\angle A + \angle B + \angle C + \angle D = 360°$

$$\Rightarrow\ \frac{\angle A}{2} + \frac{\angle B}{2} + \frac{\angle C}{2} + \frac{\angle D}{2} = \frac{360°}{2} = 180°$$

$\Rightarrow\ p + q + t + s = 180°$

($\because$ *Diagonals AC and BD of rhombus ABCD bise the angles A, C and B, D respectively*)

Also, diagonals of a rhombus are perpendicula to each other so $r = 90°$

$\therefore\ p + q + r + s + t = 180° + 90° = \mathbf{270°}$.

19. (d) Let p and q be the two parallel lines cut by tw distinct transversals l and m. The figure forme by the lines *AB, BC, CD, DA* will always be trapezium as at least one pair of given lines wi be parallel.

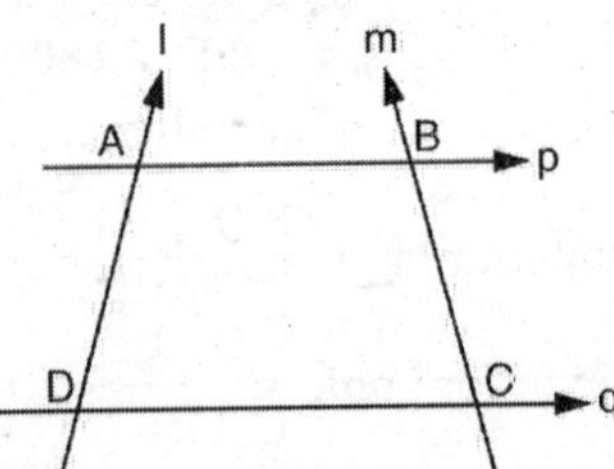

20. (c) In ΔDEF,

$\angle EDF + \angle DEF + \angle DFE = 180°$

$\Rightarrow\ \angle DEF + \angle DFE = 180° - \angle EDF$

$\Rightarrow\ \angle DEF + \angle DFE = 180° - 40° = 140°$ ...

In ΔBCF, $CB = CF \Rightarrow \angle CFB = \angle CBF$

(*Isos.* Δ *prop.*)

or $\angle EFD = \angle CBF$ ...(ii)

In $\Delta AEC, CA = CE \Rightarrow \angle CEA = \angle CAE$ (*Isos. Δ prop.*)...(iii)

or $\angle DEF = \angle CAE$

$\therefore$ From (i), (ii) and (iii)

$\angle CAE + \angle CBF = 140°$

$180° - \angle CAD + 180° - \angle CBD = 140°$

$\Rightarrow \angle CAD + \angle CBD = 360° - 140° = 220°$

In quad. *DACB*,

$\angle ADB + \angle CAD + \angle CBD + \angle ACB = 360°$

$\Rightarrow 40° + 220° + \angle ACB = 360°$

$\Rightarrow \angle ACB = 360° - 260° = \mathbf{100°}$.

21. (b) In quad. *ABCD*

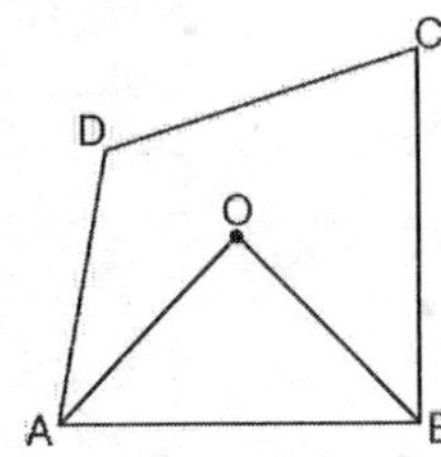

$\angle A + \angle B + \angle C + \angle D = 360°$

$\Rightarrow \angle A + \angle B = 360° - (\angle C + \angle D)$

$\Rightarrow \frac{\angle A}{2} + \frac{\angle B}{2} = 180° - \frac{1}{2}(\angle C + \angle D)$

$\Rightarrow \angle OAB + \angle OBA = 180° - \frac{1}{2}(\angle C + \angle D)$

($\therefore$ *OA* bisects $\angle A$, *OB* bisects $\angle B$)

In ΔAOB, $\angle AOB = 180° - (\angle OAB + \angle OBA)$

$= 180° - \left\{(180° - \frac{1}{2}(\angle C + \angle D)\right\}$

$= \frac{1}{2}(\angle C + \angle D)$

23. (c) Let *AB* and *CD* be the diagonals of a circle such that $AB \perp CD$.

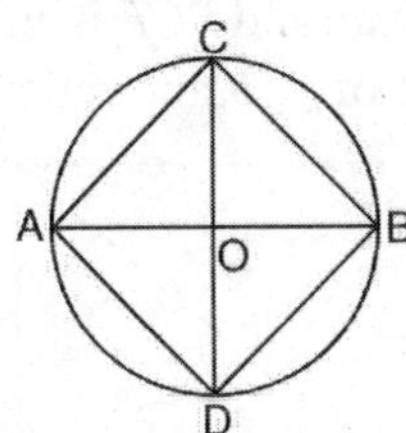

Joining points *A*, *C*, *B*, *D* in order we see that *AB* and *CD* are the equal diagonals of quad. *ACBD* which intersect at right angle.

$\therefore$ *ACBD* is a square.

24. (a) Reflex $\angle BCD = 360° - b$

In quad. *ABCD*,

$\angle A + \angle B + \text{reflex } \angle C + \angle D = 360°$

$\Rightarrow a + c + 360° - b + x = 360°$

$\Rightarrow x = 360° - 360° - a - c + b$

$= b - a - c.$

25. (a) In ΔADP,

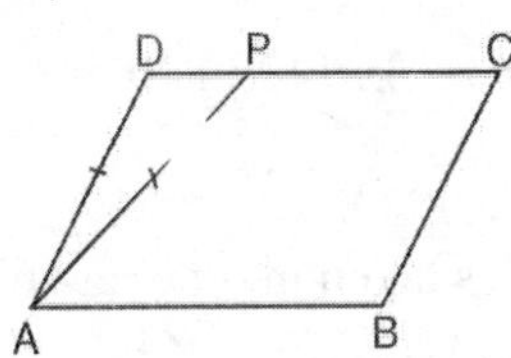

$AD = AP \Rightarrow \angle APD = \angle ADP$

Also, $\angle ADP + \angle APD + \angle PAD = 180°$

$\Rightarrow 2\angle ADP = 180° - \angle PAD$ ($\because \angle APD = \angle ADP$)...(i)

In quad. *ABCD*,

$\angle A + \angle B + \angle C + \angle D = 360°$

$\angle PAD + \angle PAB + 2\angle D + \angle C = 360°$ ($\because \angle B = \angle D$)

$\angle PAD + \angle PAB + 180° - \angle PAD + \angle C = 360°$ (From (i))

$\Rightarrow \angle PAB + \angle BCD = 360° - 180° = 180°$.

26. (a) Given, *E* and *F* are the centroids of ΔABD and ΔBCD.

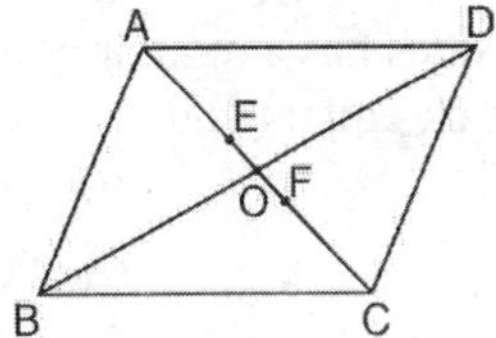

So, $\frac{AE}{EO} = \frac{CF}{FO} = \frac{2}{1}$, *i.e.*,

$AE = CF = 2x$, if $EO = FO = x$

$\Rightarrow EF = EO + FO = x + x = 2x$

$\therefore AE = EF = CF$

27. (b) $AD \| BC$, *DB* is the transversal, therefore, $\angle CBD = \angle ADB = 55°$ (*alt.* $\angle s$)

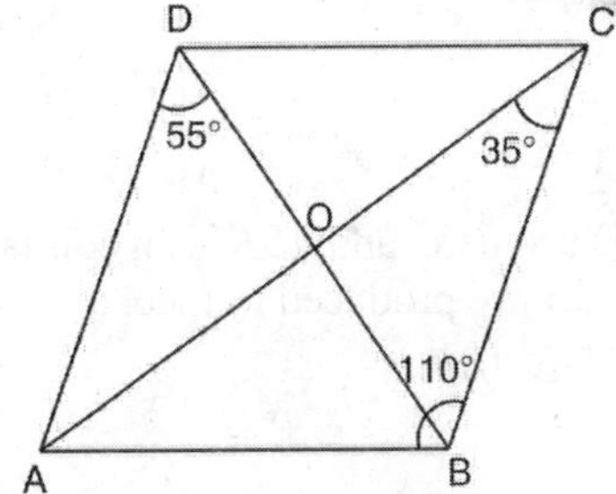

In ΔBOC,

$\angle BOC = 180° - (\angle OCB + \angle OBC)$

$= 180° - (35° + 55°) = 180° - 90° = 90°$

$\Rightarrow BD \perp AC$

Also in $\Delta ABC, \angle BAC = 180° - (110° + 35°)$

$= 180° - 145° = 35°$

Hence, $\angle ACB = \angle BAC \Rightarrow AB = BC$.

$\therefore$ *ABCD* is a parallelogram whose adjacent sides are equal and diagonals are perpendicular to each other

$\Rightarrow$ *ABCD* is a rhombus.

28. (c) Let the angles be $7x$, $13x$, $12x$ and $8x$

Then, $7x + 13x + 12x + 8x = 360°$

$\Rightarrow 40x = 360° \Rightarrow x = 9°$

$\therefore$ The angles taken in order are 63°,117°,108°,72°

This shows that two pairs of adjacent angles are supplementary (63° + 117° = 108° and 108° + 72° = 180°), but opposite angles are not equal.

Therefore, the given quadrilateral will be a trapezium.

29. (c) $\angle PCV = 135°$

(*Internal angle of a regular octagon*)

$\angle DCV = 180° - \angle PCV = 180° - 135° = 45°$

(*PCD is a straight line*)

$\therefore \angle B = \angle DCV = \mathbf{45°}$. (*Prop. of isos. trapezium*)

Self Assessment Sheet–19

1. How many sides has a polygon the sum of whose interior angles is 1980°.

(a) 10 (b) 13
(c) 11 (d) None of these

2. Find the value of each angle of a regular polygon of 15 sides

(a) 130° (b) 156°
(c) 146° (d) 160°

3. How many sides has a regular polygon each interior angle of which equals 160°.

(a) 18 (b) 16
(c) 15 (d) 20

4. Find the angles of a parallelogram if one angle is three times another.

(a) 45°, 135°, 45°, 135°
(b) 50°, 130°, 50°, 130°
(c) 40°, 140°, 40°, 140°
(d) 55°, 125°, 55°, 125°

5. The ratio of the measure of an exterior angle of a regular nonagon to the measure of one of its interior angles is :

(a) 7 : 2 (b) 2 : 7
(c) 4 : 3 (d) 3 : 4

6. *ABCD* is a square, and *ABE* is an equilateral triangle. *AC* and *EB* are produced to meet at *F*. Calculate the angles of ΔABF.

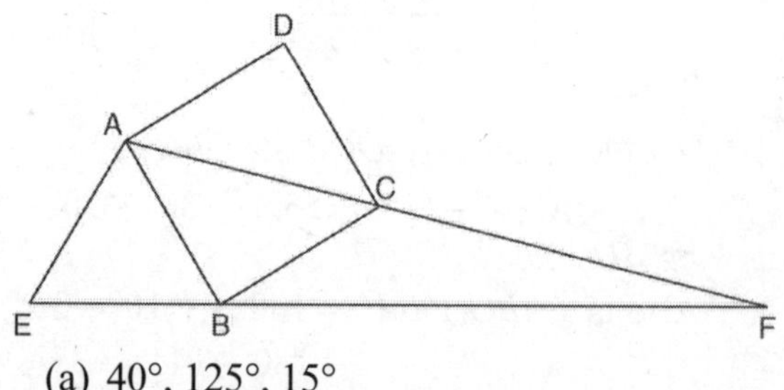

(a) 40°, 125°, 15°
(b) 50°, 115°, 15°
(c) 45°, 120°, 15°
(d) 35°, 130°, 15°

7. Find x

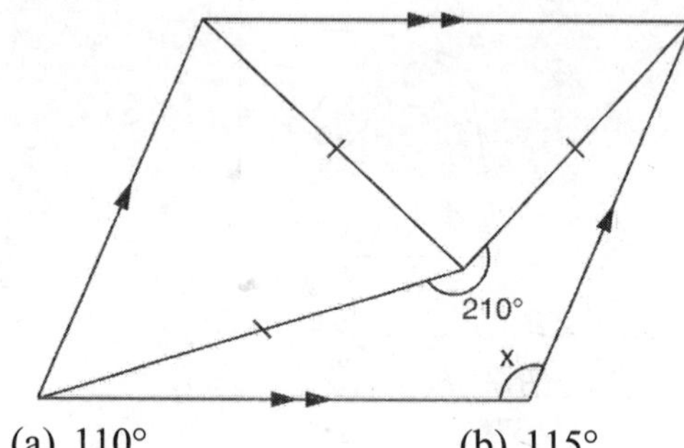

(a) 110° (b) 115°
(c) 100° (d) 105°

8. The equilateral triangle *ABP* lies inside the square *ABCD*. Find $\angle CPD$.

(a) 150° (b) 145°
(c) 155° (d) 160°

9. How many sides does a polygon have if the sum of the measures of its internal angles is five times as large as the sum of the measures of its exterior angles ?

(a) 20 (b) 12
(c) 15 (d) 10

10. *ABCD* is a square and *BCE* is an equilateral triangle. Find the value of *a*, *b* and reflex $\angle BED$.

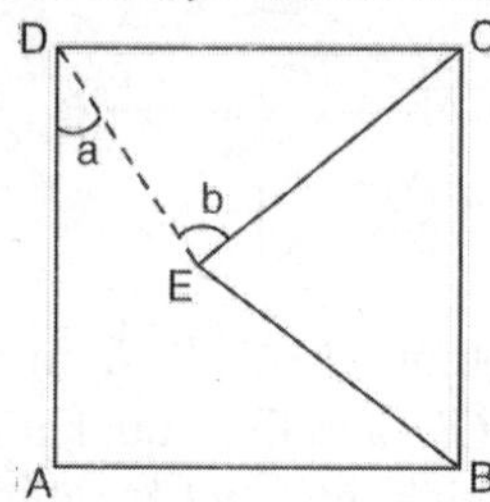

(a) 20°, 70°, 230° (b) 18°, 72°, 210°
(c) 15°, 75°, 225° (d) 25°, 65°, 210°

Answers

1. (b)	2. (b)	3. (a)	4. (a)	5. (b)	6. (c)	7. (d)	8. (a)	9. (b)	10. (c)

CIRCLES

1. A **circle** is a simple closed curve all of whose points are at a constant distance from a fixed point in the same plane. The fixed point is called the **centre** of the circle.
2. **Radius:** A line segment joining the centre of the circle with any point on it is called the radius (Plural : radii) of the circle. All radii of a circle are equal.
3. **Diameter:** A line segment which passes through the centre of a circle and has the end points on the boundary of the circle is called the **diameter** of the circle.

$$\boxed{\text{Diameter} = 2 \times \text{radius}}$$

4. **Chord:** A line segment joining any two points on a circle is called a **chord** of the circle.The diameter is the longest chord.

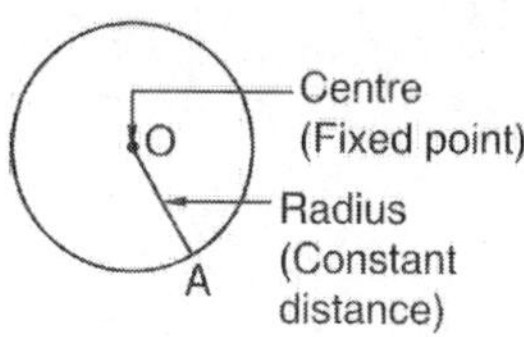

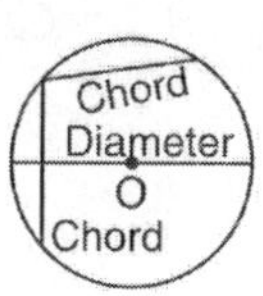

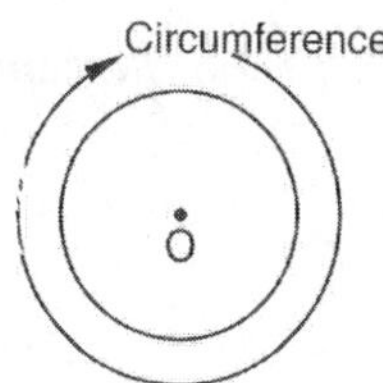

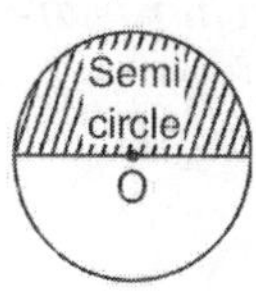

5. **Circumference:** The distance around the circle is called its **circumference.**
6. **Semi-circle:** A diameter of a circle divides the circle into two equal parts. Each part is called a semi-circle.
7. **Arc:** An arc is a part of a circle included between two points on a circle. In the figure ***APB*** and ***AQB*** are arcs, They are denoted by $\overset{\frown}{APB}$ and $\overset{\frown}{AQB} \cdot \overset{\frown}{APB}$ is less than a semi circle, and is called a **minor arc**. $\overset{\frown}{AQB}$ is greater than a semi-circle, and is called a **major arc.**

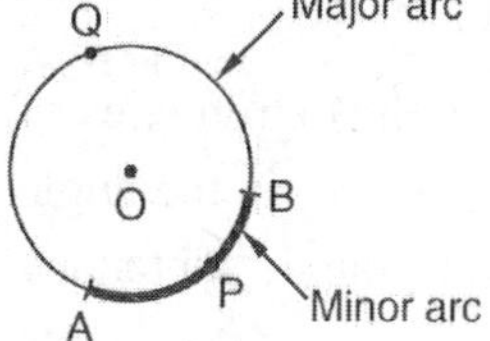

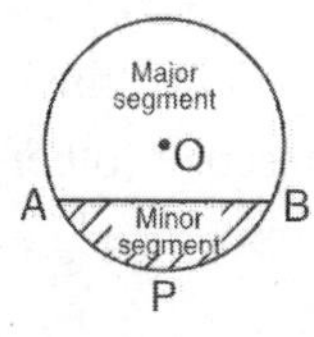

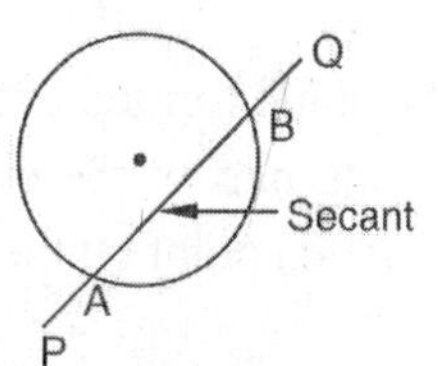

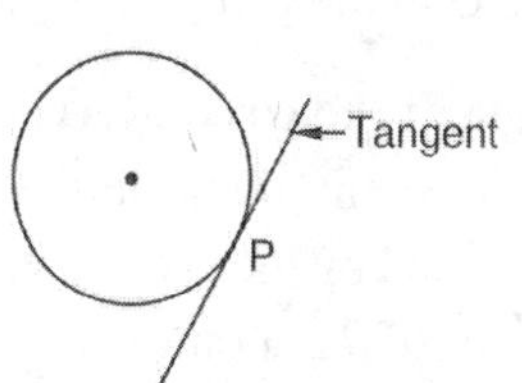

8. **Segments:** A chord *AB* of a circle divides the circular region into two parts. Each part is called a segment of the circle.The bigger part containing the centre of the circle is called the **major segment** and the smaller part which does not contain the centre is called the **minor segment** of the circle.

9. **Secant:** A line which intersects the circle at two distinct points is called a **secant.** PQ is a secant intersecting the circle at points A and B.

10. **Tangent:** A line touching a circle at a point is called a tangent.

Some important properties of a circle:

(i) ***The angle in a semi circle is a right angle.*** E.g., $\angle ACB = 90^\circ$.

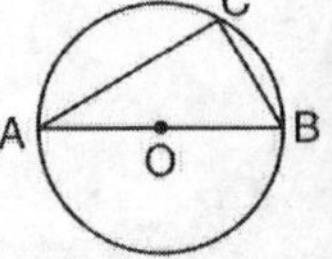

(ii) ***The line joining the centre of the circle and the mid-point of a chord (not passing through the chord) is perpendicular to the chord.***

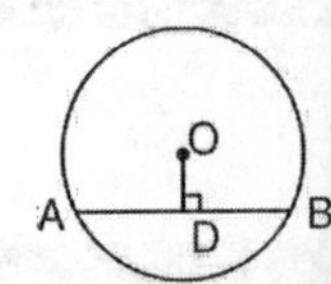

D is the mid-point of $AB \Rightarrow OD \perp AB$.

(iii) ***The perpendicular from the centre to a chord bisects the chord.***

$OC \perp AB \Rightarrow AC = CB$.

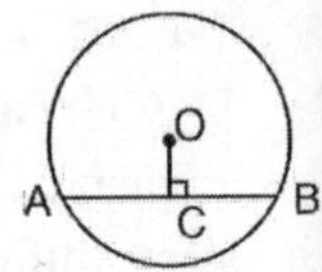

(iv) ***The perpendicular bisectors of two chords of a circle intersect at its centre.***

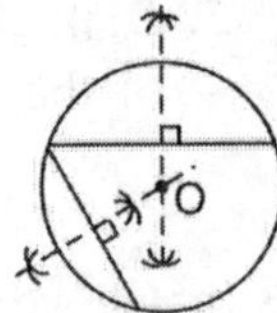

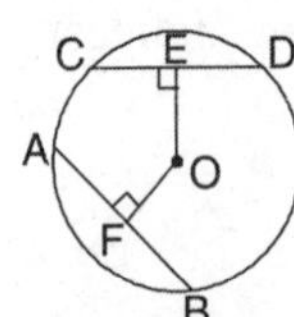

(v) ***Equal chords are equidistant from the centre*** E.g., $AB = CD \Rightarrow OE = OF$

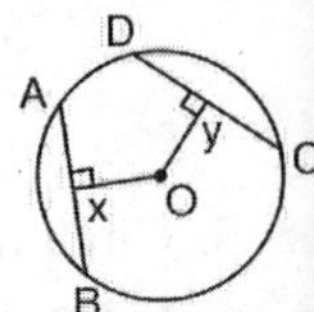

(vi) ***Chords equidistant from the centre are equal*** E.g., $OX = OY \Rightarrow AB = CD$.

(vii) There can be one and only one circle passing through three non collinear points.

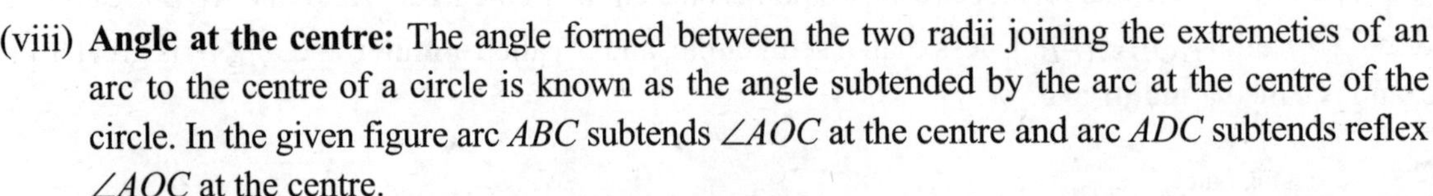

(viii) **Angle at the centre:** The angle formed between the two radii joining the extremeties of an arc to the centre of a circle is known as the angle subtended by the arc at the centre of the circle. In the given figure arc ABC subtends $\angle AOC$ at the centre and arc ADC subtends reflex $\angle AOC$ at the centre.

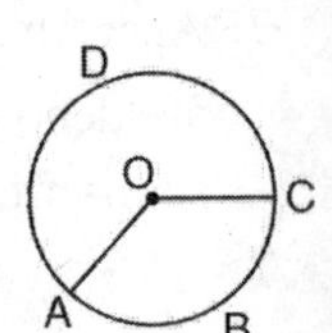

Angle at the circumference: The angle formed between the two lines joining the extremeties of an arc to any point on the remaining part of the circumference (other than the arc) is the angle subtended by the given arc, at the given point on the circumference. $\angle ADB$ is the angle subtended by arc ACB at point D.

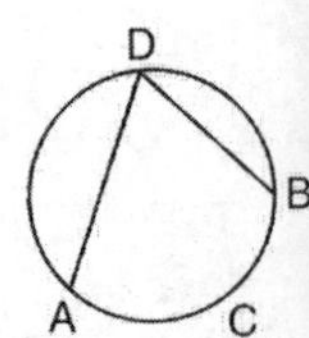

The given figure shows four angles all subtended by arc ACB on the remaining part of the circumference at four different points.

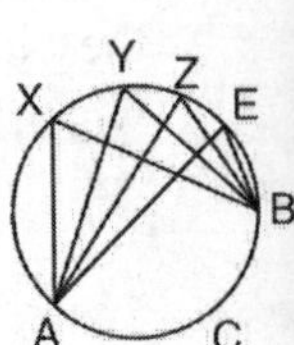

Such angles are said to be angles in the same segment. Thus, $\angle AXB$, $\angle AYB$, $\angle AZB$, $\angle AEB$ are all angles subtended by arc ACB at points X, Y, Z and E.

(ix) ***The angle subtended at the centre by an arc of a circle is double the angle which this arc subtends at any remaining part of the circumference.***

E.g., in Fig (i) and (ii)

$\angle AOB = 2\angle ACB$

In Fig (iii)

Reflex $\angle AOB = 2\ \angle ACB$

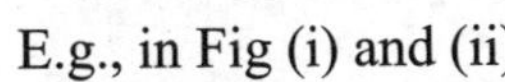

(x) ***Angles in the same segment are equal.***

E.g., $\angle APB = \angle AQB = \angle ARB$

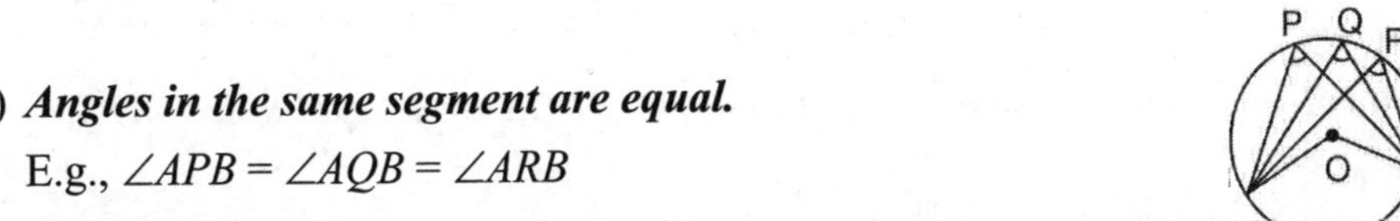

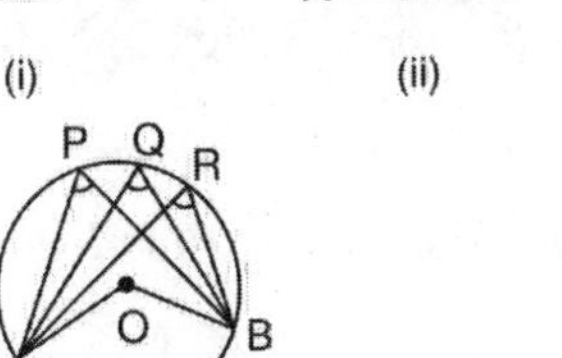

(xi) **Cyclic quadrilateral:** If the vertices of a quadrilateral lie on a circle, it is called a cyclic quadrilateral. $ABCD$ is a cyclic quadrilateral.

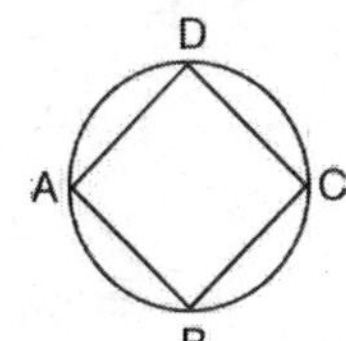

(xii) ***The opposite angles of a cyclic quadrilateral are supplementary,***

i.e., $\angle A + \angle C = 180^\circ$, $\angle B + \angle D = 180^\circ$.

(xiii) ***If a side of a cyclic quadrialteral is produced, the exterior angle so formed is equal to the interior opposite angle.***

E.g., $\angle ABE = \angle ADC$

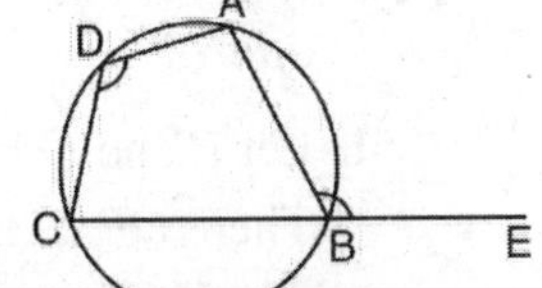

Solved Examples

Ex. 1. ***In the given circle with diameter AB find the value of x.***

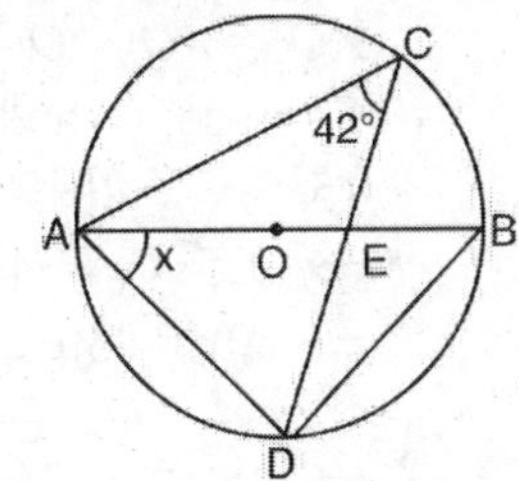

Sol. $\angle ADB = 90^\circ$ *(Angle in a semi circle is a right angle)*

Also $\angle ABD = \angle ACD = 42^\circ$ *(Angles in the same segment)*

$\therefore$ In $\Delta\ ADB$,

$x = \angle BAD = 180^\circ - (\angle ADB + \angle ABD)$

$= 180^\circ - (90^\circ + 42^\circ)$ *(Angle sum prop. of a Δ)*

$= 180^\circ - 132^\circ =$ **48°**.

Ex. 2. ***O is the centre of the circle. Find the values of p, q and r.***

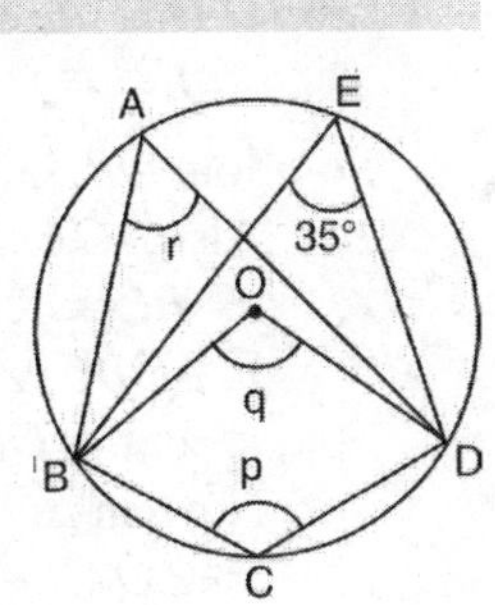

Sol. $q = \angle BOD = 2 \times \angle BED = 2 \times 35^\circ =$ **70°**.

(Angle at the centre = 2 × Angle at the remaining part of the cicumference)

$r = \angle BAD = \angle BED =$ **35°**. *(Angles in the same segment are equal)*

In cyclic quad. $BEDC$, $\angle BCD + \angle BED = 180^\circ$.

(opp. ∠s of a cyclic quad. are supplementary)

$\Rightarrow p + 35^\circ = 180^\circ$

$\Rightarrow p = 180^\circ - 35^\circ =$ **145°**.

Ex. 3. ***If the length of a chord of a circle is equal to its radius, then find the angle subtended by it at the minor arc of the circle.***

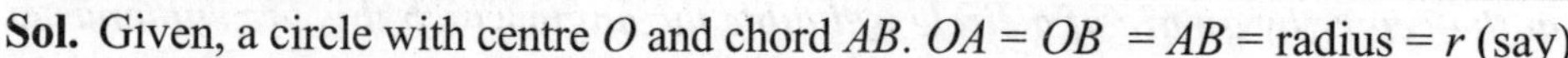

Sol. Given, a circle with centre O and chord AB. $OA = OB = AB$ = radius = r (say)

$\angle ACB$ is the angle subtended by chord AB at the minor arc.

ΔAOB is clearly an equilateral triangle

$\Rightarrow \angle AOB = 60°$.

$\therefore$ Reflex $\angle AOB = 360° - 60° = 300°$

$\therefore \angle ACB = \frac{1}{2} \times$ Reflex $\angle AOB = \frac{1}{2} \times 300° = \mathbf{150°}$.

(Angle at the centre = 2 × Angle at remaining part of the circumference)

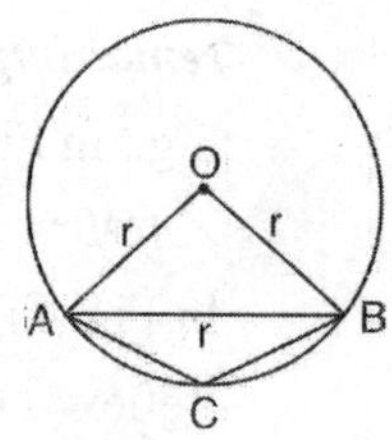

Ex. 4. ***PQ is the diameter of the given circle, whose centre is O. Given, ∠ROS = 54°, calculate ∠RTS.***

Sol. $\angle RQS = \frac{1}{2} \times \angle ROS = \frac{1}{2} \times 54° = 27° \Rightarrow \angle RQT = 27°$

(∠s RQS and RQT being the same angles)

(∠s Angle at centre = 2 × Angle at remaining part of the circumference)

Now, in ΔPRQ, $\angle PRQ = 90°$ *(Angle in a semi circle)*

$\therefore \angle QRT = 90°$ *(Linear pair)*

So, in ΔRQT, $\angle RTQ = 180° - (\angle RQT + \angle QRT) = 180° - (27° + 90°)$

$= 180° - 117° = \mathbf{63°}$.

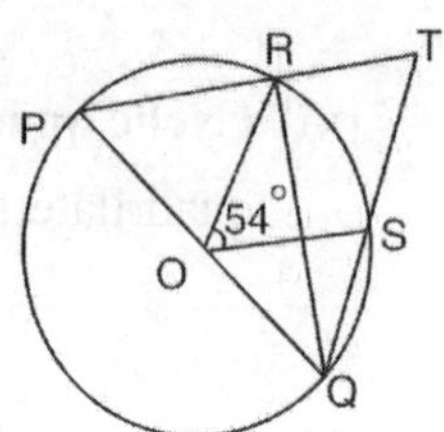

Ex. 5. ***What is the length of the common chord of two circles of radii 15 cm and 20 cm whose centres are 25 cm apart ?***

Sol. Let P and Q be the centres of the two circles respectively and AB be the common chord.

Then, $OP \perp AB$ and bisects AB, *i.e.*, $AO = OB$

Also, $OQ \perp AB$

Given, $AP = 15$ cm, $AQ = 20$ cm and $PQ = 25$ cm

Let $OP = x$ cm. Then $OQ = (25 - x)$ cm

In rt. ∠d ΔAOP, $AO^2 = AP^2 - OP^2 = 15^2 - x^2$ ...(i)

In rt. ∠d ΔAOQ,

$AO^2 = AQ^2 - OQ^2 = 20^2 - (25 - x)^2$ ...(ii)

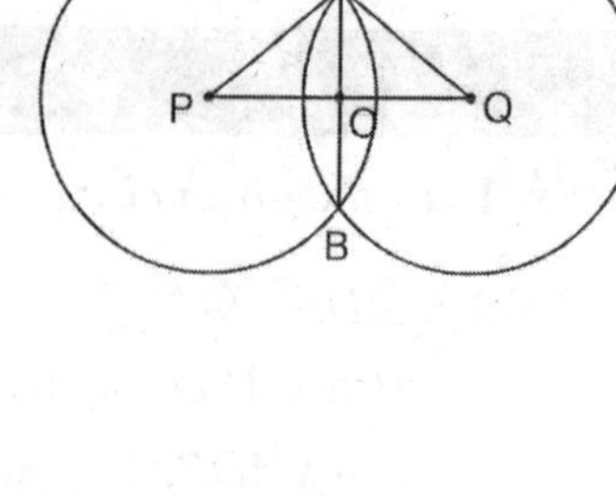

From eq. (i) and (ii)

$15^2 - x^2 = 20^2 - (25 - x)^2$

$\Rightarrow 225 - x^2 = 400 - (625 - 50x + x^2) \Rightarrow 225 - x^2 = 400 - 625 + 50x - x^2$

$\Rightarrow 50x = 450 \Rightarrow x = 9$

$\therefore AO = \sqrt{15^2 - 9^2} = \sqrt{225 - 81} = \sqrt{144} = 12$ cm (From (i))

$\Rightarrow AB = 2 \times 12$ cm = **24 cm.**

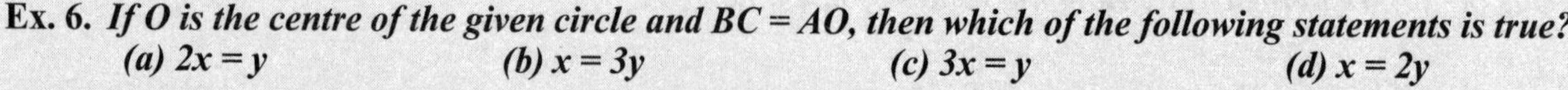

Ex. 6. ***If O is the centre of the given circle and BC = AO, then which of the following statements is true?***

(a) 2x = y ***(b) x = 3y*** ***(c) 3x = y*** ***(d) x = 2y***

Sol. Join OB. Given, $BC = OA$

Also, $OA = OB$ *(radii of same circle)*

$\therefore BC = OA = OB$

In ΔOBC, $\angle BOC = \angle BCO = y$ *(∠s equal sides are equal).*

$\therefore$ ext. $\angle OBA = \angle BOC + \angle BCO = y + y = 2y$ *(Ext. ∠ prop. of a Δ)*

Now, in ΔOAB, $OB = OA$

$\Rightarrow \angle OAB = \angle OBA = 2y$ *(isos. Δ property)*

Again for ΔAOC, applying the exterior angle property, $x = \angle OAC + \angle OCA = 2y + y = 3y$.

Option (b) is correct.

Question Bank–20

1. If the two circles C_1 and C_2 have three points in common, then which of the following is correct?

(a) C_1 and C_2 are concentric
(b) C_1 and C_2 are the same circle
(c) C_1 and C_2 have different centres
(d) None of the above

2. Which of the following pairs of lines can be parallel?

1. Two tangents to a circle.
2. Two diameters of a circle.
3. A chord of circle and a tangent to a circle.
4. Two chords of a circle.

Select the correct answer using the codes given below:

Codes:

(a) 1, 2 and 3 (b) 2, 3 and 4
(c) 1, 3 and 4 (d) 1, 2 and 4

3. Three circles with equal radii touch each other externally. The figure formed by joining the centres of these circles is

(a) an isosceles triangle
(b) an equilateral triangle
(c) a scalene triangle
(d) a right angled triangle

4. Two non-intersecting circles one lying inside another arc of diameters a and b. The minimum distance between their circumferences is c. The distance between their centres is

(a) $a-b-c$ (b) $a+b-c$
(c) $\frac{1}{2}(a-b-c)$ (d) $\frac{1}{2}(a-b)-c$

5. In the given figure, if AB is the diameter of the circle and PM the internal bisector of $\angle APB$, then the measure of angle ABM is

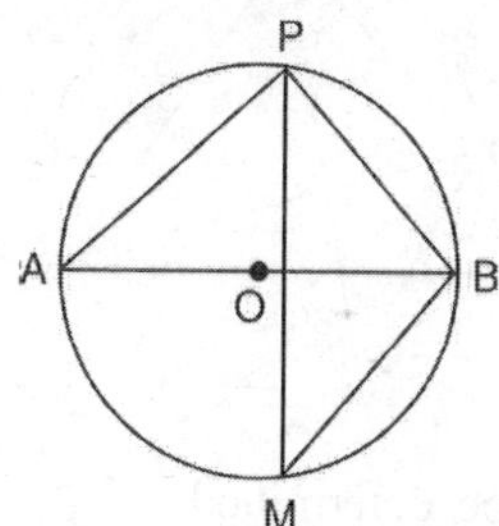

(a) 15° (b) 30°
(c) 45° (d) 60°

6. A square is inscribed in a circle with centre O. What angle does each side subtend at the centre O ?

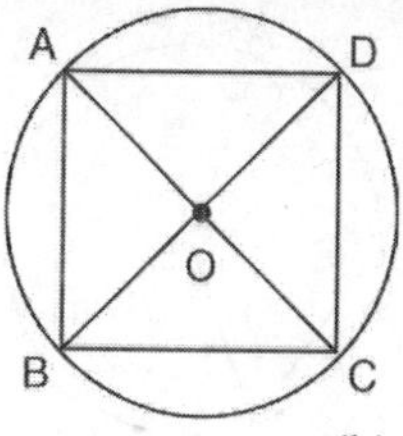

(a) 45° (b) 60°
(c) 75° (d) 90°

7. A regular polygon is inscribed in a circle. If a side subtends an angle of 45° at the centre. What is the number of sides of the polygon?

(a) 6 (b) 5
(c) 10 (d) 8

8. The length of a chord of a circle at a distance of 5 cm from the centre is 24 cm. The diameter of the circle is

(a) 26 cm (b) 24 cm
(c) 13 cm (d) 12 cm

9. In a circle of radius 25 cm, a chord is drawn at a distance of 7 cm from the centre. Find the length of the chord.

(a) 24 cm (b) 48 cm
(c) 50 cm (d) 36 cm

10. A chord 6 cm long is at a distance of 4 cm from the centre of a circle. Find the length of a chord which is at a distance of 3 cm from the centre of the circle.

(a) 10 cm (b) 6 cm
(c) 8 cm (d) 12 cm

11. In the given figure, ΔABC is inscribed in a circle and the bisector of $\angle A$ meets BC in D and the circle in E. If $\angle ECD = 30°$, what is $\angle A$?

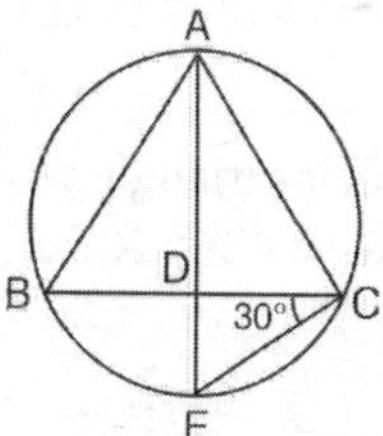

(a) 60° (b) 45°
(c) 70° (d) 150°

12. O is the centre of a circle $\angle AOB = 90°$ and $\angle BOC = 120°$. $\angle ABC$ is equal to

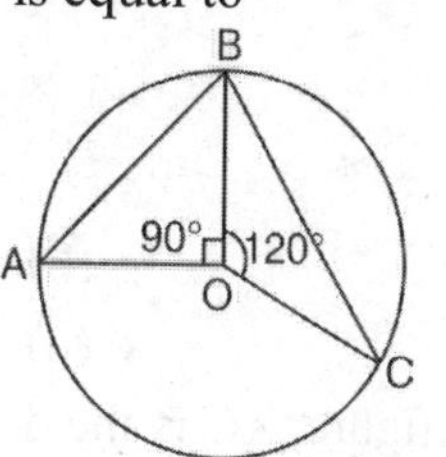

(a) 150° (b) 210°
(c) 75° (d) 105°

13. In the given figure, O is the centre of the circle. The value of x is

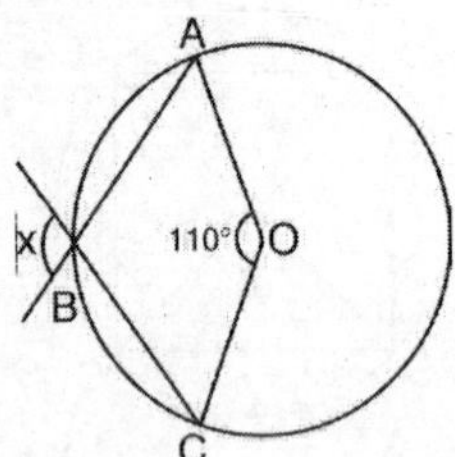

(a) 75° (b) 55°
(c) 125° (d) 110°

14. $PQRS$ is a cyclic quadrilateral. Find the measure of $\angle P$ and $\angle Q$.

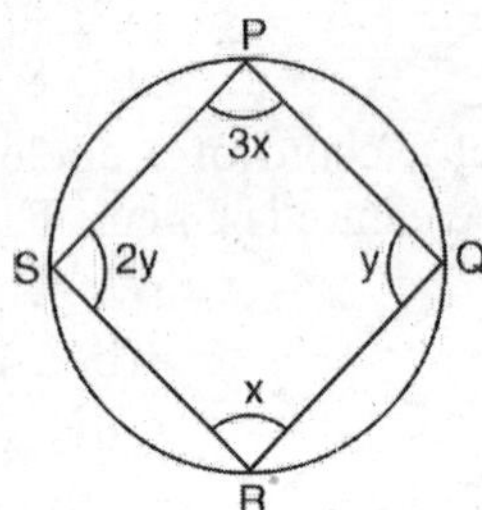

(a) 135°, 60° (b) 60°, 120°
(c) 60°, 90° (d) 100°, 120°

15. If $\angle ABO = 35°$ and $\angle ACO = 20°$, then $\angle x$ is

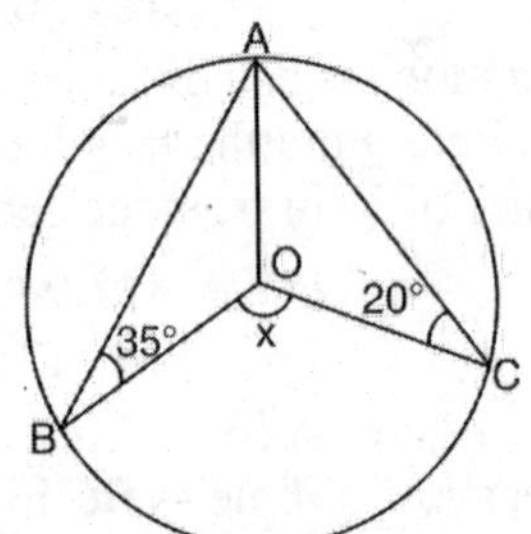

(a) 55° (b) 110°
(c) 80° (d) 70°

16. ABC is an isosceles triangle in the given circle with centre O. $\angle ABC = 42°$, $\angle CDE$ is equal to

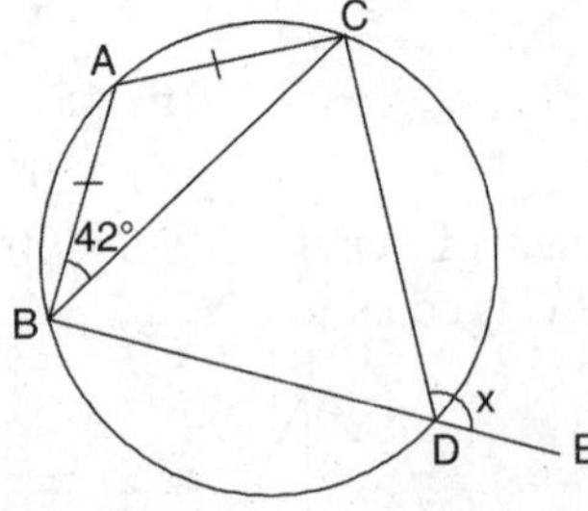

(a) 84° (b) 138°
(c) 96° (d) 148°

17. In the given figure, AC is the diameter of the circle with centre O. Chord BD is perpendicular to AC. Express p in terms of x.

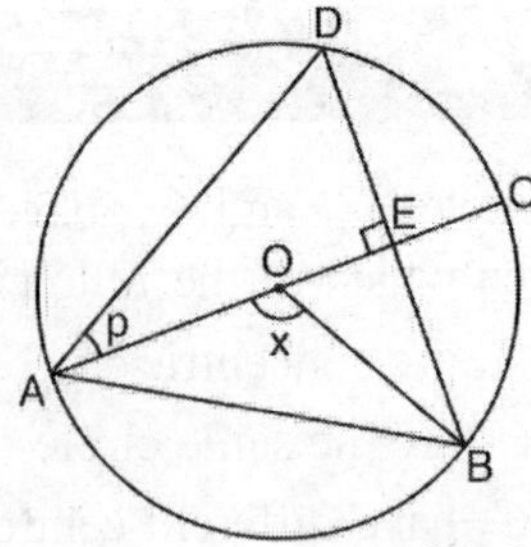

(a) $x/2$ (b) $90° + x/2$
(c) $90° - x/2$ (d) $180° - x$

18. In the given figure, AE is the diameter of the circle. Write down the numerical value of $\angle ABC + \angle CDE$.

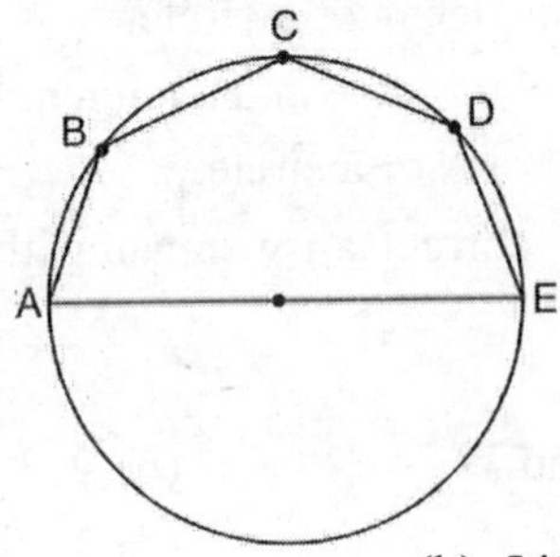

(a) 360° (b) 540°
(c) 180° (d) 270°

19. In the given figure, $\angle PAQ = 59°$, $\angle APD = 40°$, then what is $\angle AQB$?

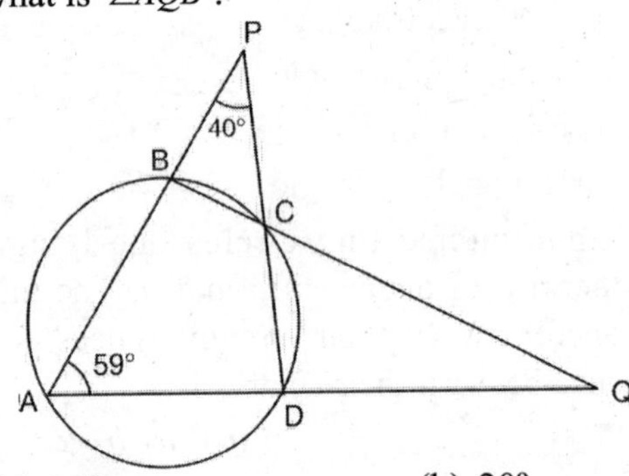

(a) 19° (b) 20°
(c) 22° (d) 27°

20. In the given figure $\angle CAB = 80°$, $\angle CBA = 55°$ and $\angle DCA = 45°$. The statement BD is the diameter is:

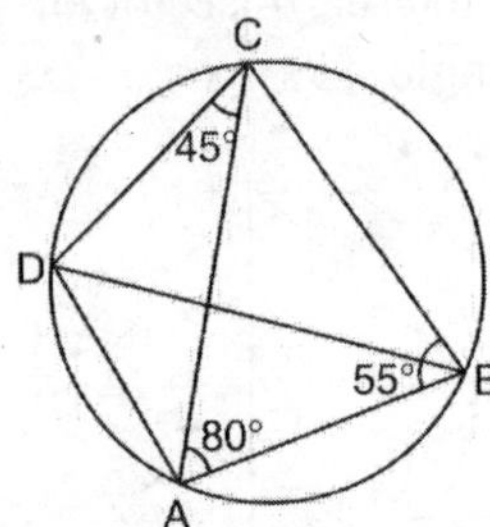

(a) False
(b) cannot be determined
(c) True
(d) Not possible

21. In the given figure, C and D are points on a semi circle described on AB as diameter. $\angle ABD = 75°$ and $\angle DAC = 35°$. What is $\angle BDC$?

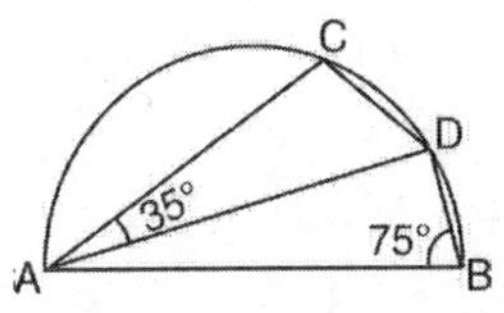

(a) 130° (b) 110°
(c) 90° (d) 100°

22. In the adjoining figure, chord *ED* is parallel to the diameter of the circle. If $\angle CBE = 65°$, then what is the value of $\angle DEC$?

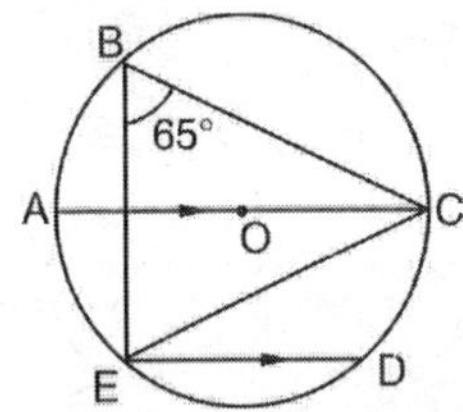

(a) 35° (b) 55°
(c) 45° (d) 25°

23. *AB* and *BC* are two equal chords of a circle with centre *O*. $OM \perp AB$ and $ON \perp BC$. *OB* is joined. State if each of the following statements is true or false. Give reasons in each case.

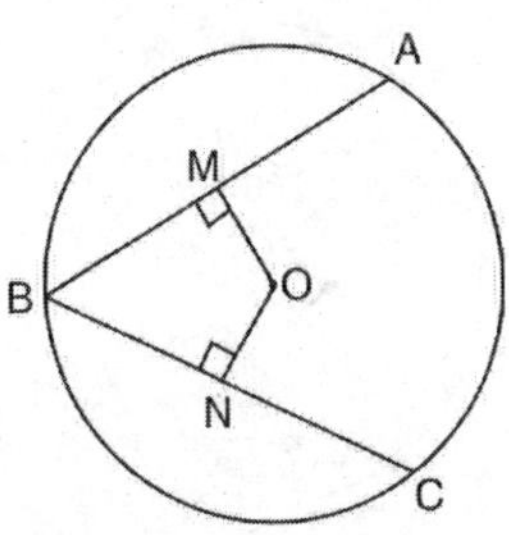

(i) $OM = ON$

(ii) $\Delta OMB \cong \Delta ONB$

(iii) *BO* bisects $\angle ABC$

24. In the given figure, find $x + y$.

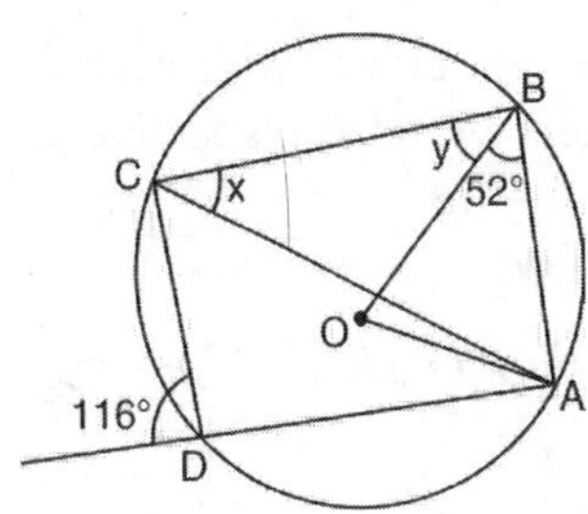

(a) 116° (b) 102°
(c) 64° (d) 76°

25. In the given figure, *O* is the centre of the circle. *ABD* is a straight line and $\angle CBD = 65°$. Find reflex $\angle AOC$ (marked $x°$).

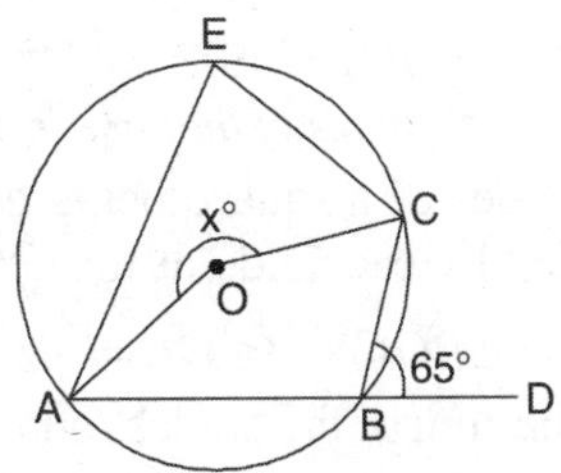

(a) 130° (b) 230°
(c) 190° (d) 65°

Answers

1. (b)	**2.** (c)	**3.** (b)	**4.** (d)	**5.** (c)	**6.** (d)	**7.** (d)	**8.** (a)	**9.** (b)	**10.** (c)
11. (a)	**12.** (c)	**13.** (c)	**14.** (a)	**15.** (b)	**16.** (c)	**17.** (c)	**18.** (d)	**19.** (c)	**20.** (c)
21. (a)	**22.** (d)	**23.** All are true statements.			**24.** (b)	**25.** (b)			

Hints and Solutions

1. (b) C_1 and C_2 are the same circle as there can be one and only one circle passing through three non collinear points.

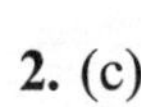

2. (c)

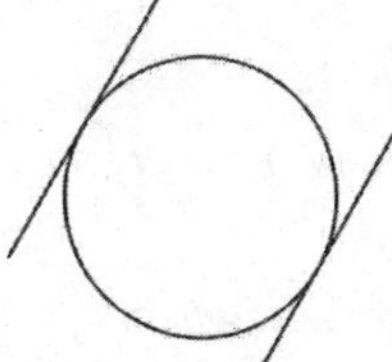

Two tangents to a circle

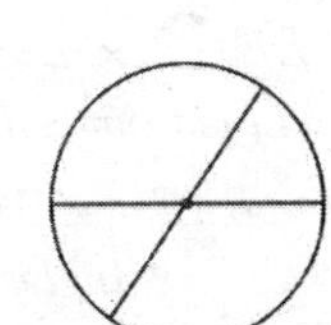

Two diameters of a circle

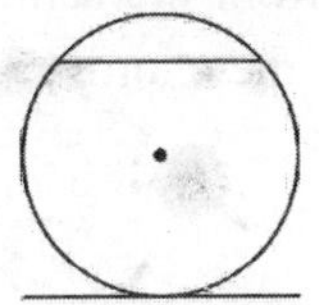

A chord and a tangent to a circle

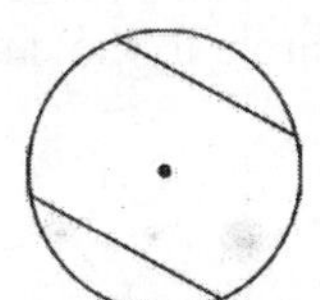

Two chords of a circle

3. (b) Let the three circles be of equal radii *r*. Then, the triangle formed by joining the centres of these circles is an equilateral triangle with each side 2*r*.

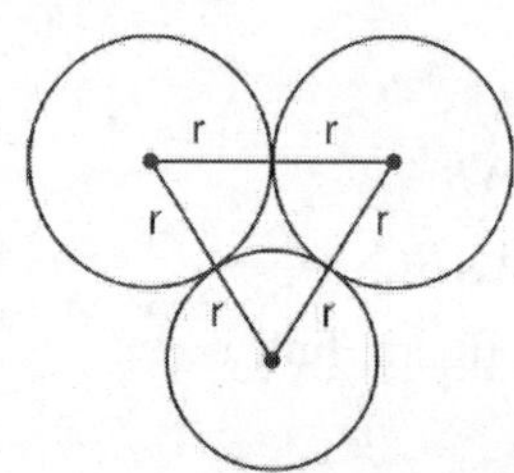

4. (d) Let the two circles with centres A and B have diameters a and b respectively. Then required distance $= BA$

$= AC - BC$

$= AC - (BD + DC)$

$= \frac{a}{2} - \left(\frac{b}{2} + c\right)$

$= \frac{a}{2} - \frac{b}{2} - c = \frac{1}{2}(a - b) - c.$

5. (c) $\angle APB = 90°$

(Angle in a semi circle is a right angle)

$\Rightarrow \quad \angle APM = \frac{1}{2}\angle APB = 45°$ *(PM bisects $\angle APB$)*

$\therefore \quad \angle ABM = \angle APM = 45°$

(Angles in the same segment are equal)

6. (d) All the sides of a square being equal, the angles subtended by each side at the centre are equal.

$\Rightarrow \angle AOB = \angle BOC = \angle COD = \angle DOA = x$ (say)

Since, the sum of the angles round a point = 360°, therefore,

$\angle AOB + \angle BOC + \angle COD + \angle DOA = 360°$

$\Rightarrow x + x + x + x = 360°$

$\Rightarrow 4x = 360° \Rightarrow x = \mathbf{90°}$.

7. (d) Number of sides of the polygon

$= \frac{\text{Sum of the angles at the centre}}{\text{Angle subtended by one side at the centre}}$

$= \frac{360°}{45°} = \mathbf{8}.$

8. (a) Given $AB = 24$ cm and $OC = 5$ cm

Since, the perpendicular from the centre of the circle to the chord bisects the chord,

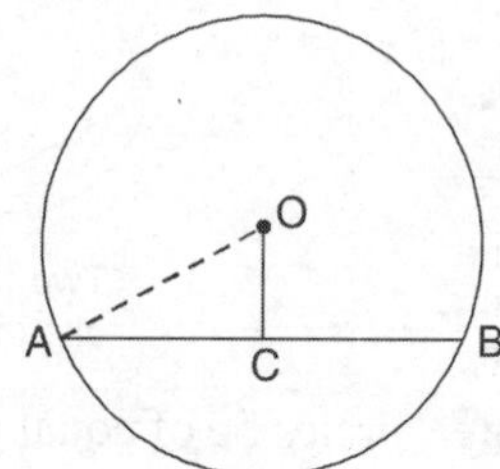

$AC = CB = 12\text{cm}$

Join, OA, in ΔAOC,

$AO^2 = AC^2 + OC^2$ *(Pythagoras Theorem)*

$= 12^2 + 5^2$

$= 144 + 25 = 169$

$\Rightarrow AO = \sqrt{169} = 13$ cm

$\therefore$ Diameter = 2 × radius = 2 × AO = 2 × 13 cm = **26 cm**.

9. (b) Given, $OD = 7$ cm, $OB = 25$ cm

Since, $OD \perp BC$, it bisects BC, *i.e.*, $BD = DC$.

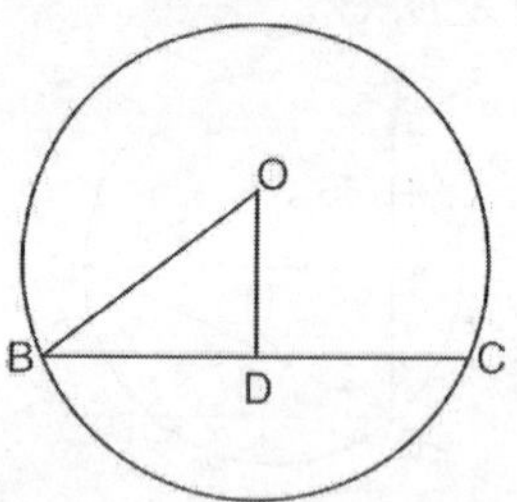

In ΔOBD,

$BD^2 = OB^2 - OD^2 = 25^2 - 7^2 = 625 - 49 = 576$

$\Rightarrow BD = \sqrt{576} = 24$

$\therefore$ Chord $BC = 2 \times 24$ cm $=$ **48 cm**.

10. (c) Give $OC = 4$ cm, Chord $AB = 6$ cm

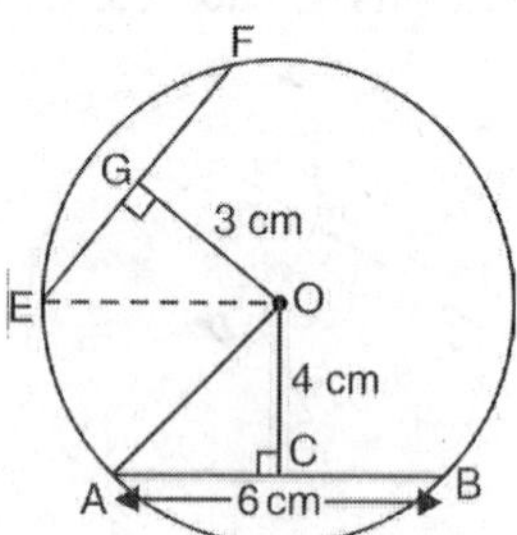

$\because OC \perp AB \Rightarrow OC$ bisects AB

$\Rightarrow AC = CB = 3\text{cm}$

$\therefore$ In ΔOCA,

$OA^2 = OC^2 + AC^2$ *(Pythagoras, Theorem)*

$= 4^2 + 3^2 = 16 + 9 = 25$

$\Rightarrow OA = 5$ cm

Let EF be the chord at a distance of 3 cm from the centre.

Given, radius $= OE = 5$ cm

$\therefore$ In ΔOGE, $EG^2 = OE^2 - OG^2$

$= 5^2 - 3^2 = 25 - 9 = 16$

$\Rightarrow EG = 4$ cm

$\therefore EF = 2 \times EG =$ **8 cm.**

(Line from centre $\perp$ to chord bisects the chord)

11. (a) $\angle BAE = \angle BCE = 30°$

(Angles in the same segment are equal)

$\therefore \angle BAC = 2 \times \angle BAE = 2 \times 30° = \mathbf{60°}$.

($\because$ AE bisects $\angle BAC$)

12. (c) Reflex $\angle AOC = \angle AOB + \angle BOC$

$= 90° + 120° = 210°$

$\therefore \angle AOC = 360° - 210° = 150°$

$\therefore \angle ABC = \frac{1}{2} \times \angle AOC = \frac{1}{2} \times 150° = \mathbf{75°}.$

(Angle subtended by an arc at the centre is twice the angle at any remaining part of the ⊙ce of the circle).

13. (c) Reflex $\angle O = 360° - 110° = 250°$

$\therefore \angle ABC = \frac{1}{2} \times 250° = 125°$

(Angle subtended by an arc at the centre is twice the angle at any remaining part of the ⊙ce of the circle)

$\therefore x = \angle ABC = \mathbf{125°}$. *(Vertically opposite angles)*

14. (a) Opposite angles of a cyclic quadrilateral are supplementary.

$\therefore 3x + x = 180°$ and $2y + y = 180°$

$\Rightarrow 4x = 180°$ and $3y = 180°$

$\Rightarrow x = 45°$ and $y = 60°$.

$\therefore \angle P = 3 \times 45° = \mathbf{135°}$ and $\angle Q = \mathbf{60°}$.

15. (b) In ΔOAB, $OA = OB$ *(Radii of same circle)*

$\Rightarrow \angle OBA = \angle OAB = 35°$

(Angles opposite to equal sides are equal)

$\therefore \angle AOB = 180° - (35° + 35°) = 180° - 70° = 110°$

In OAC, $OA = OC$ *(Radii of same circle)*

$\Rightarrow \angle OCA = \angle OAC = 20°$ *(Isosceles Δ prop.)*

$\therefore \angle AOC = 180° - (20° + 20°) = 180° - 40° = 140°$

$\therefore$ Reflex $\angle BOC = \angle AOB + \angle AOC$

$= 110° + 140° = 250°$

$\Rightarrow x = 360° - 250° = \mathbf{110°}.$

16. (c) $AB = AC \Rightarrow \angle ACB = \angle ABC = 42°$

(Isosceles Δ prop.)

$\therefore$ In ΔABC, $\angle BAC = 180° - (42° + 42°)$

$= 180° - 84° = 96°$

$\therefore \angle CDE = \angle BAC = \mathbf{96°}.$

(Exterior ∠ of a cyclic quadrilateral is equal to the interior opposite angle).

17. (c) $\angle AOB = x \Rightarrow \angle ADB = x/2$

(Angle subtended by an arc at the centre is twice the angle at the remaining part of the ⊙ce of the circle).

Since, chord $BD \perp AC$, $\angle AED = 90°$

$\therefore$ In ΔAED,

$p = \angle DAE = 180° - (\angle ADE + \angle AED)$

$= 180° - (x/2 + 90°)$

$= 90° - x/2$.

18. (d) Join AD.

In the cyclic quadrilateral $ABCD$,

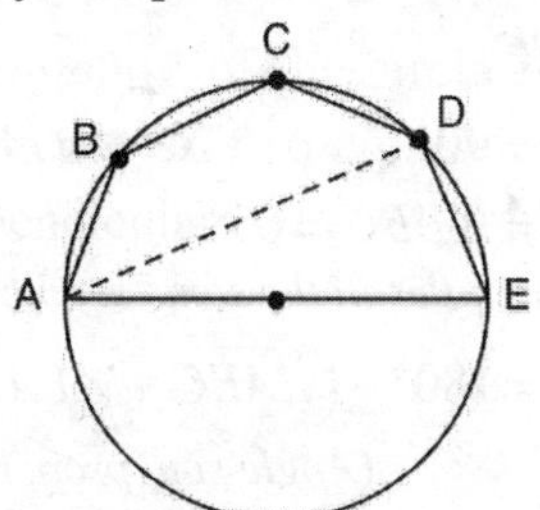

$\angle ABC + \angle ADC = 180°$ ...(i)

(opp. ∠s angles of a cyclic quadrilateral are supplementary).

Also, $\angle ADE = 90°$ ...(ii)

(Angle in a semicircle is a right angle)

$\therefore$ Adding eqns. (i) and (ii), we get,

$\angle ABC + \angle ADC + \angle ADE = 180° + 90°$

$\Rightarrow \angle ABC + \angle CDE = \mathbf{270°}.$

19. (c) $\angle DCQ = 59°$

(Exterior angle theorem of cyclic quadrant).

In ΔADP,

Exterior $\angle CDQ$ = Int. opp. $(\angle PAD + \angle APD)$

$= 59° + 40° = 99°$

$\therefore$ In ΔDCQ, $\angle DQC = 180° - (\angle DCQ + \angle CDQ)$

$= 180° - (59° + 99°)$

$= 180° - 158° = 22°$

(Angle sum property of a Δ)

$\Rightarrow \angle AQB = \mathbf{22°}.$

20. (c) $\angle BAC = \angle BDC = 80°$

(Angles in the same segment are equal)

In ΔBAC, $\angle ACB = 180° - (\angle BAC + \angle CBA)$

$= 180° - (80° + 55°)$

$= 180° - 135° = 45°$

$\therefore \angle BCD = \angle BCA + \angle ACD = 45° + 45° = 90°$

$\Rightarrow$ BD is the diameter. *(Angle in a semi-circle = 90°)*

21. (a) In ΔADB, $\angle ADB = 90°$ *(Angle in a semi-circle)*

$\therefore \angle DAB = 180° - (\angle ADB + \angle ABD)$

$= 180° - (90° + 75°)$

$= 180° - 165° = 15°$

$\therefore \angle CAB = \angle CAD + \angle DAB = 35° + 15° = 50°$

In cyclic quadrilateral $ABDC$,

$\angle BDC + \angle CAB = 180°$

(Opp. ∠s of a cyclic quadrilateral are supplementary)

$\therefore \angle BDC = 180° - \angle CAB = 180° - 50° = \mathbf{130°}.$

22. (d) Join AE.

Given, $\angle EBC = 65°$ and $AC \| ED$.

In ΔAEC,

$\angle AEC = 90°$ *(Angle in a semi circle)*

$\angle EAC = \angle EBC = 65°$

(Angles in the same segment are equal)

$\therefore \angle ACE = 180° - (\angle AEC + \angle EAC)$

(Angle sum property of a circle)

$= 180° - (90° + 65°) = 180° - 155° = 25°$.

$\angle CED = \angle ACE = 25°$

($AC \| ED$, alternate angles are equal)

23. (i) True, equal chords are equidistant from the centre.

(ii) True, In Δs OMB and ONB,

$OM = ON$ (Proved in (i))

$OB = BO$ (Common)

$\angle OMB = \angle ONB = 90°$

$\therefore \Delta OMB \cong \Delta ONB$ (RHS)

(iii) $\Delta OMB \cong \Delta ONB \Rightarrow \angle OBM = \angle OBN$ (*cpct*)

$\Rightarrow BO$ bisects $\angle ABC$.

Hence, all the statements are true statements.

24. (b) For cyclic quadrilateral $ABCD$,

$y + 52° = 116° \Rightarrow y = 116° - 52° = 64°$

(Ext. angle property of a cyclic quadrilateral)

$OB = OA$ *(radii of same circle)*

$\Rightarrow \angle OAB = \angle OBA = 52°$

$\therefore \angle BOA = 180° - (\angle OAB + \angle OBA)$

$= 180° - (52° + 52°)$

$= 180° - 104° = 76°$.

(Angle sum property of a Δ)

$\therefore x = \angle BCA = \frac{1}{2} \times \angle BOA = \frac{1}{2} \times 76° = 38°$

(Angle at the centre is twice the angle at any remaining part of circumference)

$\therefore x + y = 64° + 38° = \mathbf{102°}$.

25. (b) $AECB$ being a cyclic quadrilateral,

$\Rightarrow$ interior opposite $\angle AEC$ = exterior $\angle CBD = 65°$

$\therefore \angle AOC = \angle AEC \times 2 = 65° \times 2 = 130°$

(Angle at centre = 2 × angle at remaining part of the circumference)

$\therefore$ Reflex $\angle AOC = 360° - 130° = \mathbf{230°}$.

Self Assessment Sheet–20

1. Which of the following statements is not TRUE ?

(a) The diameter is the greatest chord that can be drawn in a circle.

(b) A straight line cannot intersect a circle in more than two points.

(c) A diameter bisects a circle.

(d) In the figure $\angle A = \angle C$ and $\angle B = \angle D$.

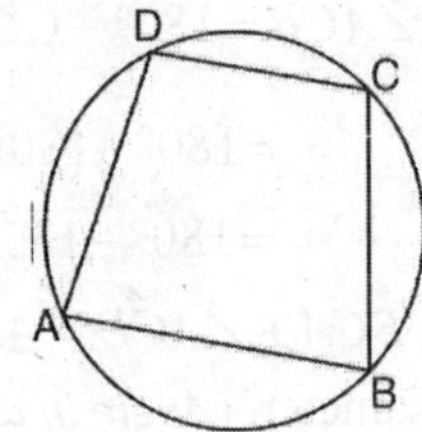

2. O is the centre of the circle, ABN is a straight line. Find $\angle AOC$.

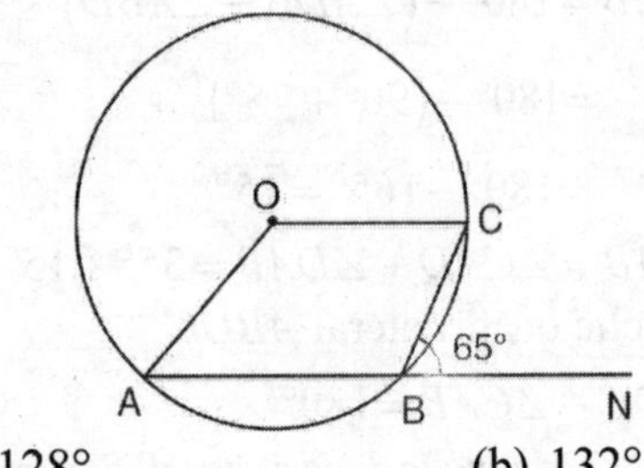

(a) 128° (b) 132°

(c) 130° (d) 135°

3. Find the size of the angle marked x.

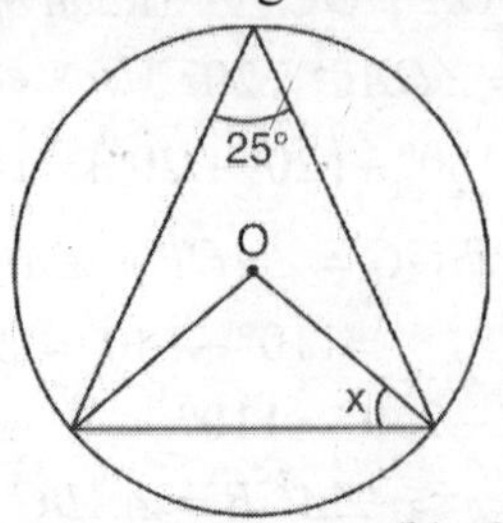

(a) 60° (b) 65°

(c) 70° (d) 55°

4. The length of a chord of a circle of radius 10 cm is 12 cm. Find the distance of the chord from the centre of the circle.

(a) 6 cm (b) 5 cm

(c) 8 cm (d) 7 cm

5. Chord $ED \|$ diameter AC. Determine $\angle CED$.

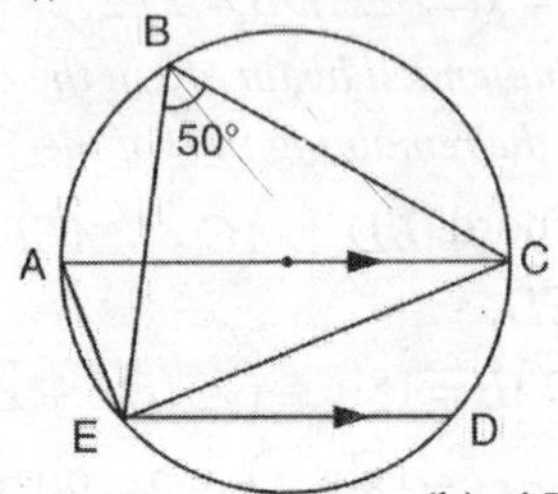

(a) 50° (b) 45°

(c) 55° (d) 40°

6. The measure of the line segment joining the centre of a circle to the mid-point of a chord is :
   (a) twice the measure of the chord
   (b) half the measure of the chord
   (c) equal to the measure of the chord
   (d) none of the above

7. *ABCD* is a cyclic quadrilateral whose side *AB* is a diameter of the circle through *A, B, C, D*. If $\angle ADC = 130°$, find $\angle BAC$.

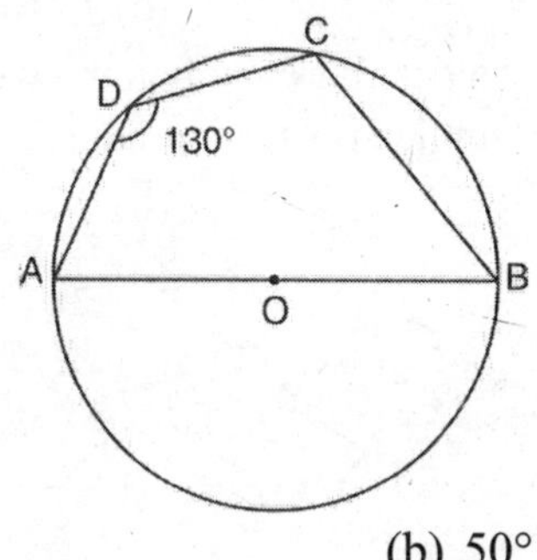

(a) 40° (b) 50°
(c) 60° (d) 30°

8. *O* is the centre of the circle *APQB; AOBR, PQR* are straight lines. Find *x* in terms of *y* and *z*.

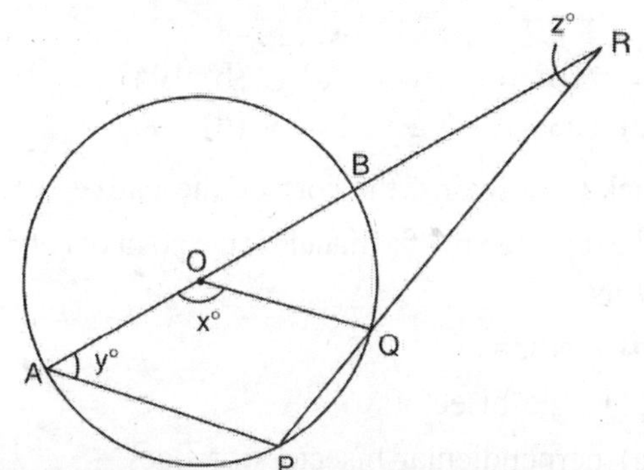

(a) $x = y + z$ (b) $x = 2y + z$
(c) $x = y + 2z$ (d) $x = 2(y + z)$

9. *O* is the centre of the circle *ABC*, radius 5 cm, *AB* = 8 cm, *AC* = 6 cm. Calculate the lengths of the perpendiculars *OP, OQ* from *O* to *AB, AC*.

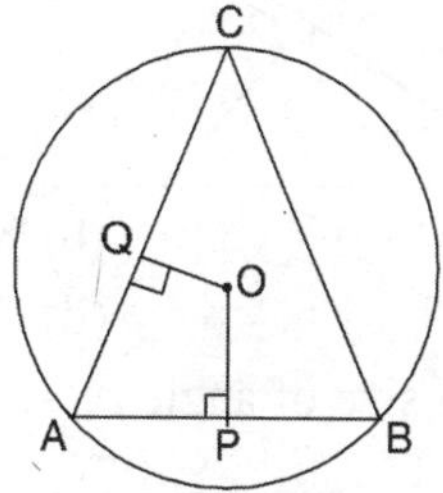

(a) 3 cm, 4 cm (b) 4 cm, 3 cm
(c) 2 cm, 3 cm (d) 3.5 cm, 2.5 cm

10. *AB, AC* are equal chords of the circle *ABCD*. Calculate $\angle BAD$.

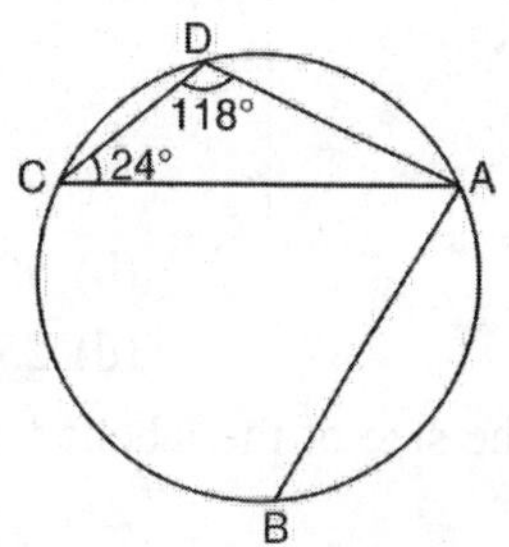

(a) 100° (b) 94°
(c) 96° (d) 80°

Answers

1. (d)	2. (c)	3. (b)	4. (c)	5. (d)	6. (d)	7. (a)	8. (d)	9. (a)	10. (b)

Unit Test–4

1. Which of the following is equidistant from the vertices of a triangle ?
   (a) circumcentre (b) centroid
   (c) orthocenter (d) incentre

2. The circumcentre in a right triangle is :
   (a) inside the triangle
   (b) outside the triangle
   (c) on one of the perpendicular sides
   (d) on the hypotenuse

3. *P* is the incentre of Δ *ABC*. Which of the following statements is true ?

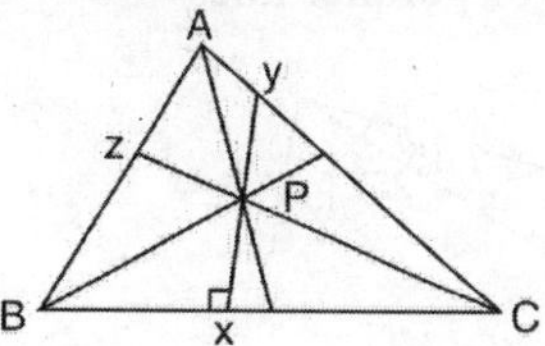

(a) $AZ = BZ$ (b) $AY = BX$
(c) $PY = PZ$ (d) $PA = PC$

4. The incentre of a triangle coincides with the circumcentre, orthocenter and centroid in case of :
   (a) an isosceles triangle
   (b) an equilateral triangle

(c) a right - angled triangle

(d) a right - angled isosceles triangle

5. Find x

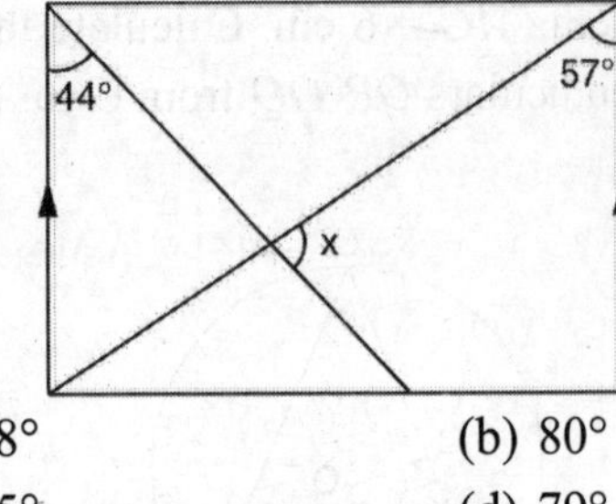

(a) 78° (b) 80°

(c) 75° (d) 79°

6. Calculate the size of angle p.

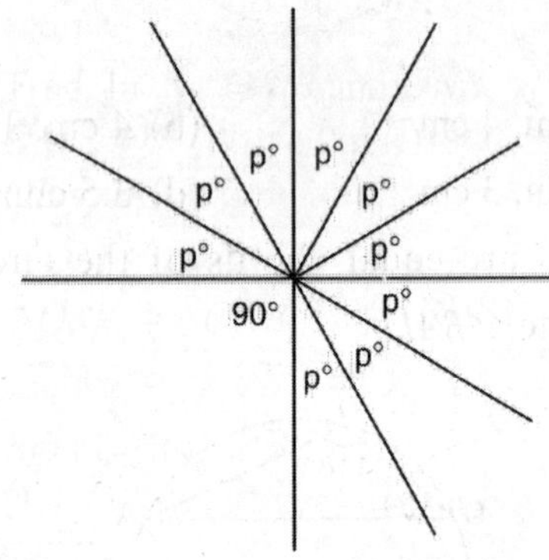

(a) 20° (b) 30°

(c) 40° (d) 25°

7. Calculate the size of the labelled angles

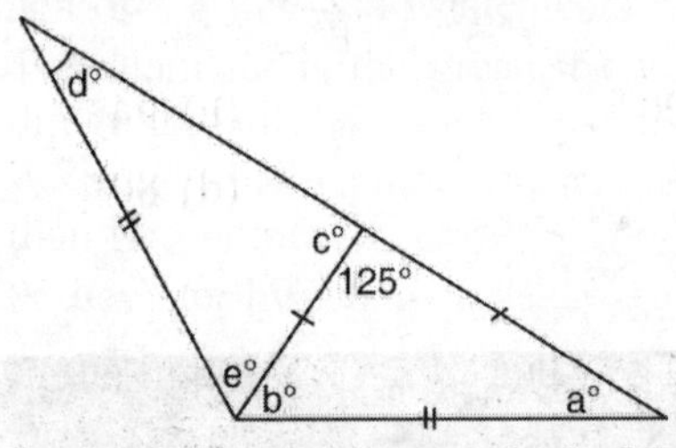

(a) $a=27.5°, b=27.5°, c=55°, d=27.5°, e=97.5°$

(b) $a = 28°, b = 27°, c = 60°, d = 28°, e = 102°$

(c) $a = 30°, b = 25°, c = 55°, d = 30°, e = 95°$

(d) None of these

8. Find pairs of parallel lines

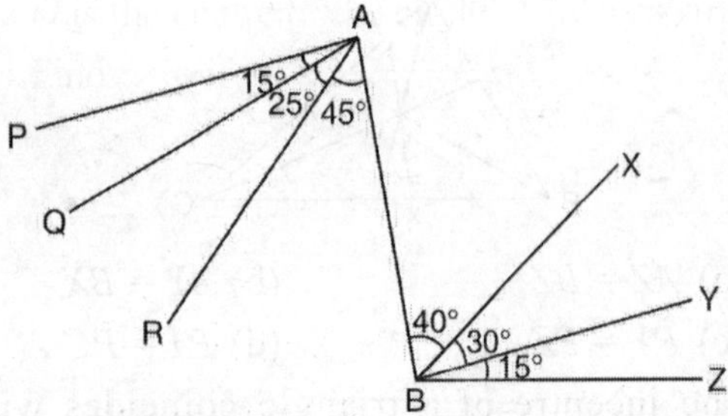

(a) $AR, BX; AP, BY$ (b) $AQ, BZ; AP, BX$

(c) $AQ, BY; AP, BZ$ (d) $AQ, BX; AR, BZ$

9. In the figure, $AB \parallel CD$, then

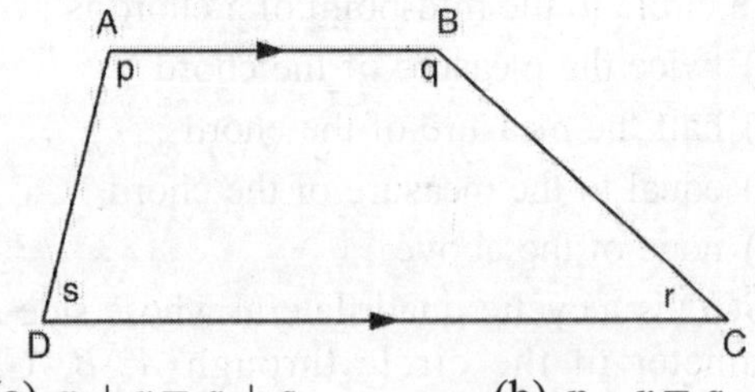

(a) $p + r = q + s$ (b) $p - r = q - s$

(c) $p + s = q + r$ (d) $p - q = s - r$

10. How many sides does a polygon have if the sum of its interior angles is 30 right angles ?

(a) 15 (b) 17

(c) 19 (d) 20

11. Find x

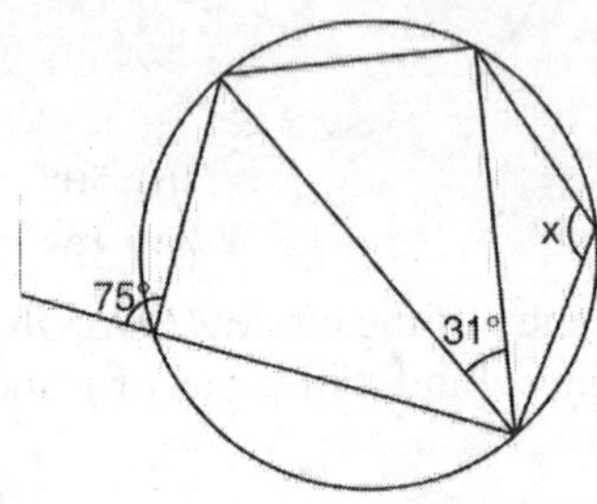

(a) 110° (b) 104°

(c) 108° (d) 106°

12. Tick (✓) against the correct alternative.

The orthocentre of a triangle is the point of concurrency of its.

(a) medians

(b) angle bisectors

(c) perpendicular bisectors of sides

(d) altitudes drawn to sides from opposite vertices.

13. Match correctly

(a) centroid	(1) medians of a Δ
(b) incentre	(2) centre of the circumcircle point of intersection of the perp. bisectors of the sides of a Δ
(c) circumcentre	(3) point of intersection of the medians of a Δ
(d) concurrent	(4) centre of the incircle point of intersection of the angle bisectors of a Δ

14. The lengths of the sides of a Δ ABC are given below. In which of these cases are angles of the triangle in the increasing order of magnitude as $\angle C, \angle B, \angle A$.

(a) $BC = 5$ cm, $CA = 6.5$ cm, $AB = 7.9$ cm
(b) $BC = 10$ cm, $CA = 6.9$ cm, $AB = 5.4$ cm
(c) $BC = 3$ cm, $CA = 4$ cm, $AB = 5$ cm
(d) $BC = 3.5$ cm, $CA = 3$ cm, $AB = 4$ cm

15. $ABCD$ is a rhombus and AED is an equilateral triangle. E and C lie on opposite sides of AD. If $\angle ABC = 78°$, calculate $\angle DCE$ and $\angle ACE$.

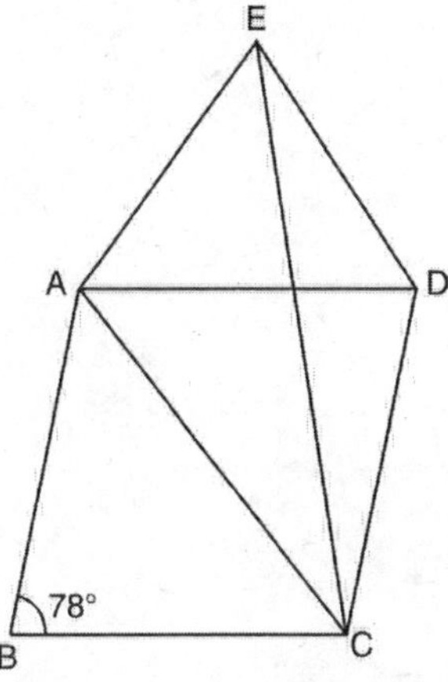

(a) 20°, 30° (b) 21°, 31°
(c) 22°, 32° (d) 19°, 29°

16. The number of sides of two regular polygons are in the ratio 1: 2 and their interior angles are in the ratio 3 : 4. Find the number of sides in each polygon.

(a) 6, 12 (b) 8, 16
(c) 5, 10 (d) 7, 14

17. A regular polygon is inscribed in a circle. If a side subtends an angle of 30° at the centre, what is the number of its sides ?

(a) 10 (b) 8
(c) 6 (d) 12

18. Answer True or False. ACB is an arc of a circle with centre O and $\angle ABC = 45°$, then, $AO \perp OC$.

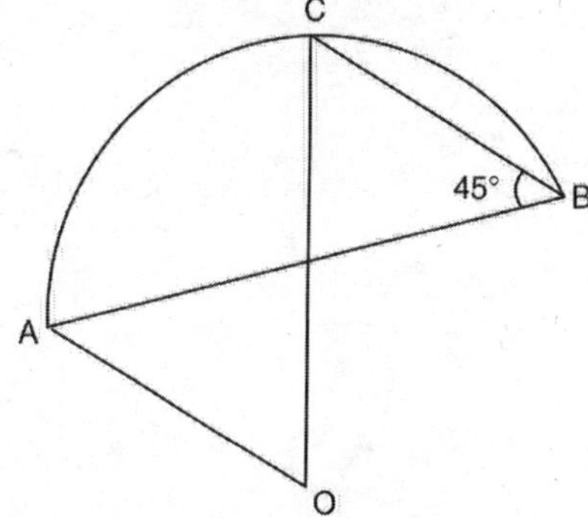

19. Consider the following statements

(1) The bisectors of all the four angles of a parallelogram enclose a rectangle.

(2) The figure formed by joining the midpoints of the adjacent sides of a rectangle is a rhombus.

(3) The figure formed by joining the midpoints of the adjacent sides of a rhombus is a square.

Which of these statements are correct ?

(a) 1 and 2 (d) 2 and 3
(c) 3 and 1 (d) 1, 2 and 3

20. If the sum of the diagonals of a rhombus is 12 cm, and its perimeter is $8\sqrt{5}$ cm, then the lengths of the diagonals are :

(a) 6 cm and 6 cm (b) 7 cm and 5 cm
(c) 8 cm and 14 cm (d) 9 cm and 3 cm

Answers

1. (a) **2.** (d) **3.** (c) **4.** (b) **5.** (d) **6.** (b) **7.** (a) **8.** (c) **9.** (b) **10.** (b)
11. (d) **12.** (d) **13.** (a) $\rightarrow$3 (b)$\rightarrow$4 (c)$\rightarrow$2 (d)$\rightarrow$1 **14.** (b) **15.** (b) **16.** (c) **17.** (d)
18. True **19.** (a) **20.** (c)

UNIT-5

MENSURATION

- *Perimeter and Area*
- *Circumference and Area of a Circle*
- *Volume and Surface Area of a Cube and a Cuboid*

Chapter 21

PERIMETER AND AREA

KEY FACTS

1. Rectangle

(i) Perimeter = $2(l + b)$ (ii) Area = $l \times b$ (iii) Diagonal = $\sqrt{l^2 + b^2}$

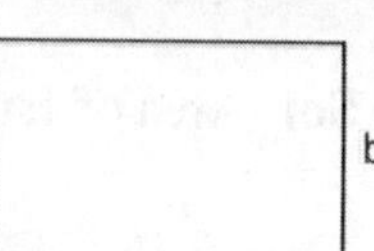

2. Square

(i) Perimeter = $4a$ (ii) Diagonal $(d) = a\sqrt{2}$ (iii) Area = $a^2 = \frac{1}{2}d^2$

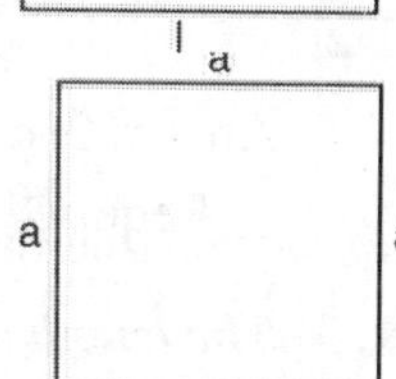

3. Right-angled triangle

(i) Perimeter = $p + b + h$ (ii) Area = $\frac{1}{2} \times b \times p$

(iii) Hypotenuse $(h) = \sqrt{(\text{perp})^2 + (\text{base})^2}$

$= \sqrt{p^2 + b^2}$

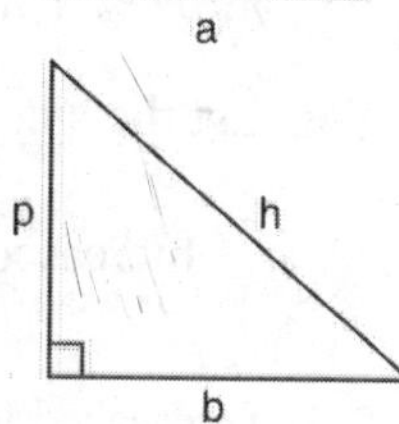

4. Right-angled isosceles Δ.

(i) Hypotenuse = $\sqrt{a^2 + a^2} = a\sqrt{2}$ (ii) Perimeter = $2a + \sqrt{2}a$

(iii) Area = $\frac{1}{2} \times a \times a = \frac{1}{2}a^2$

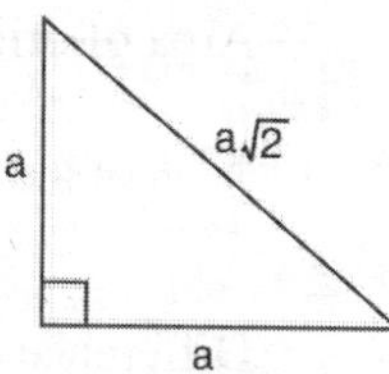

5. Equilateral triangle

(i) Perimeter = $3a$ (ii) Height = $\frac{\sqrt{3}}{2}a$ (iii) Area = $\frac{\sqrt{3}}{4}a^2$

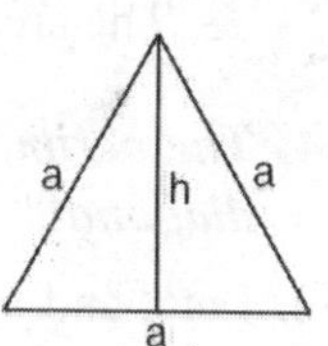

6. Scalene triangle

(i) Perimeter $(2s) = a + b + c$ (ii) Area = $\sqrt{s(s-a)(s-b)(s-c)}$,

where a, b, c are the sides of the triangle and semi-perimeter $(s) = \frac{a+b+c}{2}$.

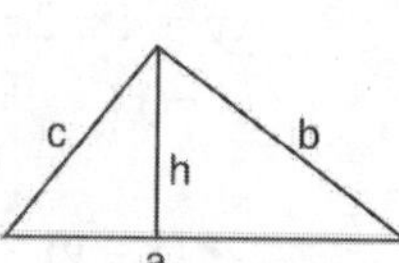

7. Parallelogram

(i) Perimeter = $2(a + b)$ (ii) Area = base × height $= ah_1 = bh_2$

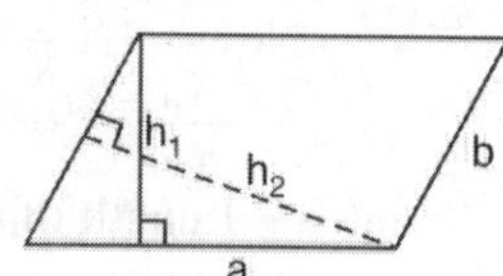

Solved Examples

Ex. 1. ***Expenditure incurred in cultivating a square field at the rate of ₹ 170 per hectare is ₹ 680. What would be the cost of fencing the field at the rate of ₹ 3 per metre?***

Sol. Area of the square field $= \dfrac{680}{170} = 4$ hectares $= (4 \times 10000)$ sq m $= 40000$ sq m

$\therefore$ Side of the square field $= \sqrt{40000 \text{ m}^2} = 200$ m

$\therefore$ Perimeter of the square field $= (4 \times 200)$ m $= 800$ m

$\therefore$ Cost of fencing the field $=$ ₹ $(3 \times 800) =$ **₹ 2400.**

Ex. 2. ***The length of the diagonal of a square and that of the side of another square are both 10 cm. What is the ratio of the area of the first square to that of the second?***

Sol. Area of 1st square $= \dfrac{1}{2}d^2 = \dfrac{1}{2} \times 10^2 \text{ cm}^2$

$= \dfrac{100}{2} \text{ cm}^2 = 50 \text{ cm}^2$

Area of 2nd square = side$^2 = (10)^2 \text{ cm}^2 = 100 \text{ cm}^2$

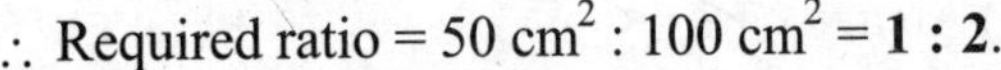

$\therefore$ Required ratio $= 50 \text{ cm}^2 : 100 \text{ cm}^2 =$ **1 : 2.**

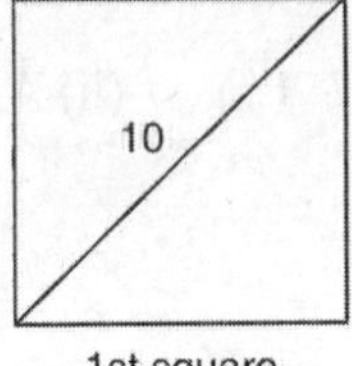

1st square

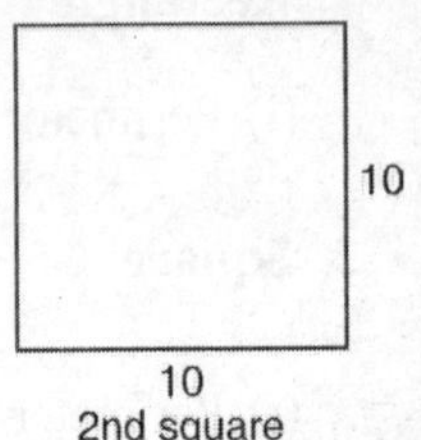

2nd square

Ex. 3. ***The length of one pair of opposite sides of a square is reduced by 10% and that of the other pair is increased by 10%. Compare the area of the new rectangle with the area of the original square.***

Sol. Let the side of the square be a

$\therefore$ Increased length $= a + 10\%$ of $a = \dfrac{11a}{10}$

Decreased length $= a - 10\%$ of $a = \dfrac{9a}{10}$

Area of original square $= a^2$

Area of the new rectangle $= \dfrac{11a}{10} \times \dfrac{9a}{10} = \dfrac{99a^2}{100}$, *i.e.*

Difference of the two areas $= a^2 - \dfrac{99a^2}{100} = \dfrac{a^2}{100}$

$\Rightarrow$ The area of the new rectangle is 1% less than the area of original square.

Ex. 4. ***The perimeter of the top of rectangular table is 28 m, whereas its area is 48 m². What is the length of its diagonal?***

Sol. Let x and y be the length and breadth of the rectangle.

$xy = 48,\ 2x + 2y = 28 \quad \Rightarrow \quad x + y = 14$

$(x - y)^2 = (x + y)^2 - 4xy = 14^2 - 4 \times 48$

$= 196 - 192 = 4$

$\therefore (x - y) = \sqrt{4} = 2$

Now $x + y = 14$ and $x - y = 2$

$\Rightarrow 2x = 16 \quad \Rightarrow x = 8 \quad \Rightarrow \quad y = 6$

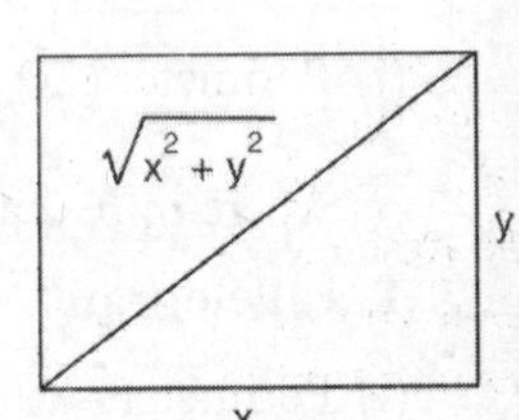

$\therefore$ Length of diagonal $= \sqrt{x^2 + y^2} = \sqrt{64 + 36}$

$= \sqrt{100} =$ **10 cm.**

Ex. 5. ***In the given diagram, ABCD is a rectangle. ADEF, CDHG, BCLM and ABNO are four squares. If the perimeter of ABCD is 16 cm and total area of the four squares is 68 cm², then what is the area of ABCD?***

Sol. Let $AD = a$ and $AB = b$. Then,

$2(a + b) = 16 \quad \Rightarrow \quad a + b = 8$ and

$2(a^2 + b^2) = 68 \quad \Rightarrow \quad a^2 + b^2 = 34$

$\therefore \ (a + b)^2 = a^2 + b^2 + 2ab$

$\Rightarrow 64 = 34 + 2ab$

$\Rightarrow 2ab = 30 \quad \Rightarrow \quad ab = 15$

Hence, area of rect. $ABCD =$ **15 cm²**.

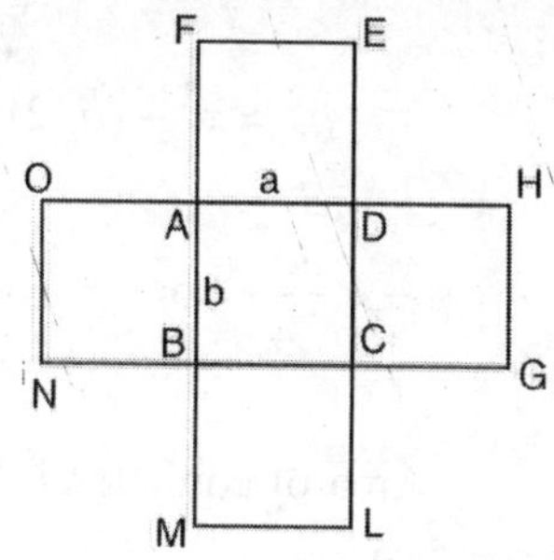

Ex. 6. ***What is the area of the square ABCD shown in the diagram?***

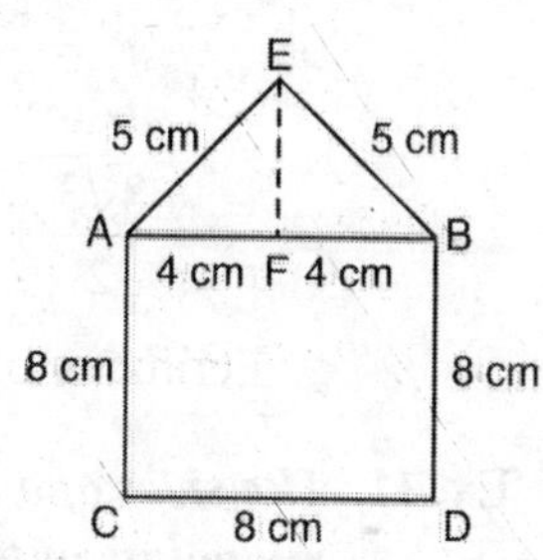

Sol. $AD = \sqrt{12^2 + 5^2}$ cm $= \sqrt{144 + 25}$ cm

$= \sqrt{169}$ cm $= 13$ cm

$\therefore$ Area of square $ABCD = (13)^2 \text{ cm}^2 =$ **169 cm²**.

Ex. 7. ***What is the area of a figure formed by a square of side 8 cm and an isosceles triangle with base as one side of the square and the perimeter as 18 cm?***

Sol. Each of the equal sides of the isosceles $\Delta = \frac{18 - 8}{2}$ cm $= 5$ cm

$\because$ In the isosceles $\Delta\, EAB$, altitude EF bisects the base AB, $AF = FB = 4$ cm

$\therefore$ In $\Delta\, EFA$, $EF = \sqrt{EA^2 - AF^2} = \sqrt{25 - 16}$ cm

$= \sqrt{9}$ cm $= 3$ cm

$\therefore$ Area of the figure = Area of $\Delta\ EAB$ + Area of square $ABCD$

$= \frac{1}{2} \times 8 \times 3 \text{ cm}^2 + (8 \times 8) \text{ cm}^2$

$= 12 \text{ cm}^2 + 64 \text{ cm}^2 =$ **76 cm²**.

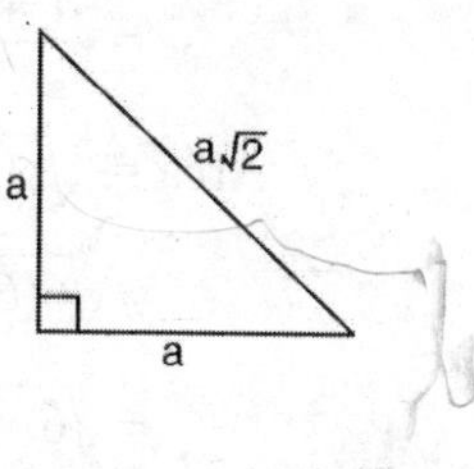

Ex. 8. ***If the perimeter of a right angled isosceles triangle is $\sqrt{2} + 1$, then what is the length of the hypotenuse?***

Sol. Given, $a + a + a\sqrt{2} = \sqrt{2} + 1$

$\Rightarrow 2a + a\sqrt{2} = \sqrt{2} + 1 \quad \Rightarrow \quad a\sqrt{2}(\sqrt{2} + 1) = \sqrt{2} + 1$

$\Rightarrow a = \frac{(\sqrt{2} + 1)}{\sqrt{2}(\sqrt{2} + 1)} \quad \Rightarrow \quad a = \frac{1}{\sqrt{2}}$

$\therefore$ Hypotenuse $= a\sqrt{2} = \frac{1}{\sqrt{2}} \times \sqrt{2}$ cm $=$ **1 cm.**

Ex. 9. ***If x is the length of a median of an equilateral triangle, then what is its area?***

Sol. In an equilateral triangle, the median and the altitude coincide. Let a be the length of each side of the equilateral triangle.

In rt. $\angle d\ \Delta\ ADB$,

$$AB^2 = AD^2 + BD^2$$

$$\Rightarrow a^2 = x^2 + (a/2)^2 \quad \Rightarrow \quad a^2 - \frac{a^2}{4} = x^2$$

$$\Rightarrow \frac{3a^2}{4} = x^2 \quad \Rightarrow \quad a^2 = \frac{4x^2}{3}$$

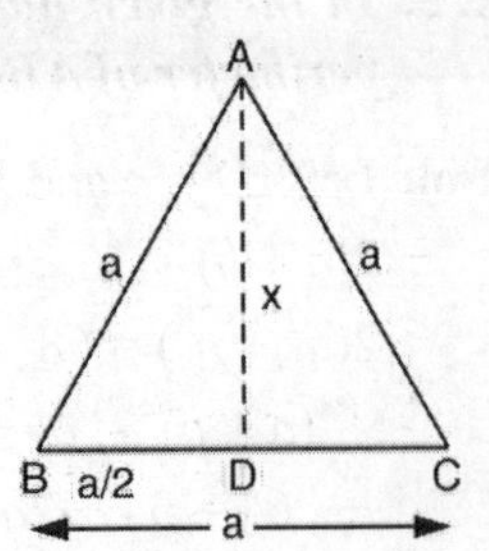

Area of equilateral $\Delta = \frac{\sqrt{3}}{4}a^2$

$$= \frac{\sqrt{3}}{4} \times \frac{4}{3} \times x^2 = \frac{\mathbf{x^2}}{\sqrt{\mathbf{3}}}.$$

Ex. 10. ***From a point in the interior of an equilateral triangle the perpendicular distances of the sides are $\sqrt{3}$ cm, $2\sqrt{3}$ cm and $5\sqrt{3}$. What is the perimeter (in cm) of the triangle?***

Sol. Let each side of the triangle be a cm.

Area of ΔABC = Area of ΔAOB + Area of ΔAOC + Area of ΔBOC

$$\Rightarrow \frac{\sqrt{3}}{4}a^2 = \frac{1}{2} \times a \times \sqrt{3} + \frac{1}{2} \times a \times 2\sqrt{3} + \frac{1}{2} \times a \times 5\sqrt{3} \qquad \Rightarrow \frac{\sqrt{3}}{4}a^2 = \frac{1}{2}a(\sqrt{3} + 2\sqrt{3} + 5\sqrt{3})$$

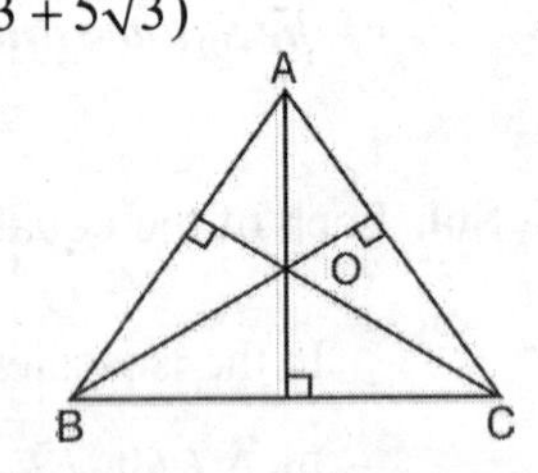

$$\Rightarrow \frac{\sqrt{3}}{4}a^2 = \frac{1}{2} \times a \times 8\sqrt{3} \Rightarrow \frac{\sqrt{3}}{4}a = \frac{1}{2} \times 8\sqrt{3}$$

$$\Rightarrow a = \frac{8\sqrt{3}}{2} \times \frac{4}{\sqrt{3}} \quad \Rightarrow \quad a = 16$$

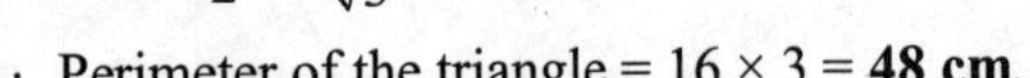

$\therefore$ Perimeter of the triangle = 16 × 3 = **48 cm.**

Ex. 11. ***The sides of a triangle are 3 cm, 4 cm and 5 cm. What is the area (in cm^2) of a triangle formed by joining the midpoints of this triangle?***

Sol. ABC is the given triangle. PQR is the triangle formed by joining the midpoints of ΔABC.

Area of $\Delta PQR = \frac{1}{4}$ (Area of ΔABC)

For area of ΔABC,

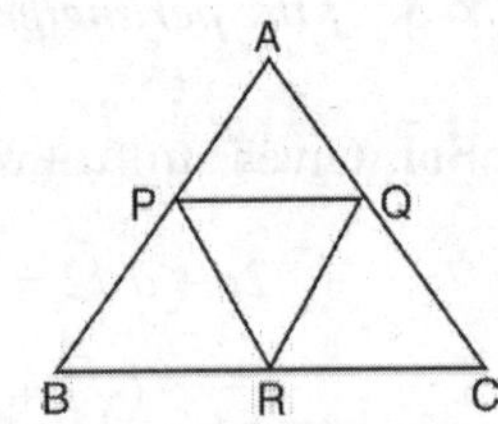

$$s = \frac{a+b+c}{2} = \frac{3\text{cm} + 4\text{cm} + 5\text{cm}}{2} = 6 \text{ cm}$$

$$\text{Area} = \sqrt{s(s-a)(s-b)(s-c)}$$

$$= \sqrt{6(6-3)(6-4)(6-5)} \text{ cm}^2$$

$$= \sqrt{6 \times 3 \times 2 \times 1} \text{ cm}^2$$

$$= \sqrt{36} \text{ cm}^2 = 6 \text{ cm}^2$$

$\therefore$ Area of $\Delta PQR = \frac{1}{4} \times 6\text{cm}^2$ = **1.5 cm²**.

Ex. 12. ***ABCD is a square of area 1 m^2. P and Q are the midpoints of AB and BC respectively. What is the area of ΔDPQ?***

Sol. Area of ΔDPQ = Area of square $ABCD$ – Area of ΔAPD – Area of ΔBPQ – Area of ΔDQC

$$= 1-\left(\frac{1}{2}\times AP\times AD\right)-\left(\frac{1}{2}\times BP\times BQ\right)-\left(\frac{1}{2}\times QC\times DC\right)$$

$$= 1\,m^2-\left(\frac{1}{2}\times\frac{1}{2}\times 1\right)m^2-\left(\frac{1}{2}\times\frac{1}{2}\times\frac{1}{2}\right)m^2-\left(\frac{1}{2}\times\frac{1}{2}\times 1\right)m^2$$

$$= \left(1-\frac{1}{4}-\frac{1}{8}-\frac{1}{4}\right)m^2 = \left(1-\frac{5}{8}\right)m^2 = \mathbf{\frac{3}{8}\,m^2}.$$

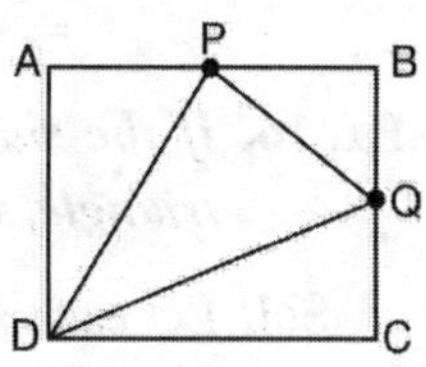

Ex. 13. ***If the height of a triangle is decreased by 40% and its base is increased by 40%, what will be the effect on its area?***

Sol. Let the original base and height of the triangle be b units and h units respectively. Then,

Original area = $\frac{1}{2}bh = 0.5bh$

New base = $b + 40\%$ of $b = 1.4b$,

New height = $h - 40\%$ of $h = 0.6h$

$\therefore$ New area $= \frac{1}{2}\times 1.4b\times 0.6h = \frac{0.84bh}{2} = 0.42bh$

Decrease in area = $0.5bh - 0.42bh = 0.08bh$

$\therefore$ % decrease = $\frac{0.08bh}{0.5bh}\times 100 = \mathbf{16\%}$.

Ex. 14. ***If an equilateral triangle of area X and a square of area Y have the same perimeter, then X is:***

(a) equal to Y ***(b) greater than Y***

(c) less than Y ***(d) less than or equal to Y***

Sol. Let each side of the equilateral triangle be x cm and that of square be y cm. Then, $X = \frac{\sqrt{3}}{4}x^2$ and $Y = y^2$

Also given, $3x = 4y$, *i.e.*, $y = \frac{3x}{4}$

$\therefore$ $X = \frac{\sqrt{3}}{4}x^2$ and $Y = \left(\frac{3x}{4}\right)^2 = \frac{9x^2}{16}$ $\Rightarrow X = \frac{1.732}{4}x^2 = 0.433x^2$ and $Y = 0.5625x^2$

$\Rightarrow$ X is less than Y.

Ex. 15. ***A lawn is in the form of an isosceles triangle. The cost of turfing it come to ₹ 1200 at ₹ 4 per m^2. If the base be 40 m long, find the length of each side.***

Sol. Area of the lawn = $\frac{₹1200}{₹4} = 300\ m^2$

$\Rightarrow$ $\frac{1}{2}\times$ base $\times$ altitude = 300

$\Rightarrow$ altitude = $\frac{300\times 2}{40}$ m = 15 m

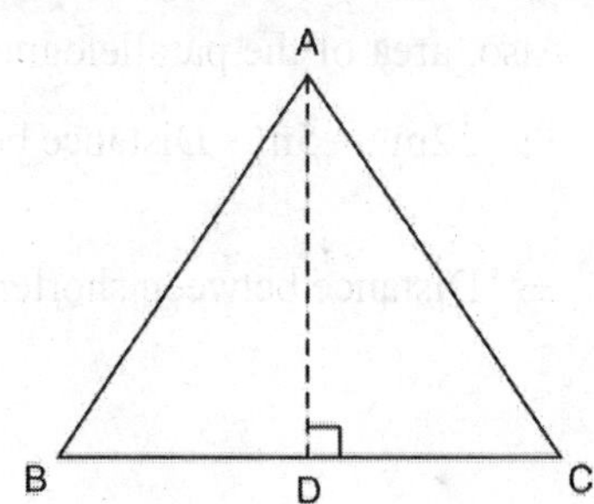

In an isosceles triangle the altitude bisects the base, so $BD = DC = 20$ cm

In rt. $\angle d\ \ \Delta ABD$, $AD = 15$ m, $BD = 20$ cm

$\therefore\ AB = \sqrt{15^2 + 20^2}$ cm $= \sqrt{225 + 400}$ cm

$= \sqrt{625}$ cm $=$ **25 cm.**

Ex. 16. ***If the sides of an equilateral triangle are increased by 20%, 30% and 50% respectively to form a new triangle, what is the percentage increase in the perimeter of the equilateral triangle?***

Sol. Let each side of the equilateral triangle be x cm. Then, after increase the three sides are

$$x + \frac{20}{100}x,\ x + \frac{30}{100}x \text{ and } x + \frac{50}{100}x,$$

i.e., $x + 0.2x$, $x + 0.3x$ and $x + 0.5x$,

i.e., $1.2x$, $1.3x$ and $1.5x$.

$\therefore$ Original perimeter $= 3x$,

Increased perimeter $= 1.2x + 1.3x + 1.5x = 4x$

$$\% \text{ increase in perimeter} = \frac{\text{Increase in perimeter}}{\text{Original perimeter}} \times 100 = \left(\frac{4x - 3x}{3x}\right) \times 100\%$$

$$= \frac{100}{3}\% = \mathbf{33\frac{1}{3}\%}.$$

Ex. 17. ***The base of a triangular field is three times its height. If the cost of cultivating the field at ₹ 26.38 per hectare is ₹ 356.13, find the base and height of the field.***

Sol. Area of the field $= \dfrac{356.13}{26.38}$ hectares $= 13.5 \times 10000\ \text{m}^2$

$= 135000\ \text{m}^2$.

Let the height of the field $= x$ m

Then, base $= 3x$ m

Given, $\dfrac{1}{2} \times 3x \times x\ \text{m}^2 = 135000\ \text{m}^2$

$\Rightarrow\quad x^2 = 90000\ \text{m}^2 \qquad \Rightarrow \qquad x = \sqrt{90000\ \text{m}^2} = 300$ m

$\therefore\quad$ Height $= 300$ m and Base $= 900$ m.

Ex. 18. ***The adjacent sides of a parallelogram are 8 m and 5 m. The distance between the longer sides is 4 m. What is the distance between the shorter sides?***

Sol. Area of the parallelogram $=$ Longer side $\times$ Distance between them

$= 8\ \text{m} \times 4\ \text{m} = 32\ \text{m}^2$

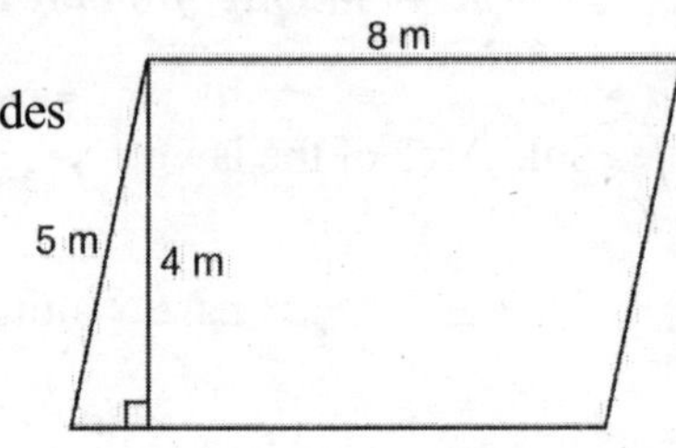

Also, area of the parallelogram $=$ Shorter side $\times$ Distance between shorter sides

$\Rightarrow\ 32\text{m}^2 = 5\text{m} \times$ Distance between shorter sides

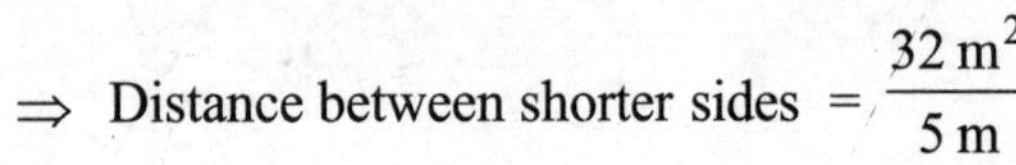

$\Rightarrow$ Distance between shorter sides $= \dfrac{32\ \text{m}^2}{5\ \text{m}}$

$=$ **6.4 m.**

Ex. 19. ***If the base of a parallelogram is (x + 4), altitude to the base is (x – 3) and the area is (x^2 – 4), then what is the actual area equal to?***

Sol. Area of the parallelogram = base × altitude

$$= (x+4) \times (x-3) = x^2 + 4x - 3x - 12$$
$$= x^2 + x - 12$$

Given, $x^2 + x - 12 = x^2 - 4 \Rightarrow x = 8.$

$\therefore$ Actual area = $(8)^2 - 4 = 64 - 4 =$ **60 sq units.**

Question Bank-21(a)

1. The length of a rectangle is increased by 60%. By what per cent would the width have to be reduced to maintain the same area?
(a) $37\frac{1}{2}\%$ (b) 60%
(c) 75% (d) 120%

2. A rectangular field has dimensions 25 m by 15 m. Two mutually perpendicular passages of 2 m width have been left in its central part and the grass has been grown in the rest of the field. The area under grass is:
(a) 295 m^2 (b) 299 m^2
(c) 300 m^2 (d) 375 m^2

3. The diagonal of a square is $4\sqrt{2}$ cm. The diagonal of another square whose area is double that of the first square is
(a) 8 cm (b) $8\sqrt{2}$ cm
(c) 16 cm (d) $4\sqrt{2}$ cm

4. If the length and breadth of a rectangular plot are each increased by 1 m, then the area of the floor is increased by 21 sq m. If the length is increased by 1 m and breadth is decreased by 1 m, then the area is decreased by 5 sq m. What is the perimeter of the floor?
(a) 30 m (b) 32 cm
(c) 36 m (d) 40 m

5. A typist uses a sheet measuring 20 cm by 30 cm lengthwise. If a margin of 2 cm is left on each side and a 3 cm margin on top and bottom, then the per cent of page used for typing is
(a) 40 (b) 60
(c) 64 (d) 72

6. A rectangular farm has to be fenced on one long side, one short side and the diagonal. If the cost of fencing is ₹ 100 per metre, the area of the farm is 1200 cm^2 and the short side is 30 m long, how much would the job cost?
(a) ₹ 14,000 (b) ₹ 12,000
(c) ₹ 7000 (d) ₹ 15,000

7. The diagonal of a rectangle is $\sqrt{41}$ cm and its area is 20 sq cm. The perimeter of the rectangle must be
(a) 9 cm (b) 18 cm
(c) 41 cm (d) 20 cm

8. The length and breadth of a rectangle are in the ratio 3:2 respectively. If the sides of the rectangle are extended on each side by 1 m, the ratio of length to breadth becomes 10:7. Find the area of the original rectangle in square metres.
(a) 2350 m^2 (b) 1150 m^2
(c) 1350 m^2 (d) 1000 m^2

9. The area of a 6 metres wide road outside a garden in all its four sides is 564 sq metres. If the length of the garden is 20 metres, what is its breadth?
(a) 18 metres (b) 16 metres
(c) 15 metres (d) 19 metres

10. The ratio between the length and breadth of a rectangular garden is 5:3. If the perimeter of the garden is 160 metres, what will be the area of 5 metre wide road around its outside?
(a) 600 m^2 (b) 1200 m^2
(c) 900 m^2 (d) 1000 m^2

11. A square S_1 encloses another square S_2 in such a manner that each corner of S_2 is at the midpoint of the side of S_1. If A_1 is the area of S_1 and A_2 is the area of S_2, then
(a) $A_1 = A_2$ (b) $A_2 = 2A_1$
(c) $A_1 = 2A_2$ (d) $A_1 = 4A_2$

12. The perimeter of a rectangle and a square are 160 m each. The area of the rectangle is less than that of the square by 100 square metre. The length of the rectangle is
(a) 30 m (b) 60 m
(c) 40 m (d) 50 m

13. The ratio between the length and perimeter of a rectangular plot is 1:3. What is the ratio between the length and breadth of the plot?
(a) 1 : 2 (b) 2 : 1
(c) 3 : 2 (d) 1 : 3

14. A rectangular paper, when folded into two congruent parts had a perimeter of 34 cm for each part folded along one set of sides and the same is 38 cm when folded along the other set of sides. What is the area of the paper?

(a) 140 cm^2 (b) 240 cm^2
(c) 560 cm^2 (d) 646 cm^2

15. 50 square stone slabs of equal size were needed to cover a floor area of 72 sq m. The length of each stone slab is

(a) 102 cm (b) 120 cm
(c) 201 cm (d) 210 cm

16. In a rectangle, the difference between the sum of adjacent sides and the diagonal is half the length of longer side. What is the ratio of the shorter to the longer side?

(a) $\sqrt{3}:2$ (b) $1:\sqrt{3}$
(c) 2 : 5 (d) 3 : 4

17. *A* took 15 seconds to cross a rectangular field diagonally walking at the rate of 52 m/min and *B* took the same time to cross the same field along its sides walking at the rate of 68 m/min. The area of the field is

(a) 30 m^2 (b) 40 m^2
(c) 50 m^2 (d) 60 m^2

18. A rectangular plank $\sqrt{2}$ m wide is placed symmetrically on the diagonal of a square of side 8 metres as shown. What is the area of the plank?

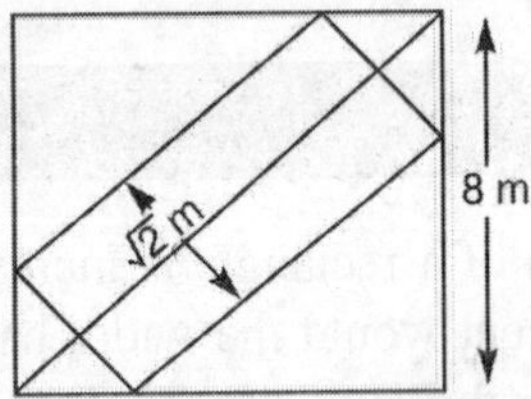

(a) $(16\sqrt{2}-3)$ sq m (b) $7\sqrt{2}$ sq m
(c) 98 sq m (d) 14 sq m

19. Four sheets of 50 cm × 5 cm are to be arranged in such a manner that a square could be formed. What will be the area of inner part of the square so formed?

(a) 2000 cm^2 (b) 2025 cm^2
(c) 1800 cm^2 (d) 2500 cm^2

Answers

1. (a)	**2.** (b)	**3.** (a)	**4.** (d)	**5.** (c)	**6.** (b)	**7.** (b)	**8.** (c)	**9.** (c)	**10.** (c)
11. (c)	**12.** (d)	**13.** (b)	**14.** (a)	**15.** (b)	**16.** (d)	**17.** (d)	**18.** (d)	**19.** (b)	

Hints and Solutions

1. (a) Let the original length and width of the rectangle be l and w respectively.

Original area = lw

New length = $l_1 = l + 60\%$ of $l = l + \frac{60l}{100} = 1.6l$

New width = b_1

Given, $l_1 w_1 = lw \Rightarrow 1.6l \times w_1 = lw$

$$\Rightarrow w_1 = \frac{lw}{1.6l} = \frac{10w}{16} = \frac{5}{8}w$$

$$\therefore \text{ \% reduction in width} = \left(\frac{w - 5/8w}{w} \times 100\right)\%.$$

$$= \left(\frac{3}{8} \times 100\right)\% = \mathbf{37\frac{1}{2}\%}.$$

2. (b) Area under grass

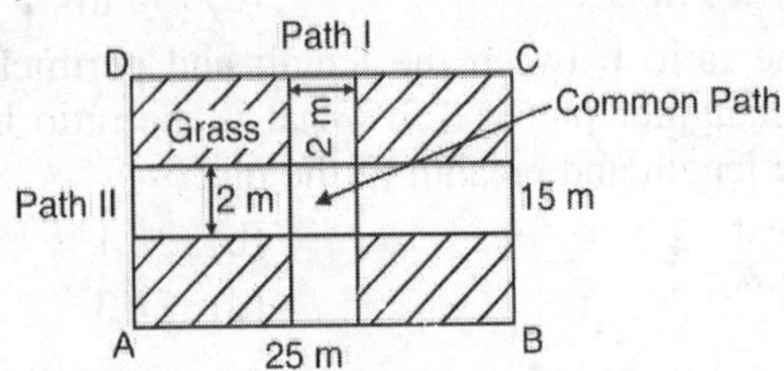

= Area of field – (Area of path I + Area of path II) + Area of common path

= 25 m × 15 m – (15 m × 2 m + 25 m × 2 m) + 2 m × 2 m

$= 375\text{ m}^2 - (30\text{ m}^2 + 50\text{ m}^2) + 4\text{ m}^2$

$= 379\text{ m}^2 - 80\text{ m}^2$

= **299 m^2**.

3. (a) Area of square with diagonal $4\sqrt{2}$ cm

$= \frac{1}{2} \times (\text{diagonal}) = \frac{1}{2} \times (4\sqrt{2})^2\text{ cm}^2 = 16\text{ cm}^2$

Area of the second square = $2 \times 16\text{ cm}^2 = 32\text{ cm}^2$

Let d cm be diagonal of the second square. Then

$\frac{1}{2}d^2\text{ cm}^2 = 32\text{ cm}^2 \Rightarrow d^2 = 2 \times 32\text{ cm}^2$

$\Rightarrow d = \sqrt{64\text{cm}^2}$ = **8 cm.**

4. (d) Let the length and breadth of the rectangle be x m and y m respectively.

Then, $(x + 1)(y + 1) = xy + 21$ ...(*i*)

$(x+1)(y-1) = xy - 5$ ...(ii)

$\Rightarrow xy + y + x + 1 = xy + 21$

$\Rightarrow x + y + 1 = 21 \Rightarrow x + y = 20$ ...(iii)

and $xy + y - x - 1 = xy - 5$

$\Rightarrow -x + y - 1 = -5 \Rightarrow x - y = 4$ ...(iv)

Adding (*iii*) and (*iv*), we get, $2x = 24 \Rightarrow x = 12$

Putting the value of x in (*iii*), we get, $12 + y = 20$

$\Rightarrow y = 8$.

$\therefore$ Perimeter $= 2(x + y) = 2 \times 20 =$ **40 m**.

5. (c) Area of the sheet $= 30 \text{ cm} \times 20 \text{ cm} = 600 \text{ cm}^2$

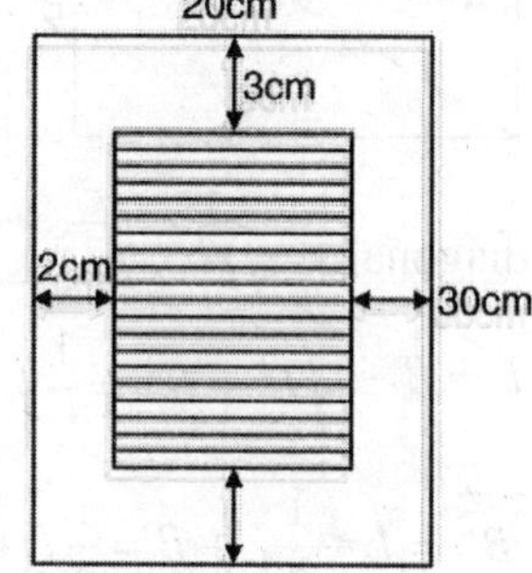

Area of the sheet used for typing

$= (30 - 6) \text{ cm} \times (20 - 4) \text{ cm}$

$= 24 \text{ cm} \times 16 \text{ cm} = 384 \text{ cm}^2$

$\therefore$ % of sheet used for typing

$= \left(\frac{384}{600} \times 100\right)\% = \mathbf{64\%}$.

6. (b) Let the length of the farm be l m.

Given breadth (b) = 30 m

Then, $l = \frac{\text{Area}}{\text{Breadth}} = \frac{1200 \text{ m}^2}{30 \text{ m}} = 40 \text{ m}$

Diagonal $= \sqrt{l^2 + b^2} = \sqrt{40^2 + 30^2}$ m

$= \sqrt{1600 + 900}$ m

$= \sqrt{2500}$ m $= 50$ m

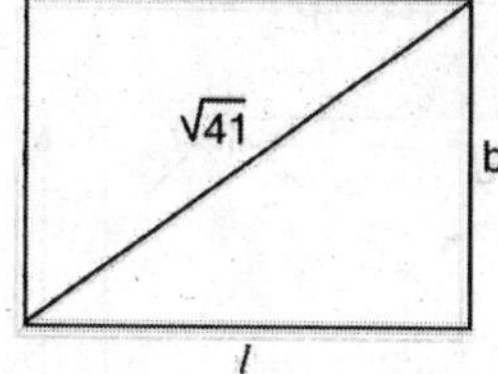

$\therefore$ Cost of fencing = ₹ {100 × Length of fence}

= ₹ {100 × (30 m + 40 m + 50 m)}

= ₹ {(100 × 120 m)} = **₹ 12000**.

7. (b) $l^2 + b^2 = (\sqrt{41})^2 = 41$. Also, $lb = 20$

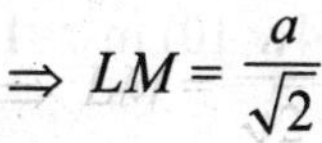

We know that

$(l + b)^2 = l^2 + b^2 + 2lb$

$= 41 + 2 \times 20 = 41 + 40 = 81$

$\Rightarrow l + b = \sqrt{81} = 9$

$\therefore$ Perimeter $= 2(l + b) = 2 \times 9 =$ **18 cm**.

8. (c) Let the length and breadth of the original rectangle be $3x$ and $2x$ respectively.

Given, $\frac{3x+1}{2x+1} = \frac{10}{7} \Rightarrow 7(3x+1) = 10(2x+1)$

$\Rightarrow 21x + 7 = 20x + 10$

$\Rightarrow 21x - 20x = 10 - 7 \Rightarrow x = 3$

Dimensions of the original rectangle are 3×3 m and 2×3 m, *i.e.*, 9 m and 6 m.

$\therefore$ Area of original rectangle = 45 m × 30 m

$= \mathbf{1350 \text{ m}^2}$.

9. (c) Let the breadth of the garden be b m. Then,

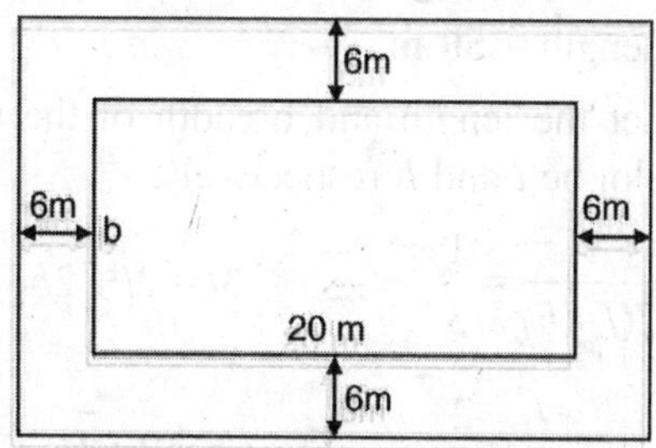

$(20 + 12) \times (b + 12) - (20) \times (b) = 564$

$32(b + 12) - 20b = 564$

$\Rightarrow 32b + 384 - 20b = 564$

$\Rightarrow 12b = 564 - 384 = 180 \Rightarrow b =$ **15 m**.

10. (c) Let the length and breadth of the rectangular garden be $5x$ metres and $3x$ metres.

Given $2(5x + 3x) = 160 \Rightarrow 16x = 160 \Rightarrow x = 10$

$\therefore$ Length of garden = 50 m and

Breadth of garden = 30 m

Area of 5 m road around its outside

$= (50 + 10) \times (30 + 10) - 50 \times 30$

$= 60 \times 40 - 50 \times 30$

$= (2400 - 1500)\text{m}^2 = \mathbf{900 \text{ m}^2}$.

11. (c) Let each side of the larger square be a units. Then,

By Pythagoras Theorem,

$LM^2 = LD^2 + DM^2$

$= (a/2)^2 + (a/2)^2$

$= \frac{2a^2}{4} = \frac{a^2}{2}$

$\Rightarrow LM = \frac{a}{\sqrt{2}}$

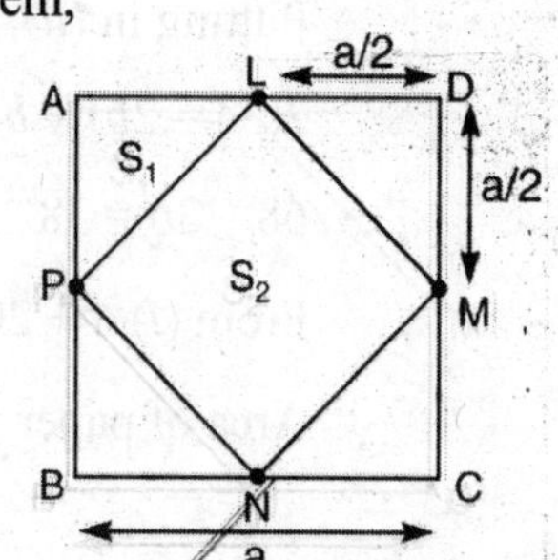

$\therefore \quad A_1 = \text{Area of } S_1 = a^2$ and

$A_2 = \text{Area of } S_2 = \left(\dfrac{a}{\sqrt{2}}\right)^2 = \dfrac{a^2}{2} = \dfrac{A_1}{2}$

$\Rightarrow \mathbf{A_1 = 2A_2}$

12. (d) Side of the square $= \dfrac{160}{4}$ m = 40 m.

Let the length and breadth of the rectangle be l and b respectively. Then,

$(40 \times 40) - (l \times b) = 100$

$\Rightarrow 1600 - lb = 100 \quad \Rightarrow \quad lb = 1500$

Also, $2(l + b) = 160 \Rightarrow l + b = 80 \quad \ldots(i)$

$\therefore \ (l - b)^2 = (l + b)^2 - 4lb = 80^2 - 4 \times 1500$
$= 6400 - 6000 = 400$

$\Rightarrow l - b = 20 \quad \ldots(ii)$

Adding (i) and (ii) $2l = 100 \quad \Rightarrow \quad l = 50$ m

$\therefore$ From (i) we get $b = 80 - 50 = 30$ m.

$\therefore$ Length = **50 m.**

13. (b) Let the length and breadth of the rectangular plot be l and b respectively.

$\dfrac{l}{2(l+b)} = \dfrac{1}{3} \quad \Rightarrow \quad 3l = 2l + 2b \Rightarrow l = 2b$

$\Rightarrow \quad \dfrac{l}{b} = \dfrac{2}{1} \quad \Rightarrow \quad l : b = \mathbf{2 : 1}.$

14. (a) When folded along the breadth, we have

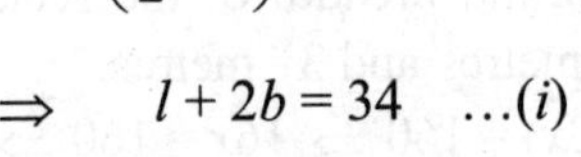

$2\left(\dfrac{l}{2} + b\right) = 34$

$\Rightarrow \quad l + 2b = 34 \quad \ldots(i)$

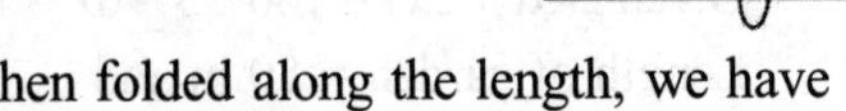

When folded along the length, we have

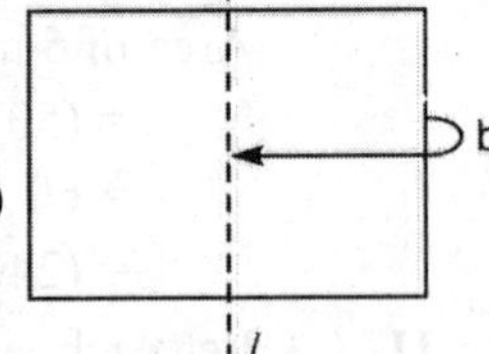

$2\left(l + \dfrac{b}{2}\right) = 38$

$\Rightarrow 2l + b = 38 \quad \ldots(ii)$

From (i) we have,

$l = 34 - 2b$

Putting in (ii),

$2(34 - 2b) + b = 38 \quad \Rightarrow \quad 68 - 4b + b = 38$

$\Rightarrow 68 - 3b = 38 \quad \Rightarrow \quad 3b = 30 \Rightarrow b = 10$

From (i), $l + 20 = 34 \quad \Rightarrow \quad l = 14$

$\therefore$ Area of paper $= l \times b = (14 \times 10)\ \text{m}^2 = \mathbf{140\ m^2}$.

15. (b) Area of each stone slab $= \dfrac{72}{50}\ \text{m}^2 = 1.44\ \text{m}^2$

$\therefore$ Length of each stone slab $= \sqrt{1.44}$ m
$= 1.2$ m $= (1.2 \times 100)$ cm
$=$ **120 cm.**

16. (d) Let the length and breadth of the rectangle be L and B respectively.

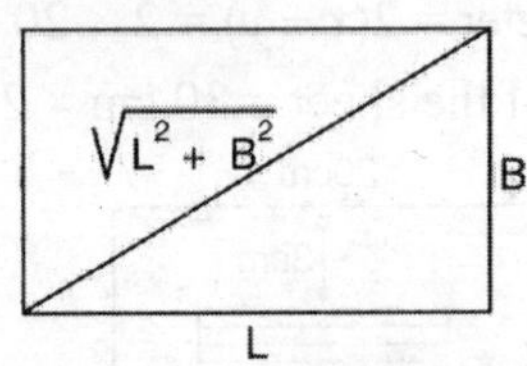

Then, diagonal $= \sqrt{L^2 + B^2}$

Given $L + B - \sqrt{L^2 + B^2} = \dfrac{1}{2}L$

$\Rightarrow \sqrt{L^2 + B^2} = L - \dfrac{1}{2}L + B = \dfrac{1}{2}L + B$

Squaring both the sides,

$L^2 + B^2 = \left(\dfrac{L}{2} + B\right)^2 \Rightarrow L^2 + B^2 = \dfrac{L^2}{4} + B^2 + LB$

$\Rightarrow \dfrac{3L^2}{4} = LB \quad \Rightarrow \quad \dfrac{LB}{L^2} = \dfrac{3}{4} \quad \Rightarrow \quad \dfrac{B}{L} = \dfrac{\mathbf{3}}{\mathbf{4}}.$

17. (d) Distance = Speed × Time

$\therefore$ Length of diagonal $= \left(52 \times \dfrac{15}{60}\right)$ m $= 13$ m

Also, Length + Breadth $= \left(68 \times \dfrac{15}{60}\right)$ m $= 17$ m

$\Rightarrow \sqrt{l^2 + b^2} = 13$ and $l + b = 17$ and Area $= lb = ?$

We know $(l + b)^2 = l^2 + b^2 + 2lb$

$\Rightarrow 2lb = (l + b)^2 - l^2 - b^2 = (l + b)^2 - (l^2 + b^2)$
$= (l + b)^2 - (\sqrt{l^2 + b^2})^2$
$= 17^2 - 13^2 = 289 - 169$
$= 120.$

$\therefore \quad lb = \dfrac{1}{2} \times 120 = \mathbf{60\ m^2}.$

18. (d) Let $AF = AE = a$

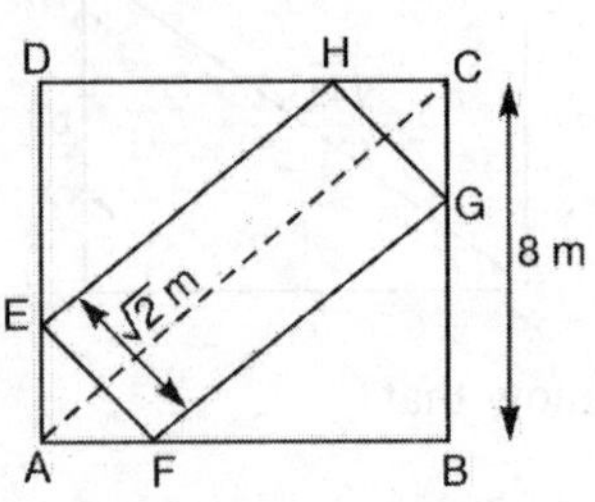

$\Rightarrow AF^2 + AE^2 = EF^2$ (By Pythagoras Th.)

$\Rightarrow a^2 + a^2 = (\sqrt{2})^2 \Rightarrow 2a^2 = 2 \Rightarrow a = 1$

$\therefore AF = CG = 1$ m

$\Rightarrow FB = BG = 8$ m $- 1$ m $= 7$ m

$\Rightarrow$ In ΔFGB,

$FG^2 = FB^2 + BG^2$ (Pythagoras Theorem)

$\Rightarrow FG^2 = 7^2 + 7^2 = 98 \Rightarrow FG = 7\sqrt{2}$

$\therefore$ Area of the plank $= FG \times FE$

$= (7\sqrt{2} \times \sqrt{2})$ m^2 = **14 m^2**.

19. (b) The four sheets are *MBNR*, *SNCO*, *PODL* and *QLAM*.

Side of the new square sheet *ABCD* $= (50 + 5)$ cm $= 55$ cm.

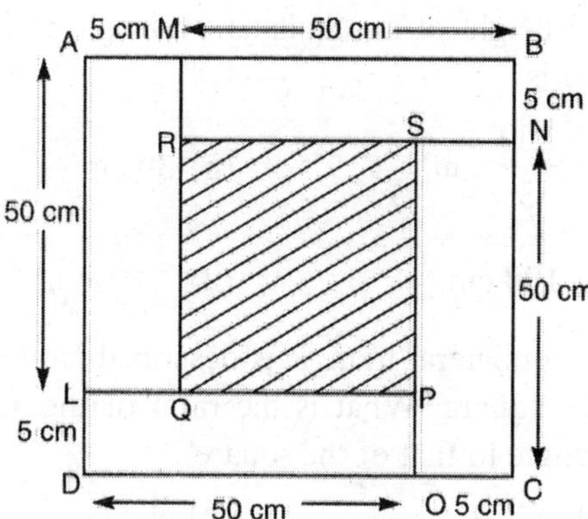

Side of the square sheet $PQRS = (55 - 10)$ cm $= 45$ cm

$\therefore$ Area of $PQRS = (45 \times 45)$ cm^2 = **2025 cm^2**.

Question Bank–21(b)

1. The base of a triangle is 15 cm and height is 12 cm. The height of another triangle of double the area having the base 20 cm is:

(a) 8 cm (b) 9 cm
(c) 12.5 cm (d) 18 cm

2. If the area of a triangle is 1176 m^2 and base : corresponding altitude is 3 : 4, then the altitude of the triangle is:

(a) 42 m (b) 52 m
(c) 54 m (d) 56 m

3. The hypotenuse of a right-angled isosceles triangle is 5 cm. The area of the triangle is

(a) 5 cm^2 (b) 6.25 cm^2
(c) 6.5 cm^2 (d) 12.5 cm^2

4. What is the area of the given figure? *ABCD* is a rectangle and *BDE* is an isosceles right triangle.

(a) ab (b) ab^2
(c) cab (d) $b(a + b/2)$

5. The ratio of the bases of two triangles is $x : y$ and that of their areas is $a : b$. The ratio of their corresponding altitudes will be

(a) $\frac{a}{x} : \frac{b}{y}$ (b) $ax : by$
(c) $ay : bx$ (d) $\frac{x}{a} : \frac{b}{y}$

6. If *D* and *E* are the midpoints of the sides *AB* and *AC* respectively of ΔABC in the figure given here, the shaded region is what per cent of the whole triangular region?

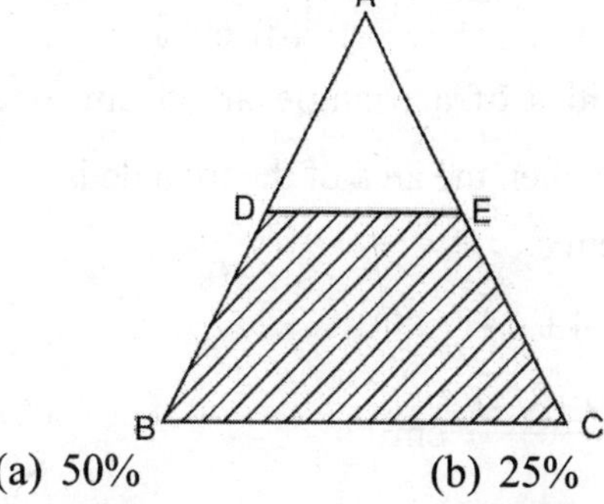

(a) 50% (b) 25%
(c) 75% (d) 60%

7. The perimeter of a right angled triangle is 60 cm. Its hypotenuse is 26 cm. The area of the triangle is

(a) 120 cm^2 (b) 240 cm^2
(c) 390 cm^2 (d) 780 cm^2

8. The area of an equilateral triangle is $400\sqrt{3}$ m^2. Its perimeter is

(a) 120 m (b) 150 m
(c) 90 m (d) 135 m

9. The areas of two equilateral triangles are in the ratio 25 : 36. Their altitudes will be in the ratio

(a) 36 : 25 (b) 25 : 36
(c) 5 : 6 (d) $\sqrt{5} : \sqrt{6}$

10. From a point within an equilateral triangle, perpendiculars drawn to the three sides are 6 cm, 7 cm and 8 cm respectively. The length of the side of the triangle is

(a) 7 cm (b) 10.5 cm
(c) $14\sqrt{3}$ cm (d) $\frac{14\sqrt{3}}{3}$ cm

11. The height of an equilateral triangle is 10 cm. Its area is

(a) $\frac{100}{3}$ cm^2 (b) 30 cm^2

(c) 100 cm^2 (d) $\frac{100}{\sqrt{3}}$ cm^2

12. An equilateral triangle is described on the diagonal of a square. What is the ratio of the area of the triangle to that of the square?

(a) $2:\sqrt{3}$ (b) $4:\sqrt{3}$

(c) $\sqrt{3}:2$ (d) $\sqrt{3}:4$

13. A square and an equilateral triangle have the same perimeter. If the diagonal of the square is $12\sqrt{2}$ cm, then the area of the triangle is

(a) $24\sqrt{3}$ cm^2 (b) $24\sqrt{2}$ cm^2

(c) $64\sqrt{3}$ cm^2 (d) $32\sqrt{3}$ cm^2

14. If the side of an equilateral triangle is decreased by 20%, its area is decreased by

(a) 36% (b) 64%

(c) 40% (d) 60%

15. If the sides of a triangle are 5 cm, 4 cm and $\sqrt{41}$ cm, then the area of the triangle is

(a) 20 cm^2

(b) $(5+4+\sqrt{41})$ cm^2

(c) $\frac{5+4+\sqrt{41}}{2}$ cm^2

(d) 10 cm^2

16. The area of a triangle is 216 cm^2 and its sides are in the ratio 3 : 4 : 5. The perimeter of the triangle is

(a) 6 cm (b) 12 cm

(c) 36 cm (d) 72 cm

17. In a triangular field having sides 30 m, 72 m and 78 m, the length of the altitude to the side measuring 72 m is

(a) 25 m (b) 28 m

(c) 30 m (d) 35 m

18. If every side of an equilateral triangle is doubled, the area of the new triangle is K times the area of the old one. K is equal to

(a) $\sqrt{2}$ (b) 2

(c) 3 (d) 4

19. If the perimeter of a right angled isosceles triangle is $(6+3\sqrt{2})$ m, then the area of the triangle will be

(a) 4.5 m^2 (b) 5.4 m^2

(c) 9 m^2 (d) 81 m^2

20. If A be the area of a right angled triangle and b is the length of one of the sides containing the right angle, then the length of the altitude on the hypotenuse is

(a) $\frac{2Ab}{\sqrt{b^2+4A^2}}$ (b) $\frac{2Ab}{b^2+4A^2}$

(c) $\frac{2Ab}{\sqrt{b^4+4A^4}}$ (d) $\frac{2Ab}{\sqrt{b^4+4A^2}}$

21. Inside an equiangular triangular park, there is a flower bed forming a similar triangle. Around the flower bed runs a uniform path of such a width that the sides of the park are exactly double the corresponding sides of the flower bed. The ratio of the areas of the path to the flower bed is

(a) 1 : 1 (b) 1 : 2

(c) 1 : 3 (d) 3 : 1

Answers

1. (d)	**2.** (d)	**3.** (b)	**4.** (d)	**5.** (a)	**6.** (c)	**7.** (a)	**8.** (a)	**9.** (c)	**10.** (c)
11. (d)	**12.** (c)	**13.** (c)	**14.** (a)	**15.** (d)	**16.** (d)	**17.** (c)	**18.** (d)	**19.** (a)	**20.** (d)
21. (d)									

Hints and Solutions

1. (d) Area of the given triangle = $\frac{1}{2}$ × base × height

$= \frac{1}{2} \times 15 \text{ cm} \times 12 \text{ cm}$

$= 90 \text{ cm}^2$.

Area of another triangle = 180 cm^2,

Base = 20 cm

$\therefore\ A = \frac{1}{2}bh \Rightarrow h = \frac{2A}{b} = \frac{2\times 180 \text{ cm}^2}{20 \text{ cm}} = \mathbf{18 \text{ cm}}.$

2. (d) Let the base and altitude of the triangle be $3x$ and $4x$ respectively. Then,

$\frac{1}{2}\times 3x\times 4x = 1176 \Rightarrow 6x^2 = 1176$

$\Rightarrow x^2 = 196 \quad \Rightarrow \quad x = 14$

$\therefore$ Altitude $= 4 \times 14$ m $=$ **56 m.**

3. (b) Let the two equal sides containing the right angle be a cm each. Then,

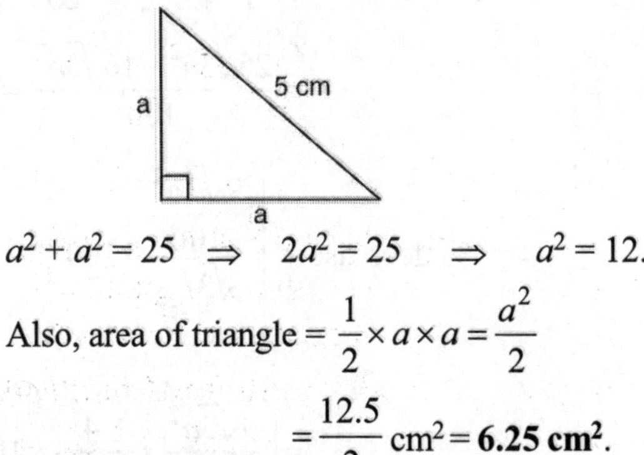

$a^2 + a^2 = 25 \Rightarrow 2a^2 = 25 \Rightarrow a^2 = 12.5$

Also, area of triangle $= \frac{1}{2} \times a \times a = \frac{a^2}{2}$

$= \frac{12.5}{2}$ cm$^2 =$ **6.25 cm^2.**

4. (d) Area of the given figure

$=$ Area of rectangle $ABCD +$ Area of ΔBDE

$= ab + \frac{1}{2} \times b \times b = \mathbf{b\left(a + \frac{b}{2}\right)}$.

5. (a) Let the corresponding altitudes be h_1 and h_2. Then,

$\frac{1}{2} \times h_1 \times x = a \Rightarrow h_1 = \frac{2a}{x}$ and $\frac{1}{2} \times h_2 \times y = b$

$\Rightarrow h_2 = \frac{2b}{y}$

$\therefore h_1 : h_2 = \frac{2a}{x} : \frac{2b}{y} = \mathbf{\frac{a}{x} : \frac{b}{y}}$.

6. (c) For ΔABC, let height $= h$ and base $BC = x$

$\therefore$ Area $= \frac{1}{2} hx$

For ΔADE, height $= h/2$ and base $DE = x/2$

$\therefore$ Area $= \frac{1}{2} \times \frac{h}{2} \times \frac{x}{2} = \frac{hx}{8}$

Area of shaded portion $= \frac{hx}{2} - \frac{hx}{8} = \frac{3hx}{8}$

$\therefore$ Required % $= \left(\frac{\frac{3}{8}hx}{\frac{hx}{2}} \times 100\right)\% =$ **75%.**

7. (a) Let the other two sides be x and y.

Then, $x + y + 26 = 60$

$\Rightarrow x + y = 34$

Also, hypotenuse $= \sqrt{x^2 + y^2} = 26$

$\Rightarrow x^2 + y^2 = 26^2 = 676$

Now,

$(x + y)^2 = x^2 + y^2 + 2xy \Rightarrow 34^2 = 676 + 2xy$

$\Rightarrow 2xy = 1156 - 676 = 480 \Rightarrow xy = 240$

Area of the $\Delta = \frac{1}{2} xy = \frac{1}{2} \times 240 =$ **120 m^2.**

8. (a) Let each side of the equilateral triangle be a m. Then,

$\frac{\sqrt{3}}{4} a^2 = 400\sqrt{3} \Rightarrow a^2 = 1600 \Rightarrow a = 40$ m

$\therefore$ Perimeter $= 3a =$ **120 m.**

9. (c) Let the sides of the equilateral triangle be a and b respectively. Then,

$\frac{\frac{\sqrt{3}}{4} a^2}{\frac{\sqrt{3}}{4} b^2} = \frac{25}{36} \Rightarrow \frac{a}{b} = \frac{5}{6}$

If h_1 and h_2 be their altitudes, then $\frac{\frac{1}{2} \times a \times h_1}{\frac{1}{2} \times b \times h_2} = \frac{25}{36}$

$\Rightarrow \frac{5h_1}{6h_2} = \frac{25}{36} \Rightarrow \frac{h_1}{h_2} = \frac{25}{36} \times \frac{6}{5} = \frac{5}{6}$

$\Rightarrow h_1 : h_2 =$ **5 : 6.**

10. (c) Type Solved Example 11.

Let the length of each side of the Δ be x cm.

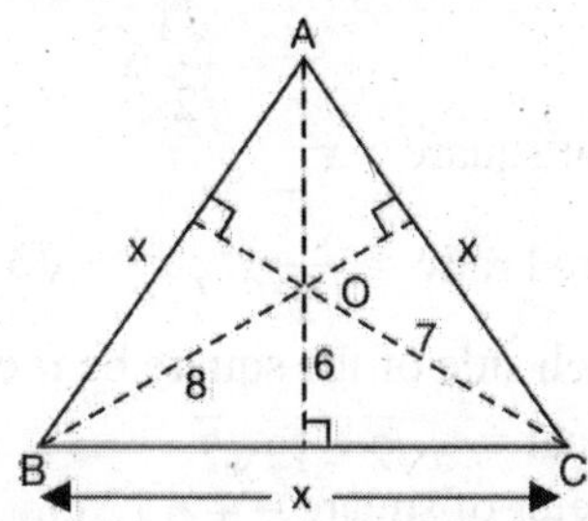

Then,

$\frac{1}{2} \times x \times 6 + \frac{1}{2} \times x \times 7 + \frac{1}{2} \times x \times 8 = \frac{\sqrt{3}}{4} x^2$

$\Rightarrow x(3 + 3.5 + 4) = \frac{\sqrt{3}}{4} x^2$

$\Rightarrow x = \frac{10.5 \times 4}{\sqrt{3}} = \frac{42}{\sqrt{3}} \times \frac{\sqrt{3}}{\sqrt{3}} = \frac{42\sqrt{3}}{3} =$ **$14\sqrt{3}$ cm.**

11. (d) Let each side of the Δ be x cm. Then, as altitude bisects the base,

$QS = SR = x/2$

In ΔPQS,

$$\left(\frac{x}{2}\right)^2 + 10^2 = x^2$$

$$\Rightarrow \frac{x^2}{4} + 100 = x^2$$

$$\Rightarrow \frac{3x^2}{4} = 100 \Rightarrow x^2 = \frac{400}{3}$$

Area of the equilateral $\Delta = \frac{\sqrt{3}}{4}x^2$

$$= \frac{\sqrt{3}}{4} \times \frac{400}{3} = \frac{\mathbf{100}}{\sqrt{3}} \textbf{ cm}^2.$$

12. (c) Let each side of the square be x units. Then diagonal of square = $x\sqrt{2}$ units.

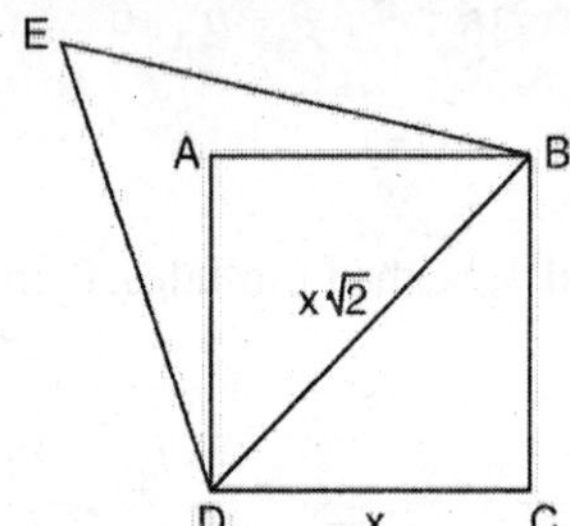

$\Rightarrow$ Each side of the equilateral $\Delta = x\sqrt{2}$ units

Area of equilateral $\Delta = \frac{\sqrt{3}}{4}(x\sqrt{2})^2$

$$= \frac{\sqrt{3}}{2}x^2$$

Area of square = x^2

$\therefore$ Required ratio = $\frac{\sqrt{3}}{2}x^2 : x^2 = \mathbf{\sqrt{3} : 2}$.

13. (c) Let each side of the square be a cm. Then,

Diagonal = $a\sqrt{2} = 12\sqrt{2} \Rightarrow a = 12$ cm

$\therefore$ Perimeter of square = 4 × 12 cm = 48 cm

$\Rightarrow$ Perimeter of equilateral $\Delta = 48$ cm

$\Rightarrow$ Each side of equilateral $\Delta = \frac{48}{3}$ cm = 16 cm

$\therefore$ Area of $\Delta = \frac{\sqrt{3}}{4}$ (side)^2 = $\frac{\sqrt{3}}{4} \times 16 \times 16$ cm^2

$$= \mathbf{64\sqrt{3}} \textbf{ cm}^2.$$

14. (a) Let each side of the original $\Delta = a$ cm. Then,

its area = $\frac{\sqrt{3}}{4}a^2$

Each side of the Δ after 20% decrease

$$= 80\% \text{ of } a = \frac{80}{100}a \text{ cm} = \frac{4}{5}a \text{ cm}$$

$\therefore$ Area of Δ after decrease = $\frac{\sqrt{3}}{4} \times \left(\frac{4}{5}a\right)^2$

$$= \frac{4\sqrt{3}a^2}{25}$$

$\therefore$ Decrease in area = $\left(\frac{\sqrt{3}}{4}a^2 - \frac{4\sqrt{3}a^2}{25}\right)$

$$= \frac{25\sqrt{3}a^2 - 16\sqrt{3}a^2}{100} = \frac{9\sqrt{3}a^2}{100}$$

$\therefore$ % decrease = $\left(\frac{\frac{9\sqrt{3}a^2}{100}}{\frac{\sqrt{3}}{4}a^2} \times 100\right)\%$

$$= \left(\frac{9\sqrt{3}a^2}{100} \times \frac{4}{\sqrt{3}a^2} \times 100\right)\%$$

$$= \mathbf{36\%}.$$

15. (d) Since $5^2 + 4^2 = (\sqrt{41})^2$, the given triangle is a right angled Δ.

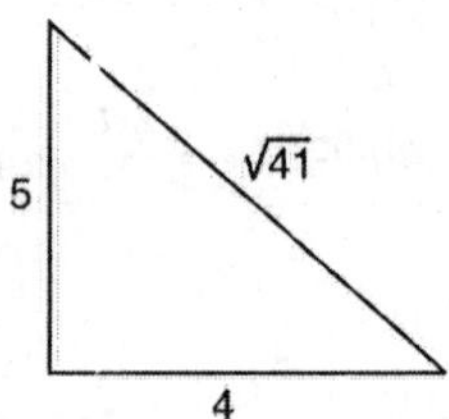

$\therefore$ Area of $\Delta = \frac{1}{2} \times 5 \times 4$ cm^2 = **10 cm^2**.

16. (d) Let the sides of the triangle be $3x$ cm, $4x$ cm and $5x$ cm. Then,

$$s = \frac{3x + 4x + 5x}{2} = \frac{12x}{2} = 6x$$

$\therefore$ Area of Δ = $\sqrt{s(s-a)(s-b)(s-c)}$

$$= \sqrt{6x(6x-3x)(6x-4x)(6x-5x)}$$

$$= \sqrt{6x \times 3x \times 2x \times x} = 6x^2$$

Given, $6x^2 = 216 \Rightarrow x^2 = \frac{216}{6} = 36 \Rightarrow x = 6$.

$\therefore$ The sides of the triangle are (3 × 6) cm, (4 × 6) cm and (5 × 6) cm, *i.e*, 18 cm, 24 cm and 30 cm.

Perimeter of the Δ = 18 cm + 24 cm + 30 cm

= **72 cm.**

17. (c) $s = \frac{30 + 72 + 78}{2} = \frac{180}{2} = 90$

$\therefore$ Area of $\Delta = \sqrt{90(90-30)(90-72)(90-78)}$ m^2

$$= \sqrt{90 \times 60 \times 18 \times 12} \text{ m}^2 = 1080 \text{ m}^2$$

Area of the Δ with base (72 m)

$$=\frac{1}{2}\times 72\times \text{altitude}=1080$$

$\therefore$ Required length of altitude $=\frac{1080\times 2}{72}=$ **30 m.**

18. (d) $A_1=\frac{\sqrt{3}}{4}a^2$ and $A_2=\frac{\sqrt{3}}{4}(2a^2)=4\times\frac{\sqrt{3}}{4}a^2=4A_1$

$\Rightarrow$ **$K=4$.**

19. (a) Let the two equal sides be x metres, then hypotenuse $= x\sqrt{2}$ metres

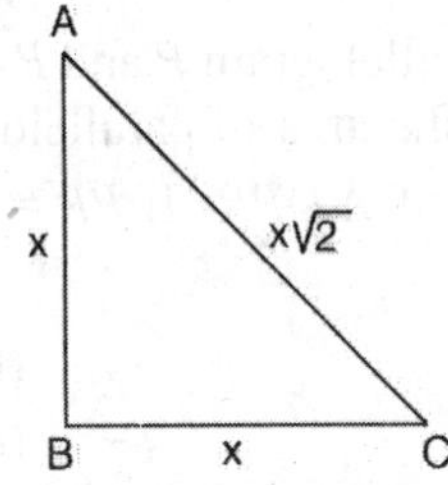

$$\Rightarrow x+x+x\sqrt{2}=6+3\sqrt{2}$$

$$\Rightarrow 2x+x\sqrt{2}=6+3\sqrt{2}$$

$$\Rightarrow x(2+\sqrt{2})=3(2+\sqrt{2}) \Rightarrow x=3.$$

$\therefore$ Area of the triangle $=\left(\frac{1}{2}\times 3\times 3\right)$ m² $=$ **4.5 m².**

20. (d) Let the length of the other side and the altitude on the hypotenuse be x and p respectively. Then,

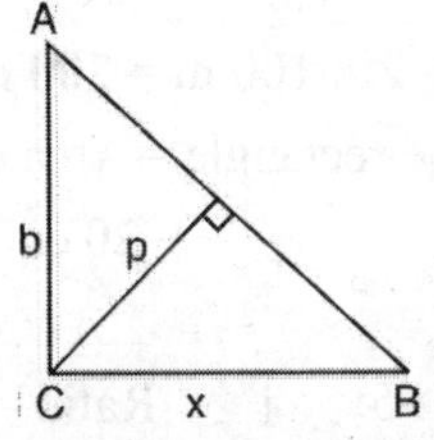

$$A=\frac{1}{2}\times b\times x \text{ and also } A=\frac{1}{2}\times p\times AB$$

$$\Rightarrow x=\frac{2A}{b} \text{ and } p=\frac{2A}{AB}$$

$$AB^2=x^2+b^2$$

$$\Rightarrow AB=\sqrt{x^2+b^2}=\sqrt{\left(\frac{2A}{b}\right)^2+b^2}$$

$$\therefore\ p=\frac{2A}{\sqrt{\left(\frac{2A}{b}\right)^2+b^2}}=\frac{2A}{\sqrt{\frac{1}{b^2}(4A^2+b^4)}}$$

$$=\mathbf{\frac{2Ab}{\sqrt{b^4+4A^2}}}.$$

21. (d) Let each side of the park be $2x$.
Then, each side of the flower bed $= x$
Area of path
= Area of park – Area of flower bed

$$=\frac{\sqrt{3}}{4}(2x)^2-\frac{\sqrt{3}}{4}(x)^2=\frac{\sqrt{3}}{4}\times 3x^2$$

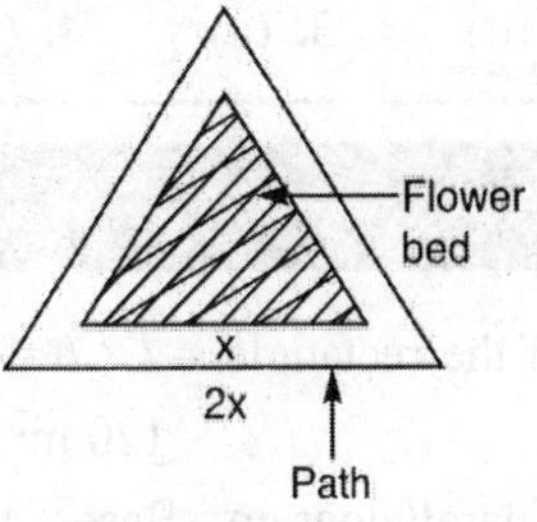

$\therefore$ Ratio of area of path: Flower bed

$$=\frac{\frac{\sqrt{3}}{4}\times 3x^2}{\frac{\sqrt{3}}{4}\times x^2}=\frac{3}{1}=\mathbf{3:1}.$$

Question Bank–21(c)

1. A rectangle and a parallelogram have equal areas. If the sides of a rectangle are 10 m and 12 m and the base of the parallelogram is 20 m, then the altitude of the parallelogram is
   (a) 7 m (b) 6 m
   (c) 5 m (d) 3 m
2. If a parallelogram with area P, a rectangle with area R and a triangle with area T are all constructed on the same base and all have the same altitude, then a false statement is
   (a) $P=2T$ (b) $T=\frac{1}{2}R$
   (c) $P=R$ (d) $P+T=2R$
3. The base of a parallelogram is three times its height. If the area of the parallelogram is 75 sq cm, then its height is
   (a) 5 cm (b) $5\sqrt{2}$ cm
   (c) $3\sqrt{2}$ cm (d) 15 cm

4. A triangle and a parallelogram are constructed on the same base such that their areas are equal. If the altitude of the parallelogram is 100 m, then the altitude of the triangle is

(a) 100 m (b) 200 m

(c) $100\sqrt{2}$ m (d) $10\sqrt{2}$ m

5. A rectangle and a parallelogram have equal areas. The base of the parallelogram is 20 cm and the altitude is 6 cm. Which one of the following cannot be the ratio of dimensions of the rectangle?

(a) 7 : 5 (b) 40 : 3

(c) 15 : 2 (d) 30 : 1

6. A parallelogram has sides 30 m, 70 m and one of its diagonals is 80 m long. Its area will be

(a) 600 m² (b) $1200\sqrt{3}$ m²

(c) 1200 m² (d) $600\sqrt{3}$ m²

7. One diagonal of a parallelogram is 40 cm and the perpendicular distance of this diagonal from either of the outlying vertices is 19 cm. The area of the parallelogram (in sq cm) is

(a) 700 cm² (b) 380 cm²

(c) 760 cm² (d) 1140 cm²

8. The ratio of two adjacent sides of a parallelogram is 3 : 4. Its perimeter is 105 cm. Find its area if altitude corresponding to the larger side is 15 cm.

(a) 900 cm² (b) 600 cm²

(c) 300 cm² (d) 450 cm²

9. The area of a rhombus is 128 cm² and its perimeter is 32 cm. The altitude of the rhombus is

(a) 7 cm (b) 8 cm

(c) 16 cm (d) 12 cm

10. *ABCD* is a parallelogram *P* and *R* are two points on *AB* such that the area of parallelogram *ABCD* is 8 times the area of ΔDPR. If $PR = 5$ cm, then *CD* is equal to

(a) 10 cm (b) 5 cm

(c) 20 cm (d) 12 cm

Answers

1. (b)	2. (d)	3. (a)	4. (b)	5. (a)	6. (b)	7. (c)	8. (d)	9. (c)	10. (c)

Hints and Solutions

1. (b) Area of the rectangle $= l \times b = 10 \text{ m} \times 12 \text{ m} = 120 \text{ m}^2$

Area of parallelogram = Base × Altitude = 120 m²

$\Rightarrow$ Altitude $= \dfrac{\text{Area}}{\text{Base}} = \dfrac{120 \text{ m}^2}{20 \text{ m}} = $ **6 m.**

2. (d) Let the base of each of them be *b* units and the height = *h* units. Then,

$P = b \times h, R = b \times h$ and $T = \dfrac{1}{2} \times b \times h$.

Then, $P = R, P = 2T$ and $T = \dfrac{1}{2}R$ are all correct statements.

3. (a) Let the height of the parallelogram be *x* m.

Then, base = 3*x*

$\therefore$ Area $= b \times h = x \times 3x = 75$

$\Rightarrow 3x^2 = 75 \quad \Rightarrow \quad x^2 = \dfrac{75}{3}$

$\Rightarrow x^2 = 25 \quad \Rightarrow \quad x = 5$

$\therefore$ Height = **5 cm.**

4. (b) Let the altitude of the $\Delta = h_1$, altitude of the parallelogram $= h_2$ and base of both Δ and parallelogram = *b*.

Then, $b \times h_2 = \dfrac{1}{2} \times b \times h_1$ where $h_2 = 100$ m

$\therefore \ h_1 = 2h_2 = 2 \times 100 \text{ m} = $ **200 m.**

5. (a) Area of the rectangle = Area of parallelogram $= 20 \text{ cm} \times 6 \text{ cm} = 120 \text{ cm}^2$

Now,

Ratio = 7 : 5	Ratio = 40 : 3
Product = 35	Product = 120
Ratio = 15 : 2	Ratio = 30 : 1
Product = 30	Product = 30

$\therefore$ Ratio = 7 : 5 does not match with the condition as 120 is not divisible by 35.

6. (b) The diagonal of a parallelogram divides it into two congruent triangle, so

Area (parallelogram *ABCD*) = 2 × Area (ΔABC)

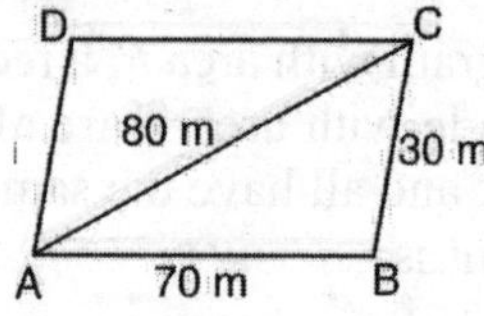

For ΔABC,

$$s = \frac{80\text{m}+30\text{m}+70\text{m}}{2} = \frac{180\text{m}}{2} = 90\text{m}$$

$\therefore$ Area $= \sqrt{90(90-80)(90-30)(90-70)}\ \text{m}^2$

$= \sqrt{90\times10\times60\times20}\ \text{m}^2$

$= 600\sqrt{3}\ \text{m}^2$

$\therefore$ Area of parallelogram $ABCD = 2$(area of $\Delta\ ABC$)

$= (2\times600\sqrt{3})\ \text{m}^2$

$=$ **$1200\sqrt{3}\ \text{m}^2$**.

7. (c)

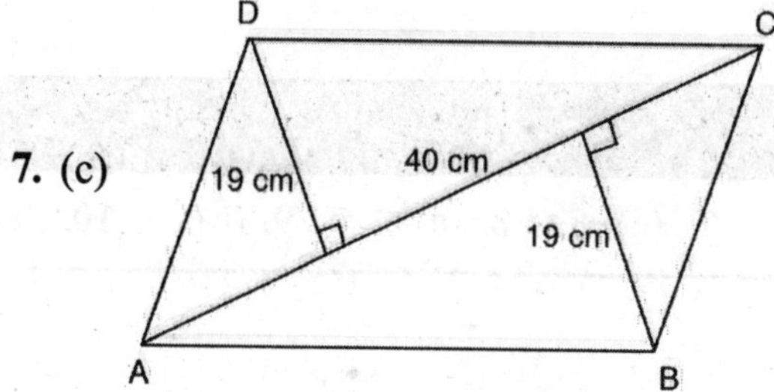

Area of the parallelogram

= Area of ΔABC + Area of ΔACD

$= \frac{1}{2}\times40\times19\ \text{cm}^2 + \frac{1}{2}\times40\times19\text{cm}^2$

$= 380\ \text{cm}^2 + 380\ \text{cm}^2 =$ **$760\ \text{cm}^2$**.

8. (d) Let the two adjacent sides of the parallelogram be $3x$ and $4x$. Then,

$2(3x + 4x) = 105$

$\Rightarrow 14x = 105$

$\Rightarrow x = 7.5$

$\therefore$ The two sides are 3×7.5 cm and 4×7.5 cm, *i.e*, 22.5 cm and 30 cm.

Area of the parallelogram = base × altitude

$= 30\ \text{cm}\times15\ \text{cm} =$ **$450\ \text{cm}^2$**.

9. (c) A rhombus is a parallelogram with all four sides equal.

$\therefore$ Each side of the rhombus $= \frac{32}{4}$ cm = 8 cm,

Area = 128 cm^2

$\Rightarrow$ base × altitude = 128 cm^2

$\Rightarrow$ altitude $= \frac{128}{8}$ cm = **16 cm**.

10. (c) Area of parallelogram $ABCD$

$= 8\times$ Area of ΔDPR

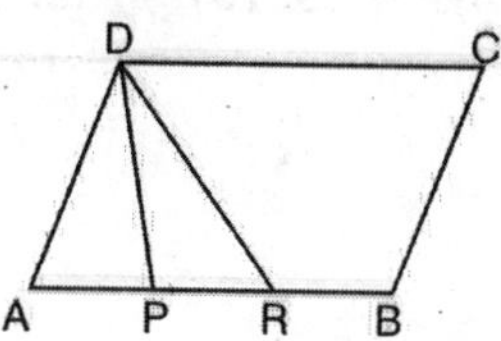

$\Rightarrow AB\times\text{height} = 8\times\left(\frac{1}{2}\times PR\times\text{height}\right)$

(**Note:** Height will be same for both the Δ and parallelogram)

$\Rightarrow AB = 4\times PR = 4\times5\ \text{cm} = 20\ \text{cm}$

$\therefore\ CD = 20$ cm

(Opp. sides of a parallelogram are equal)

Self Assessment Sheet–21

1. Two sides of a parallelogram are 10 cm and 15 cm. If the altitude corresponding to the side of length 15 cm is 5 cm, then what is the altitude to the side of length 10 cm ?

(a) 5 cm (b) 7. 5 cm

(c) 10 cm (d) 15 cm

2. What is the area of a right angled isosceles triangle whose hypotenuse is $6\sqrt{2}$ cm?

(a) 12 cm^2 (b) 18 cm^2

(c) 24 cm^2 (d) 36 cm^2

3. If A is the area of a triangle in cm^2, whose sides are 9 cm, 10 cm and 11 cm, then which one of the following is correct ?

(a) $A < 40\ \text{cm}^2$

(b) $40\ \text{cm}^2 < A < 45\ \text{cm}^2$

(c) $45\ \text{cm}^2 < A < 50\ \text{cm}^2$

(d) $A > 50\ \text{cm}^2$

4. The cost of turfing a triangular field at the rate of ₹ 45 per 100 m^2 is ₹ 900. If double the base of the triangle is 5 times the height, then the height is :

(a) 50 m (b) 45 m

(c) 60 m (d) 40 m

5. The triangular side - wall of a flyover have been used for advertisements. The sides of the walls are 122 m, 22 m and 120 m. The advertisements yield an earning of Rs 5000 per m^2 per year. A company hired one of its walls for 3 months. How much rent did it pay ?

(a) 1750000 (b) 1600000

(c) 1650000 (d) None of these

6. The perimeter of a square is 48 m. The area of a rectangle is 4 sq m less than the area of given square. If the length of the rectangle is 14 m, find the breadth.

(a) 8 m (b) 9 m

(c) 10.5 m (d) 10 m

7. A rectangular lawn 80 m × 60 m has two roads each 10 m wide running in the middle of it, one parallel to the length and the other parallel to the breadth. The cost of gravelling them at ₹ 30 per square metre is

(a) ₹ 38000 (b) ₹ 40000
(c) ₹ 39000 (d) ₹ 39500

8. A diagonal of a rhombus is 80% of the other diagonal. Then, area of the rhombus is how many times the square of the length of the longer diagonal?

(a) $\frac{2}{5}$ (b) $\frac{4}{5}$
(c) $\frac{3}{4}$ (d) $\frac{1}{4}$

9. The area of a rhombus, one of whose diagonals measures 8 cm and the side is 5 cm, is :

(a) 25 cm^2 (b) 24 cm^2
(c) 24. 5 cm^2 (d) 26 cm^2

10. A parallelogram has two sides 60 m and 25 m and a diagonal 65 m long. The area of the parallelogram is :

(a) 1000 m^2 (b) 1400 m^2
(c) 1600 m^2 (d) 1500 m^2

Answers

1. (b)	2. (b)	3. (b)	4. (d)	5. (c)	6. (d)	7. (c)	8. (a)	9. (b)	10. (d)

Chapter

22 CIRCUMFERENCE AND AREA OF A CIRCLE

KEY FACTS

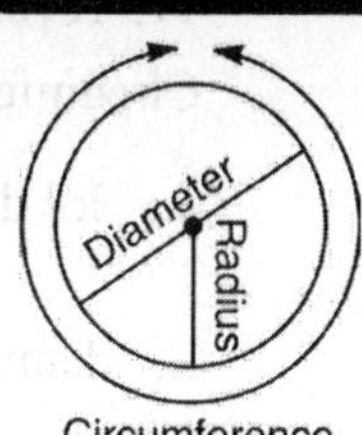

1. The length of the **diameter 'd'** is twice the length of the **radius 'r'**.

2. The distance around a circle is called the **circumference** of the circle.
   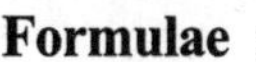
   Formulae :
   Circumference $= \pi \times \text{diameter}$
   $= 2\pi \times \text{radius}$, *i.e.*,
   $C = 2\pi r$

   Note: Some reasonably useful approximations of π are $3\frac{1}{7}$, $\frac{22}{7}$, 3.14 etc.

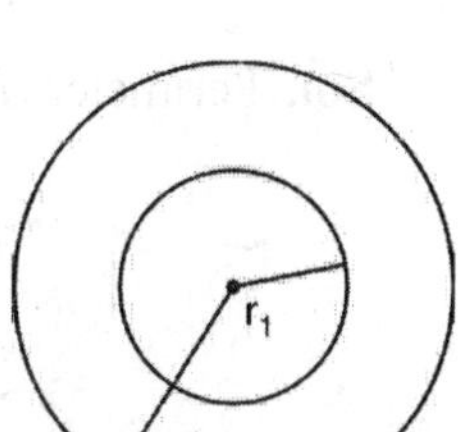

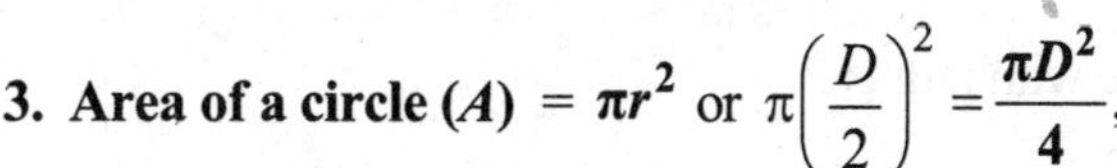

3. **Area of a circle (A)** $= \pi r^2$ or $\pi\left(\frac{D}{2}\right)^2 = \frac{\pi D^2}{4}$,
   where r is the radius and D is the diameter of the circle.

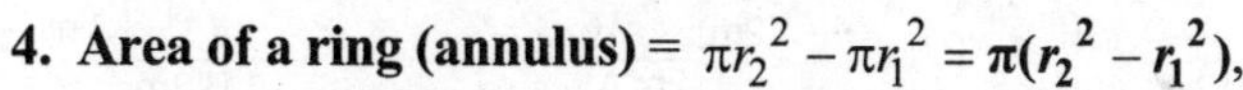

4. **Area of a ring (annulus)** $= \pi r_2^2 - \pi r_1^2 = \pi(r_2^2 - r_1^2)$,
   where r_1 is the radius of the inner circle and r_2 is the radius of the outer circle.

Solved Examples

Ex. 1. ***The area of a circular plot is 3850 square metres. What is the circumference of the plot ?***

Sol. Let r be the radius of the plot. Then,

$$\pi r^2 = 3850 \Rightarrow r^2 = \frac{3850 \times 7}{22}$$

$$\Rightarrow r^2 = 1225 \Rightarrow r = 35 \text{ m}$$

$\therefore$ Circumference of the plot $= 2\pi r = 2 \times \frac{22}{7} \times 35$ m
$= \mathbf{220\ m.}$

Ex. 2. ***What is the length (in metres) of a rope, by which a cow must be tethered in order that it may graze over an area of 616 m² ?***

Sol. Let x be the length of the rope. Then,
The cow can graze over a circular area of radius x.

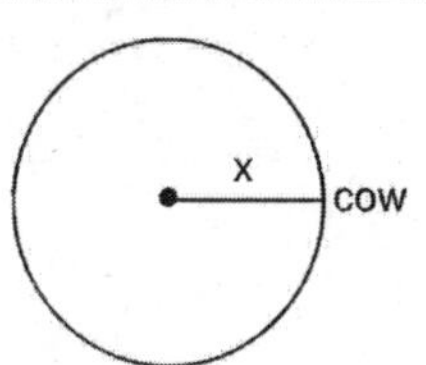

So, $\pi x^2 = 616 \Rightarrow x^2 = \frac{616 \times 7}{22} = 196 \Rightarrow x = \mathbf{14}$ **metres.**

Ex. 3. ***The radius of a wheel is 0.25 m. How many rounds will it take to complete the distance of 11 km ?***

Sol. Distance covered in one round = Circumference of wheel $= 2\pi r = \left(2\times\frac{22}{7}\times 0.25\right)\text{m} = \frac{11}{7}\text{ m}$

$\therefore$ Number of rounds to cover 11 km $= \dfrac{\text{Total distance covered}}{\text{Distance covered in one round}} = \dfrac{11\times 1000}{\frac{11}{7}} = \mathbf{7000}.$

Ex. 4. ***Find the diameter of a wheel that makes 113 revolutions to go 2 km 26 decametres*** $\left(\pi = \frac{22}{7}\right)$.

Sol. 2 km 26 dam = (2 × 1000 + 26 × 10) m
= 2260 m

113 revolutions are equivalent to 2260 m $\Rightarrow$ 1 revolution is equivalent to $\frac{2260}{113}$ m = 20 m

Circumference of wheel = 20 m

$\Rightarrow \pi\times\text{diameter} = 20\text{ m} \Rightarrow \frac{22}{7}\times\text{diameter} = 20\text{ m}$

$\Rightarrow \text{diameter} = \frac{20\times 7}{22} = \frac{70}{11}\text{ m} = \mathbf{6\frac{4}{11}\ m}.$

Ex. 5. ***What is the perimeter of the given figure correct to one decimal place ?***

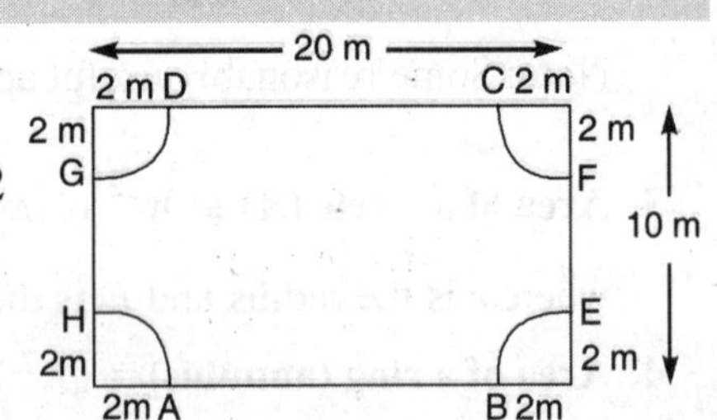

Sol. Perimeter $(P) = AB + EF + CD + GH + 4 \times \text{Arc } AH$

$= 2 \times AB + 2 \times EF$ + Circumference of circle with radius 2

$= 2 \times (20 - 4)\text{ m} + 2 \times (10 - 4)\text{ m} + 2 \times \frac{22}{7} \times 2$

$= 32\text{ m} + 12\text{ m} + 12.57\text{ m} = \mathbf{56.6\ m.}$

Ex. 6. ***In the given figure PQRS is a rectangle of 8 cm × 6 cm inscribed in a circle. What is the area of the shaded portion ?***

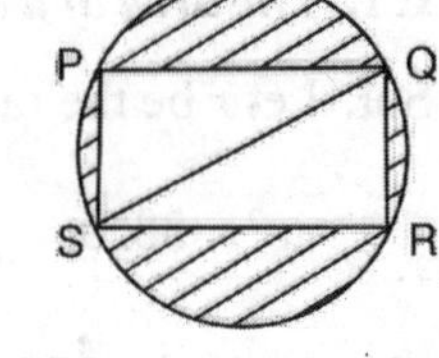

Sol. Diameter of the circle = Diagonal of the rectangle

$\Rightarrow SQ^2 = SR^2 + QR^2 = 8^2 + 6^2 = 64 + 36 = 100 \Rightarrow SQ = \sqrt{100}\text{ cm} = 10\text{ cm}$

$\therefore$ Area of shaded portion = Area of circle – Area of rectangle $PQRS$

$= \frac{22}{7}\times 5\times 5\text{ cm}^2 - 8\times 6\text{ cm}^2$

$= 78.57\text{ cm}^2 - 48\text{ cm}^2 = \mathbf{30.57\ cm^2}.$

Ex. 7. ***In the given figure, ABC is a right angled triangle with B as right angle. Three semicircles are drawn with AB, BC and AC as diameters. What is the area of the shaded portion, if the area of △ ABC is 12 square units ?***

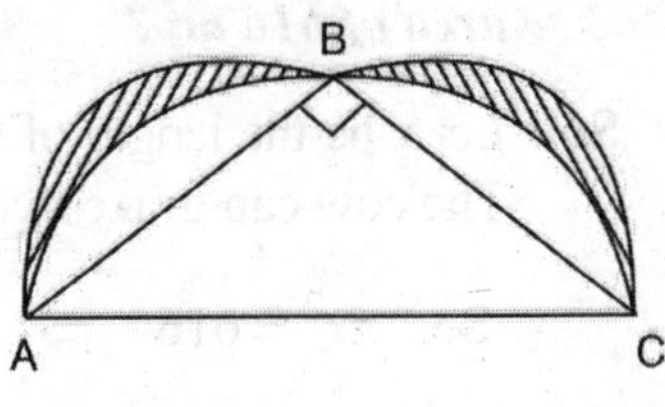

Sol. Area of shaded portion = (Area of two semi-circles – Area of biggest semicircle) + Area of ΔABC

$= \left\{\frac{1}{2}\pi\left(\frac{AB}{2}\right)^2 + \frac{1}{2}\pi\left(\frac{BC}{2}\right)^2 - \frac{1}{2}\pi\left(\frac{AC}{2}\right)^2\right\} + 12$

$= \frac{\pi}{8}(AB^2 + BC^2 - AC^2) + 12$

$= \frac{\pi}{8}(AC^2 - AC^2) + 12 = 0 + 12 = \mathbf{12\ sq\ units.}$

($\because AB^2 + BC^2 = AC^2$, by Pythagoras Theorem)

Ex. 8. ***PQRS is a diameter of a circle. The lengths of PQ, QR and RS are equal. Semi circles are drawn on PQ and QS to create the shaded figure or given. What is the perimeter of the shaded figure ?***

Sol. $PS = 2r \quad \Rightarrow \quad PQ = QR = RS = \dfrac{2r}{3}$

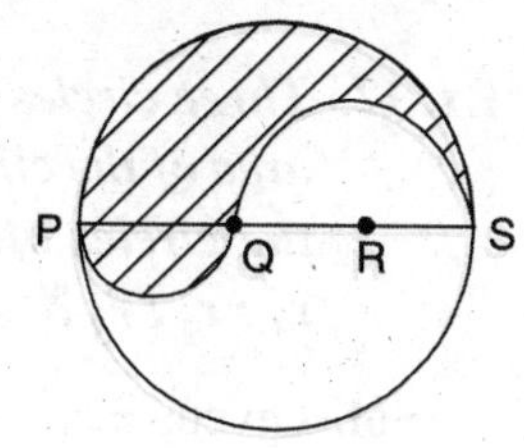

$\therefore$ Perimeter of the shaded portion = Smaller arc PQ + Arc QS + Arc PS

$$= \frac{1}{2}\left(2\pi\times\frac{r}{3}\right)+\frac{1}{2}\left(2\pi\times\frac{2r}{3}\right)+\frac{1}{2}\times 2\pi r$$

$$= \frac{\pi r}{3}+\frac{2\pi r}{3}+\pi r = \mathbf{2\pi r}.$$

Ex. 9. ***If the wire is bent into the shape of a square, the area of the square is 81 sq cm. When the wire is bent into a semicircular shape, what is the area of the semicircle ?*** $\left(\pi = \dfrac{22}{7}\right)$

Sol. Let 'a' be the length of each side of the square. Then, $a^2 = 81 \quad \Rightarrow \quad a = 9$ cm

$\therefore$ Length of wire = Perimeter of square = $4a$

$= 36$ cm

When the wire is bent into a semi-circular shape, then

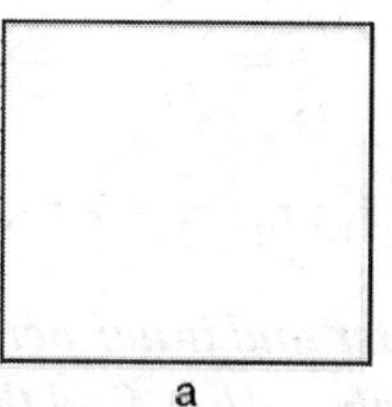

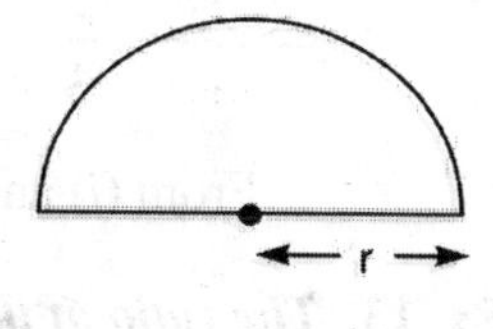

$$\pi r + 2r = 36$$

$$\Rightarrow r = \frac{36}{\pi+2} = \frac{36}{\frac{22}{7}+2} = \frac{36\times 7}{(22+14)}$$

$$= \frac{36\times 7}{36}\text{ cm} = 7 \text{ cm}$$

$\therefore$ Area of the semicircle $= \dfrac{1}{2}\pi r^2 = \dfrac{1}{\not{2}_1}\times\dfrac{\not{22}^{11}}{\not{7}_1}\times \not{7}^1 \times 7\text{cm}^2$

$= \mathbf{77 cm^2}$

Ex. 10. ***The quadrants shown in the given figure are each of diameter 12 cm. What is the area of the shaded portion?***

Sol. Area of shaded portion = Area of square – 4 × Area of one quadrant

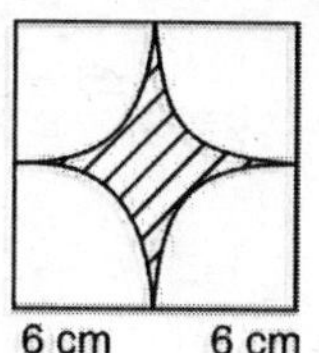

$$= (12 \times 12)\text{ cm}^2 - \left(4\times\frac{1}{4}\times\pi\times 6^2\right)\text{cm}^2$$

$$= (144 - 36\pi)\text{ cm}^2 = \mathbf{36\,(4-\pi)\text{ cm}^2}.$$

Ex. 11. ***If the radius of the circle is increased by 50%, then by how much per cent the area of the circle is increased?***

Sol. Let the radius of the circle = r units

Then original area = πr^2

Increased radius = r + 50% of r $= r + \dfrac{50r}{100} = \dfrac{150r}{100} = 1.5r$

New area = $\pi(1.5r)^2 = 2.25\,\pi r^2$

$$\therefore \quad \% \text{ increase} = \left(\frac{2.25\pi r^2 - \pi r^2}{\pi r^2} \times 100\right)\% = \left(\frac{1.25\pi r^2}{\pi r^2} \times 100\right)\%$$

$$= (1.25 \times 100)\% = \mathbf{125\%}.$$

Ex. 12. ***Three circles of radii r_1, r_2 and r_3 are drawn concentric to each other. The radii r_1 and r_2 are such that the area of the circle with radius r_1 is equal to the area between the circles of radius r_2 and r_1. The area between the circles of radii r_3 and r_2 is equal to area between the circles of radii r_2 and r_1. What is the value of $r_1 : r_2 : r_3$?***

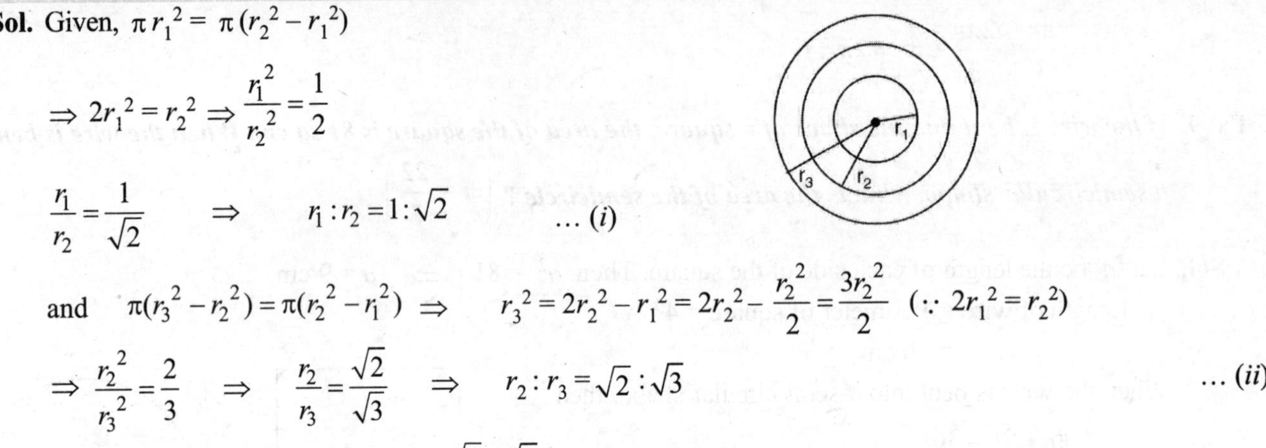

Sol. Given, $\pi r_1^2 = \pi (r_2^2 - r_1^2)$

$$\Rightarrow 2r_1^2 = r_2^2 \Rightarrow \frac{r_1^2}{r_2^2} = \frac{1}{2}$$

$$\frac{r_1}{r_2} = \frac{1}{\sqrt{2}} \quad \Rightarrow \quad r_1 : r_2 = 1 : \sqrt{2} \qquad \dots (i)$$

and $\pi(r_3^2 - r_2^2) = \pi(r_2^2 - r_1^2) \Rightarrow \quad r_3^2 = 2r_2^2 - r_1^2 = 2r_2^2 - \frac{r_2^2}{2} = \frac{3r_2^2}{2} \quad (\because 2r_1^2 = r_2^2)$

$$\Rightarrow \frac{r_2^2}{r_3^2} = \frac{2}{3} \quad \Rightarrow \quad \frac{r_2}{r_3} = \frac{\sqrt{2}}{\sqrt{3}} \quad \Rightarrow \quad r_2 : r_3 = \sqrt{2} : \sqrt{3} \qquad \dots (ii)$$

$\therefore$ From (*i*) and (*ii*) $r_1 : r_2 : r_3 = \mathbf{1 : \sqrt{2} : \sqrt{3}}$.

Ex. 13. ***The ratio of the outer and inner perimeters of a circular path is 23 : 22. If the path is 5 m wide, what is the diameter of the circle ? Also, find the area of path enclosed between the two circles.***

Sol. Let the inner and outer radii be r_1 and r_2 respectively. Also, let the inner and outer perimeters be $22x$ and $23x$ respectively.

Given, $\quad r_2 = r_1 + 5\,;\; 2\pi r_1 = 22x$

and $\quad 2\pi r_2 = 23x \;\Rightarrow\; 2\pi(r_1 + 5) = 23x$

$\Rightarrow \quad 2\pi r_1 + 10\pi = 23x \Rightarrow 22x + 10\pi = 23x$

$\Rightarrow \quad x = 10\pi \Rightarrow 2\pi r_1 = 22 \times 10\pi = 220\pi$

$\Rightarrow \quad r_1 = \frac{220\pi}{2\pi} = 110$ m and $2\pi r_2 = 23 \times 10\pi = 230\pi$

$\Rightarrow \quad r_2 = \frac{230\pi}{2\pi} = 115$ m

$\therefore \quad$ Area of path $= \pi r_2^2 - \pi r_1^2 = \pi (r_2^2 - r_1^2)$

$$= \frac{22}{7} \times (115^2 - 110^2)\text{ m}^2 = \frac{22}{7}(115 + 110)(115 - 110)\text{ m}^2$$

$$= \left(\frac{22}{7} \times 225 \times 5\right)\text{m}^2$$

$$= \frac{22}{7} \times 1125\text{ m}^2 = \mathbf{3535.71\text{ m}^2}.$$

Ex. 14. ***A square circumscribes a circle and another square is inscribed in this circle with one vertex at the point of contact. What is the ratio of the areas of the circumscribed and inscribed squares ?***

Sol. Let each edge of the square $ABCD = x$ units

Then, area of square $ABCD = x^2$ sq units

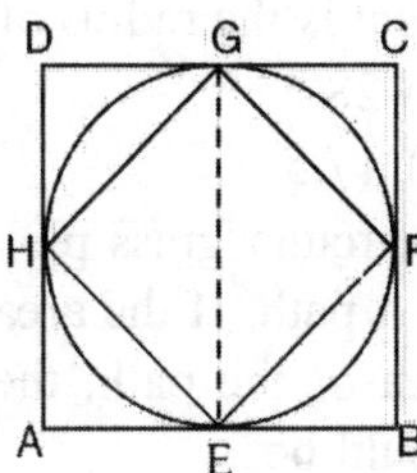

For square $EFGH$, diagonal $EG = x$

$\therefore$ Side of square $EFGH = \dfrac{x}{\sqrt{2}}$

$\therefore$ Area of square $EFGH = \dfrac{1}{2} \times (\text{diagonal})^2 = \dfrac{1}{2} \times \left(\dfrac{x}{\sqrt{2}}\right)^2 = \dfrac{x^2}{4}$.

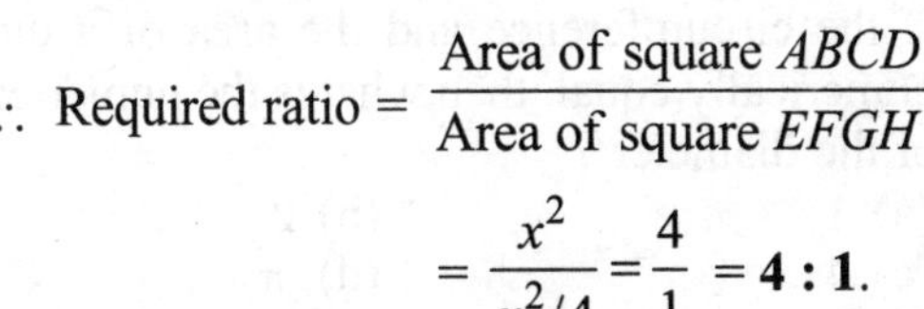

$\therefore$ Required ratio $= \dfrac{\text{Area of square } ABCD}{\text{Area of square } EFGH}$

$= \dfrac{x^2}{x^2/4} = \dfrac{4}{1} = \mathbf{4 : 1}$.

Ex. 15. ***An equilateral triangle, a square and a circle have equal perimeters. If T denotes the area of the triangle. S is the area of the square and C, the area of the circle, then:***

(a) S < T < C ***(b) T < C < S*** ***(c) T < S < C*** ***(d) C < S < T***

Sol. Let the perimeter of each figure be P cm.

Then, side of equilateral triangle $= \dfrac{P}{3}$

Side of square $= \dfrac{P}{4}$, Radius of circle $= \dfrac{P}{2\pi}$

$\therefore \quad T = \dfrac{\sqrt{3}}{4}\left(\dfrac{P}{3}\right)^2 = \dfrac{\sqrt{3}P^2}{36}$, $S = \left(\dfrac{P}{4}\right)^2 = \dfrac{P^2}{16}$, $C = \pi \times \left(\dfrac{P}{2\pi}\right)^2 = \dfrac{P^2}{4\pi} = \dfrac{7P^2}{88} = 0.0795\,P^2$

$\Rightarrow T = \dfrac{\sqrt{3}P^2}{36} = \dfrac{1.732}{36}P^2 = 0.0481P^2$, $S = 0.0625\,P^2$, $C = 0.0795\,P^2$

$\therefore \; C > S > T$.

Question Bank–22

1. If the perimeter of a semicircle is 18 cm, what will be its diameter ?
 (a) 9 cm (b) 28 cm
 (c) 14 cm (d) 7 cm
2. What is the area of a circle with circumference 88 cm ?
 (a) 616 sq cm (b) 546 sq cm
 (c) 600 sq cm (d) 615 sq cm
3. The area of the biggest circle that can be drawn inside a square of side 21 cm is $\left(\text{Take } \pi = \dfrac{22}{7}\right)$
 (a) 344.5 sq cm (b) 364.5 sq cm
 (c) 346.5 sq cm (d) 366.5 sq cm
4. The number of revolutions a wheel of diameter 40 cm makes in travelling a distance of 176 m is $\left(\pi = \dfrac{22}{7}\right)$
 (a) 140 (b) 150
 (c) 160 (d) 166
5. The diameter of a bullock cart wheel is $\dfrac{14}{11}$ metres. This wheel makes 10 complete revolutions per minute. What would be the speed of the cart in kilometres per hour ?
 (a) 4.8 (b) 9.6
 (c) 8.8 (4) 2.4
6. Diameter of a wheel is 3 m. The wheel revolves 28 times in a minute. To cover the distance of 5.280 km, the wheel will take $\left(\pi = \dfrac{22}{7}\right)$
 (a) 10 minutes (b) 20 minutes
 (c) 30 minutes (d) 40 minutes
7. A circular park has a path of uniform width around it. The difference between outer and inner circumference of the circular path is 132 m. Its width is $\left(\pi = \dfrac{22}{7}\right)$
 (a) 22 m (b) 20 m
 (c) 21 m (d) 24 m
8. The difference between the radii of the smaller circle and the bigger circle is 7 cm and the difference

between the areas of the two circles is 1078 sq cm. What is the radius of the smaller circle in cm ?

(a) 28 (b) 21
(c) 17.5 (d) 35

9. A circular grass plot 40 m in radius is surrounded by a path. If the area of the grass plot is twice the area of the path, the width of the path in metres would be

(a) $40\left(1+\sqrt{\frac{2}{3}}\right)$ (b) $40\left(1-\sqrt{\frac{2}{3}}\right)$
(c) $40\left(\sqrt{\frac{3}{2}}-1\right)$ (d) $40\left(\sqrt{\frac{3}{2}}+1\right)$

10. A circular disc of area A_1 is given. With its radius as diameter, a circular disc of area A_2 is cut out of it. The area of the remaining disc is denoted by A_3. Then,

(a) $A_1A_3 < 16A_2^2$ (b) $A_1A_3 > 16A_2^2$
(c) $A_1A_3 = 16A_2^2$ (d) $A_1A_3 > 2A_2^2$

11. If the difference between the circumference and diameter of a circle is 30 cm, then what is the radius of the circle ?

(a) 14 cm (b) 3.5 cm
(c) 7 cm (d) 6 cm

12. If the circumference of a circle is reduced by 50%, then by how much per cent is the area of the circle reduced ?

(a) 25% (b) 50%
(c) 65% (d) 75%

13. A wire is looped in the form of a circle of radius 21 cm. It is rebent in the shape of an equilateral triangle. Find the area of the triangle. (Take $\pi = \frac{22}{7}$)

(a) 484 cm² (b) $484\sqrt{3}$ cm²
(c) 308 cm² (d) $308\sqrt{3}$ cm²

14. A circular wire of diameter 42 cm is bent in the form of a rectangle whose sides are in the ratio 6 : 5. The area of the rectangle is $\left(\pi = \frac{22}{7}\right)$

(a) 540 cm² (b) 1080 cm²
(c) 2160 cm² (d) 4320 cm²

15. If the area of a circle and a square are equal, then the ratio of their perimeters is

(a) 1 : 1 (b) 2 : π
(c) π : 2 (d) $\sqrt{\pi}$: 2

16. A right angled isosceles triangle is inscribed in a circle of radius r. What is the area of the remaining portion of the circle ?

(a) $\frac{\pi r^2}{2}$ (b) $\left(\pi-\frac{1}{2}\right)r^2$
(c) $(\pi - 1)r^2$ (d) $(\pi - 2)r^2$

17. The largest possible square is inscribed in a circle of circumference 2 π units. The area of the square in square units is

(a) $4\pi\sqrt{2}$ (b) $2\pi\sqrt{2}$
(c) $\sqrt{2}$ (d) 2

18. If the circumference and the area of a circle are numerically equal, then what is the numerical value of the diameter ?

(a) 1 (b) 2
(c) 4 (d) π

19. What is the perimeter of a square whose area is equal to that of a circle with perimeter $2\pi x$?

(a) $2\pi x$ (b) $\sqrt{\pi}\,x$
(c) $4\sqrt{\pi x}$ (d) $4x\sqrt{\pi}$

20. If the circumference of a circle increases from 4 π to 8 π, what change occurs in its area ?

(a) Change = 2 × Original area
(b) Change = 3 × Original area
(c) No change
(d) Change = 4 × Original area

21. This figure is made up of a semicircle and quarter of a circle. The length of AB = 9.1 cm. The distance from A to the centre of the semicircle is 3.5 cm. The area of the figure is

(a) 25 cm² (b) 28.26 cm²
(c) 22.715 cm² (d) 88 cm²

22. The figures in the rectangle are 8 identical quadrants. What is the area of the shaded part ? (Take π = 3.14)

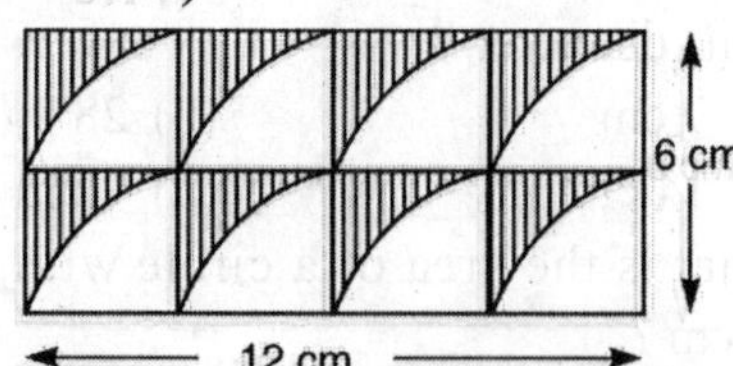

(a) 22.52 cm² (b) 15.48 cm²
(c) 56.52 cm² (d) 28.36 cm²

23. X and Y are the centres of 2 circles on a square as shown in the figure. What is the area of the shaded portion ? (Take π = 3.14)

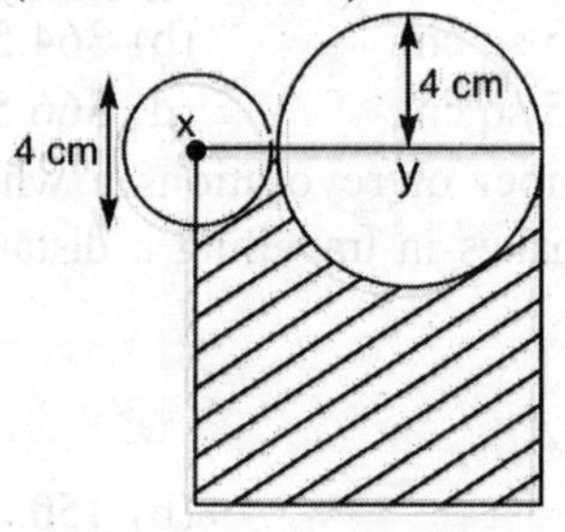

(a) 25.12 cm² (b) 28.26 cm²
(c) 71.74 cm² (d) 88 cm²

24. A wire is bent into the shape as shown. It is made up of 5 semi-circles. What is the length of the wire? (Take $\pi = 3.14$)

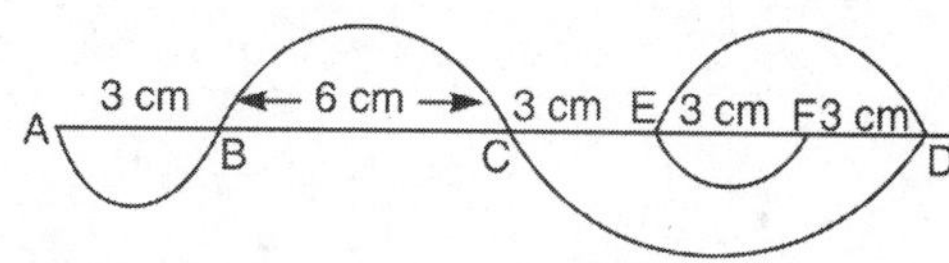

(a) 27 cm (b) 42.39 cm
(c) 45 cm (d) 27.92 cm

25. In the given figure, *ABCD* is a square with side 10 cm. *BFD* is an arc of a circle with centre *C* and *BGD* is an arc of a circle with centre *A*. What is the area of the shaded region in square centimetres ?

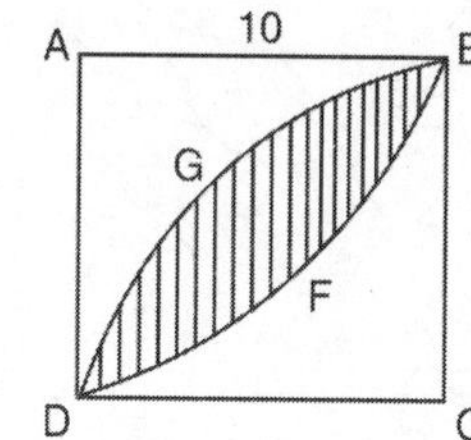

(a) $100 - 50\pi$ (b) $100 - 25\pi$
(c) $50\pi - 100$ (d) $25\pi - 100$

26. If a triangle of base 6 cm has the same area as that of a circle of radius 6 cm, then the altitude of the triangle is

(a) 6π cm (b) 8π cm
(c) 10π cm (d) 12π cm

27. In the given figure, *P* is the centre of the circle. The area and perimeter respectively of the figure are

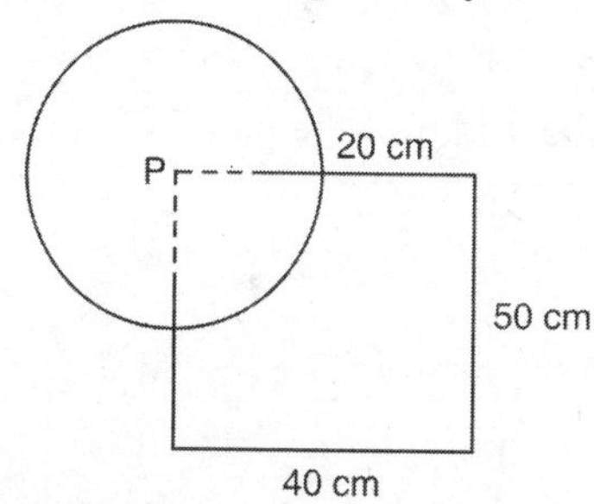

(a) 3256 cm^2, 234.2 cm
(b) 2492 cm^2, 229.2 cm
(c) 2942 cm^2, 234.2 cm
(d) 3256 cm^2, 229.2 cm

28. Four identical coins are placed in a square. For each coin, the ratio of area to circumference is the same as the ratio of circumference to area. The area of the square not covered by the coins is

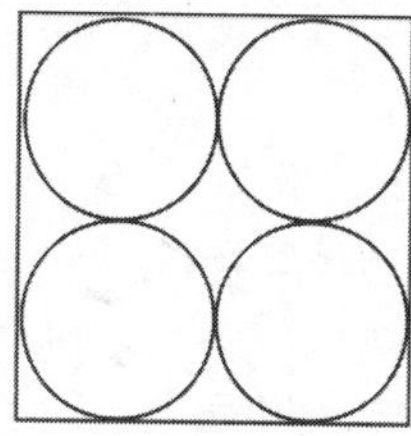

(a) $16(\pi - 1)$ (b) $16(8 - \pi)$
(c) $16(4 - \pi)$ (d) $16(4 - \frac{\pi}{2})$

29. *ABCD* is a square of side 5 cm. At the four corners, four circular arcs each of radius 1 cm are drawn. A circle of radius 2.5 cm with centre *O* is drawn inside the square. What is the approximate area of the shaded portion ?

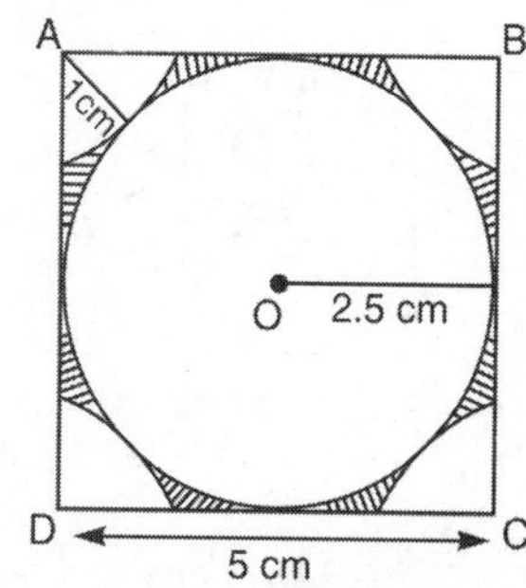

(a) 1 cm^2 (b) 1.4 cm^2
(c) 1.8 cm^2 (d) 2.2 cm^2

30. In the given figure, $MN = x$. What is the area of the shaded region?

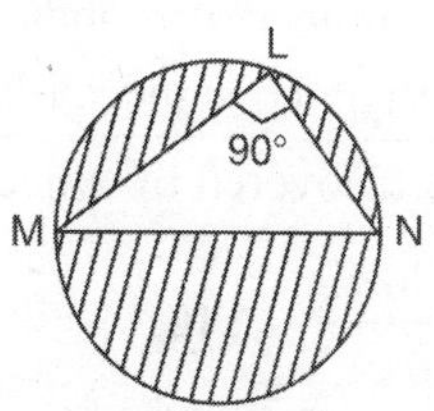

(a) $\frac{\pi x^2}{2}$ (b) $\frac{\pi x^2}{4}$
(c) πx^2 (d) $4\pi x^2$

Answers

1. (d)	**2.** (a)	**3.** (c)	**4.** (a)	**5.** (d)	**6.** (b)	**7.** (c)	**8.** (b)	**9.** (c)	**10.** (a)
11. (c)	**12.** (d)	**13.** (b)	**14.** (b)	**15.** (d)	**16.** (c)	**17.** (d)	**18.** (c)	**19.** (d)	**20.** (b)
21. (c)	**22.** (b)	**23.** (c)	**24.** (b)	**25.** (c)	**26.** (d)	**27.** (c)	**28.** (c)	**29.** (d)	**30.** (b)

Hints and Solutions

1. (d) Given, $\pi r + 2r = 18$

$\Rightarrow r(\pi + 2) = 18 \quad \Rightarrow \quad r = \dfrac{18}{\frac{22}{7}+2}$

$\Rightarrow r = \dfrac{18\times 7}{36} = 3.5$ cm

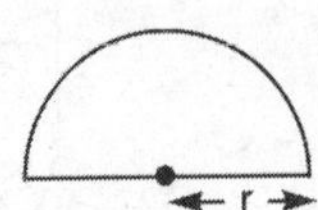

$\Rightarrow$ Diameter = **7 cm.**

2. (a) $2\pi r = 88$ cm $\quad \Rightarrow \quad r = \dfrac{88\times 7}{2\times 22} = 14$ cm

$\therefore$ Area = $\pi r^2 = \left(\dfrac{22}{7}\times 14\times 14\right)$ cm² = **616 cm².**

3. (c) Radius = $\dfrac{\text{Diameter}}{2} = \dfrac{\text{Side of square}}{2}$

$= \dfrac{21}{2}$ cm = 10.5 cm

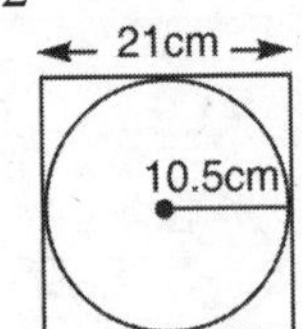

$\therefore$ Area of circle = $\left(\dfrac{22}{7}\times 10.5\times 10.5\right)$ cm²

= **346.5 cm².**

4. (a) Distance covered by the wheel in 1 round

= $\pi \times$ diameter

$= \dfrac{22}{7}\times 40 \text{ cm} = \dfrac{880}{7}$ cm

$\therefore$ Required number of rounds

$= \dfrac{\text{Total distance covered}}{\text{Distance covered by wheel in 1 round}}$

$= \dfrac{176\times 100\times 7}{880}$ = **140.**

5. (d) Radius of the wheel = $\dfrac{1}{2}\times\dfrac{14}{11}$ m $= \dfrac{7}{11}$ m

Distance covered by wheel in 1 revolution

= Circumference of wheel

$= 2\times\dfrac{22}{7}\times\dfrac{7}{11} = 4$ metres

$\therefore$ Distance covered in 10 revolutions

= (4×10) metres

= 40 metres

Thus, distance of 40 metres is covered in one minute, *i.e.*, 2400 metres in 60 minutes, *i.e.*, 2.4 km in one hour.

$\therefore$ Speed = **2.4 km / hour.**

6. (b) Distance covered by the wheel in 1 minute

= 28 × Circumference of wheel

$= \left(28\times 2\times\dfrac{22}{7}\times\dfrac{3}{2}\right)$ m = 264 m

$\therefore$ Time taken to cover a distance of 5.280 km, *i.e.*,

(5280 m) = $\dfrac{5280}{264}$ min = **20 min.**

7. (c) Given, $2\pi r_2 - 2\pi r_1 = 132$

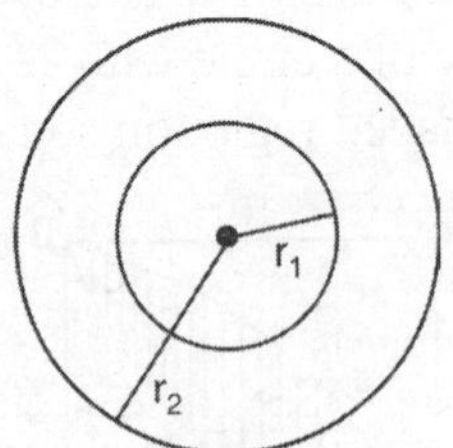

$\Rightarrow r_2 - r_1 = \dfrac{132}{2\pi} = \dfrac{132\times 7}{44}$ m = 21 m

i.e., width of the path = **21 m.**

8. (b) Let the radius of the smaller circle be r cm. Then, radius of bigger circle = $(r + 7)$ cm.

Given, $\pi (r + 7)^2 - \pi (r)^2 = 1078$

$\Rightarrow \pi (r^2 + 14r + 49) - \pi r^2 = 1078$

$\Rightarrow \pi r^2 + 14\pi r + 49\pi - \pi r^2 = 1078$

$\Rightarrow 14\times\dfrac{22}{7}\times r + 49\times\dfrac{22}{7} = 1078$

$\Rightarrow 44r + 154 = 1078 \Rightarrow 44r = 924 \Rightarrow r =$ **21 cm.**

9. (c) Let the width of the path be x metres.

Given, Area of plot = 2 × Area of path

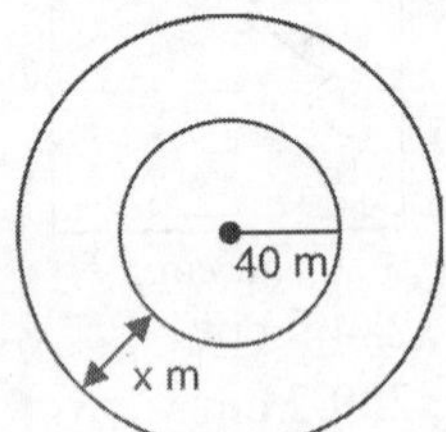

$\pi\times(40)^2 = 2\times\left\{\pi\times(40+x)^2 - \pi\times(40)^2\right\}$

$\Rightarrow \pi\times(40)^2 = 2\times\pi\times(40+x)^2 - 2\times\pi\times(40)^2$

$\Rightarrow 3\pi\times(40)^2 = 2\pi\times(40+x)^2$

$\Rightarrow (40+x)^2 = \dfrac{3}{2}\times(40)^2$

$\Rightarrow 40 + x = 40\times\sqrt{\dfrac{3}{2}}$

$\Rightarrow x = 40\sqrt{\dfrac{3}{2}} - 40 = \mathbf{40\left(\sqrt{\dfrac{3}{2}} - 1\right)}$ **m.**

10. (a) Let the radius of the disc of area A_1 be R. Then,

$$A_1 = \pi R^2;\ A_2 = \pi\left(\frac{R}{2}\right)^2 = \frac{\pi R^2}{4}$$

$$\left(\because \text{Radius of } A_2 = \frac{R}{2}\right)$$

$$A_3 = A_1 - A_2 = \pi R^2 - \frac{\pi R^2}{4} = \frac{3\pi R^2}{4}$$

$$\therefore\ A_1 \times A_3 = \pi R^2 \times \frac{3\pi R^2}{4}$$

$$= 3\pi R^2 \times \frac{\pi R^2}{4} = 3 \times 4A_2 \times A_2 = 12A_2^2$$

$$\Rightarrow A_1A_3 < 16A_2^2.$$

11. (c) Given, $2\pi r - 2r = 30 \Rightarrow 2r(\pi - 1) = 30$

$$\Rightarrow r = \frac{15}{(\pi-1)} = \frac{15}{\left(\frac{22}{7}-1\right)} = \frac{15\times 7}{15} = \mathbf{7\ cm.}$$

12. (d) Let the radius of the circle be r. Then, its circumference $= 2\pi r$, Area $= \pi r^2$

Circumference reduces by 50%

$\Rightarrow$ New circumference $= \pi r$

$\Rightarrow$ New radius $= \frac{r}{2}$

$\Rightarrow$ New area $= \pi\left(\frac{r}{2}\right)^2 = \frac{\pi r^2}{4}$

$$\therefore\ \%\text{ decrease} = \frac{\left(\pi r^2 - \frac{\pi r^2}{4}\right)}{\pi r^2}\times 100\%$$

$$= \left(\frac{3\pi r^2}{4\pi r^2}\times 100\right)\% = \mathbf{75\%.}$$

13. (b) Perimeter of the triangle

= Circumference of the circle

Let each side of the triangle be a cm. Then,

$$3a = 2\times\frac{22}{7}\times 21 \text{ cm} \Rightarrow a = 44 \text{ cm}$$

$$\therefore\ \text{Area of the triangle} = \frac{\sqrt{3}}{4}a^2 = \frac{\sqrt{3}\times 44\times 44}{4}$$

$$= \mathbf{484\sqrt{3}\ cm^2}.$$

14. (b) Let the sides of the rectangle be $6x$ cm and $5x$ cm.

Perimeter of the rectangle

= Circumference of circle

$$\Rightarrow 2(6x + 5x) = 2\times\frac{22}{7}\times 21$$

$$\Rightarrow 22x = 132 \quad\Rightarrow\quad x = 6$$

$\therefore$ The sides of the rectangle are 6×6 cm and 5×6 cm, *i.e.*, 36 cm and 30 cm.

$\therefore$ Area of the rectangle = 36 cm × 30 cm

= **1080 cm²**.

15. (d) Let the length of each side of the square be a cm and radius of the circle = r cm. Then,

$$a^2 = \pi r^2 \quad\Rightarrow\quad a = r\sqrt{\pi}$$

$$\therefore\ \text{Required ratio} = \frac{2\pi r}{4a} = \frac{2\pi r}{4r\sqrt{\pi}} = \frac{\sqrt{\pi}}{2} = \boldsymbol{\sqrt{\pi} : 2}.$$

16. (c) The hypotenuse of the right angled triangle is the diameter of the circle.

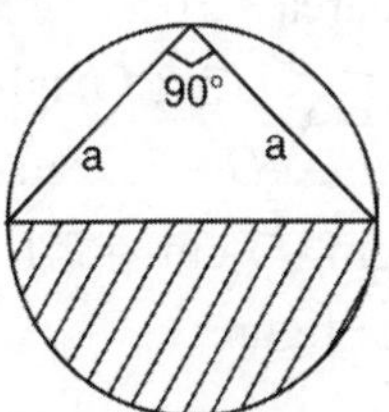

Let the sides containing the right angle be a units each. Then,

$a^2 + a^2 = (2r)^2$, where r is the radius of the circle.

$$\Rightarrow 2a^2 = 4r^2 \quad\Rightarrow\quad a = r\sqrt{2}$$

$\therefore$ Remaining area

= Area of circle – Area of triangle

$$= \pi r^2 - \frac{1}{2}(r\sqrt{2})(r\sqrt{2})$$

$$= \pi r^2 - r^2 = \mathbf{r^2(\pi - 1)}.$$

17. (d) Circumference $= 2\pi r = 2\pi \Rightarrow r = 1$

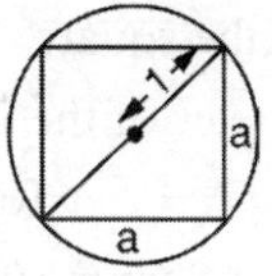

$\therefore$ Diagonal of the square = diameter

= 2 units

$$\therefore\ \text{Area of the square} = \frac{1}{2}(\text{diagonal})^2$$

$$= \frac{1}{2}\times 4 = \mathbf{2\ sq\ units.}$$

18. (c) Given, $2\pi r = \pi r^2$, where r is the radius of the circle.

$\Rightarrow r = 2$

$\therefore$ Diameter $= 2 \times 2 =$ **4.**

19. (d) Radius of the circle with perimeter $2\pi x$

$$= \frac{2\pi x}{2\pi} = x$$

$\therefore$ Its area $= \pi x^2$

Area of square with side $a = a^2 = \pi x^2 \Rightarrow a = x\sqrt{\pi}$

$\therefore$ Perimeter of square $= 4a = \mathbf{4x\sqrt{\pi}}$.

20. (b) Radius of the circle before change $= \frac{4\pi}{2\pi} = 2$ units

$\therefore$ Area of the circle before change $= \pi(2)^2$ sq units $= 4\pi$ sq units

Radius of the circle after change $= \frac{8\pi}{2\pi} = 4$ units

$\therefore$ Area of the circle after change $= \pi(4)^2$ sq units $= 16\pi$ sq units

Change in area $= 16\pi - 4\pi = 12\pi = 3 \times 4\pi$

$=$ **3 × Original area.**

21. (c) $AB = 9.1$ cm, $AO = 3.5$ cm $\Rightarrow AC = 7$ cm

$\therefore$ $BC = 9.1$ cm $- 7$ cm $= 2.1$ cm

Area of the figure

$$= \frac{1}{2} \times \pi \times AO^2 + \frac{1}{4} \times \pi \times BC^2$$

$$= \frac{1}{2} \times \frac{22}{7} \times (3.5)^2 + \frac{1}{4} \times \frac{22}{7} \times (2.1)^2$$

$= 19.25$ cm² $+ 3.465$ cm²

$=$ **22.715 cm².**

22. (b) 8 identical quadrants = 2 circles of radius 3 cm each

$\therefore$ Area of shaded part

= Area of rectangle – Area of 2 circles

$= 12$ cm $\times 6$ cm $- 2 \times 3.14 \times 3 \times 3$ cm²

$= 72$ cm² $- 56.52$ cm² $=$ **15.48 cm².**

23. (c) Each side of the square

= Diameter of the larger circle + Radius of smaller circle

$= 8$ cm $+ 2$cm $= 10$ cm

$\therefore$ Area of shaded portion

= Area of the square – Area of semicircle with centre Y – Area of quarter circle with centre X

$$= 100 \text{ cm}^2 - \left(\frac{1}{2} \times 3.14 \times 16\right) \text{cm}^2 - \left(\frac{1}{4} \times 3.14 \times 4\right) \text{cm}^2$$

$= 100$ cm² $- 25.12$ cm² $- 3.14$ cm² $=$ **71.74 cm².**

24. (b) Length of the wire

= Length of (Semicircle AB + Semicircle BC + Semicircle CD + Semicircle DE + Semicircle EF)

$= \pi \times 1.5$ cm $+ \pi \times 3$ cm $+ \pi \times 4.5$ cm $+ \pi \times 3$ cm $+ \pi \times 1.5$ cm

$= \pi \times (1.5 + 3 + 4.5 + 3 + 1.5)$ cm $= 3.14 \times 13.5$ cm

$=$ **42.39 cm.**

25. (c) Area of the portion $DEBC$

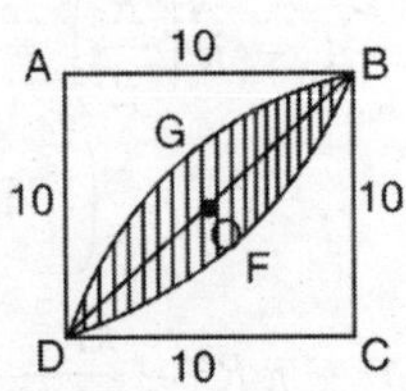

= Area of the quadrant of a circle with centre C and radius 10 cm

$$= \frac{1}{4} \times \pi \times (10)^2 = 25\pi \text{ cm}^2$$

Area of $\Delta BCD = \frac{1}{2} \times 10 \times 10$ cm² $= 50$ cm²

$\therefore$ Area of portion $BEDOB$

= Area of quad. – Area of ΔBCD

$= (25\pi - 50)$ cm²

$\therefore$ Area of shaded portion $= 2 \times (25\pi - 50)$ cm²

$=$ **(50 π – 100)cm².**

26. (d) Let the altitude of the triangle be h cm. Then,

$$\frac{1}{2} \times \text{base} \times \text{height} = \pi \times (\text{radius})^2$$

$\frac{1}{2} \times b \times h = \pi \times 6 \times 6 \Rightarrow$ **$h = 12\pi$ cm.**

27. (c) P being the centre of the circle, radius $= PE = PF = 20$ cm

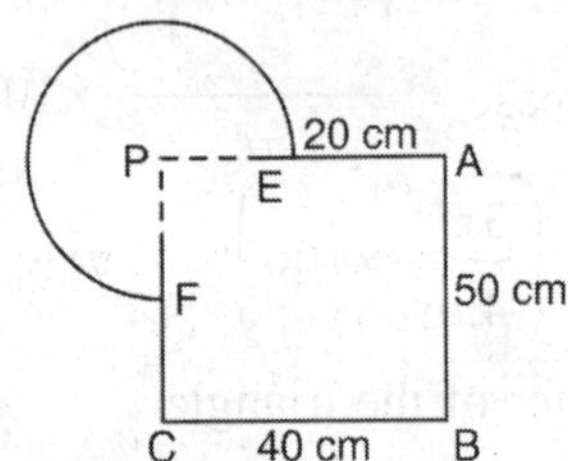

$\therefore$ Area of the figure

$= \frac{3}{4} \times$ Area of circle + Area of rectangle

$$= \left(\frac{3}{4} \times 3.14 \times 400\right) \text{cm}^2 + (50 \times 40) \text{ cm}^2$$

$= (942 + 2000)$ cm²

$= 2942$ cm²

Perimeter of the figure

$= EA + AB + BC + CF + \frac{3}{4} \times$ Circumference of circle with centre P

$$= 20 \text{ cm} + 50 \text{ cm} + 40 \text{ cm} + 30 \text{ cm} + \left(\frac{3}{4} \times 3.14 \times 40\right) \text{cm}$$

$= 140$ cm $+ 94.2$ cm $=$ **234.2 cm.**

28. (c) Let the radius of each coin = r

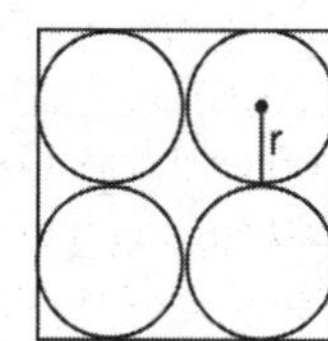

Then, $\frac{\pi r^2}{2\pi r} = \frac{2\pi r}{\pi r^2} \Rightarrow r^2 = 4 \Rightarrow r = 2$

∴ Each side of the square = 4 × 2 cm = 8 cm

∴ Area of the square not covered by coins

= Area of square – Area covered by 4 coins

= $64 - 4 \times \pi \times (2)^2 = 64 - 16\pi = \mathbf{16(4-\pi)}$.

29. (d) Area of shaded portion

= Area of square – (Area of circle with radius 2.5 cm) – 4 (Area of a quadrant of circle with radius 1 cm)

$= (5 \times 5)\,\text{cm}^2 - \pi(2.5)^2\,\text{cm}^2 - 4 \times \frac{1}{4} \times \pi \times (1)^2\,\text{cm}^2$

$= (25 - 6.25\pi - \pi)\,\text{cm}^2 = (25 - 7.25\pi)\,\text{cm}^2$

$= \{25 - (7.25 \times 3.14)\}\,\text{cm}^2$

$= 25\,\text{cm}^2 - 22.765\,\text{cm}^2 = \mathbf{2.235\,cm^2}$.

30. (b) Shaded area

$= \frac{1}{2} \times \pi \times \left(\frac{LM}{2}\right)^2 + \frac{1}{2} \times \pi \times \left(\frac{LN}{2}\right)^2 + \frac{1}{2} \times \pi \times \left(\frac{MN}{2}\right)^2$

$= \frac{\pi}{8}(LM^2 + LN^2 + MN^2) = \frac{\pi}{8}(MN^2 + MN^2)$

$= \frac{2\pi}{8} MN^2 = \frac{2\pi x^2}{8} = \mathbf{\frac{\pi x^2}{8}}$.

(By Pythagoras Theorem $LM^2 + LN^2 = MN^2$)

Self Assessment Sheet–22

1. A horse is tied to a pole fixed at one corner of a 50 m × 50 m square field of grass by means of a 20 m long rope. What is the area to the nearest whole number of that part of the field which the horse can graze ?

(a) 1256 m^2 (b) 942 m^2
(c) 628 m^2 (d) 314 m^2

2. From a rectangular metal sheet of sides 25 cm and 20 cm, a circular sheet as large as possible is cut off. What is the area of the remaining sheet ? ($\pi = 3.14$)

(a) 186 cm^2 (b) 144 cm^2
(c) 93 cm^2 (d) 72 cm^2

3. A circle and a rectangle have the same perimeter. The sides of the rectangle are 18 cm and 26 cm. What is the area of the circle ?

(a) 88 cm^2 (b) 154 cm^2
(c) 616 cm^2 (d) 1250 cm^2

4. The area of a circle is 24.64 m^2. The circumference of the circle is :

(a) 14.64 m (b) 16.36 m
(c) 17.60 m (d) 18.40 m

5. The area of the largest circle that can be drawn inside a rectangle with sides 18 cm and 14 cm is :

(a) 49 cm^2 (b) 154 cm^2
(c) 378 cm^2 (d) 1078 cm^2

6. The perimeters of a circular field and a square field are equal. If the area of the square field is 12100 m^2, the area of the circular field will be :

(a) 15500 m^2 (b) 15400 m^2
(c) 15200 m^2 (d) 15300 m^2

7. A circular disc of area (0.49π) m^2 rolls down a length of 1.76 km. The number of revolutions it makes, is :

(a) 300 (b) 400
(c) 600 (d) 4000

8. To make a marriage tent, poles are planted along the perimeter of a square field at a distance of 5 metres from each other and the total number of poles used is 20. What is the area (in sq metres) of the square field ?

(a) 500 (b) 400
(c) 900 (d) None of these

9. A track is in the form of a ring whose inner circumference is 352 m and the outer circumference is 396 m. The width of the track is :

(a) 44 m (b) 14 m
(c) 22 m (d) 7 m

10. A copper wire when bent in the form of an equilateral triangle encloses an area of $121\sqrt{3}$ cm^2. If the same wire is bent in the form of a circle, then the area of the circle is :

(a) 121 cm^2 (b) 342 cm^2
(c) 346.5 cm^2 (d) 154.8 m^2

Answers

1. (d) **2.** (a) **3.** (c) **4.** (c) **5.** (b) **6.** (b) **7.** (b) **8.** (d) **9.** (d) **10.** (c)

Chapter

23 VOLUME AND SURFACE AREA OF A CUBE AND A CUBOID

KEY FACTS

1. Cuboid

Let length = l, breadth = b and height = h units. Then,

(i) Volume = $(l \times b \times h)$ cubic units (ii) Surface area = $2(lb + bh + lh)$ sq units

(iii) Diagonal = $\sqrt{l^2 + b^2 + h^2}$ units

2. Cube

Let each edge of a cube be of length a. Then,

(i) Volume = a^3 cubic units (ii) Surface Area = $6a^2$ sq units (iii) Diagonal = $\sqrt{3}a$ units

Solved Examples

Ex. 1. *A wax cube of edge 6 cm is melted and formed into three smaller cubes. If the edges of two smaller cubes are 3 cm and 4 cm, what is the edge of the third smaller cube?*

Sol. Let the edge of the third cube be a cm. Then,

$a^3 + 3^3 + 4^3 = 6^3 \Rightarrow a^3 + 27 + 64 = 216$

$\Rightarrow a^3 = 125 \Rightarrow a = \sqrt[3]{125} = \mathbf{5\ cm}$.

Ex. 2. *The breadth of a cuboid is twice its height and half its length. If the volume of the cuboid is 512 m³, then what is the length of the cuboid?*

Sol. Let x be the breadth of the cuboid. Then,

Height of the cuboid = $\frac{x}{2}$; Length of the cuboid = $2x$

$\therefore$ Volume of the cuboid = $x \times \frac{x}{2} \times 2x = x^3$

Given $x^3 = 512 \Rightarrow x = \sqrt[3]{512} = 8$

$\therefore$ Length of the cuboid = 2×8 cm = **16 cm**.

Ex. 3. *The dimensions of a rectangular box are in the ratio 1 : 2 : 4 and the difference between the costs of covering it with the cloth and sheet at the rate of ₹ 20 and ₹ 20.50 per square metres respectively is ₹ 126. Find the dimensions of the box?*

Sol. Let, length = x, breadth = $2x$ and height = $4x$. Then,

area of the box = $2(lb + bh + lh) = 2(2x^2 + 8x^2 + 4x^2) = 28x^2$

Given, $20.5 \times 28x^2 - 20 \times 28x^2 = 126 \Rightarrow 28x^2 \times 0.5 = 216 \Rightarrow x^2 = 9 \Rightarrow x = 3$

$\therefore$ The dimensions are **3 cm, 6 cm** and **12 cm**.

Ex. 4. ***Three cubes of sides 4 cm each are joined end to end to form a cuboid. What is the ratio of the surface area of the resulting cuboid to the total surface area of the three cubes?***

Sol. The dimensions of the resulting cuboid are:

length = 12 cm, breadth = 4 cm, height = 4 cm

$\therefore$ Surface area of the cuboid $= 2(12 \times 4 + 4 \times 4 + 12 \times 4)$ cm^2

$= 2 \times 112$ cm^2 $= 224$ cm^2

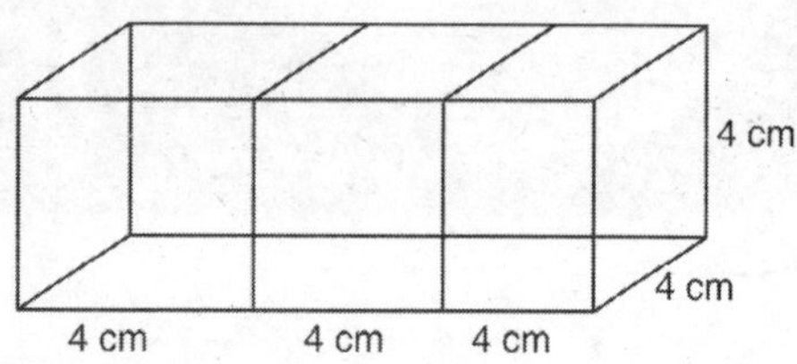

Surface area of the three cubes $= 3 \times 6 \times (4)^2$ cm$^2 = 288$ cm^2

$\therefore$ Required ratio = 224 : 288 = **7 : 9**.

Ex. 5. ***What is the volume of a cube (in cubic cm) whose diagonal measures $4\sqrt{3}$ cm?***

Sol. Let the length of an edge of the cube = x cm. Then, diagonal of the cube = $x\sqrt{3}$

$\Rightarrow x\sqrt{3} = 4\sqrt{3} \quad \Rightarrow \quad x = 4$ cm.

$\therefore$ Volume of the cube = $(4)^3$ cm^3 = **64 cm³**.

Ex. 6. ***By what per cent does the volume of a cube increase, if the length of each edge is increased by 50%?***

Sol. Let each edge of the cube = x cm

Then, volume of the cube = x^3 cm^3

Length of edge after increase $= \frac{150}{100} x$ cm $= 1.5x$ cm

$\therefore$ Increased volume $= (1.5x)^3$ cm^3 $= 3.375x^3$ cm^3

$\therefore$ % increase $= \frac{(3.375x^3 - x^3)}{x^3} \times 100\% = (2.375 \times 100)\% =$ **237.5%**.

Ex. 7. ***The areas of three adjacent faces of a cuboid are a, b and c. If the volume of the cuboid is V, then what is V^2 equal to?***

Sol. Let the length, breadth and height of the cuboid be x, y and z units respectively.

Then, $x \times y = a,\ y \times z = b,\ x \times z = c$

$\therefore x \times y \times y \times z \times x \times z = a \times b \times c$

$\Rightarrow x^2y^2z^2 = abc$

$\Rightarrow (xyz)^2 = abc$

$\Rightarrow V^2 =$ **abc**.

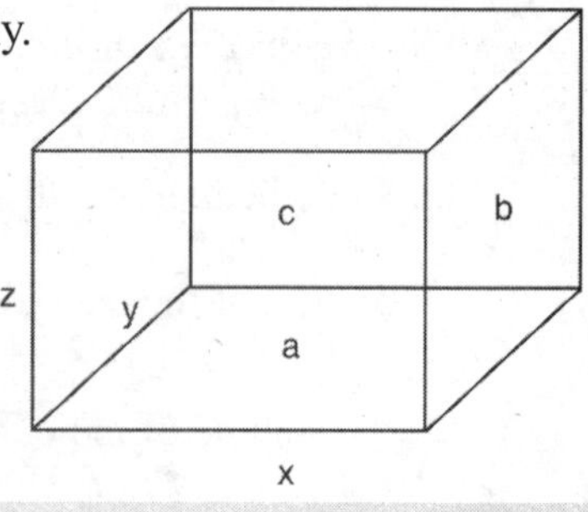

Ex. 8. ***Find the length of the longest rod that can be placed in a hall of 10 m length, 6 m breadth and 4 m height?***

Sol. Length of the longest rod

$= \sqrt{l^2 + b^2 + h^2}$

$= \sqrt{100 + 36 + 16} = \sqrt{152}$

$= \sqrt{4 \times 38} = \mathbf{2\sqrt{38}}$ **m**.

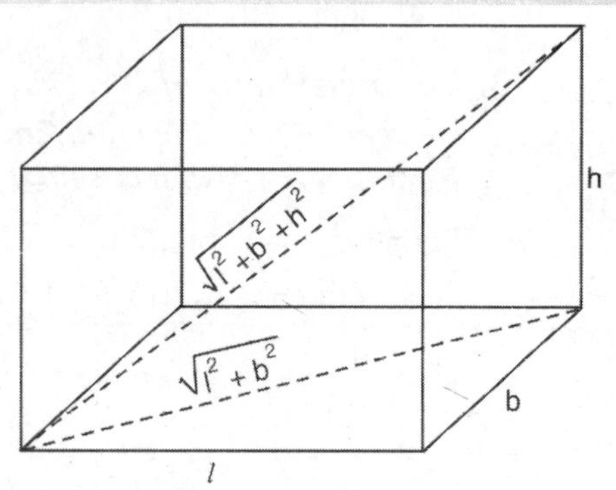

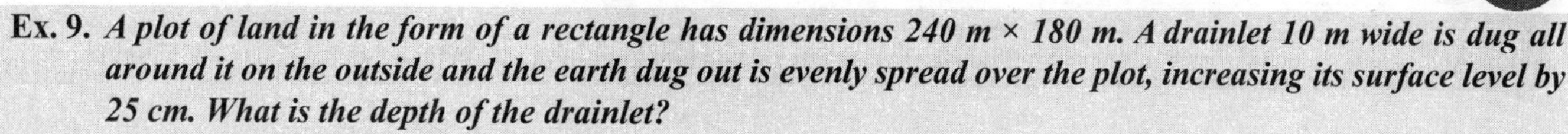

Ex. 9. ***A plot of land in the form of a rectangle has dimensions 240 m × 180 m. A drainlet 10 m wide is dug all around it on the outside and the earth dug out is evenly spread over the plot, increasing its surface level by 25 cm. What is the depth of the drainlet?***

Sol. Outer length = (240 + 20) m = 260 m

Outer breadth = (180 + 20) m = 200 m

∴ Area of the drainlet = $(260 \times 200)\ m^2 - (240 \times 180)\ m^2$

$= 52000\ m^2 - 43200\ m^2 = 8800\ m^2$

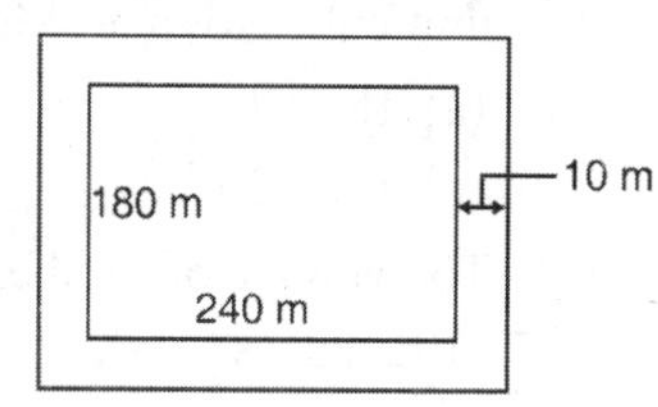

Let d be the depth of the drainlet. Then,

Volume of earth dug out = Volume of earth spread on the land

$8800 \times x = 180 \times 240 \times 0.25$

$$\Rightarrow x = \frac{180 \times 240 \times 0.25}{8800} = \frac{27}{22} = \mathbf{1.227\ m.}$$

Ex. 10. ***A hall is 15 m long and 12 m broad. If the sum of the areas of the floor and the ceiling is equal to the sum of the areas of the 4 walls, what is the volume of the hall?***

Sol. Area of floor + Area of ceiling = Area of four walls

$\Rightarrow lb + lb = 2 \times h \times (l + b)$, where l, b, h are the length, breadth and height of the hall.

$\Rightarrow lb = h \times (l + b)$

$\Rightarrow 15\ m \times 12\ m = h \times (15\ m + 12\ m)$

$$\Rightarrow h = \frac{15 \times 12}{27}\ m = \frac{20}{3}\ m$$

$$\therefore \text{Volume of the hall} = l \times b \times h = 15\ m \times 12\ m \times \frac{20}{3}\ m = \mathbf{1200\ m^3}.$$

Ex. 11. ***A wooden box measures 20 cm × 12 cm × 10 cm. Thickness of the wood is 1 cm. What is the volume of wood required to make the box?***

Sol. External dimensions are:

Length = 20 cm, Breadth = 12 cm, Height = 10 cm

Internal dimensions are:

Length = (20 – 2) cm = 18 cm

Breadth = (12 – 2) cm = 10 cm

Height = (10 – 2) cm = 8 cm

∴ Volume of wood = External volume – Internal volume = $(20 \times 12 \times 10)\ cm^3 - (18 \times 10 \times 8)\ cm^3$

$= 2400\ cm^3 - 1440\ cm^3 = \mathbf{960\ cm^3}$.

Ex. 12. ***The cost of painting the walls of a room at the rate of ₹1.35 per square metre is ₹340.20 and the cost of matting the floor at the rate of ₹0.85 per m² is ₹91.80. If the length of the room is 12 m, then what is the height of the room?***

Sol. Area of four walls of a room $= 2(l + b)h = \frac{340.20}{1.35} = 252\ m^2$ ...(*i*)

Also, $l \times b = \frac{91.8}{0.85} = 108\ m^2 \Rightarrow b = \frac{108}{12}\ m = 9\ m$ $(\because l = 12\ m)$

∴ From (*i*)

$$2(12 + 9) \times h = 252 \Rightarrow h = \frac{252}{42} = \mathbf{6\ m.}$$

Question Bank-23

1. A cuboid (3 cm × 4 cm × 5 cm) is cut into unit cubes. The ratio of the total surface area of all the unit cubes to that of the cuboid is
   (a) 180 : 3 (b) 180 : 9
   (c) 180 : 36 (d) 180 : 47
2. The volume of a cube is V. What is the total length of its edges?
   (a) $6V^{1/3}$ (b) $8\sqrt{V}$
   (c) $12V^{2/3}$ (d) $12V^{1/3}$
3. The volume of a cube is numerically equal to the sum of its edges. What is the total surface area in square units?
   (a) 66 (b) 183
   (c) 36 (d) 72
4. If the surface area of a cube is 216 cm^2, then the length of its diagonal is
   (a) $6\sqrt{3}$ cm (b) 6 cm
   (c) 8 cm (d) $7\sqrt{3}$ cm
5. If 6 cubes each of 10 cm edge are joined end to end, the surface area of the resulting solid will be
   (a) 3600 cm^2 (b) 3000 cm^2
   (c) 2600 cm^2 (d) 2400 cm^2
6. A closed box made of wood of uniform thickness has length, breadth and height 12 cm, 10 cm and 8 cm respectively. If the thickness of the wood is 1 cm, the inner surface area is
   (a) 456 cm^2 (b) 376 cm^2
   (c) 264 cm^2 (d) 696 cm^2
7. Three cubes of metal of edges 6 cm, 8 cm and 10 cm are melted to form a new cube. The diagonal of this new cube is
   (a) 8 cm (b) 12 cm
   (c) 20.78 cm (d) 21.8 cm
8. Each edge of a cube is increased by 50%. The percentage increase in the surface area of the cube is
   (a) 50 (b) 125
   (c) 150 (d) 225
9. A cuboidal water tank contains 216 litres of water. Its depth is $\frac{1}{3}$ of its length and breadth is $\frac{1}{2}$ of $\frac{1}{3}$ of the difference between length and depth. The length of the tank is
   (a) 72 dm (b) 18 dm
   (c) 6 dm (d) 2 dm
10. A cistern of capacity 8000 litres measures externally 3.3 m by 2.6 m by 1.1 m and its walls are 5 cm thick. The thickness of the bottom is
    (a) 1 m (b) 1.1 m
    (c) 1 dm (d) 90 cm
11. Water is distributed to a town of 50000 inhabitants from a rectangular reservoir consisting of three equal compartments. Each compartment has a length and breadth 200 m and 100 m respectively and 12 m depth of water in the beginning. The allowance is 20 litres per head per day. For how many days will the supply of water hold out?
    (a) 240 days (b) 720 days
    (c) 800 days (d) 900 days
12. The length of a room is $\frac{4}{3}$ times its breadth. The total area of the four walls of this room is $\frac{28}{15}$ times the area of the floor of the room. What is the ratio of the height of the room to the length of the room?
    (a) 2 : 3 (b) 1 : 2
    (c) 2 : 5 (d) 8 : 15
13. The height of a room is 'a' and the areas of the two adjacent walls of a room are 'b' and 'c'. The area of the roof will be
    (a) $\frac{bc}{a}$ (b) bc
    (c) $\frac{ac}{b^2}$ (d) $\frac{bc}{a^2}$
14. If V be the volume of the cuboid of dimensions a, b and c and S its total surface area, then $\frac{4}{S}\left(\frac{1}{a}+\frac{1}{b}+\frac{1}{c}\right)$ in terms of V is equal to
    (a) $\frac{8}{V}$ (b) $\frac{2}{V}$
    (c) $\frac{4}{V}$ (d) $\frac{1}{V}$
15. The height of a wall is six times its width and the length of the wall is seven times its height. If volume of the wall be 16128 m^3, its width is
    (a) 5 m (b) 4 m
    (c) 4.5 m (d) 6 m

16. A swimming pool is 24 m long and 15 m broad. When a number of men dive into the bath, the height of water rises by 1 cm. If the average volume of water displaced by each man be 0.1 m^3, how many men are there in the bath?
(a) 32 (b) 36
(c) 42 (d) 46

17. A cistern open at the top is to be lined with a sheet of lead which weighs 27 kg/m^2. The cistern is 4.5 m long and 3 m wide and holds 50 m^3 of water. The weight of lead required is
(a) 1660.62 kg (b) 1764.60 kg
(c) 1864.62 kg (d) 1860.62 kg

18. How many small cubes each of 96 cm^2 surface area can be formed from the material obtained by melting a larger cube of 384 cm^2 surface area.
(a) 8 (b) 5
(c) 800 (d) 8000

19. The length, breadth and height of a cuboid are in the ratio 1 : 2 : 3. If they are increased by 100%, 200% and 200% respectively, then compared to the original volume, the increase in the volume of the cuboid will be
(a) 5 times (b) 18 times
(c) 12 times (d) 17 times

20. The volume of a cuboid whose sides are in the ratio 1 : 2 : 4 is same as that of a cube. What is the ratio of the length of diagonal of the cuboid to that of the cube?
(a) $\sqrt{1.25}$ (b) $\sqrt{1.75}$
(c) $\sqrt{2}$ (d) $\sqrt{3.5}$

21. A cuboid of size 8 cm × 4 cm × 2 cm is cut into cubes of equal size of 1 cm side. What is the ratio of the surface area of the original cuboid to the surface area of all the unit cubes so formed?
(a) 13 : 14 (b) 8 : 3
(c) 7 : 24 (d) 7 : 12

22. The height of a room is 40% of its semi-perimeter. It costs ₹ 260 to paper the walls of the room with paper 50 cm wide at ₹ 2 per metre allowing an area of 15 m^2 for doors and windows. The height of the room is
(a) 2.6 m (b) 3.9 m
(c) 4 m (d) 4.2 m

23. The piece of chocolate biscuit given is filled with a thin layer of vanilla cream. What per cent of the biscuit is vanilla cream ? Assume the layer of vanilla cream forms a cuboid.

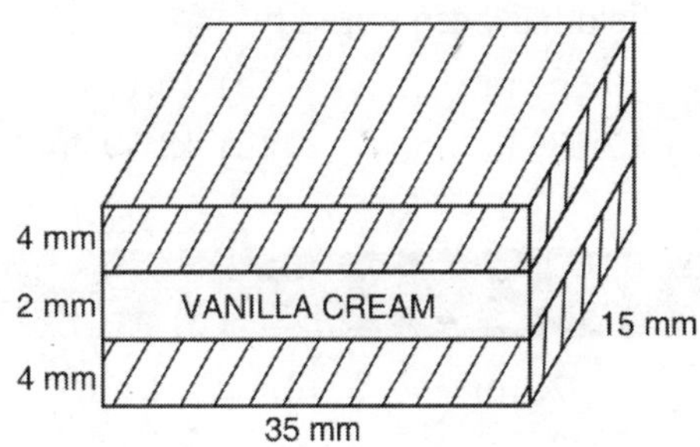

(a) 20% (b) 25%
(c) 33% (d) 50%

24. 2 cm of rain has fallen on a square kilometer of land. Assuming that 50% of the rain drops could have been collected and contained in a pool having a 100 m × 10 m base by what level would the water level in the pool have increased?
(a) 15 m (b) 20 m
(c) 10 m (d) 25 m

25. What is the number of bricks each measuring 25 cm × 12.5 cm × 7.5 cm required to construct a wall 6 m long, 5 m high and 0.5 m thick, if mortar occupies 5% of the volume of the wall.
(a) 5740 (b) 6080
(c) 3040 (d) 8120

26. A large cube is formed from the material obtained by melting three smaller cubes of 3, 4 and 5 cm size. What is the ratio of the total surface areas of the smaller cube to the large cube?
(a) 2 : 1 (b) 3 : 2
(c) 27 : 20 (d) 25 : 18

27. The length of a room is double the breadth. The cost of colouring the ceiling at ₹ 25 per sq m is ₹ 5000 and the cost of painting the four walls at ₹ 240 per sq m is ₹ 64,800. Find the height of the room?
(a) 4.5 m (b) 4 m
(c) 3.5 m (d) 5 m

28. A wooden box (open at the top) of thickness 0.5 cm, length 21 cm, width 11 cm and height 6 cm is painted on the inside. The expenses of painting are ₹ 70. What is the rate of painting per square centimetre?
(a) ₹ 0.7 (b) ₹ 0.5
(c) ₹ 0.1 (d) ₹ 0.2

29. An agricultural field is in the form of a rectangle of length 20 m and width 14 m. A pit 6 m long, 3 m wide and 2.5 m deep is dug in a corner of the field

and the earth taken out of the pit is spread uniformly over the remaining area of the field. The level of the field has been raised by

(a) 15.16 cm (b) 16.17 cm
(c) 17.18 cm (d) 18.19 cm

30. Water flows into a tank 200 m × 150 m through a rectangular pipe 1.5 m × 1.25 m at the rate of 20 km per hour. In what time will the water rise by 2 metres?

(a) 76 min (b) 80 min
(c) 90 min (d) 96 min

Answers

1. (d)	**2.** (d)	**3.** (d)	**4.** (a)	**5.** (c)	**6.** (b)	**7.** (c)	**8.** (b)	**9.** (b)	**10.** (c)
11. (b)	**12.** (d)	**13.** (d)	**14.** (b)	**15.** (b)	**16.** (b)	**17.** (c)	**18.** (a)	**19.** (d)	**20.** (b)
21. (c)	**22.** (c)	**23.** (a)	**24.** (c)	**25.** (b)	**26.** (d)	**27.** (a)	**28.** (c)	**29.** (c)	**30.** (d)

Hints and Solutions

1. (d) Number of unit cubes $= \dfrac{\text{Volume of the cuboid}}{\text{Volume of the unit cube}}$

$= \dfrac{3\text{cm} \times 4\text{cm} \times 5\text{cm}}{1\text{cm} \times 1\text{cm} \times 1\text{cm}} = 60$

∴ Total surface area of the cuboid

$= 2(3 \times 4 + 4 \times 5 + 3 \times 5)$ cm^2

$= 2(12 + 20 + 15)$ cm^2

$= 94$ cm^2

Total surface area of 60 unit cubes

$= 60 \times 6 \times (1)^2$ cm^2

$= 360$ cm^2

∴ Required ratio = 360 : 94 = **180 : 47**.

2. (d) $V = (\text{edge})^3 \Rightarrow \text{Edge} = \sqrt[3]{V}$ or $V^{1/3}$

Since a cube has 12 edges,

Total length of its edges $= 12 \times V^{1/3} = \mathbf{12V^{1/3}}$.

3. (d) Let each edge of the cube = x units. Then,

$x^3 = 12x \Rightarrow x^2 = 12$

Total surface area of the cube $= 6(x)^2 = 6x^2$

$= 6 \times 12 =$ **72 units.**

4. (a) Given, $6 \times (\text{edge})^2 = 216$

$\Rightarrow \text{edge}^2 = \dfrac{216}{6} = 36$

$\Rightarrow$ edge = 6 cm

∴ Length of the diagonal of a cube = edge × $\sqrt{3}$

$= \mathbf{6\sqrt{3}}$ **cm.**

5. (c) Length of resulting solid = 6 × 10 cm = 60 cm

Breadth = 10 cm, Height = 10 cm.

∴ Surface area of the solid

$= 2(60 \times 10 + 10 \times 10 + 10 \times 60)$ cm^2

$= 2 \times (600 + 100 + 600)$ cm^2 = **2600 cm^2**.

6. (b) Internal dimensions of the box are:

Length = (12 – 2) cm = 10 cm,

Breadth = (10 – 2) cm = 8 cm

Height = (8 – 2) cm = 6 cm

∴ Inner surface area

$= 2 \times (10 \times 8 + 8 \times 6 + 6 \times 10)$ cm^2

$= 2 \times (80 + 48 + 60)$ cm^2

$= 2 \times 188$ cm^2

= **376 cm^2**.

7. (c) Volume of the new cube $= 6^3 + 8^3 + 10^3$

$= 216 + 512 + 1000$

$= 1728$ cm^3

∴ Each edge of the new cube $= \sqrt[3]{1728} = 12$ cm

∴ Diagonal of the new cube $= (12 \times \sqrt{3})$ cm

$= (12 \times 1.732)$ cm

= **20.78 cm.**

8. (b) Let each side of the cube = a units

Then, total surface area = $6a^2$

Increased side of the cube $= \dfrac{150}{100} a$ cm $= 1.5a$ cm

∴ Increased surface area $= 6 \times (1.5a)^2$ cm^2

$= 6 \times 2.25a^2 = 13.5a^2$ cm^2

∴ % increase $= \dfrac{(13.5a^2 - 6a^2)}{6a^2} \times 100\%$

$= \left(\dfrac{7.5}{6} \times 100\right)\% = \mathbf{125\%}$.

9. (b) Let the length, breadth and height of the water in the tank be l, b, h respectively. Then,

$h = \dfrac{1}{3}l$ and $b = \dfrac{1}{2}$ of $\dfrac{1}{3}(l - h)$

$= \dfrac{1}{2} \times \dfrac{1}{3}\left(l - \dfrac{1}{3}l\right) = \dfrac{1}{6} \times \dfrac{2}{3}l = \dfrac{1}{9}l$

Given, volume of water = 216

$\Rightarrow l \times \frac{1}{9} l \times \frac{1}{3} l = 216$

$\Rightarrow l^3 = \frac{216}{1000} \times 27$ $\quad (\because 1 \text{ m}^3 = 1000\ l)$

$\Rightarrow l = 0.6 \times 3 = 1.8$ m = (1.8×10) dm = **18 dm.**

10. (c) 1 litre = 1000 cm³ $\Rightarrow 8000 l = 8000000 \text{ cm}^3$.
Let the thickness of bottom be x cm. Then,
$(330 - 10) \times (260 - 10) \times (110 - x) = 8000000$

$\Rightarrow 320 \times 250 \times (110 - x) = 8000000$

$\Rightarrow (110 - x) = \frac{8000000}{320 \times 250} = 100$

$\Rightarrow x = 110 - 100 = 10$ cm = **1 dm.**

11. (b) Volume of water present in reservoir in the beginning
$= 3 \times (200 \times 100 \times 12)$
$= 720000 \text{ m}^3 = 720000000\ l$ $\quad (\because 1 \text{ m}^3 = 1000\ l)$

Consumption of water per day = 50000×20 litres
= 1000000 litres

$\therefore$ Number of days $= \frac{720000000}{1000000} = \mathbf{720.}$

12. (d) Let breadth = x units. Then, length = $\frac{4}{3}x$ units

$\Rightarrow$ Area of floor $= x \times \frac{4}{3}x = \frac{4}{3}x^2$ square units

Let the height of the room = y units. Then,

Area of four walls $= 2\left[xy + \frac{4}{3}xy\right] = \frac{14}{3}xy$

Now, $\frac{14}{3}xy = \frac{28}{15}\left(\frac{4}{3}x^2\right) \Rightarrow xy = \frac{2}{15} \times 4x^2$

$\Rightarrow y = \frac{8}{15}x \quad \Rightarrow \quad \frac{y}{x} = \frac{8}{15}$

$\Rightarrow y : x = \mathbf{8 : 15.}$

13. (d) Let length = x, breadth = y and height = a.

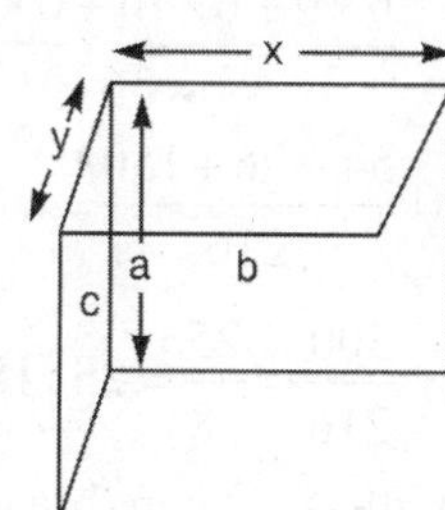

Then, $x \times a = b \Rightarrow x = \frac{b}{a}$

$y \times a = c \quad \Rightarrow \quad y = \frac{c}{a}$

$\therefore$ Area of roof $= x \times y = \frac{b}{a} \times \frac{c}{a} = \frac{bc}{a^2}$.

14. (b) Volume (V) = abc
Surface area (S) = $2(ab + bc + ca)$

$\therefore \frac{4}{S}\left(\frac{1}{a} + \frac{1}{b} + \frac{1}{c}\right) = \frac{4}{S}\left(\frac{bc + ca + ab}{abc}\right)$

$= \frac{4}{S} \times \frac{S/2}{V} = \frac{\mathbf{2}}{\mathbf{V}}.$

15. (b) Let the width of the wall be x m. Then,
Height = $6x$ m and Length = $(7 \times 6x)$ m = $42x$ m.
Given, volume of wall = 16128

$\Rightarrow 42x \times x \times 6x = 16128$

$\Rightarrow x^3 = \frac{16128}{42 \times 6} = 64 \quad \Rightarrow \quad x = 4$

$\therefore$ Width = **4 m.**

16. (b) Let the number of men be n. Then,

$n \times 0.1 = 24 \times 15 \times \frac{1}{100}$

$\Rightarrow n = 24 \times 15 \times \frac{1}{100} \times \frac{10}{1} = \mathbf{36.}$

17. (c) Let h be the height of the cistern. Then,

$4.5 \times 3 \times h = 50 \quad \Rightarrow \quad h = \frac{500}{45 \times 3} \text{ m} = \frac{100}{27} \text{ m}$

$\therefore$ Area of the cistern lined with lead
$= 2(lh + bh) + lb$

$= 2(4.5 + 3) \times \frac{100}{27} + 4.5 \times 3 = 2 \times 7.5 \times \frac{100}{27} + 13.5$

$= 55.56 + 13.5 = 69.06 \text{ m}^2$

$\therefore$ Weight of lead = (27×69.06) kg = **1864.62 kg.**

18. (a) Let the length of the edge of the larger cube be x_1 cm and that of the small cubes be x_2 cm. Then,

$6x_1^2 = 96 \Rightarrow x_1^2 = \frac{96}{6} = 16 \quad \Rightarrow \quad x_1 = 4$

$6x_2^2 = 384 \Rightarrow x_2^2 = \frac{384}{6} = 64 \quad \Rightarrow \quad x_2 = 8$

$\therefore$ Number of small cubes $= \frac{\text{Volume of larger cube}}{\text{Volume of small cube}}$

$= \frac{(x_2)^3}{(x_1)^3} = \frac{8 \times 8 \times 8}{4 \times 4 \times 4} = \mathbf{8.}$

19. (d) Let the length, breadth and height of the cuboid be x, $2x$ and $3x$ units.

The dimensions after increase are $x + \frac{100}{100}x$,

$2x + \frac{200}{100} \times 2x,\ 3x + \frac{200}{100} \times 3x,$ *i.e*, $2x, 6x$ and $9x$.

Original volume $= x \times 2x \times 3x = 6x^3$

Increased volume $= 2x \times 6x \times 9x = 108x^3$

$\therefore$ Increase in volume $= 108x^3 - 6x^3 = 102x^3$

$= 17 \times 6x^3$

= **17 times** the original volume.

20. (b) Let the sides of the cuboid be x, $2x$ and $4x$ units.

Let the length of each edge of the cube $= a$ units.

Then, $a^3 = (x)(2x)(4x)$

$\Rightarrow a^3 = 8x^3 \Rightarrow a = 2x$.

Length of diagonal of the cuboid

$= \sqrt{x^2 + (2x)^2 + (4x)^2}$

$= \sqrt{x^2 + 4x^2 + 16x^2} = \sqrt{21x^2}$

Length of diagonal of the cube $= a\sqrt{3}$

$= 2x\sqrt{3} = \sqrt{12x^2}$

$\therefore$ Required ratio $= \frac{\sqrt{21x^2}}{\sqrt{12x^2}} = \sqrt{\frac{21}{12}} = \mathbf{\sqrt{1.75}}$.

21. (c) Number of cubes

$= \frac{\text{Volume of cuboid}}{\text{Volume of cube}}$

$= \frac{8 \times 4 \times 2}{1 \times 1 \times 1} = 64$

Surface area of cuboid

$= 2 \times (8 \times 4 + 4 \times 2 + 2 \times 8)\ \text{cm}^2$

$= 2 \times (32 + 8 + 16)\ \text{cm}^2 = 112\ \text{cm}^2$

Surface area of 64 cubes $= 64 \times 6\ \text{cm}^2 = 384\ \text{cm}^2$

$\therefore$ Required ratio $= \frac{112}{384} = \frac{7}{24} = \mathbf{7 : 24}$.

22. (c) Let the length, breadth and height of the room be l, b and h respectively. Then, $h = 0.4\,(l + b)$

Area of four walls $= 2(l + b)h$

$= 2(l + b) \times 0.4(l + b)$

$= 0.8(l + b)^2$

$\therefore$ Area which is papered

$= 0.8(l + b)^2 - 15$

Now, area of paper = Area of wall

$\Rightarrow$ Length $\times$ breadth $= 0.8(l + b)^2 - 15$

$\Rightarrow$ Length $= \frac{0.8(l + b)^2 - 15}{0.5}$

($\because$ Width = 50 cm = 0.5 m)

Given, ₹ $\frac{\left(0.8(l + b)^2 - 15\right)}{0.5} \times 2 =$ ₹ 260

$\Rightarrow 0.8\,(l + b)^2 - 15 = \frac{260 \times 0.5}{2}$

$\Rightarrow 0.8\,(l + b)^2 - 15 = 65$

$\Rightarrow 0.8(l + b)^2 = 65 + 15 = 80$

$\Rightarrow (l + b)^2 = \frac{80}{0.8} = 100 \Rightarrow l + b = 10$

$\therefore h = 0.4 \times 10 = \mathbf{4\ m}$.

23. (a) Volume of vanilla cream $= 2 \times 35 \times 15 = 1050\ \text{mm}^3$

Volume of chocolate $= 2 \times 4 \times 35 \times 15 = 4200\ \text{mm}^3$

$\therefore$ Total volume $= (4200 + 1050)\ \text{mm}^3 = 5250\ \text{mm}^3$

% of vanilla $= \left(\frac{\text{Vol. of vanilla}}{\text{Total volume}} \times 100\right)\%$

$= \left(\frac{1050}{5250} \times 100\right)\% = \mathbf{20\%}$.

24. (c) Total volume of raindrops = 2 cm × 1 sq km

$= \frac{2}{100}\ \text{m} \times 1000\ \text{m} \times 1000\ \text{m} = 20000\ \text{m}^3$

Water collected in the pool $= 50\%$ of $20000\ \text{m}^3$

$= 10000\ \text{m}^3$

$\therefore$ Increase in level $= \frac{10000}{100 \times 10} = \mathbf{10\ m}$.

25. (b) Number of bricks $= \frac{95\%\ \text{of volume of wall}}{\text{Volume of a brick}}$

$= \frac{0.95 \times 600 \times 500 \times 50}{25 \times 12.5 \times 7.5}$

$= \mathbf{6080}$.

26. (d) Let x be the edge of the large cube. Then,

$x^3 = 3^3 + 4^3 + 5^3 \Rightarrow x^3 = 27 + 64 + 125 = 216$

$\Rightarrow x = \sqrt[3]{216} = 6$ cm.

$\therefore$ Required ratio

$= \frac{\text{Total surface area of smaller cubes}}{\text{Surface area of larger cube}}$

$= \frac{6(3)^2 + 6(4)^2 + 6(5)^2}{6(6)^2}$

$= \frac{6 \times 9 + 6 \times 16 + 6 \times 25}{6 \times 36}$

$= \frac{54 + 96 + 150}{216}$

$= \frac{300}{216} = \frac{25}{18} = \mathbf{25 : 18}$.

27. (a) Let length $= l$. Then, breadth $= l/2$

Given, area $= \frac{₹\ 5000}{₹\ 25} \Rightarrow l \times l/2 = 200$

$\Rightarrow l^2 = 400 \Rightarrow l = 20$.

Also, area of four walls = $2h(l + b)$

$\Rightarrow 2h(l + l/2) = \frac{64800}{240} \Rightarrow 3lh = 270$

$\Rightarrow h = \frac{270}{3 \times 20} = \mathbf{4.5\ m.}$

28. (c) As the box has to be painted from inside, the internal dimensions are:

Length $= (21 - 0.5 - 0.5) = 20$ cm,

Breadth $= (11 - 0.5 - 0.5) = 10$ cm,

Height $= (6 - 0.5) = 5.5$ cm

Total area to be painted

= Area of four walls + Area of base

$= 2 \times (10 \times 5.5 + 20 \times 5.5) + (20 \times 10)$

$= 2 \times (55 + 110) + 200 = 530\ cm^2$

Since the total expense of painting this area is ₹ 70, the rate of painting $= ₹\ \frac{70}{530} = ₹\ 0.13$

= **₹ 0.1 approx.**

29. (c) Volume of the earth dug out $= (6 \times 3 \times 2.5)\ m^3$
$= 45\ m^3$

Area of the remaining field

= Area of field – Area of pit

$= (20 \times 14)\ m^2 - (6 \times 3)\ m^2$

$= (280 - 18)\ m^2 = 262\ m^2$

Let the level of the field raised be x m. Then,

$262 \times x = 45 \Rightarrow x = \frac{45}{262} = 0.17175\ m = \mathbf{17.18\ cm.}$

30. (d) Volume of water flown through the pipe in 1 hour

$= (1.5 \times 1.25 \times 20 \times 1000)\ m^3$

$= 37500\ m^3$

Volume of water flown in the tank

$= (200 \times 150 \times 2)\ m^3$

$= 60000\ m^3$

$\therefore$ Time taken $= \frac{60000}{37500}$ hrs $= \frac{8}{5}$ hrs

$= \left(\frac{8}{5} \times 60\right)$ min = **96 min.**

Self Assessment Sheet–23

1. The volume of a rectangular block of stone is 10368 dm^3. Its dimensions are in the ratio 3 : 2 : 1. If its entire surface is polished at 2 paise per dm^2, then the total cost will be

(a) ₹ 31.50 (b) ₹ 31.68
(c) ₹ 63 (d) ₹ 63.36

2. The volume of a cuboid is twice the volume of a cube. If the dimensions of the cuboid are 9 cm, 8 cm and 6 cm, the total surface area of the cube is :

(a) 72 cm^2 (b) 216 cm^2
(c) 432 cm^2 (d) 108 cm^2

3. The edges of a cuboid are in the ratio 1 : 2 : 3 and its surface area is 88 cm^2. The volume of the cuboid is :

(a) 120 cm^3 (b) 64 cm^3
(c) 48 cm^3 (d) 24 cm^3

4. A small indoor greenhouse (herbarium) is made entirely of glass panes (including base) held together with tape. It is 30 cm long, 25 cm wide and 25 cm high. The area of glass is :

(a) 5000 cm^2 (b) 4800 cm^2
(c) 4250 cm^2 (d) 4500 cm^2

5. Two cubes each with 14 cm edge are joined face to face, thus farming a cuboid. What is the surface area of the resulting cuboid ?

(a) 3528 cm^2 (b) 2352 cm^2
(c) 1960 cm^2 (d) 1568 cm^2

6. A rectangular room measures 10 m × 10 m × 5 m. What is the maximum length of the stick it can accommodate ?

(a) 10 m (b) 15 m
(c) 20 m (d) 25 m

7. A godown measures 40 m × 25 m × 10 m. The maximum number of wooden crates each measuring 1.0 m × 1.25 m × 0.5 m that can be stored in the godown is :

(a) 14000 (b) 18000
(c) 15000 (d) 16000

8. A solid cube of side 12 cm is cut into eight cubes of equal volume. The side of the new cube and the ratio between the surface areas of the two cubes respectively are :

(a) 6 cm, 5 : 1 (b) 4 cm, 4 : 1
(c) 6 cm , 4 : 1 (d) 5 cm, 6 : 1

9. A rectangular cardboard sheet measures 48 cm × 36 cm. From each of its corners a square of 8 cm is cut off. An open box is made of the remaining sheet. The volume of the box is :

(a) 5120 cm^3 (b) 6400 cm^3
(c) 8960 cm^3 (d) 2560 cm^3

10. The volume (in m^3) of a cube whose diagonal is 2.5 metre, is :

(a) $\frac{125\sqrt{3}}{72}$ (b) $\frac{625}{8}$
(c) $\frac{125\sqrt{2}}{32}$ (d) $\frac{125}{8}$

Answers

1. (d)	2. (b)	3. (c)	4. (c)	5. (c)	6. (b)	7. (d)	8. (c)	9. (a)	10. (a)

Unit Test–5

1. If the longer side of a rectangle is doubled and the other is reduced to half, then the area of the new rectangle goes up by :

(a) 50% (b) 100%
(c) 0% (d) 150%

2. The wheel of a cycle covers 660 metres by making 500 revolutions. What is the diameter of the wheel (in cm) ?

(a) 42 (b) 21
(c) 30 (d) 60

3. A village, having a population of 4000 requires 150 litres of water per head per day. It has a tank measuring 20 m × 15 m × 6 m. For how many days will the water of this tank last ?

(a) 5 days (b) 3 days
(c) 4 days (d) 2 days

4. In exchange for a square plot of land, one of whose sides is 84 m, a man wants to buy a rectangular plot 144 m long and of the same area as the square plot. Determine the width of the rectangular plot.

(a) 48 m (b) 50 m
(c) 49 m (d) 49.5 m

5. What will be the perimeter of a rectangle if its length is 3 times its width and the length of the diagonal is $8\sqrt{10}$ cm ?

(a) $16\sqrt{10}$ cm (b) $15\sqrt{10}$ cm
(c) 64 cm (d) $24\sqrt{10}$ cm

6. The area of a circle drawn with its diameter as the diagonal of a cube of side of length 1 cm each is :

(a) $\frac{4\pi}{3}$ sq cm (b) $\frac{3\pi}{2}$ sq cm
(c) $\frac{3\pi}{4}$ sq cm (d) $\frac{2\pi}{3}$ sq cm

7. The perimeter of an equilateral triangle and a square are same, then $\frac{\text{area of }\Delta}{\text{area of }\square} =$

(a) $\frac{4}{3}$ (b) 1
(c) 1.5 (d) < 1

8. The quadrants shown in the figure given here are each of diameter 12 cm. What is the area of the shaded portion ?

(a) $12(12-\pi)cm^2$
(b) $144(4-\pi)cm^2$
(c) $36\pi cm^2$
(d) $36(4-\pi)cm^2$

9. A plastic box 1.5 m long, 1.25 m wide and 65 cm deep is to be made. It is to be opened at the top. Ignoring the thickness of the plastic the cost of sheet for it, if a sheet measuring 1 m^2 costs ₹ 20, is :

(a) ₹ 100 (b) ₹ 109
(c) ₹ 115 (d) ₹ 110

10. A wooden box measures 20 cm by 10 cm by 9 cm. Thickness of wood is 1 cm. Volume of wood to make the box (in cubic cm) is :

(a) 792 (b) 519
(c) 2400 (d) 1120

11. Front side wall of a house consists of a rectangle of 6 m × 4 m surrounded by a semicircle with base 4 m. It has two isosceles triangles made with vertical sides of the rectangle. The net area of the wall in sq m is

(a) $4(15+\frac{\pi}{2})$
(b) $4(18+\pi)$
(c) $4(24+\pi)$
(d) $64+4\pi$

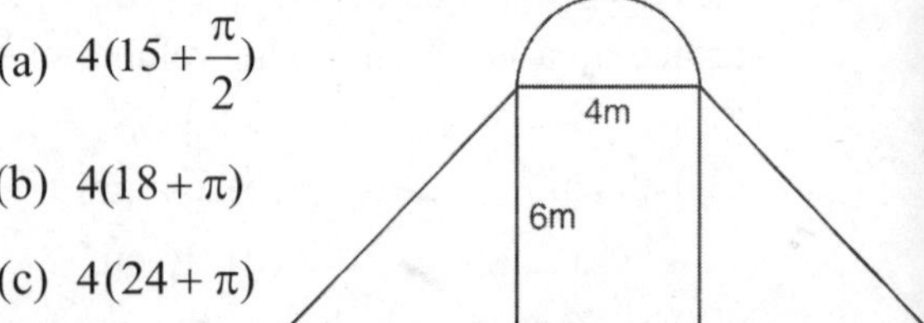

12. If the sum of diagonals of a rhombus is 10 cm and its area is 12 cm^2, then the lengths of its diagonal are :

(a) 5, 5 (b) 9, 1
(c) 8, 2 (d) 6, 4

13. Four equal-sized maximum circular plates are cut off from a square paper of area 784 cm^2. The circumference of each plate is

(a) 22 cm (b) 44 cm
(c) 66 cm (d) 88 cm

14. The area of the given figure *ABCDEF* is :

(a) 22.82 cm^2
(b) 25.82 cm^2
(c) 26.82 cm^2
(d) 28.82 cm^2

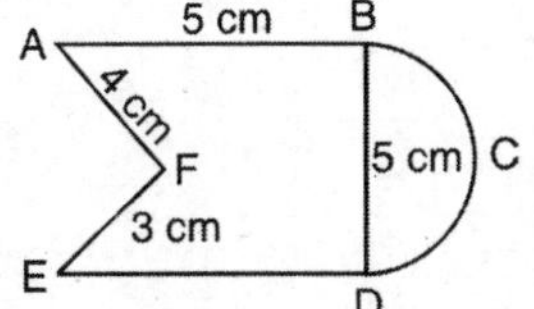

15. The volume of a cube of 60 cm side is the same as that of a cuboid one of whose sides is 36 cm. If the ratio of the other two sides is 15 : 16, then the largest side of the cuboid is :

(a) 60 cm (b) 75 cm
(c) 80 cm (d) 90 cm

16. The area of a square and a rectangular field is equal and is 900 m^2. If the perimeter of the rectangular field is 2 m more than that of the square field, calculate the dimensions of the rectangular field.

(a) 24 m, 10 m (b) 30 m, 25 m
(c) 36 m, 25 m (d) 36 m, 24 m

17. The sum of length, breadth and height of a cuboid is 12 cm long and its diagonal is 8 cm. The total surface area of the cuboid is :

(a) 88 sq cm (b) 85 sq cm
(c) 90 sq cm (d) 80 sq cm

18. The sum of length, breadth and height of a room is 19 m. The length of the diagonal is 11 m. The cost of painting the total surface area of the room at the rate of ₹ 10 per m^2 is :

(a) ₹ 240 (b) ₹ 2400
(c) ₹ 430 (d) ₹ 4200

19. Three cubes with sides in the ratio 3 : 4 : 5 are melted to form a single cube whose diagonal is $12\sqrt{3}$ cm. The sides of the cubes are :

(a) 6 cm, 8 cm, 10 cm (b) 3 cm, 4 cm, 5 cm
(c) 9 cm, 12 cm, 15 cm (d) None of these

20. *ABCD* is a parallelogram with sides *AB* = 12 cm, *BC* = 10 cm and diagonal *AC* = 16 cm. The area of the parallelogram and the distance between the shorter sides are respectively.

(a) 120.5 cm^2 ; 12.05 cm
(b) 119.8 cm^2 ; 11.98 cm
(c) 118.71 cm^2 ; 11.87 cm
(d) 117.9 cm^2 ; 11.79 cm

Answers

1. (c) **2.** (a) **3.** (b) **4.** (c) **5.** (c) **6.** (c) **7.** (d) **8.** (d) **9.** (b) **10.** (a)
11. (a) **12.** (d) **13.** (b) **14.** (d) **15.** (c) **16.** (c)
17. (d) [**Hint.** Use the formula $2(ab + bc + ca) = (a + b + c)^2 - (a^2 + b^2 + c^2)$] **18.** (b) **19.** (a) **20.** (b)

UNIT–6

STATISTICS

- *Data Handling*
  - *Mean, Median and Mode*
  - *Graphs (Bar graph, Pictograph, Pie charts, Line graphs)*
  - *Probability*

DATA HANDLING

KEY FACTS

1. The collection of a particular type of information in numbers is called **data**. For example, the temperatures of different cities, amount of rainfall (in cm) in a particular place at different times, marks of student etc.
2. Arranging this set of data in ascending or descending order is called an **array**.
3. The **range** of a set of data is the difference between the highest and lowest value.
4. The number of times a particular observation occurs is called its **frequency** (denoted by f).
5. Mean of ungrouped data $= \dfrac{\text{Sum of values}}{\text{Number of values}}$
6. The mean, $\bar{x}$, from a frequency table is calculated using the rule,

$$\bar{x} = \frac{\text{The sum of (the frequency} \times \text{value of item)}}{\text{Sum of frequency}} = \frac{\Sigma fx}{\Sigma f}$$

7. The **mode** of a set of data is the value which occurs most often.
8. The **median** of a set of values is the middle value when the data is arranged in order of size, *i.e.,* either ascending or descending.
9. In case, the set of values has even number of values, the median is the mean of two middle values.
10. (i) Data that can be counted is called **discrete data**.
    For example, the marks of a student, number of pupils using different modes of transport to school etc.
    (ii) Data that is measured is called **continuous data**. For example, height, weight, age, time etc.
11. A bar chart can be used to display data that can be counted.

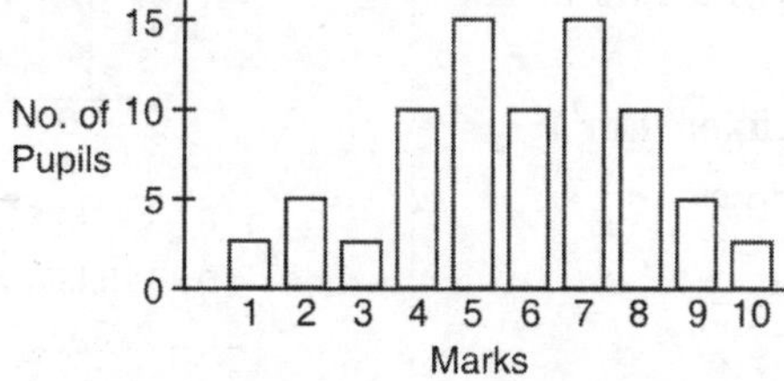

12. A pictogram can be used to illustrate data that can be counted using symbols to represent amounts.

Bus	☺
Car	☺☺☺
Walk	☺◖
Cycle	☺☺
Train	☺

Key ☺ = 20 pupils

13. **Pie charts** are usually used to display discrete data. Each sector, called a pie represents, part of a whole. The angles at the centre of a pie chart add up to 360°. The given pie chart shows how Rishi spends his day.

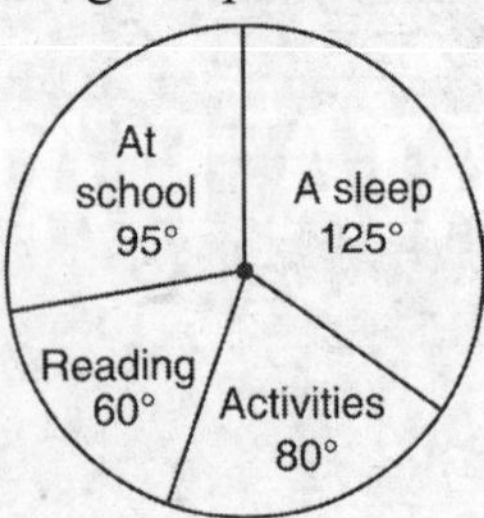

14. **Line graphs** can be used to display continuous data. They are used to show trends of temperatures, rainfall etc., over a period of times.

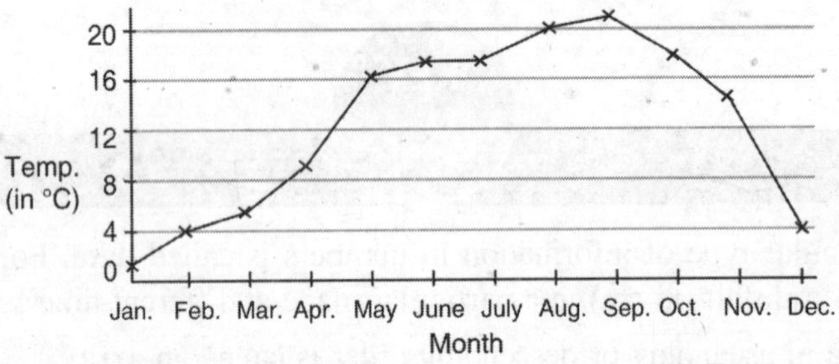

Question Bank–24(a)

1. Which of the following represents statistical data ?
   (a) The names of owners of shops located in a shopping complex.
   (b) A list giving the names of all states of India.
   (c) A list of all European countries and their respective capital cities.
   (d) The volume of rainfall in a certain geographical area recorded every month for 24 consecutive months.

2. In statistics, a suitable graph for representing the partitioning of total into sub parts is :
   (a) A bar graph (b) A picto graph
   (c) A pie chart (d) A line graph

3. The following table gives the areas of the oceans of the world :

Ocean	Area (million km²)
Pacific	70.8
Atlantic	41.2
Indian	28.5
Antarctic	7.6
Arctic	4.8

   A pie diagram of this data is to be drawn. What is the angle which the sector representing Pacific Ocean subtends at the centre ?
   (a) 167° (b) 97°
   (c) 67° (d) 18°

4. Which of the following statements is not correct for a bar graph ?
   (a) All bars have different thickness.
   (b) Distance between two consecutive bars is the same.
   (c) The bars can touch each other.
   (d) The thickness has no significance.

5. The given double graph shows the average heights of boys and girls at specific ages. Which conclusion is supported by the graph ?

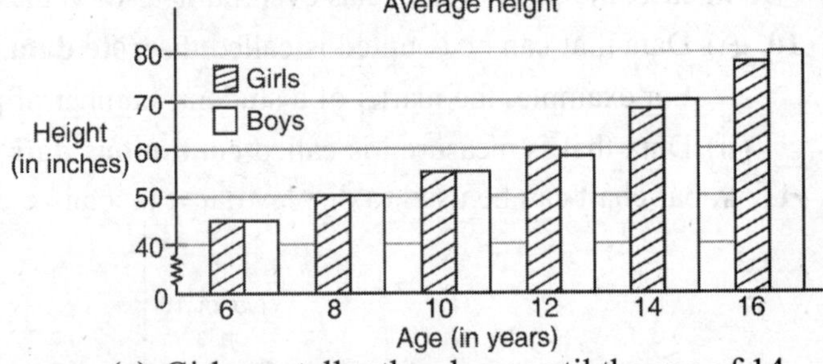

   (a) Girls are taller than boys until the age of 14.
   (b) Girls and boys grow the same amount each year.
   (c) After the age of 14, boys grow faster than girls.
   (d) Boys are always taller than girls.

6. You want to display a set of data showing the number of students in the line for lunch in the school canteen every 15 minutes during the lunch-break. Which graph shall display the data most appropriately ?
   (a) Bar Graph (b) Line Graph
   (c) Pie Graph (d) Pictograph

7. The given line graph shows the growth rate of a kitten. During which 2-month period, the kitten's weight increased the most.

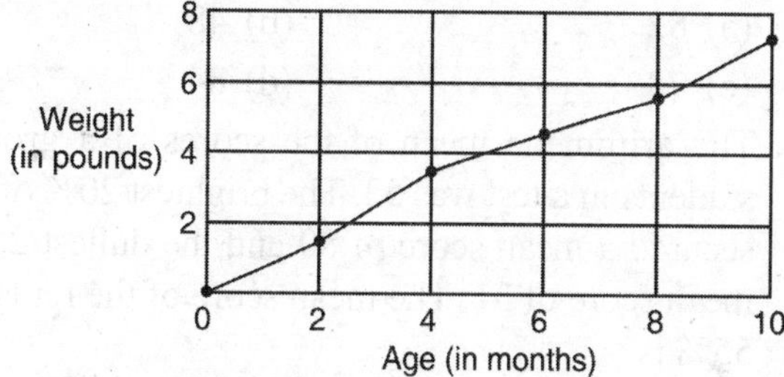

(a) 0 to 2 months (b) 2 to 4 months
(c) 4 to 6 months (d) 6 to 8 months

8. From the pie graph shown alongside. Find the per cent of students that are seventh grades.

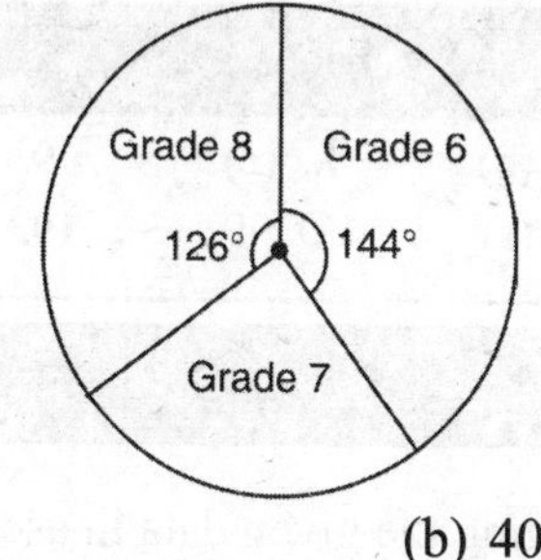

(a) 35% (b) 40%
(c) 25% (d) 28%

9. The given pie-chart shows the marks scored by a student in five different subjects – English, Hindi, Mathematics, Science and Social Science. Assuming that the total marks obtained for the examination are 540, find the subject in which the student scored 22.2% marks.

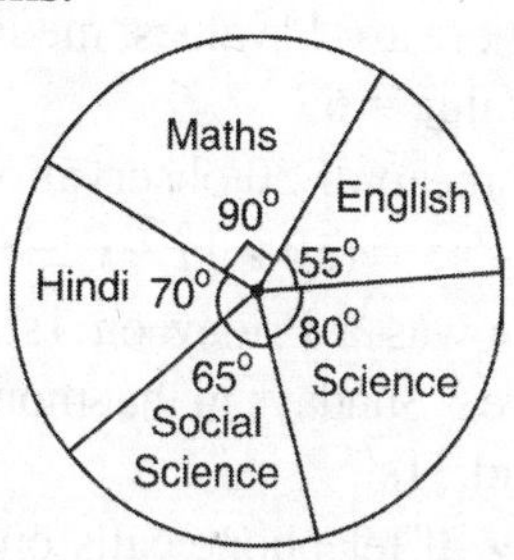

(a) Hindi (b) Science
(c) Social Science (d) English

10. The following bar graph shows the rainfall at selected locations in certain months.

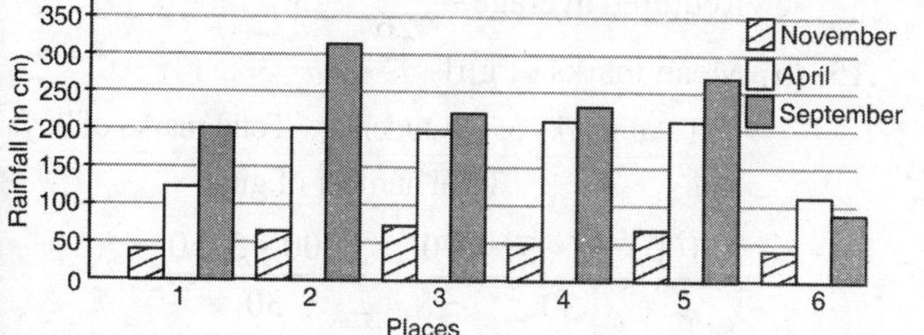

Which of the following statements is correct :

(a) November rainfall exceeds 100 cm in each location.
(b) September rainfall exceeds April rainfall by 50 cm in each location.
(c) November rainfall is lower than April rainfall in each location.
(d) None of the above.

11. The data given below are the times in minutes, it takes seven students to go to school from their homes. Which statement about the data is false ? 11, 6, 22, 7, 10, 6, 15.

(a) The median is 11 (b) The mean is 11
(c) The range is 16 (d) The mode is 6

12. A variate takes 11 values which are arranged in ascending order of their magnitudes. It is found that 4th, 6th and 8th observations are 8, 6 and 4 respectively. What is the median of the distribution?

(a) 4 (b) 6
(c) 8 (d) 10

13. From a series of 50 observations, an observation with the value of 45 is dropped, but the mean remains the same. What was the mean of 50 observations ?

(a) 50 (b) 49
(c) 45 (d) 40

14. A person made 165 telephone calls in the month of May in a year. It was Friday on 1st May of the year. The average of telephone calls on Sundays of the month was 7. What was the average of the telephone calls per day on the rest days of the month.

(a) $\frac{165}{31}$ (b) 5
(c) 7 (d) $\frac{137}{27}$

15. The mean of the marks in Statistics of 100 students in a class was 72. The mean of marks for boys was 75, while their number was 70. The mean of marks of girls in the class was

(a) 35 (b) 65
(c) 68 (d) 86

16. The numbers 4 and 9 have frequencies x and $(x - 1)$ respectively. If their arithmetic mean is 6, then x is equal to :

(a) 2 (b) 3
(c) 4 (d) 5

17. If the median of $\frac{x}{5}, x, \frac{x}{4}, \frac{x}{2}$ and $\frac{x}{3}$ (where $x > 0$) is 8, then the value of x would be :

(a) 24 (b) 32
(c) 8 (d) 16

18. In the frequency distribution of discrete data given below, the frequency x against value 0 is missing.

Variable x :	0	1	2	3	4	5
Frequency f :	x	20	40	40	20	4

If the mean is 2.5, then the missing frequency x will be

(a) 0 (b) 1
(c) 3 (d) 4

19. The mean of 10 numbers is 7. If each number is multiplied by 12, then the mean of new set of numbers is

(a) 82 (b) 48
(c) 78 (d) 84

20. The arithmetic mean of the scores of a group of students in a test was 52. The brightest 20% of them secured a mean score of 80 and the dullest 25%, a mean score of 31. The mean score of the remaining 55% is

(a) 45% (b) 50%
(c) 51.4% (approx) (d) 54.6% (approx)

Answers

1. (d)	**2.** (c)	**3.** (a)	**4.** (a)	**5.** (c)	**6.** (b)	**7.** (b)	**8.** (c)	**9.** (b)	**10.** (c)
11. (a)	**12.** (b)	**13.** (c)	**14.** (b)	**15.** (b)	**16.** (b)	**17.** (a)	**18.** (d)	**19.** (d)	**20.** (c)

Hints and Solutions

1. (d) Since all the other sets of data do not have numerical figures, they do not represent statistical data.

3. (a) Angle representing Pacific Ocean

$$= \left(\frac{70.8}{70.8+41.2+28.5+7.6+4.8}\right) \times 360°$$

6. (b) Line Graphs are used to illustrate data collected at intervals in time.

7. (b) The line graph shown is steepest from the period of 2 months to 4 months, showing that the growth is maximum.

8. (c) Angle at the centre representing Grade 7 $= 360° - (126° + 144°) = 360° - 270° = 90°$

$$\therefore \text{ Reqd. \%} = \frac{90°}{360°} \times 100 = \mathbf{25\%}.$$

9. (b) Calculating the percentage of marks in different subjects, we get

$$\text{Hindi} = \frac{70°}{360°} \times 100 = 19.4\%$$

$$\text{Science} = \frac{80°}{360°} \times 100 = 22.2\%$$

$$\text{Social Science} = \frac{65°}{360°} \times 100 = 18.1\%$$

$$\text{English} = \frac{55°}{360°} \times 100 = 15.28\%$$

11. (a) Arranging the given data in ascending order, we get, 6, 6, 7, 10, 11, 15, 22

$$\text{Mean} = \frac{6+6+7+10+11+15+22}{7} = \frac{77}{7} = 11$$

Mode = 6
Median = 4th value = 10
Range = 22 – 6 = 16

12. (b) Since there are 11values, median = middle value = 6th value = **6.**

13. (c) Let the mean of 50 observations be x. Then, $50 \times x - 45 = (50 - 1) \times x \Rightarrow x = 45$.

14. (b) Since it was a Friday on 1st May of the year, there are 5 Sundays in that month, 3rd, 10th, 17th, 24th and 31st.

$\therefore$ Number of telephone calls on Sundays $= 7 \times 5 = 35$

$\therefore$ Remaining number of telephone calls $= 165 - 35 = 130$

$$\text{Required average} = \frac{130}{26} = \mathbf{5}.$$

15. (b) Mean marks of girls

$$= \frac{\text{Total marks of all students} - \text{Total marks of boys}}{\text{Total number of girls}}$$

$$= \frac{100 \times 72 - 75 \times 70}{30} = \frac{7200 - 5250}{30}$$

$$= \frac{1950}{30} = \mathbf{65}.$$

16. (b) Given, $\dfrac{4x+9(x-1)}{x+(x-1)} = 6$

$\Rightarrow \dfrac{13x-9}{2x-1} = 6 \Rightarrow 13x - 9 = 12x - 6 \Rightarrow x = \mathbf{3}.$

17. (a) Arranging in ascending order, the values are $\dfrac{x}{5}, \dfrac{x}{4}, \dfrac{x}{3}, \dfrac{x}{2}, x$

Middle value $= \dfrac{x}{3} \Rightarrow \dfrac{x}{3} = 8 \Rightarrow x = \mathbf{24}.$

18. (d) Mean $= \dfrac{\Sigma fx}{\Sigma f}$

$\Rightarrow \dfrac{0\times x + 1\times 20 + 2\times 40 + 3\times 40 + 4\times 20 + 5\times 4}{x + 20 + 40 + 40 + 20 + 4} = 2.5$

$\Rightarrow \dfrac{20 + 80 + 120 + 80 + 20}{124 + x} = 2.5$

$\Rightarrow 320 = 2.5\,(124 + x)$

$\Rightarrow 320 = 310 + 2.5x \Rightarrow 2.5x = 10 \Rightarrow x = \mathbf{4}.$

19. (d) Total of 10 numbers = $10 \times 7 = 70$

If each number is multiplied by 12,

New Total = 70×12

$\therefore$ New mean $= \dfrac{70\times 12}{10} = \mathbf{84}.$

20. (c) Mean score of remaining 55%

$= \dfrac{[100\times 52 - (20\times 80 + 25\times 31)]}{55}$

$= \dfrac{5200 - (1600 + 775)}{55}$

$= \dfrac{5200 - 2375}{55} = \dfrac{2825}{55} = 51.36\%$

$=$ **51.4%** (approx)

PROBABILITY

1. Probability means the chance of occurrence of an event. It is the measure of uncertainty.
2. A **trial** is a random experiment repeated under same conditions. For example, tossing a coin, throwing a dice are trials.
3. An **event** is a possible outcome of the trial. For example, getting a head or tail on tossing a coin, getting a '6' on the upper face of a dice etc., are events.
4. When two events have an even chance of happening the two events are **equally likely**. For example, while throwing a dice, chance of occurring of Head or Tail are equally likely events.
5. Probability uses numbers to measure the chance of an outcome happening.
6. A likelihood scale runs from impossible to certain, with an 'even chance' in the middle.
   So, all probabilities have a value between 0 and 1.

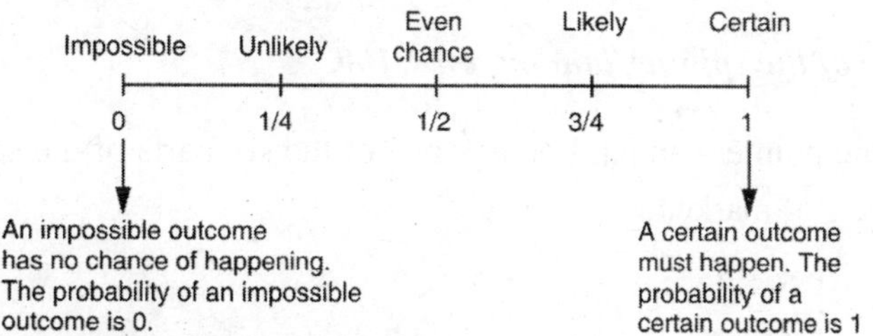

> **Note:** Probability is never greater than 1 or less than zero.

7. Probability of something happening $= \dfrac{\text{Number of favourable outcomes}}{\text{Total number of outcomes}}$

   (i) Here, favourable outcome is an outcome that matches the event. For example, in throwing of a dice, the number of favourable outcomes corresponding to the appearance of a prime number is 3, *i.e.,* 2, 3 and 5.

(ii) The set of the total number of outcomes is also described by the word **sample space.** Hence the sample space for throwing a dice is $S = \{1, 2, 3, 4, 5, 6\}$.

(iii) One of the most common examples used in probability is that of a deck of 52 playing cards.
A pack of 52 cards is divided in four suits :
Hearts (red), Spades (black), Diamond (red), Clubs (black). Each suit consists of 13 cards bearing the values 2, 3, 4, 5, 6, 7, 8, 9, 10, Jack, Queen, King and Ace. The Jack, Queen and King are called 'Face Cards'. The total number of outcomes is 52.

8. Probability of an event not occurring = 1 – Probability of an event occurring

Solved Examples

Ex. 1. ***A dice is rolled once. What is the probability of getting :***
(a) a 3 ? ***(b) an even number ?***

Sol. (a) Number of 3's on a dice = 1
Total number of possible outcomes = 6
($\because$ Sample space = {1, 2, 3, 4, 5, 6})
$\therefore$ Probability of getting 3 = $\frac{1}{6}$.

(b) Number of even numbers on a dice = 3 (2, 4, 6)
Total number of possible outcomes = 6
$\therefore$ Probability of getting an even number = $\frac{3}{6} = \frac{1}{2}$.

Ex. 2. ***A bag contains 4 red, 6 blue and 7 yellow balls. One ball is selected at random. What is the probability that it is (a) a red ball ? (b) blue or yellow ball ? (c) not blue ball ?***

Sol. (a) Number of red balls = 4
Total number of balls = 4 + 6 + 7 = 17
$\therefore$ Probability of a red ball = $\frac{4}{17}$.

(b) Number of the blue and yellow balls = 6 + 7 = 13
Total number of balls = 17
$\therefore$ Probability of a blue or a yellow ball = $\frac{13}{17}$.

(c) Total number of balls that are not blue = Red balls + Yellow balls = 4 + 7 = 11
Total number of balls = 17
$\therefore$ Probability of not a blue ball = $\frac{11}{17}$.

Ex. 3. ***What is the probability of the spinner landing on ₹ 100.***

Sol. In the spinner shown, the pointer can land on any one of the six parts of the same size.
There are 3 parts with ₹ 100 marked.

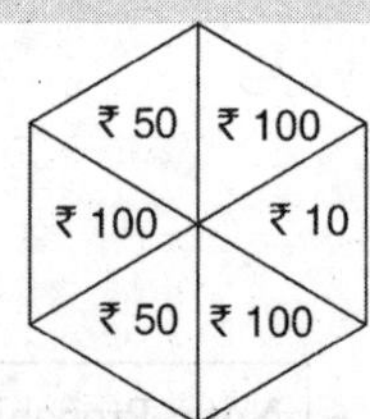

$\therefore$ Required probability = $\frac{3}{6} = \frac{1}{2}$.

Ex. 4. ***One of 26 letter keys on a type writer is pressed. What is the probability that the key prints a letter other than 'a'?***

Sol. Total number of letters other than 'a' = 25
Total number of possible out comes = 26
$\therefore$ Required probability = $\frac{25}{26}$.

Ex. 5. ***A survey of the local neighbourhood showed that 15% of the population was under 12 years old and 25% of the population was over 60 years. What is the probability that a person selected at random was between 12 years old and 60 years old ?***

Sol. The percentage of population between 12 years old and 60 years old is (100 – (15 + 25))% = 60%.

$\therefore$ Required probability $= \frac{60}{100} = \mathbf{\frac{3}{5}}$.

Ex. 6. ***The probability of selecting a queen from a standard pack of cards is $\frac{1}{13}$. Find the probability of not selecting a queen ?***

Sol. P (event not happening) = 1 – P (event happening)

P (not selecting a queen)$= 1 - \frac{1}{13} = \mathbf{\frac{12}{13}}$.

Ex. 7. ***A card is drawn from a pack of 52 cards. What is the probability that it is one of the following?***
(a) a black card (b) a seven (c) a red Queen

Sol. There are 52 cards, so total number of possible outcomes = 52

(a) There are 26 black cards, so P(black card) $= \frac{26}{52} = \mathbf{\frac{1}{2}}$.

(b) There are 4 sevens, so P(seven) $= \frac{4}{52} = \mathbf{\frac{1}{13}}$.

(c) There are 2 red Queens, so P(red Queen) $= \frac{2}{52} = \mathbf{\frac{1}{26}}$.

Ex. 8. ***What is the probability of drawing a red face card from a pack of 52 playing cards.***

Sol. There are 3 face cards in each suit.

$\therefore$ There are 6 red face cards in all.

Total number of possible outcomes = 52

P (red face card) $= \frac{6}{52} = \mathbf{\frac{3}{26}}$.

Question Bank–24(b)

1. The event of drawing a red card from a pack of blue, white and black cards is
(a) unlikely (b) certain
(c) impossible (d) likely

2. On the probability line, we would describe the event – *A new born child will be a girl* as
(a) unlikely (b) even chance
(c) certain (d) impossible

3. In a class there are 14 boys and 10 girls. If one child is absent, the probability that it is a boy is
(a) $\frac{5}{12}$ (b) $\frac{7}{12}$
(c) $\frac{10}{14}$ (d) $\frac{1}{3}$

4. A box contains 8 slips of paper which are numbered 0 to 7. If one slip of paper is drawn unseen, the probability of drawing a number greater than 4 is
(a) $\frac{0}{8}$ (b) $\frac{1}{8}$
(c) $\frac{1}{4}$ (d) $\frac{3}{8}$

5. A dice is rolled once. What is the probability of rolling a prime number.
(a) $\frac{2}{3}$ (b) $\frac{1}{2}$
(c) $\frac{1}{6}$ (d) $\frac{5}{6}$

6. A bag contains red, white and blue marbles. The probability of selecting a red marble is $\frac{2}{15}$ and that of selecting a blue marble is $\frac{4}{15}$. The probability of selecting a white marble is

(a) $\frac{13}{15}$ (b) $\frac{11}{15}$

(c) $\frac{3}{5}$ (d) $\frac{2}{5}$

7. A letter is chosen at random from the word 'PROBABILITY'. The probability that it is a vowel is

(a) $\frac{3}{11}$ (b) $\frac{6}{11}$

(c) $\frac{4}{11}$ (d) $\frac{7}{11}$

8. The probability of drawing a red 9 from a standard pack of 52 playing cards is

(a) $\frac{1}{13}$ (b) $\frac{1}{26}$

(c) $\frac{1}{2}$ (d) $\frac{1}{4}$

9. Eight sided dice are used in adventure games. They are marked with the numbers 1 to 8. The score is the upper most face. The probability of scoring a square number is

(a) $\frac{3}{8}$ (b) $\frac{1}{2}$

(c) $\frac{1}{8}$ (d) $\frac{1}{4}$

10. The probability of drawing a face card from a standard pack of 52 cards is

(a) $\frac{4}{13}$ (b) $\frac{1}{13}$

(c) $\frac{3}{13}$ (d) $\frac{10}{13}$

11. A box contains 50 coloured stones. What is the total number of orange stones in the box if the probability of selecting an orange stone is 0.4.

(a) 20 (b) 0

(c) 10 (d) 40

12. Akshay spins a spinner that is split into 10 equal sections. The sections are labelled 1, 3, 2, 1, 2, 2, 3, 1, 2, 1. What is the probability that the spinner will land on the number 2.

(a) $\frac{1}{10}$ (b) $\frac{2}{5}$

(c) $\frac{1}{2}$ (d) $\frac{1}{5}$

13. A chocolate gift box contains 15 chocolates. Six are 'Five Stars', four are 'Fruit 'n' Nut', five are 'Dairy Milk'. After I have eaten the first chocolate, a 'Fruit 'n' Nut'. I pick another one. The probability that I pick a 'Fruit 'n' Nut' again is

(a) $\frac{4}{15}$ (b) $\frac{1}{3}$

(c) $\frac{1}{5}$ (d) $\frac{3}{14}$

14. This frequency table shows the results of a small class quiz. If a student is selected at random, the probability that he or she scored 2, 4 or 5 is

Marks	Frequency
1	4
2	3
3	2
4	6
5	5

(a) $\frac{3}{20}$

(b) $\frac{1}{4}$

(c) $\frac{7}{10}$

(d) $\frac{9}{20}$

15. The king, queen and jack of hearts are removed from a deck of 52 playing cards and well shuffled. One card is selected from the remaining cards. The probability of drawing a '10' of hearts is

(a) $\frac{10}{49}$ (b) $\frac{13}{49}$

(c) $\frac{3}{49}$ (d) $\frac{1}{49}$

Answers

1. (c)	**2.** (b)	**3.** (b)	**4.** (c)	**5.** (b)	**6.** (c)	**7.** (c)	**8.** (b)	**9.** (d)	**10.** (c)
11. (a)	**12.** (b)	**13.** (d)	**14.** (c)	**15.** (d)					

Hints and Solutions

1. (c) Since there is no red card in the pack, the event is impossible.

2. (b) A new born child will either be a girl or a boy, so the event has an even chance.

3. (b) Probability of boy being absent

$= \frac{14}{14+10} = \frac{14}{24} = \mathbf{\frac{7}{12}}.$

4. (c) Number of slips with numbers greater than 4 = 3, *i.e.,* 5, 6, 7

$\therefore$ Reqd. probability $= \mathbf{\frac{3}{8}}.$

5. (b) Prime numbers on the faces of a dice are 2, 3 and 5.

$\therefore P(\text{rolling a prime no.}) = \frac{3}{6} = \mathbf{\frac{1}{2}}.$

6. (c) P (white marble) = Total probability – [P (red marble) + P(blue marble)]

$= 1 - \left[\frac{2}{15} + \frac{4}{15}\right] = 1 - \frac{6}{15} = \frac{9}{15} = \mathbf{\frac{3}{5}}.$

7. (c) The vowels in the word 'PROBABILITY' are *o, a, i, i.*

$\therefore$ Reqd. probability

$= \frac{\text{Number of vowels}}{\text{Number of letters in 'PROBABILITY'}}$

$= \mathbf{\frac{4}{11}}.$

8. (b) There are two red nines, one of hearts and one of diamonds. In all these are 52 cards.

$\therefore$ Reqd. probability $= \frac{2}{52} = \mathbf{\frac{1}{26}}.$

9. (d) Square numbers between 1 and 8 = 1 and 4

$\therefore$ Reqd. probability $= \frac{2}{8} = \mathbf{\frac{1}{4}}.$

10. (c) Face cards or more commonly picture cards are 12 in number, three from each suit.

$\therefore P(\text{face card}) = \frac{12}{52} = \mathbf{\frac{3}{13}}.$

11. (a) Let the total number of orange stones in the box be x.

Then, $P(\text{orange}) = \frac{x}{50}$

Given, $\frac{x}{50} = 0.4 \Rightarrow x = \mathbf{20}.$

12. (b) The number of sections labelled '2' = 4

Total number of sections = 10

$\therefore P(\text{number 2}) = \frac{4}{10} = \mathbf{\frac{2}{5}}.$

13. (d) Total number of chocolates remaining after I pick and eat one 'Fruit 'n' Nut' chocolate = 15 – 1 = 14

Total no. of 'Fruit 'n' Nut' remaining = 4 – 1 = 3

$\therefore$ Required Probability $= \mathbf{\frac{3}{14}}.$

14. (c) Total number of students in the class

= 4 + 3 + 2 + 6 + 5 = 20

Number of students scoring 2, 4 or 5

= 3 + 6 + 5 = 14

$\therefore P(2, 4 \text{ or } 5) = \frac{14}{20} = \mathbf{\frac{7}{10}}.$

15. (d) Number of cards remaining after the king, queen and jack of hearts have been removed

= 52 – 3 = 49

There is only one '10' of hearts.

$\therefore P(\text{'10' of hearts}) = \mathbf{\frac{1}{49}}.$

Self Assessment Sheet–24

1. Which data set has a median of 16 ?
   (a) 16, 21, 26, 29, 32
   (b) 0, 4 , 7, 10, 16, 16
   (c) 0, 9, 31, 17, 18, 22
   (d) 25, 14, 7, 16, 21
2. A batsman scores 80 runs in his sixth innings and thus increases his average by 5. What is his average after six innings?
   (a) 50 (b) 55
   (c) 60 (d) 65
3. The average salary of male employees in a firm is ₹ 520 and that of female employees is ₹ 420. The mean salary of all the employees is ₹ 500. What is the percentage of female employees ?
   (a) 40% (b) 30%
   (c) 25% (d) 20%
4. The mean, median and mode of given data of scores are 21, 23 and 22 respectively. If 3 is added to each score. What are the new values of mean, median and mode respectively.
   (a) 21, 23, 22 (b) 24, 26, 25
   (c) 24, 23, 22 (d) 23, 21, 24
5. A student represents his scores in Mathematics. Statistics and Economics in a pie-chart. The central angle for Mathematics is 120°. He scored 96 in Statistics and 84 in Economics. The central angle for Statistics is :
   (a) 116° (b) 128°
   (c) 192° (d) 212°
6. You spin the spinner at right, which is divided into equal parts. Match the event with the letter on the

number line that indicates the probability of the event.

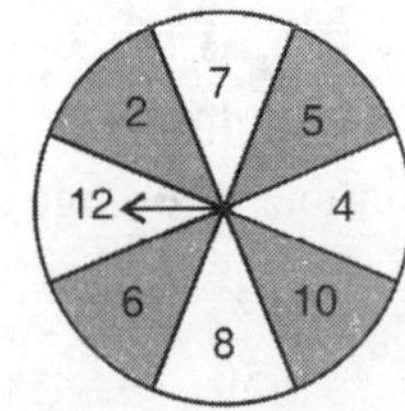

(i) Pointer lands on the shaded part — *A*
(ii) Pointer lands on an even number — *B*
(iii) Pointer lands on 6 — *C*
(iv) Pointer lands on a prime number — *D*

A B C D
0 0.25 0.5 0.75 1

7. Jenny draws a card from a standard pack of 52 cards. What is the probability that she draws an Ace or a King.

(a) $\frac{4}{13}$ (b) $\frac{1}{13}$

(c) $\frac{9}{13}$ (d) $\frac{2}{13}$

8. The probability that Vikram is late for school in the morning is 0.1. How many times would you expect Vikram to be on time in 20 mornings ?

(a) 2 (b) 9

(c) 18 (d) 1

9. Soni has digit cards 1, 4 and 7. She makes 2-digit numbers using each card only once. The probability that a 2-digit number chosen at random is divisible by 2 is :

(a) $\frac{1}{3}$ (b) $\frac{2}{3}$

(c) $\frac{1}{6}$ (d) $\frac{5}{6}$

10. Twelve sides dice are used in adventure games. They are marked with the numbers 1 to 12. The score is the upper most face. If this dice is thrown, what is the probability that the score is a factor of 12 ?

(a) $\frac{1}{4}$ (b) $\frac{2}{3}$

(c) $\frac{1}{2}$ (d) $\frac{1}{3}$

Answers

1. (d) **2.** (b) **3.** (d) **4.** (b) **5.** (b) **6.** (i) → C, (ii) → D, (iii) → A, (iv) → B

7. (d) **8.** (c) **9.** (a) **10.** (c)

UNIT-7

SETS

- *Sets*

Chapter 25

SETS

KEY FACTS

1. A **set** is a well defined collection of objects.
2. Each object in a set is called an **element** of the set. The symbol $\in$ is used to denote 'a member of' or 'belongs to'.
   $x \in A$ means x is an element of set A.
   $x \notin A$ means x is not an element of set A.
3. Sets are specified by any of the following methods :
   (i) **Description method :** By writing the description in braces, *i.e.,*
   A = {consonants}, B = { letters of the word SCHOOL}
   (ii) **Roster method :** By listing the names of the elements in braces, *i.e.,* the set of days of week in roster form is {Sunday, Monday, Tuesday, Wednesday, Thursday, Friday, Saturday}
   (iii) **Set builder form :** A set is specified in the form { x : statement of property x satisfies}.
   For example , $\{x : x = n^2,\ n \in N \text{ and } n \leq 8\}$
4. A set containing only one element is called a **singleton set** and a set containing no element is called the **empty set.** The **empty set** is denoted by { } or ϕ.
5. Sets containing equal number of elements are called **equivalent sets.** Set containing same elements are called **equal sets.** For example, if
   A = {letters of the word 'pat'}
   B = {letters of the word 'tap'}
   C = {first three natural numbers}
   then, $A = B$ and $A \Leftrightarrow C$, $B \Leftrightarrow C$ [$\Leftrightarrow$ $\Rightarrow$ equivalent]
6. A set containing a definite number of objects is a **finite set**, otherwise it is an **infinite set**.
7. The number of elements in a set A is called its **cardinal number** and is denoted by $n(A)$.
   For example, if A = {vowels}, then $n(A) = 5$.
8. Two sets are **overlapping** if they have at least one element in common.
   For example, A = { letters of 'SET'} and B = { letters of 'SON'} are overlapping as they have letter S common.
9. If two sets have no element in common then they are called **disjoint sets**. For example, A = {even numbers} and B = {odd numbers} are disjoint sets.
10. If every member of a set P is also a member of a second set Q, then P is a **subset** of Q.

Note that :
(i) ***Every set is a subset of itself.***
(ii) ***Null set is a subset of every set.***

   The symbol '$\subseteq$' denotes '***is a subset of***'.
11. If P is a subset of Q and $n(P) < n(Q)$ then P is called a **proper subset** of Q.
   The symbol '$\subset$' denotes '***is a proper subset of***'.

For example,

If $A = \{1, 3, 4, 5, 6\}$, $B = \{3, 4, 5\}$, $C = \{\}$ or ϕ

Then, $B \subseteq A$, $C \subseteq A$ and $C \subseteq B$

but $B \subset A$, $C \subset A$, $C \subset B$

'$\not\subseteq$' denotes 'is not a subset of'.

'$\not\subset$' denotes 'is not a proper subset of'.

Thus, all the subsets of a set other than the set itself are proper subsets.

12. If $A \subseteq B$ then B is a superset of A, which is denoted by $B \supseteq A$, *i.e.*, B contains A.

13. If there are n elements in a set, the total number of subsets is $\mathbf{2^n}$.

The total number of proper subset is $2^n - 1$.

For example, if $A = \{p, q, r\}$, then the number of subsets of $A = 2^3 = 8$ and number of proper subsets of $A = 2^3 - 1 = 8 - 1 = 7$.

14. The superset of all the sets for a particular discussion is called an **universal set.** It is denoted by U. **For example,**

(i) The set of whole numbers is the universal set for the sets of even numbers and odd numbers.

(ii) The set of alphabets is the universal set for the sets of vowels or consonants.

15. The set of elements of the universal set, which are not in a given set (say A) is the complement of A, denoted by A' *or* $\bar{A}$. Thus, $A' = \{x \in U : x \notin A\}$

For example, $U = \{$ set of alphabets$\}$, $A = \{$ set of consonants$\}$

Then, $A' = \{$ set of vowels $\}$

16. Union of sets : The **Union of two sets** is the set whose elements occur either in one set or in the other (or in both the sets). The common elements of both the sets are included only once in the union of sets.

The symbol '$\cup$' denotes 'union of '

For example, $A = \{1, 3, 4, 5\}$, $B = \{2, 3, 6, 7\}$

Then, $A \cup B = \{1, 2, 3, 4, 5, 6, 7\}$

Note: If A is any set then, (i) $A \cup \phi = A$ (ii) $A \cup U = U$ (iii) $A \cup A = A$ (iv) $A \cup A' = U$

17. Intersection of sets : The intersection of two sets whose elements are common to both the sets.

The symbol '$\cap$' denotes ' intersection of '.

For example,

$A = \{$ non zero multiples of 2 less than 10$\}$ $B = \{$ non zero multiples of 3 less than 10$\}$

Then, $A = \{2, 4, 6, 8\}$ and $B = \{3, 6, 9\}$

$\therefore A \cap B = \{6\}$

Note: If A is any set, then (i) $A \cap \phi = \phi$ (ii) $A \cap U = A$ (iii) $A \cap A = A$ (iv) $A \cap A' = \phi$

18. By taking a few examples, it can be shown that

(*i*) $\mathbf{n(A) + n(B) = n(A \cap B) + n(A \cup B)}$ (*ii*) $\mathbf{(A \cup B)' = A' \cap B'}$ (*iii*) $\mathbf{(A \cap B)' = A' \cup B'}$

19. Venn diagrams : Sets are pictorially represented by venn diagrams. The universal set is generally represented by a rectangle and its subsets are shown as circles, ovals etc., within the rectangle, the elements of the set are written inside the curve.

For example, the set $U = \{a, b, c, d, e, f)$ and $A = \{a, e\}$ can be represented on a Venn diagram as shown in Fig.

U b c A a e d f

Some of common relationships amongst sets can be represented by the Venn diagrams as follows:

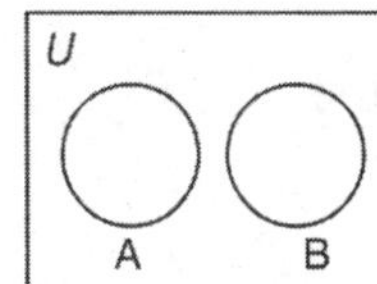

Disjoint sets

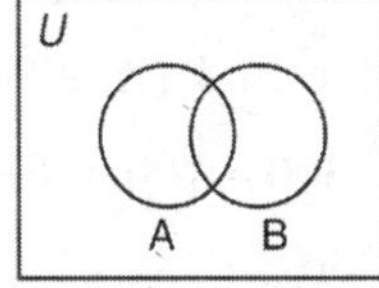

Overlapping sets

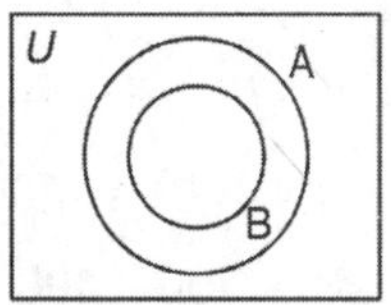

A universal set, its subsets where $B \subset A$

The shaded regions of the each of the diagrams given below describes the set given below the diagram.

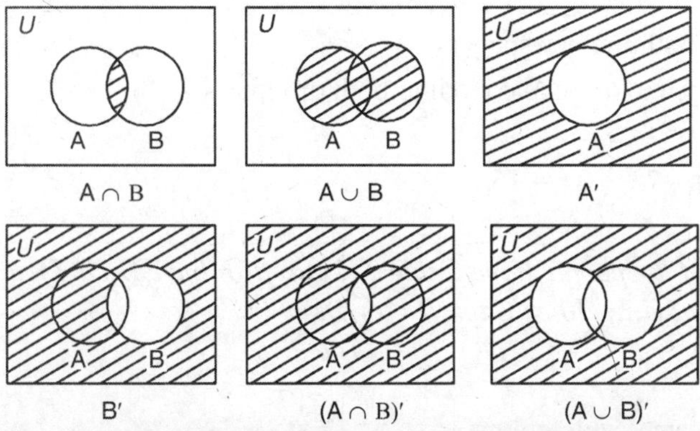

$A \cap B$ $\quad$ $A \cup B$ $\quad$ A'

B' $\quad$ $(A \cap B)'$ $\quad$ $(A \cup B)'$

Solved Examples

Ex. 1. ***Which of the following sets is non - empty ?***

(a) Set of odd prime numbers less than 3. *(b)* $A = \{x \mid x + 4 = 0,\ x \in N\}$

(c) $B = \{x \mid 4 < x < 5, x \in W\}$ *(d)* C = *Set of even prime numbers*

Sol. (d)

(a) Odd number less than 3 is 1, which is not a prime number. Hence the set is empty.

(b) No natural number satisfies the equation $x + 4 = 0$, hence $A = \phi$.

(c) Again, there is no whole number between 4 and 5, hence $B = \phi$.

(d) $C = \{2\}$, hence it is a non - empty set.

Ex. 2. ***Which of the following pairs of sets are not equal ?***

(a) $A = \{1, 3, 3, 1\}, B = \{1, 4\}$ *(b)* $A = \{x \mid x + 4 = 4\},\ B = \{0\}$

(c) $A = \{m, a, t, h, e, m, a, t, i, c, s\}, B = \{a, m, t, h, e, i, c, s\}$ *(d)* $A = \{1, 4, 9, 16, 25, ...\}$ $B = \{x/x = n^2, n \in N\}$

Sol. Ans. (a)

(a) $A = \{1, 3\}, B = \{1,4\} \Rightarrow A \neq B$

(b) $A = \{x \mid x + 4 = 4\} \Rightarrow A = \{0\}, B = \{0\}$

(c) $A = \{m,a,t,h,e,m,a,t,i,c,s\} = \{m,a,t,h,e,i,c,s\}$ $\quad B = \{a, m, t, h, e, i, c, s\} \Rightarrow A = B$

(d) $A = \{1, 4, 9, 16, ...\}, B = \{1^2, 2^2, 3^2, 4^2, ...\} = \{1, 4, 9, 16, ...\} \Rightarrow A = B$

Ex. 3. ***If S and T are two sets such that S has 21 elements. T has 32 elements and $S \cap T$ has 11 elements, then find the number of elements in $S \cup T$.***

Sol. Given, $n(S) = 21,\ n(T) = 32,\ n(S \cap T) = 11$

Now $n(S) + n(T) = n(S \cap T) + n(S \cup T)$

$\Rightarrow n(S \cup T) = 21 + 32 - 11 = \mathbf{42}$.

Ex. 4. ***If A and B are subsets of U such that n(U) = 700, n(A) = 200, n(B) = 300, n(A ∩ B) = 100, then find n(A' ∩ B').***

Sol. We know that $(A \cup B)' = A' \cap B'$, so we need to find $(A \cup B)$ first.

$n(A \cup B) = n(A) + n(B) - n(A \cap B) = 200 + 300 - 100 = 400$

$n(A \cup B)' = n(U) - n(A \cup B) = 700 - 400 = 300$

$\Rightarrow n(A' \cap B') = \mathbf{300}$.

Ex. 5. ***In a community of 175 persons, 40 read TOI, 50 read the Samachar Patrika and 100 do not read any. How many persons read both the papers.***

Sol. Let x no. of people read both the newspapers.

Since 100 do not read any, no. of people reading both or either of newspapers = 175 – 100 = 75

$\therefore (40 - x) + x + (50 - x) = 75 \Rightarrow \mathbf{x = 15}$.

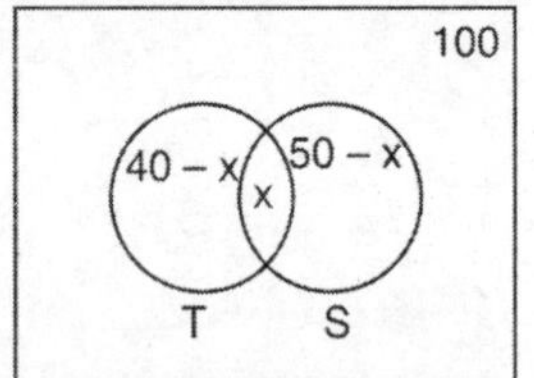

Ex. 6. ***In a locality, two-thirds of the people have Cable TV, one-fifth have Dish TV and one-tenth have both. What is the fraction of people having either Cable TV or Dish TV ?***

Sol. Fraction of people who watch cable only $= \left(\frac{2}{3} - \frac{1}{10}\right) = \frac{17}{30}$

Fraction of people who watch Dish TV only $= \left(\frac{1}{5} - \frac{1}{10}\right) = \frac{1}{10}$

$\therefore$ Fraction of people who watch either Cable or Dish TV $= \frac{17}{30} + \frac{1}{10} = \frac{20}{30} = \mathbf{\frac{2}{3}}$.

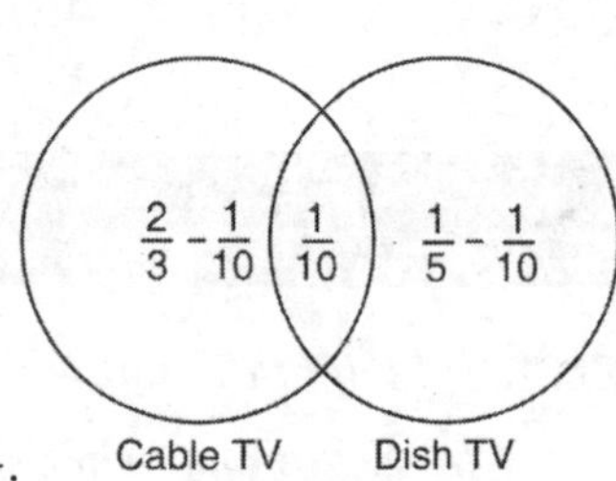

Question Bank–25

1. A = {an integer whose square is a negative value} is

(a) singleton set (b) null set
(c) infinite set (d) disjoint set

2. Which of the following statements is true ?

(a) $\phi \subset \{a, b, c\}$ (b) $\phi \in \{a, b, c\}$
(c) $0 \in \phi$ (d) $\{a\} \in \{a, b, c\}$

3. The sets $A = \{x \mid x \text{ is prime}, x < 20\}$ and $B = \{x \mid x = n^2, x \in N \text{ and } x < 5\}$ are:

(a) overlapping sets (b) equal sets
(c) equivalent sets (d) disjoint sets

4. Let A = { multiples of 3 less than 20 }
B = {multiples of 5 less than 20}.

Then, $A \cap B$ =

(a) {3, 5} (b) 15
(c) {15} (d) ϕ

5. Let U = {all digits of our number system}
A = {prime numbers}, B = {factors of 36}.

Then, $A \cup B$ =

(a) { 2, 3, 4, 6, 9, 12, 18}
(b) {1, 2, 3, 4, 5, 6, 7, 8, 9}
(c) {0, 1, 2, 3, 4, 5, 6, 7, 8, 9}
(d) {1, 2, 3, 4, 5, 6, 7, 9}

6. Let A = {even numbers}, B = {prime numbers}. Then, $A \cap B$ equals

(a) {odd numbers} (b) {composite numbers}
(c) {2} (d) {whole numbers}

7. If $A = \{1, 2, 3, 4\}$, $B = \{2, 4, 5, 6\}$ and $C = \{1, 2, 5, 7, 8\}$, then $(A \cup C) \cap B$ is equal to:

(a) {1, 2, 5} (b) {2, 4, 5}
(c) {1, 2, 4, 5, 7, 8} (d) {1, 2, 3, 4, 5, 7, 8}

8. A and B are two sets such that $A \cup B$ has 18 elements. If A has 8 elements and B has 15 elements, then the number of elements in $A \cap B$ will be :

(a) 5 (b) 8
(c) 7 (d) 4

9. If X and Y are two sets, then $X \cap (Y \cup X)'$ equals :

(a) X (b) Y

(c) ϕ (d) $\{0\}$

10. Let $P = \{x \mid x$ is a multiple of 3 and less than 100, $x \in N\}$

$Q = \{x \mid x$ is a multiple of 10 and less than 100, $x \in N\}$

Then, which of the following statements is true ?

(a) $Q \subset P$

(b) $P \cup Q = \{x \mid x$ is multiple of 30; $x \in N\}$

(c) $P \cap Q = \phi$

(d) $P \cap Q = \{x \mid x$ is a multiple of 30 ; $x \in N\}$

11. Let P = Set of all integral multiples of 3

Q = Set of all integral multiples of 4

R = Set of all integral multiples of 6

Consider the following relations:

1. $P \cup Q = R$ **2.** $P \subset R$

3. $R \subset (P \cup Q)$

Which of the relations given above is/are correct ?

(a) only 1 (b) only 2

(c) only 3 (d) 2 and 3

12. If $P = \{2m : m \in N\}$ and $Q = \{2^m : m \in N\}$, where m is positive, then:

(a) $Q \subset P$ (b) $P \subset Q$

(c) $P = Q$ (d) $P \cup Q = N$

13. The shaded region in the diagram is :

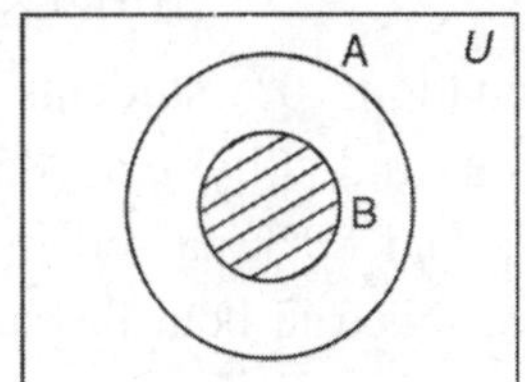

(a) $A \cup B$ (b) A'

(c) B' (d) $A \cap B$

14. The given Venn diagram represents four sets of students who have opted for Mathematics (M), Physics (P), Chemistry (C) and Electronics (E).

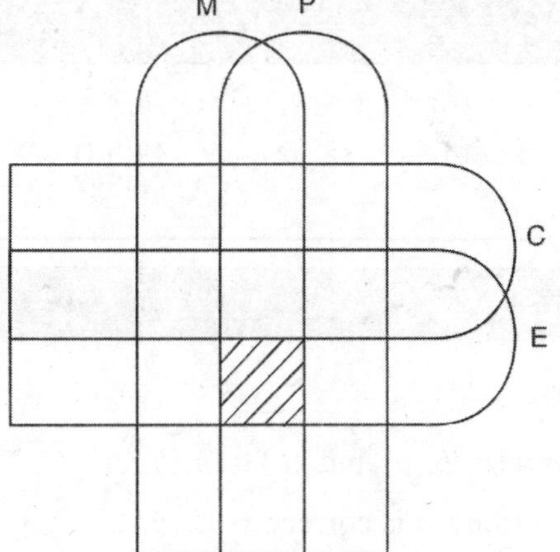

What does the shaded region represent?

(a) Students who opted for Physics, Chemistry and Electronics.

(b) Students who opted for Mathematics, Physics, Chemistry.

(c) Students who opted for Mathematics, Physics and Electronics.

(d) Students who opted for Mathematics, Chemistry and Electronics.

15. In the given diagram, circle A represents teachers who can teach Physics, circle B represents teachers who can teach Chemistry and circle C represents teachers who can teach Mathematics. Among the regions marked p, q, r, s, t, u, v the one that represents teachers who can teach Physics and Mathematics but not Chemistry is

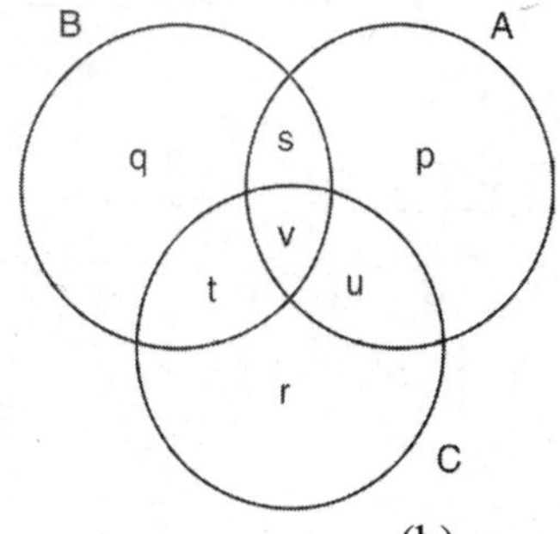

(a) v (b) u

(c) s (d) t

16. Let A, B, C be the subsets of the universal set X. Let A', B', C' denote their complements in X. Then, which of below corresponds to the shaded portion in the given figure.

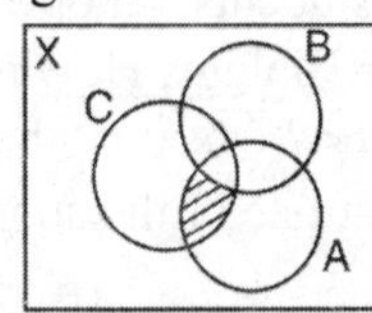

(a) $A' \cap B \cap C$ (b) $A \cap B' \cap C$

(c) $A \cap B \cap C'$ (d) $A' \cap B' \cap C$

17. If $A \cap B = A$ and $B \cap C = B$, then $A \cap C$ is equal to :

(a) B (b) C

(c) $B \cup C$ (d) A

18. The Venn diagram showing the relationship $X \subset Y \subset U$ is :

(a)

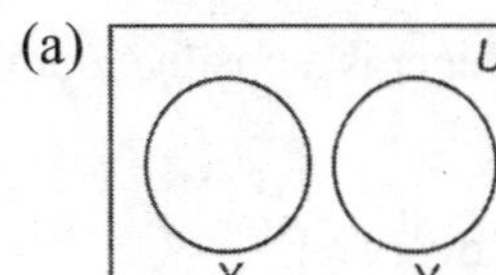

(b)

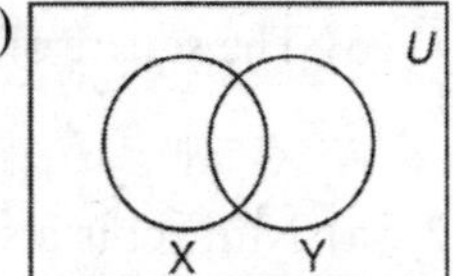

(c) (d)

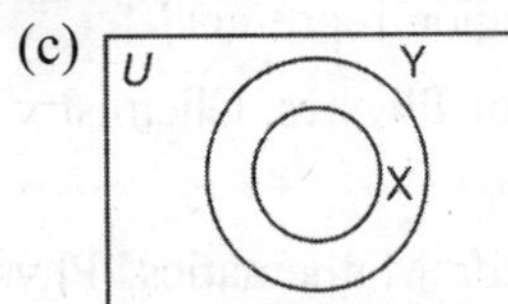

19. Let $A = \{x \mid x \in N,\ x \text{ is a multiple of } 2\}$

$B = \{x \mid x \in N,\ x \text{ is a multiple of } 5\}$

$C = \{x \mid x \in N,\ x \text{ is a multiple of } 10\}$

The set $(A \cap B) \cap C$ is equal to :

(a) A (b) $A \cap C$
(c) B (d) C

20. In a group of 15, 7 have studied German, 8 have studied French, and 3 have not studied either. The Venn diagram showing the number of students who have studied both is :

(a)

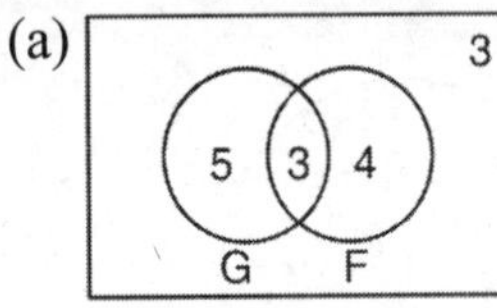

(b)

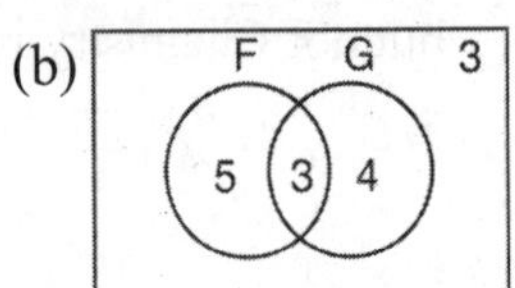

(c)

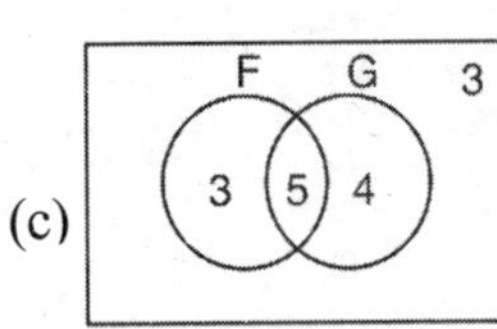

(d)

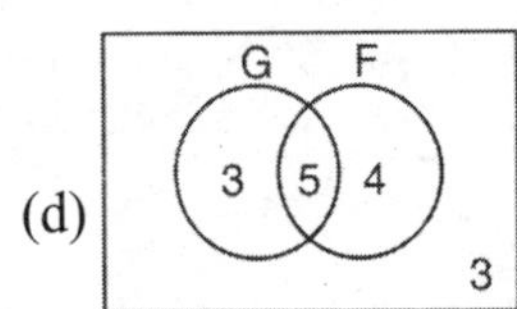

21. In a class of 45 students, 22 can speak Hindi only, 12 can speak English only. The number of students who can speak both Hindi and English is :

(a) 11 (b) 23
(c) 33 (d) 34

22. In a class of 50 students, 35 opted for Mathematics and 37 opted for Biology. How may have opted for only Mathematics ? (Assume that each student has to opt for at least one of the subjects)

(a) 15 (b) 17
(c) 13 (d) 19

23. There are 19 hockey players in a club. On a particular day 14 were wearing the prescribed hockey shirts, while 11 were wearing the prescribed hockey pants. None of them was without a hockey pant or a hockey shirt. How many of them were in complete hockey uniform ?

(a) 8 (b) 6
(c) 9 (d) 7

24. Consider the given diagram 500 candidate appeared in an examination conducted for the tests in English, Hindi and Maths represented by the circle E, H and M respectively. The diagram M gives the number of candidates who failed in different tests ?

What is the percentage of candidates who failed in atleast two subjects ?

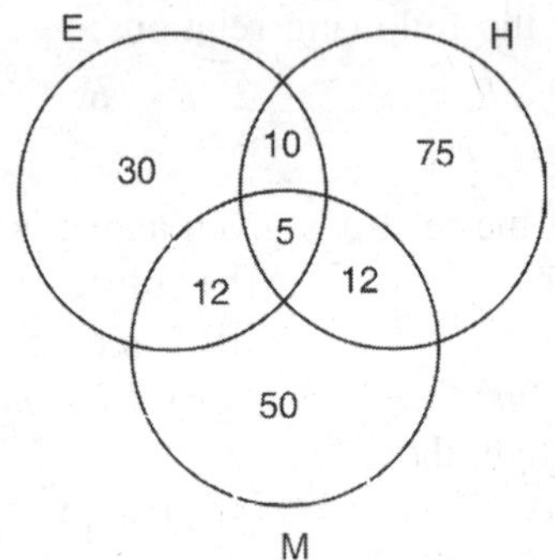

(a) 0.078 (b) 1.0
(c) 6.8 (d) 7.8

25. In an examination 70% students passed both in Mathematics and Physics, 85% passed in Mathematics and 80% passed in Physics. If 30 students have failed in both the subjects, then the total number of students who appeared in the examination is equal to :

(a) 900 (b) 600
(c) 150 (d) 100

Answers

1. (b)	2. (a)	3. (d)	4. (c)	5. (d)	6. (c)	7. (b)	8. (a)	9. (c)	10. (d)
11. (c)	12. (a)	13. (d)	14. (c)	15. (b)	16. (b)	17. (d)	18. (c)	19. (d)	20. (b)
21. (a)	22. (c)	23. (b)	24. (d)	25. (b)					

Hints and Solutions

1. (b) The square of every integer is a positive number .

$\therefore\ A = \phi$

2. (a) Null set is a subset of every set.

(b) Null set cannot belong to a set.

(c) $\phi \neq \{0\}$

(d) $a \in \{a, b, c\}$ but $\{a\} \notin \{a, b, c\}$

$\therefore$ Option (a) is correct.

3. (d) $A = \{2, 3, 5, 7, 11, 13, 17, 19\}$
$B = \{1, 4, 9, 16\}$
$\Rightarrow A \cap B = \phi$

4. (c) $A = \{3, 6, 9, 12, 15, 18\}$
$B = \{5, 10, 15\}$
$\Rightarrow A \cap B = \{15\}$

5. (d) $\xi = \{0, 1, 2, 3, 4, 5, 6, 7, 8, 9\}$
$A = \{2, 3, 5, 7\}$
$B = \{1, 2, 3, 4, 6, 9\}$
$\Rightarrow A \cup B = \{1, 2, 3, 4, 5, 6, 7, 9\}$

6. (c) $A = \{2, 4, 6, ...\}$
$B = \{2, 3, 5, ...\}$
$\therefore$ 2 is the only even prime number, $A \cap B = \{2\}$.

7. (b) $A \cup C = \{1, 2, 3, 4\} \cup \{1, 2, 5, 7, 8\}$
$= \{1, 2, 3, 4, 5, 7, 8\}$
$(A \cup C) \cap B = \{1, 2, 3, 4, 5, 7, 8\} \cap \{2, 4, 5, 6\}$
$= \{2, 4, 5\}$.

8. (a) $n(A) + n(B) = n(A \cup B) + n(A \cap B)$
$\Rightarrow 8 + 15 = 18 + n(A \cap B)$
$\Rightarrow n(A \cap B) = 23 - 18 = 5$

9. (c) $X \cap (Y \cup X)' = X \cap (Y' \cap X')$
$= X \cap Y' \cap X' = X \cap X' \cap Y'$
$= \phi \cap Y' = \phi$.

10. (d) $P = \{3, 6, 9, 12, 15, 18, 21, 24, 27, 30 \ldots, 60, \ldots, 90, \ldots, 99\}$
$Q = \{10, 20, 30, 40, 50, 60, 70, 80, 90\}$
Then looking at the multiple choices we get
$P \cap Q = \{30, 60, 90\}$
$\Rightarrow P \cap Q = \{x \mid x$ is a multiple of 30, $x \in N\}$

11. (c) $P = (3, 6, 9, 12, 15, 18, 21, \ldots\}$
$Q = \{4, 8, 12, 16, 20, \ldots\}$
$R = \{6, 12, 18, \ldots\}$
Considering the choices given :
1. $P \cup Q = \{3, 4, 6, 8, 9, 12, 15, 16, 18, \ldots\} \neq R$
2. All the elements of P are not in R, so $P \not\subset R$
3. $P \cup Q = \{3, 4, 6, 8, 9, 12, 15, 16, 18, \ldots\}$

$\Rightarrow$ All the elements of R are in $P \cup Q$
$\Rightarrow R \subset (P \cup Q)$

12. (a) $P = \{2, 4, 6, 8, 10, 12, 14, \ldots\}$
$Q = \{2^1, 2^2, 2^3, \ldots\} = \{2, 4, 8, \ldots\}$
$\Rightarrow Q \subset P$

13. (d) $B \subset A, A \cap B = B \therefore$ Shaded region $= A \cap B$

15. (b) The regions common to circles A (Physics) and circle C (Mathematics) are v and u, where v represents teachers who can teach all subjects and u represents the teachers who can teach Physics and Mathematics but not Chemistry.

17. (d) $A \cap B = A$ can be represented by a Venn diagram as :

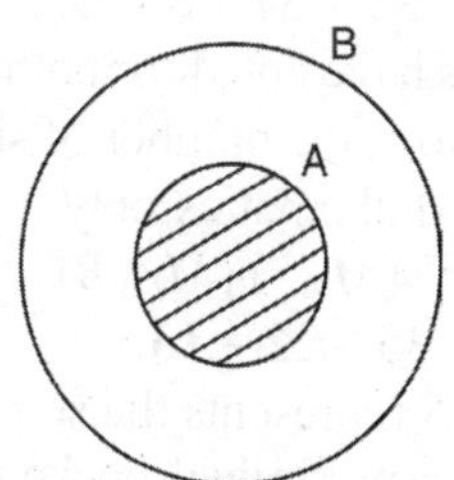

and $B \cap C = B$ can be represented by a Venn diagram as :

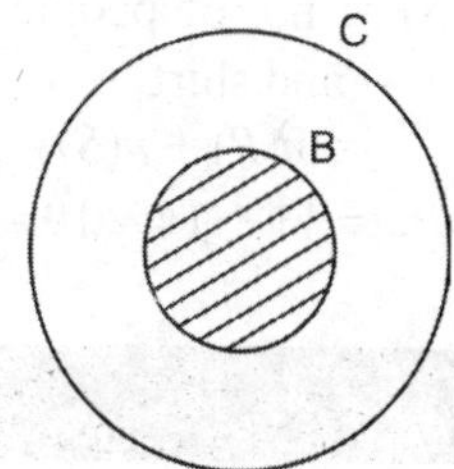

Combining both the Venn diagrams, we get

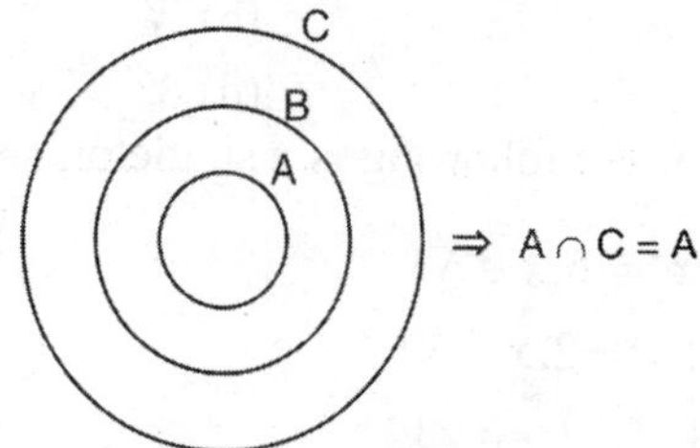

19. (d) $A = \{2, 4, 6, 8, 10, 12, 14, \ldots\}$
$B = \{5, 10, 15, 20, 25, \ldots\}$
$C = \{10, 20, 30, 40, \ldots\}$
$\Rightarrow A \cap B = \{10, 20, 30, \ldots\}$
$(A \cap B) \cap C = \{10, 20, 30, \ldots\} = C$

20. (b) No. of students in the group who have studied either or both the languages, *i.e.*, $n(G \cup F) = 15 - 3 = 12$
Given, $n(G) = 7, n(F) = 8$. Now we have to find the number of students who have studied both, *i.e.*, $n(G \cap F)$.
$n(G) + n(F) = n(G \cup F) + n(G \cap F)$
$\Rightarrow 7 + 8 = 12 + n(G \cap F) \Rightarrow n(G \cap F) =$ **3.**
$\therefore$ (b) is the best suited Venn diagram.

21. (a) Here $n(H) = 22 + x$
$n(E) = 12 + x$
$n(H \cap E) = x$
$n(H \cup E) = 45$
$\Rightarrow 45 + x = 22 + x + 12 + x \Rightarrow x =$ **11.**

H E
22 x 12

22. (c) Here, $n(M \cup B) = 50$, $n(M) = 35$, $n(B) = 37$

$\therefore\ n(M \cap B) = n(M) + n(B) - n(M \cup B)$
$= 35 + 37 - 50 = 22$

$\Rightarrow$ 22 students have opted for both Mathematics and Biology. Now, the number of students who have opted for Mathematics only
$= n(M) - n(M \cap B)$
$= 35 - 22 = \mathbf{13}$

23. (b) Let P and S represents the sets of hockey player wearing the prescribed hockey pants and shirts respectively.

Then, $n(P) = 11$, $n(S) = 14$, $n(P \cup S) = 19$

$\Rightarrow\ n(P \cap S)$ = no. of people wearing both pant and shirt
$= n(P) + n(S) - n(P \cup S)$
$= 11 + 14 - 19 = \mathbf{6}$

24. (d) Number of candidates who failed in at least two subjects $= n(E \cap H) + n(H \cap M) + n(E \cap M) + n(E \cap H \cap M)$
$= 10 + 12 + 12 + 5 = 39$

$\therefore$ Required percentage $= \left(\frac{39}{500} \times 100\right)\% = \mathbf{7.8\%}$.

25. (b) Students passed in atleast one subject
$= n(P \cup M) = n(P) + n(M) - n(P \cup M)$
$= 80 + 85 - 70 = 95$

$\therefore$ 5% students failed in both the subjects

$\Rightarrow$ 5% of total students = 30

$\Rightarrow$ Total students $= \frac{30 \times 100}{5} = \mathbf{600}$.

Self Assessment Sheet–25

1. If $X' = Y$, then $(X \cap Y)'$ is equal to

(a) ϕ (b) X
(c) U (d) Y

2. Which of the following is a singleton set ?

(a) $\{x : x^2 = 5, x \in N\}$
(b) $\{x : |x| = 2, x \in N\}$
(c) $\{x : |x| = 7, x \in Z\}$
(d) $\{x : x^2 + 4x + 4 = 0, x \in N\}$

3. If A has 5 elements and B has 8 elements such that $A \subset B$, then the number of elements in $A \cap B$ and $A \cup B$ are respectively :

(a) 8, 5 (b) 3, 3
(c) 5, 8 (d) 5, 13

4. If $aN = \{ax : x \in N\}$, then the set $3N \cap 7N$ is

(a) $\{21x : x \in N\}$
(b) $\{42x : x \in N\}$
(c) $\{63x : x \in N\}$
(d) None of these

5. In a group of 100 persons, 85 take tea, 20 take coffee and 25 take both tea and coffee. Number of persons who take neither tea nor coffee is

(a) 5 (b) 15
(c) 25 (d) 20

6. Which of the following cannot have a proper subset ?

(a) $\{x : x \in Q, 5 < x < 7\}$
(b) $\{x : x \in Z, -4 < x < 4\}$
(c) $\{x : x \in N, 5 < x < 6\}$
(d) $\{x : x + 1 = 0, x \in Z\}$

7. The relationship illustrated by the given Venn diagram is :

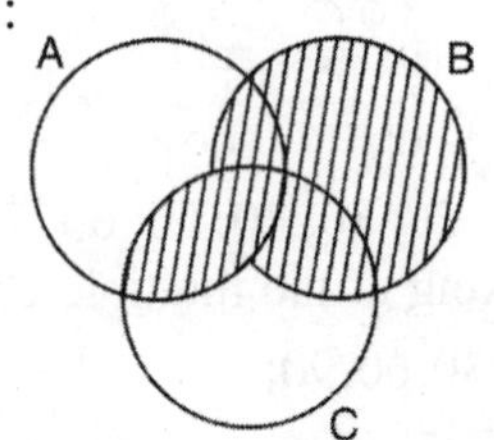

(a) $(A \cup B) \cap C$ (b) $(A \cap B) \cap C$
(c) $(A \cap C) \cup B$ (d) $(A \cup B)' \cap C$

8. If A, B, C be three sets such that $A \cup B = A \cup C$ and $A \cap B = A \cap C$, then

(a) $A = B$ (b) $B = C$ (c) $A = C$ (d) $A = B = C$

9. If $U = \{3, 4, 5, 6, 7, 8, 9\}$, $X = \{3, 4\}$, $Y = \{5, 6\}$ and $Z = \{7, 8, 9\}$, then $Y' \cap (X \cap Z)'$ is equal to

(a) $X \cup Y$ (b) $Y \cup Z$ (c) $(X \cup Z)'$ (d) $X' \cap Y'$

10. Out of 450 students in a school, 193 students read Science Today, 200 students read Science Refresher, while 80 students read neither. How many students read both the magazines ?

(a) 137 (b) 80 (c) 57 (d) 23

Answers

1. (c)	2. (b)	3. (c)	4. (a)	5. (d)	6. (c)	7. (c)	8. (b)	9. (c)	10. (d)